SYMMETRY OF CRYSTALS AND MOLECULES

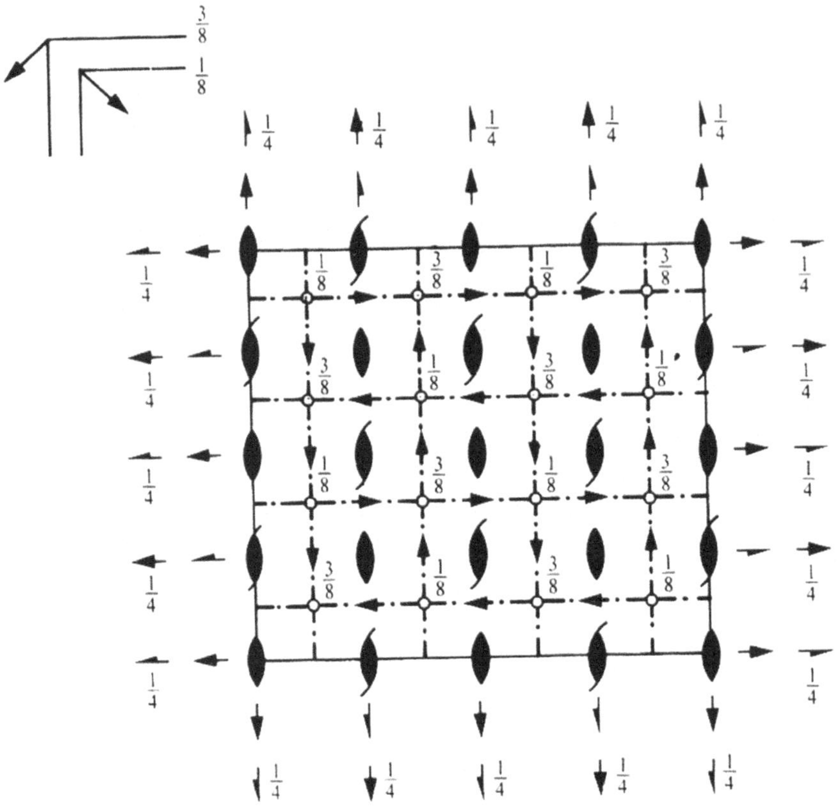

The orthorhombic space group *Fddd* ($F\frac{2}{d}\frac{2}{d}\frac{2}{d}$); origin on 222, at −1/8, −1/8, −1/8 from $\bar{1}$.

Symmetry of Crystals and Molecules

MARK LADD

Formerly Head of Chemical Physics, University of Surrey

"A thing is symmetrical if there is something you can do to it so that after you have finished doing it, it looks the same as before."

Hermann Weyl (1983): *Symmetry*

OXFORD
UNIVERSITY PRESS

OXFORD
UNIVERSITY PRESS

Great Clarendon Street, Oxford, OX2 6DP,
United Kingdom

Oxford University Press is a department of the University of Oxford.
It furthers the University's objective of excellence in research, scholarship, and education by publishing worldwide. Oxford is a registered trade mark of Oxford University Press in the UK and in certain other countries

First Edition published in 2014

Published in the United States of America by Oxford University Press
198 Madison Avenue, New York, NY 10016, United States of America

British Library Cataloguing in Publication Data
Data available

Library of Congress Control Number: 2013944301

ISBN 978–0–19–967088–8

Foreword

The observation that stone-age people produced objects with a measure of symmetry is not as unexpected as often made out. The opposite would actually be much more surprising. Totally asymmetric objects in the real world are extremely rare, if not altogether non-existent. Some objects are admittedly more symmetrical than others, and this rule applies without exception. Perfectly symmetrical objects are therefore as rare as the totally asymmetrical. The most perfectly fashioned sphere or the most exquisite crystal contains some blemish when examined in sufficient detail. The distinction drawn by chemists between chiral and achiral molecules is meaningless. I consider it all a matter of resolution; any molecule has a degree of chirality between zero and one.

The powerful concept of symmetry applied with such good effect in science is an abstract idealization with precise definition only in the sense of logic. Like the absolute zero of temperature or the relativistic limit to material velocity it can be described mathematically, but remains physically unreachable. The mathematical tool that describes symmetry is known as group theory. In terms of Klein's famous Erlangen program any geometry is characterized by a group of transformations under which its propositions remain valid. In this sense the Euclidean group is more restrictive than the projective group, and both of these are less general than the continuous topological group. Topology allows any amount of stretching or compression, without tearing; in this most general case the popular notion of symmetry disappears, but remains embedded in the group structure.

To produce a monograph on a topic as wide as symmetry needs some courage. In doing so Mark Ladd wisely stipulated at the outset that the book is restricted to crystal and molecular structure, for which he is perhaps uniquely qualified. Molecular structure in this context strictly refers to the nuclear framework as revealed in crystallographic analyses. The appropriate symmetry group in this case is the three-dimensional discrete Euclidean group, commonly referred to as space-group symmetry.

As a seasoned crystallographer, albeit of forgotten vintage, I do not find much new in this presentation, but I marvel at the comprehensive detail and rigour. In the days of visually estimated x-ray intensities and Beevers–Lipson strips, crystal geometry and space-group symmetry were experienced as practical necessities. It is not clear where and how the new generation, who twiddle a few knobs to produce a crystal structure, acquires this know-how. Working through the numerous examples could well provide the necessary hands-on experience. This is where I see this book becoming essential reading, if not prescribed for intensive study. Too many papers are being published around dubious symmetry arguments.

Up to a point the analysis is based on Newton's approach that relies only on conclusions which can be formulated mathematically—without the need of hypotheses. I notice in particular how the author wisely refrains from offering physical interpretations for observed aspects such as chirality, optical activity and other symmetry related phenomena, identified and manipulated purely as point group variables. Unfortunately, the temptation to reinterpret molecular symmetry in terms of hybrid orbitals, a concept popular in quantum chemistry, was clearly irresistible.

It is common chemical practice to visualize Lewis type electron pair covalent interaction in terms of real orthogonal classical functions, called atomic orbitals. These functions, such as the so-called p_x, p_y and p_z orbitals are often confused with quantum mechanical wave functions, which they resemble after a fashion. However, these functions cannot be described by quantum numbers, conflict with the exclusion principle, do not represent surface harmonics, which are complex and hence cannot be used to describe electronic orbital angular momentum nor spin. Atomic orbitals are routinely used in linear combination, in the same way as Euclidean unit vectors, to visualize directed covalent interaction, inferred from crystallographic analyses, to derive point group symmetry of molecular species. These concepts are, admittedly, not easily avoided, being so thoroughly entrenched in chemical group theory, but a word of warning is perhaps not misplaced. From the symmetry point of view, real p_x and p_y functions, which represent linear vibrations, can never be considered equivalent to the rotation defined by a complex function. It is not the point group description of molecules that is at issue, but the way it is used to represent hybrid orbitals. Being of fundamental importance, it could serve as an encouragement for the reader to make an independent assessment, one way or the other.

Professor Jan Boeyens
University of Pretoria

Preface

The study of crystal and molecular symmetry is often considered difficult by those meeting it for the first time. I believe that there are two particular reasons for this problem. First, crystals and molecules, unlike the more familiar plane figures, are three-dimensional bodies and it can be difficult to take in simultaneously all parts of such a body and see them in relation to the whole. In two dimensional figures, such as a rectangle or a regular hexagon, it can be done readily and we appreciate the symmetry without difficulty. The second reason, not unrelated to the first, lies in the unfamiliarity of three-dimensional relationships, notwithstanding we encounter daily the three-dimensional spatial aspects of our space-time continuum.

The purpose of this book is to present crystal and molecular symmetry in a straightforward and practical manner. It evolves from lecture courses given over many years, at undergraduate and postgraduate levels, to those pursuing crystallography, chemistry and materials science. It makes liberal use of stereoviews and computer aids to understanding the textual material. Most of the stereoscopic illustrations were drawn with the program PLUTO, by courtesy of Dr WDS Motherwell, University of Cambridge, and aids to stereoviewing are discussed in an appendix. Each chapter concludes with a set of problems and detailed tutorial solutions are provided. Computer programs relevant to the text and the problems have been devised, and are available via the publisher's web site, <www.oup.co.uk/companion/ladd>.

It is informative to study everyday objects in relation to symmetry: a glass and a beer jug; a chair and a table; a tiled floor and a brick wall; a Dalmatian and a Dobermann. In this way, symmetry can become clearer—it is not only about crystals and molecules. Two notations for symmetry are in use, the Hermann–Mauguin and Schönflies notations; each has a place in crystal and molecular symmetry, and each is discussed in detail.

The study of the symmetry of crystals and molecules may be begun either with lattices or with a description of the external symmetry of crystals. The former has advantages in a short lecture course aimed at getting to grips with X-ray diffraction. However, a morphological approach has been chosen for this book because, in this way, the book should have a more general appeal and its pace a little slower. Time is needed in order to assimilate the concepts of symmetry and to consolidate them into a working knowledge of the subject. It is hoped that this book will be helpful to all those meeting symmetry for the first time, whatever their specialization, and will prepare the reader for study of the current definitive texts on symmetry, the *International tables for crystallography*, Volumes A and A1.

Pre-knowledge is confined mainly to mathematics and physical sciences, but no more than would be studied at A-level. A little more mathematics, along lines indicated, is an advantage in certain contexts, and appropriate guidance is provided in appendices in order not to interrupt the flow of the text.

The main notation used in the book is indicated in the front matter. In addition, equations are indicated in the form Eq. (1.2) wherever they occur, or as Eq. (A1.2) if in Appendix A1; for compactness, particularly in matrices, a negative number, –1, for example, is frequently written as $\bar{1}$; symmetry elements are written as 2 or *m*, the corresponding symmetry operations or operators as ***2*** or ***m*** and the matrices representing these operations as **2** or **m**.

I would like to express my thanks to those authors and publishers for permission to reproduce the figures which bear an appropriate acknowledgement. In particular, many illustrations have been taken from four particular sources: (a) *Structure determination by X-ray crystallography*. 4th ed. Kluwer Academic/Plenum Publishing, 2003; (b) *Crystal structures: lattices and solids in stereoview*, 1999; (c) *Structure and bonding in solid state chemistry*, 1979, both Ellis Horwood Ltd, UK; (d) *International tables for X-ray crystallography*, Vol. 1, 1965; (e) *International tables for crystallography*, Vol. A; and (f) *Symmetry aspects of M. C. Escher's periodic drawings*, the latter three from the International Union of Crystallography (IUCr). Where used, the copyright holders will be acknowledged as 'Reproduced by courtesy of Springer Science+Business Media, New York © Kluwer Academic/Plenum Publishing' for (a), 'Reproduced by courtesy of Woodhead Publishing, UK' for (b) and (c), and 'Reproduced by courtesy of IUCr', followed by (1), (A) or (E), for (d), (e) and (f), as appropriate.

It is my great pleasure to acknowledge Professor Mike Glazer of the Clarendon Laboratory, University of Oxford, for both reading this book in its draft stage and offering valuable suggestions. My thanks also go to Professor Jan Boeyens of the University of Pretoria for writing the Foreword and for timely comments. I am indebted to Sönke Adlung, Senior Commissioning Editor, Physical Sciences, Oxford University Press for encouraging me to proceed with this book, the publication of which coincides with the **International Year of Crystallography**, celebrating 100 years of X-ray crystallography, to Jessica White, Assistant Commissioning Editor, Physical Sciences and to Erin Pearson, Production Editor, Law and Academic Science & Medicine, Oxford University Press, for advice and technical assistance at all stages of the preparation of the book.

Mark Ladd
Bramshott, 2014

Disclaimer

Every effort has been made to ensure the correct functioning of the software associated with this book. However, the reader planning to use the software should note that, from a legal point of view, there is no warranty, expressed or implied, that the programs are free from error or will prove suitable for a particular application. By using the software the reader accepts full responsibility for all the results produced, and the author and publisher disclaim all liability from any consequences arising from the use of the software. The software should not be relied upon for solving a problem, the incorrect solution of which could lead to injury to a person or loss of property. If you do use the programs in such a manner, it is at your own risk. The author and publisher disclaim all liability for direct or consequential damages resulting from your use of the programs.

Contents

Physical data, notation, and online materials

Physical constants

Atomic mass unit	m_u	1.6605×10^{-27} kg
Avogadro constant	L	6.0221×10^{23} mol^{-1}
Bohr radius for hydrogen	a_0	5.2918×10^{-11} m
Elementary charge	e	1.6022×10^{-19} C
Planck constant	h	6.6261×10^{-34} J s
Speed of light in a vacuum	c	2.9979×10^{8} m s^{-1}

Prefixes to units

femto	pico	nano	micro	milli	centi	deci	kilo	mega	giga
f	p	n	μ	m	c	d	k	M	G
10^{-15}	10^{-12}	10^{-9}	10^{-5}	10^{-3}	10^{-2}	10^{-1}	10^{3}	10^{6}	10^{9}

Notation

These notes indicate the main symbols used throughout the book. Inevitably, some of them have more than one application, partly from general usage, and partly from a desire to preserve a mnemonic character in the notation wherever possible. Two or more uses of one and the same symbol are separated by a semicolon in the presentation hereunder.

$A'(hkl)$ $B'(hkl)$	Components of the crystal unit cell structure factor along the real and imaginary axes, respectively, of an Argand diagram
$A(hkl)$ $B(hkl)$	Components of the crystal unit cell geometrical structure factor along the real and imaginary axes, respectively, of an Argand diagram
A	A face-centered unit cell; irreducible representation symmetric with respect to the principal C_n axis
Å	Ångström unit $(1\ \text{Å} = 10^{-10}\text{m})$
a, b, c	Unit cell edge lengths parallel to the x, y and z axes, respectively; intercepts made by the parametral plane on the x, y and z axes, respectively; glide planes with translational components of $a/2$, $b/2$ and $c/2$, respectively
a, b, c	Unit cell translation vectors along the x, y and z axes, respectively
a^*, b^*, c^*	Reciprocal unit cell edge lengths along the x^*, y^* and z^* axes, respectively
$\mathbf{a}^*, \mathbf{b}^*, \mathbf{c}^*$	Reciprocal unit cell translation vectors along x^*, y^* and z^* axes, respectively
B	B face-centered unit cell; irreducible representation asymmetric with respect to the principal C_n axis
C	C face-centered unit cell
¢	'Not constrained by symmetry to equal'

c	Speed of light
csu	Combined standard uncertainty
D_c	Calculated crystal density
d	Interplanar spacing in real space
d^*	Distance in reciprocal space
$d\,(hkl)$	Interplanar spacing of the *hkl* family of planes
$d^*\,(hkl)$	Distance from the origin to the *hkl* reciprocal lattice point
e	Electron charge
e, exp	Exponential function
esd	Estimated standard deviation
F(*hkl*)	Crystal unit cell structure factor for the *hkl* reflection
F*(*hkl*)	Complex conjugate of F(*hkl*)
| F(*hkl*) |	Amplitude of the structure factor F(*hkl*)
f	Atomic scattering factor
g	Glide line in two-dimensional space groups
H	Hexagonal (triply primitive) unit cell
(*hkl*) *or* (*hkil*)	Miller or Miller–Bravais indices of planes associated with the x, y and z axes or the x, y, u and z axes, respectively (a single index of two digits has a comma placed *after* it); family of planes
{*hkl*} *or* {*hkil*}	Form of (*hkl*) or {*hkil*} planes
hkl	Reciprocal lattice point corresponding to the (*hkl*) family of planes
h	Vector with components h, k, l in reciprocal space
h	Miller index parallel to the x axis; Planck constant; order of a group
I	Body-centered unit cell
$\mathcal{J}$	Imaginary axis on an Argand diagram
i, j, k	Unit vectors in the directions x, y, z, respectively
i	$\sqrt{-1}$; an operator that rotates a quantity on an Argand diagram anticlockwise through 90°
k	Miller index parallel to the y axis; symmetry number of a group
l	Miller index parallel to the z axis
M_r	Relative molecular mass
m	Mirror plane; mass
N	Number of atoms per unit cell; number density
n	Glide plane, with translational component of $(a+b)/2$, $(b+c)/2$ or $(c+a)/2$
R	Symmetry element—normal italic font (numbers remain Roman); rhombohedral unit cell; rotation axis of degree R
R	Symmetry operator or operation—bold, italic font (also with numbers, for example ***2*** (the symbolism ***R2*** implies the combination of these operators)
R	Matrix representing symmetry operation—bold, Roman font (also with numbers, for example, **2**
$\overline{R}$	Inversion axis of degree R
$\mathcal{R}$	Real axis on an Argand diagram
r	Radial coordinate
r	Vector distance, as in $\mathbf{r}(x, y, z)$
rms	Root mean square
RU	Reciprocal lattice unit
[*U V W*]	Zone axis or direction symbol
<*U V W*>	Form of zone axes or directions
V	Volume of a crystal unit cell
x, y, u, z	Crystallographic reference axes descriptors (u corresponds with Miller–Bravais indices)
X, Y, Z	Spatial coordinates, in absolute measure, of a point with respect to the x, y and z axes
x, y, z	Spatial fractional coordinates in a unit cell parallel to x, y, z respectively
x^*, y^*, z^*	Coordinates of a point in reciprocal space
$[x, \beta, \gamma]$	Line parallel to the x axis and intersecting the y and z axes at β and γ, respectively, with respect to the origin
(x, y, γ)	Plane normal to the z axis and intersecting it at γ with respect to the origin
$\pm\{x, y, z; \ldots\}$	$x, y, z;\ \bar{x}, \bar{y}, \bar{z} \ldots$
Z	Number of formula entities of mass M_r per unit cell; atomic number

α, β, γ	Angles between the pairs of unit-cell edges *bc*, *ca* and *ab*, respectively
$\alpha^*, \beta^*, \gamma^*$	Angles between the pairs of reciprocal unit cell edges b^*c^*, c^*a^* and a^*b^*, respectively
$\phi(hkl)$	Phase angle of the F(*hkl*) structure factor
θ	Bragg angle; spherical coordinate
λ	Wavelength
ϕ	Spherical coordinate
Φ	Interfacial (internormal) angle; molecular wave function
ρ	Electron density
χ, ψ, ω	Direction cosines of a line with respect to the *x*, *y* and *z* axes, as in $\cos\chi,\ \cos\psi,\ \cos\omega$
ψ	Atomic wave function
$^\circ$	Degree, as in 90°
$\times$	Vector (cross) product
$\cdot$	Scalar (dot) product
$\otimes$	Direct product, as in $C_i \otimes C_{3v} = D_{3d}$
$\bar{x}$	–*x* (minus *x*), in symmetry operations, matrices and coordinates lists
$< x >$	Average value of *x*

Online materials

Computer programs relevant to the text and the problems have been devised and are available via the publisher's website, <www.oup.co.uk/companion/ladd>.

Symmetry everywhere

1

SYNOPSIS

- Symmetry in science and arts
- Visualizing symmetry
- Symmetry in everyday objects
- Defining symmetry
- Introducing symmetry notation

1.1 Introduction

When I was lecturing to chemistry students on crystal symmetry, I would usually begin with the question: 'what is common to the National Westminster Bank logo, a Mercedes-Benz car emblem and a molecule of 2,4,6-triazidotriazine?' Generally, there was no reply: maybe they banked at Lloyds, drove Fords and skipped the lectures on triazines—but when I showed a slide of these entities (Fig. 1.1) there was always someone who saw that it had something to do with the number 'three'. And that is a good way to begin the study of symmetry. Of course, it can all be done mathematically and there is a place for that, as will be shown in later chapters. But for the moment, a visual expression of symmetry will suffice. This chapter sets out to introduce ideas on symmetry, and to show that it is experienced by everyone in some way every day: breakfast at a table having reflection symmetry, a midday pint in a glass of cylindrical symmetry, an evening stroll with the Dalmatian—even no symmetry is a form of symmetry.

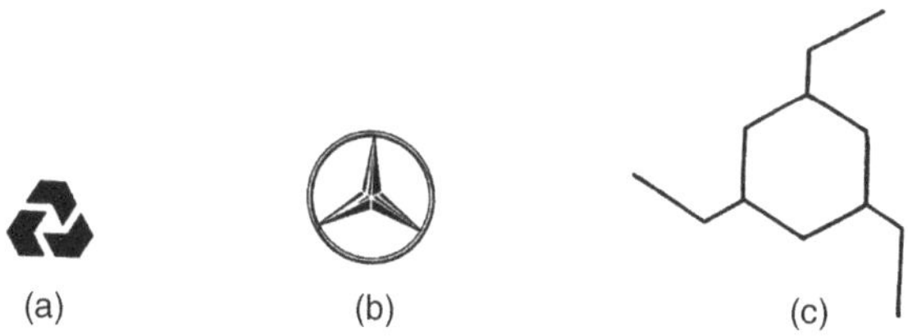

Fig. 1.1
(a) Logo of the National Westminster Bank. (b) Emblem of Mercedes-Benz cars. (c) Molecular skeleton of 2,4,6-triazidotriazine, $C_3N_3(N_3)_3$.

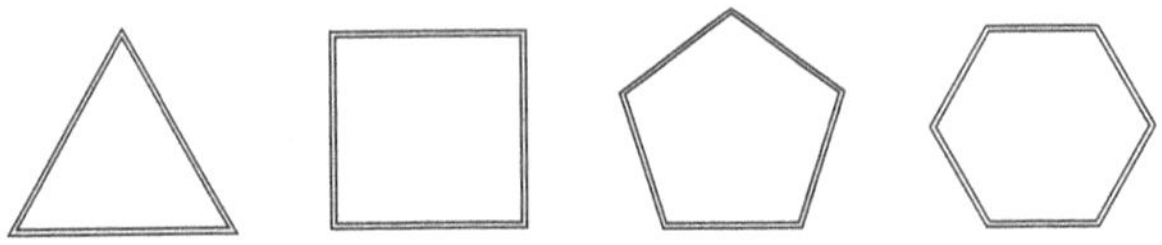

Fig. 1.2
Regular polygons.

1.2 Looking at symmetry

'"*Beauty is truth, truth beauty"—that is all ye know on earth, and all ye need to know*' [1]. Few of us have difficulty in recognizing symmetry in the plane geometrical shapes shown in Fig. 1.2. But it is a rather different matter when considering more complex three-dimensional objects (Fig. 1.3 and Fig. 1.4).

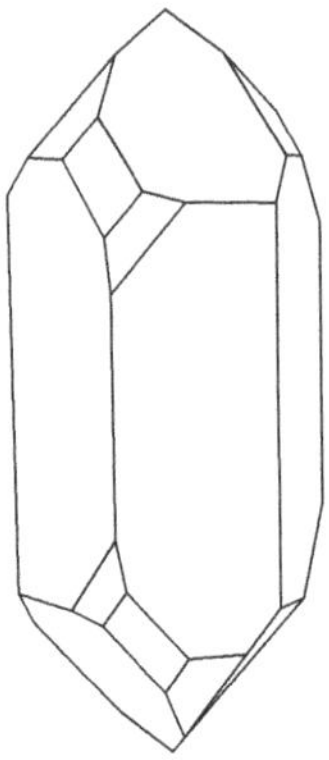

Fig. 1.3
Crystal of quartz, SiO_2.

Why should this be? I believe that the problem arises first from the fact that while one can see all parts of a two-dimensional object simultaneously, and thus take in the relationships of the parts to the whole, that cannot be done so easily in three dimensions. Secondly, although some three-dimensional objects, such as flowers, pencils and glass tumblers are simple enough to be rotated and examined visually, the natural gift for mentally perceiving and manipulating more complex objects may not be possessed by everyone. Nevertheless, the facility of doing so can be developed with suitable aids and with patience. If, initially, you have problems, take heart. You are not alone and, like many before you, you will be surprised at how swiftly the required facility can be acquired. Architects and sculptors may be blessed with a natural three-dimensional visualization aptitude, but they have learned to develop it—particularly by making and handling models.

Standard practice is always to reduce three-dimensional objects to two dimensions, in drawings such as projections and elevations: it is cheap, well suited to reproduction in books and less cumbersome than handling three-dimensional models. In this book, such two-dimensional representations still have a value, but to rely on them exclusively only delays the acquisition of a three-dimensional visualization ability. Fortunately, stereoscopic image pairs may be employed, such as that shown in Fig. 1.4. This type of illustration is a considerable help but, because it provides a view from only one standpoint, it is not always quite the equal of a model that can be examined by hand.

This illustration shows the crystal structure of oxalic acid dihydrate [2]. One half of the figure may be covered, whereupon the structure is viewed as a two-dimensional representation. Although the figure has been drawn carefully, with tapered bonds and hydrogen bonding indicated by double thin lines, it does not

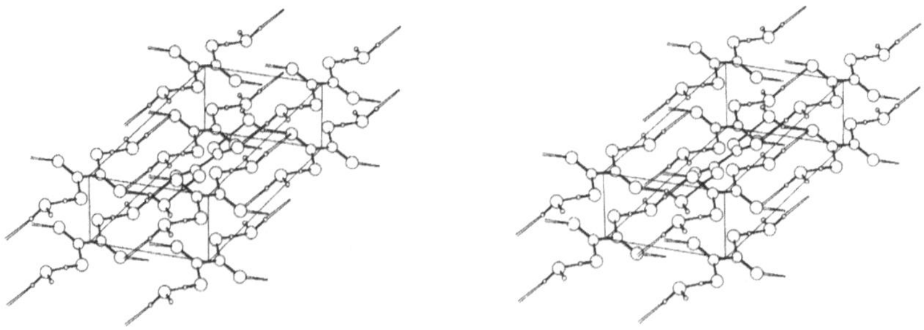

Fig. 1.4
Stereoview of the crystal structure of oxalic acid dihydrate, $(CO_2H)_2.2H_2O$; the circles in decreasing order of size represent O, C and H atoms. The double lines indicate hydrogen bonds, with the $H - O \cdots H$ bond distance of *ca.* 2.50 Å; the sum of the van der Waals radii for hydrogen and oxygen is 2.72 Å, which is good evidence of hydrogen bonding in this structure. [Reproduced by courtesy of Woodhead Publishing, UK.]

convey nearly as much information as the stereoview which, when viewed correctly, presents a convincing three-dimensional image. Notes on stereoviewers and stereoviewing are given in Appendix A1.

1.3 Some symmetrical objects

Four different objects are presented in Fig. 1.5. At first, there may not appear to be any connection between a Dobermann bitch, a Grecian urn, a molecule of 3-fluorochlorobenzene and a crystal of potassium tetrathionate. Yet each is an example of reflection symmetry: a symmetry plane (mirror plane, symbol m) can be imagined to divide each object into halves that are related

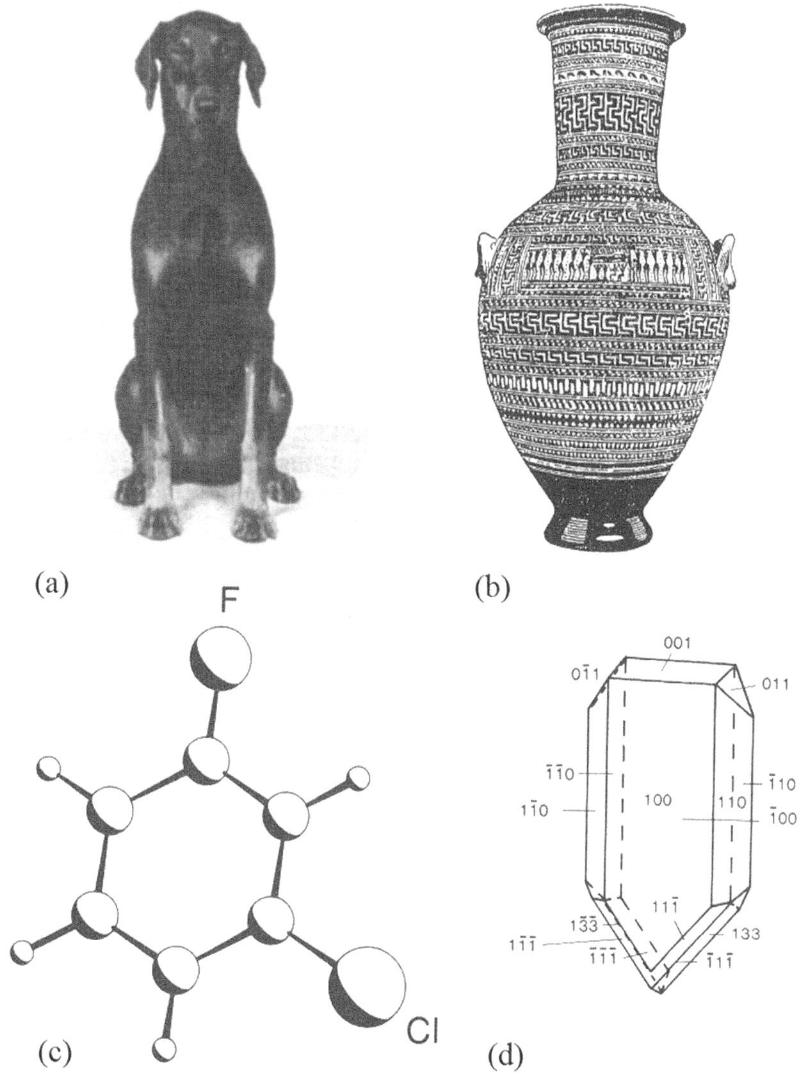

Fig. 1.5
Examples of reflection (mirror) symmetry. (a) Dobermann, *Vijentor Seal of Approval at Valmara*, JW. (b) Grecian urn. (c) Molecule of 3-chlorofluorobenzene; circles in decreasing order of size represent Cl, F, C and H atoms. (d) Crystal of potassium tetrathionate, $K_2S_2O_4$; with the crystal faces indexed by their Miller indices, (*hkl*). [Reproduced by courtesy of Woodhead Publishing, UK.]

exactly as an object is to its mirror image, like a right hand to its left hand. However, if the examples are perused in more detail, it would be noticed that the Dobermann bitch, beautiful animal that she is, does not have perfect *m* symmetry; the urn is not absolutely symmetrical; the molecule may not be totally planar; and the real crystal may have minute flaws that degrade perfect mirror symmetry.

Fig. 1.6
Portion of a floor with ideally square tiles.

In seeking symmetry around us, repeating patterns are soon encountered, such as tiled floors and brick walls (Fig. 1.6 and Fig. 1.7). Examine such structures in your locality at leisure, but do not be too critical about stains on the tiles or chips off the bricks. Perfect tiled floors and perfect brick walls are, like perfect crystals, conceptual. So what is the use of symmetry if the real objects that are to be studied are not strictly symmetrical?

The symmetry of objects can be studied both as finite bodies—the Grecian urn, the chemical molecule and as parts of larger, conceptually infinite bodies—the brick wall, the crystal. When investigating the internal structure of a crystal, which is one of the main reasons for studying crystal symmetry, it is discovered that a crystal of finite size is composed of a myriad of building blocks, or unit cells. Are all unit cells exactly alike? No; but to consider a specific case: sodium chloride crystallizes normally as cubes, and the basic unit of this crystal is also a cube, with an edge length of 0.564 nm[1] and volume of approximately 1.8×10^{-28} m^3.

[1] 1 nm = 10 Å = 10^{-9} m

Digressing for a moment: it is good practice to write numerical quantities as numbers from 0 to 9 with the appropriate 10^n multiplier; it makes for easy checks on calculations.

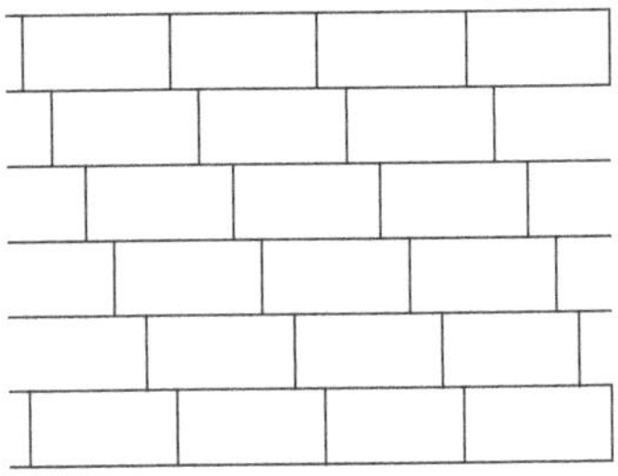

Fig. 1.7
Portion of a brick wall with ideally regular, rectangular bricks.

A crystal of sodium chloride of experimental size in an X-ray diffraction experiment could be of the order of 0.2, 0.2, 0.2 mm, a volume of 8×10^{-12} m^3. Thus, the number of unit cubes in the whole crystal is approximately 4×10^{16}. Symmetry concepts may be applied to real crystals because, although individual building units may exhibit sub-microscopic differences, the complete crystal behaves statistically towards physical observations as though it were perfect and infinite, and the results of treating the crystal in this way are found to be scientifically rewarding.

1.4 Defining symmetry

Symmetry is not an absolute property of a body: the result of a test for symmetry may depend on the nature of the examining probe. For example, the crystal structure of chromium appears different under X-ray and neutron diffraction examinations. Figure 1.8 illustrates the unit cell of chromium as seen both by X-rays and by neutrons. X-rays see a *body-centred* unit cell (a), but a neutron examination gives the different result, (b).

Elemental chromium has the electronic configuration $(Ar)(4s)^1(3d)^5$; it is antiferromagnetic below 38 °C and paramagnetic above this temperature. In the neutron diffraction experiment, although the individual atoms are similarly placed, magnetic interactions with the neutron beam reverse the direction of the magnetic moment of the central atom in the unit cell, so that it is no

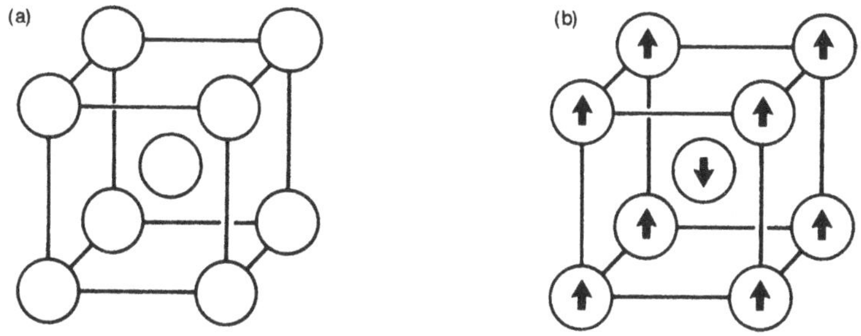

Fig. 1.8
Unit cell of the crystal structure of elemental chromium: (a) as seen by x-rays, and (b) as seen by neutrons; the arrows represent the direction of the magnetic moment vectors of the chromium atoms.

longer body-centred but *primitive*. Symmetry differences may arise also with optical and photoelastic properties.

This book will be concerned with the symmetry of the positional distribution of the parts of a molecule, crystal, body or pattern, as is revealed by visual inspection, microscopic examination or diffraction techniques. A definition of symmetry is *that spatial property of an assemblage by which the parts of the assemblage can be brought from an initial state to another indistinguishable state by means of a certain operation—a symmetry operation*. The term 'assemblage' is useful here because it can be used to describe the distribution of faces on a crystal, of bonds radiating from a central atom and of diffractions spectra from crystalline materials.

1.5 Symmetry in science

Manifestations of symmetry abound in most areas of science and, indeed, throughout nature; they are not confined to molecules and crystals. In botany, for example, the symmetry inherent in the structures of flowers and reproductive systems plays an important part in plant taxonomy. Figure 1.9 illustrates a white orchid, *Cattleya walkeriana*, var. *Alba*; its bilateral, or *m*, symmetry is clearly apparent.

Fig. 1.9
A bloom of the orchid *Cattleya walkeriana*, variation *Alba*, showing bilateral reflection symmetry. [Reproduced by courtesy of Greg Allikas.] (**See Plate 1**)

Examples of symmetry arise also in mathematics and physics. Consider first the equation

$$X^4 = 16 \tag{1.1}$$

Its roots are $X = \pm 2$ and $X = \pm 2\mathrm{i}$, where i is the square root of –1, and it may be seen immediately that these solutions have a symmetrical distribution about $X = 0$; the fourth power of each of these roots is 16.

The differential equation

$$\frac{\mathrm{d}^2 Y}{\mathrm{d}X^2} + k^2 Y = 0 \tag{1.2}$$

represents a type encountered, for example, in the physics of the simple pendulum or of a mass attached to a spring. Its general solution is

$$Y = A\exp(\mathrm{i}kX) + B\exp(-\mathrm{i}kX) \tag{1.3}$$

where A, B and k are constants. If reflection symmetry is introduced across the point $X = 0$, so that X is converted into $-X$, then the solution would become

$$Y = A\exp(-\mathrm{i}kX) + B\exp(\mathrm{i}kX) \tag{1.4}$$

Differentiating Eq. (1.4) twice with respect to X, shows that Eq. (1.4) is also a solution of Eq. (1.2). If a translation constant t is applied such that X becomes $X + t$, then again, the equation and its solution have similar symmetry properties.

The French mathematician and physicist Jean-Baptiste Joseph Fourier (1768–1830) solved problems on thermal conduction by means of a series consisting of cosine and sine terms [3]. Such series, termed Fourier series, are of fundamental importance in many areas of science. For example, in crystal structure analysis, X-ray diffraction data from a crystal structure are synthesized mathematically by Fourier summation to form an image of the electron density of the structure. A Fourier series is a single-valued, continuous, periodic function that can be represented by a series of cosine and sine terms. For a periodic, continuous function $Y(X)$, defined on the interval $\pm\pi$, the function may be formulated as the Fourier series

$$Y = A_0 + \sum_{\substack{k=-\infty \\ (k\neq 0)}}^{\infty} A_k \cos 2\pi kX + B_k \sin 2\pi kX \tag{1.5}$$

A typical cosine function, shown in Fig. 1.10, exhibits reflection symmetry about the line $X = 0$, and is an *even* function, $Y(-X) = Y(X)$. Contrast it with Fig. 1.11, a typical sine function, which shows inversion symmetry about the origin and is an *odd* function, $Y(-X) = -Y(X)$.

Example 1.1

Is the function $Y(X) = \sin X - X^3 \cos X$ even or an odd?

To answer this question, it is generally sufficient to consider the value of the function for a few values of X around the point $X = 0$. Thus, the table

X	Y(X)	X	Y(X)
–3	–26.8709	3	26.8709
–2	–4.23847	2	4.23847
–1	–0.301169	1	0.301169
0	0		

shows that the function is antisymmetric across the origin, which is indicative of an odd function.

Finally, in this section, consider a cube constructed from twelve 1 Ω resistors, as shown by Fig. 1.12. Let an electrical current I be set up across the points A and G. What would be the effective resistance of this assembly of resistors for the path of the current from A to G? The problem is simplified by considering the symmetry of the cube: the planes $ACGE$ (broken lines), $AFGD$ and $BGHA$ are all mirror planes, of the type already discussed. Thus, the currents flowing along AB, AD, AE, CG, FG and HG are all equal to $I/3$.

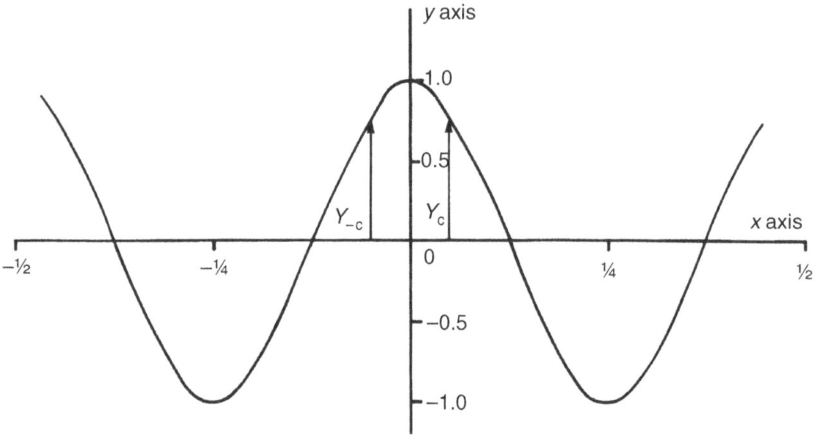

Fig. 1.10
Graph of an even function, $f(-X) = f(X)$: $Y = \cos 2\pi kX$, $k = 2$; $-1/2 \leq X \leq 1/2$.

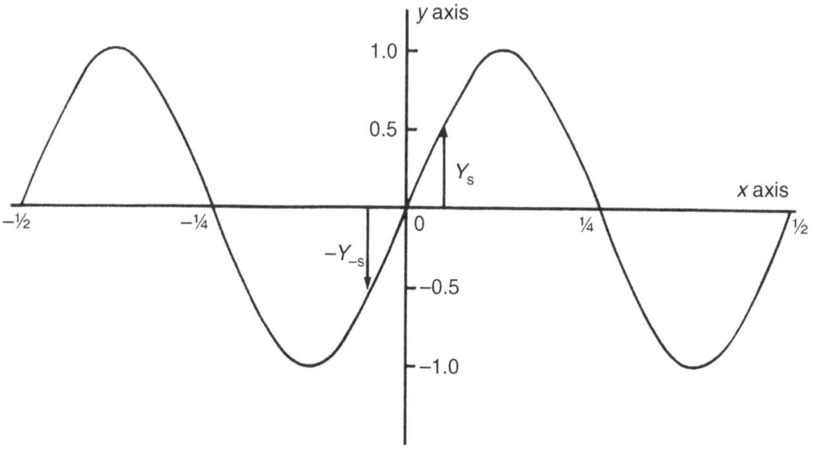

Fig. 1.11
Graph of an odd function, $f(-X) = -f(X)$: $Y = \sin 2\pi kX$, $k = 2$; $-1/2 \leq X \leq 1/2$.

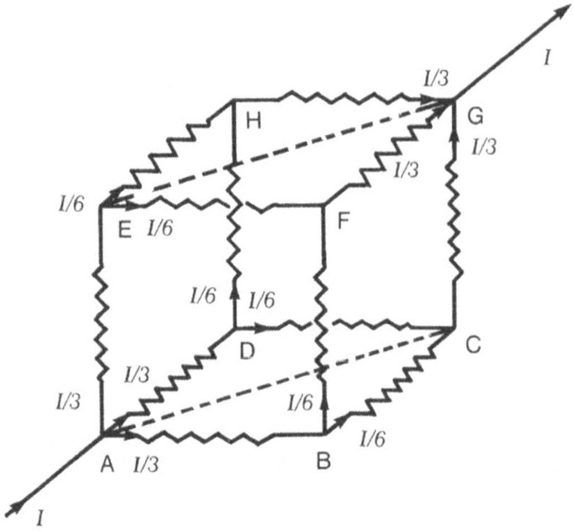

Fig. 1.12
Cubic network of 1Ω resistors; electric current I flows through the resistors from A to G.

The m symmetry requires that the currents in EF and EH are equal, and each current here is $I/6$. Similarly, the currents in BC and DC are also $I/6$, from which it follows that those in BF and DH are $I/6$. Hence, any path through the assembly of resistors from A to G has an effective resistance of 5/6 Ω.

1.6 Symmetry in music

Symmetry has featured in music from Bach to Bartók and beyond, an early example being the so-called *Crab Canon* from Bach's *Musical Offering*, which was written in a palindromic form, such that its second part is the same as the first part but written backwards (*see* Problem 1.11). Figure 1.13 illustrates an extract from the well-known classic, *Étude* No. 12, Opus 10 by Chopin, which shows approximate reflection symmetry, whereas the opening bars of Beethoven's piano sonata in C sharp minor, Opus 27, No. 2, '*Moonlight*' (Fig. 1.14) is an illustration of translational symmetry. Of course, music has to progress and to end, so that true translational symmetry is not possible. Often the symmetry in music is apparent rather than true, but can be discovered in many forms [4]. Even the piano itself exhibits an example of symmetry (Fig. 1.15).

Fig. 1.13
Reflection symmetry in music: an extract from *Étude* No. 12, Opus 10, by Chopin; the thin line marked m is an approximate reflection line.

Fig. 1.14
Opening bars of Beethoven's '*Moonlight*' sonata; an example of translational symmetry, though not of infinite extent.

Fig. 1.15
Symmetry of the piano keyboard; the keys of the triads A–C and E–G are symmetrical about the key D.

1.7 Symmetry in architecture

Architecture of all cultural periods has made extensive use of symmetry. As there are many kinds of symmetry, so many kinds of architecture have evolved. A type of symmetry is chosen so as to achieve a particular objective in terms of beauty and utility. Unlike most other arts, architecture has spatiality. While two-dimensional composition is relatively straightforward, in three dimensions there is a greater call on the imagination. The symmetry of an object is fixed but the perception of it changes with the relative position of the observer. One can not only move around but also through architectural designs, thus providing experiences of symmetry.

Architectural elements comprise solid and void components, and an architectural body may be characterized by the nature of its elements and their symmetry. Thus, differing types of symmetry occur in architecture, but there is space here to consider only a few of them.

Reflection symmetry is probably the most common architectural form (Fig. 1.16). Rotation symmetry occurs in domes, and cylindrical symmetry exists in towers and columns; even spherical symmetry occurs, as in Boullée's *Cenotaph for Isaac Newton*, projected but never brought to fruition. Chiral

Fig. 1.16
Bilateral symmetry of the Parthenon; the Greeks avoided placing the reflection line on a pillar by having an even number of pillars.

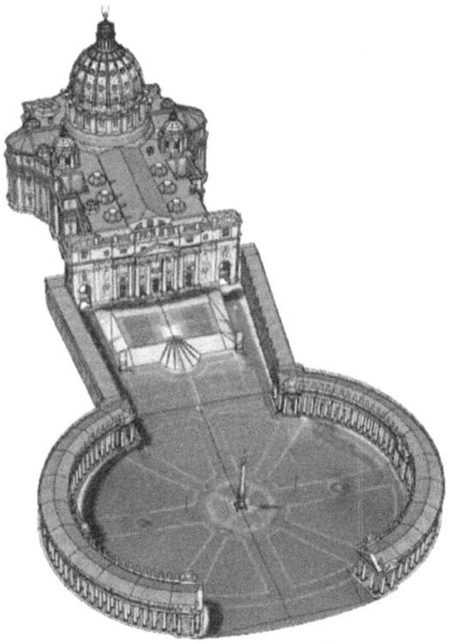

Fig. 1.17
Model of Bernini's St. Peter's Colonnade. [Reproduced by courtesy of its author, Enrico Dalbosco; <http://www.enricodalbosco.it>] (**See Plate 2**)

symmetry makes use of design based on mirror images, and is well illustrated by St. Peter's Colonnade (Fig. 1.17).

Whereas the above examples reflect point group symmetry, translational symmetry in architecture is a pseudo space group symmetry, but it falls seriously short of the ideal as its manifestations in rows of pillars or arches of viaducts are of very limited extent,.

1.8 Summary and notation

Symmetry, then, is a feature that is encountered in both scientific and everyday life. In the following chapters, crystals and molecules will be studied first as finite, non-repeating bodies, and then the expression of symmetry through symmetry functions and point groups will be developed. Subsequently, the symmetry of ideally infinite patterns will be examined, together with their application to crystal structures. On the basis of the understanding of symmetry so gained, it is complementary to discuss the mathematical approach of group theory and its applications. A brief encounter with black–white and colour symmetry will also be presented.

It may seem that the important application of symmetry in chemistry has been overlooked in this introduction. However, this topic will be effectively addressed through the work of several of the subsequent chapters, as the study of symmetry is pursued. But first a word about notation.

1.8.1 Introducing symmetry notation

There exist two notations for describing symmetry, both of which are in common use. The Hermann–Mauguin notation [5,6] is highly mnemonic, and desirable in the description of crystals and crystal structures. For work with molecules, the Schönflies notation [7] is also in general use; with space groups, however, this notation is less elegant and has little to recommend it in that application. The Hermann–Mauguin, or international, notation will be used at first; when the principles of symmetry have been grasped, there should be little problem with an alternative symmetry notation.

References 1

[1] Keats J. *Ode on a Grecian urn*. 1820.
[2] Ahmed FR and Cruickshank DWJ. *Acta Crystallogr*. 1953; 6: 385.
[3] Fourier J-BJ. *Théorie analytique de la chaleur*, Firmin, Didot [first published in 1822]. Cambridge, UK: Cambridge University Press, 2009.
[4] Dorrell P. *What is music?: solving a scientific mystery*. <http://whatismusic.info/> (2006).
[5] Hermann C. *Z. Kristallogr*. 1928; 68: 257; ibid. 69: 226; ibid. 69: 533; ibid. 76: 559.
[6] Mauguin C. *Z. Kristallogr*. 1931; 76: 542.
[7] Schönflies AM. *Krystallsysteme und Krystallstruktur*. Leipzig, 1891.

Problems 1

1.1 Seek out the following objects in the home, or elsewhere, and list their mirror symmetry. (a) Plain teacup. (b) Plain rectangular table. (c) Outer sleeve of a matchbox, ignoring the label. (d) Plain building brick. (e) Inner tray of a matchbox, ignoring colour. (f) Gaming die.

1.2 Study the patterns of Fig. 1.6 and Fig. 1.7, considering them to be extended. Illustrate each pattern by a minimum number of representative points.

1.3 Twelve identical 1 Ω resistors are connected so as to form a regular octahedron. Use the symmetry of the octahedron, which is the same as that of the cube, to evaluate its effective resistance to an electrical current between any pair of opposite apices.

1.4 Consider Fig. 1.5c. (a) What single atom change would double the number of *m* planes in the molecule? (b) Where do the planes lie in the more symmetrical molecule?

1.5 Write in upper case those letters of the alphabet that cannot exhibit *m* symmetry across a line. Your answer may depend on how you form the letters.

1.6 State whether the following functions of X have even or odd symmetry.
(a) X^3 (b) $\sin^2 X$ (c) $\cos^3(X)$ (d) $X^1 \sin X$ (e) $X^3 – X$ (f) $X \cos X$

1.7 Try out the Symmetry Game to be found at the following website: <http://www.innovationslearning.co.uk/subjects/maths/activities/year3/symmetry/shape_game.asp>.

1.8 Refer to Fig. 1.2. (a) Find the number of *m* lines in each of these figures. (b) Name the figures. (c) Deduce a relationship between the number *n* of the sides of a polygon and the number *M* of symmetry lines that it presents? (c) Why does a rectangle not follow the rule so derived?

1.9 Figure. P1.1 shows two views of the C–H bond directions in the molecule of ethane, C_2H_6, as seen in a Newman projection; the C–C bond lies normal to the plane of the diagram. What differences are there in reflection symmetry in these two conformations?

1.10 Consider the parabola $Y = (X + 3)^2 – 4$. Determine the position of the vertical symmetry line of the parabola.

1.11 Figure. P1.2 shows a portion of Bach's *Musical Offering* (*Crab Canon*). What symmetry exists in this extract of music?

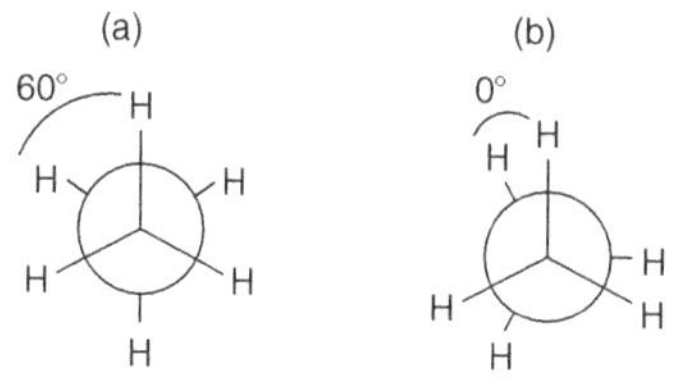

Fig. P1.1
Newman projections of the molecule of ethane, C_2H_6: (a) staggered, (b) eclipsed. The dihedral angle is 60° for (a) and 0° for (b); in the latter diagram, a twist of a few degrees has been applied to the C–C bond in order to make the eclipsed hydrogen atom apparent. The eclipsed form is approximately 12 kJ mol^{-1} higher in energy (less stable) than the staggered form.

Fig. P1.2
Portion of Bach's *Musical Offering* (*The Crab Canon*).

1.12 What aspects of symmetry are evident in the illustration of the Taj Mahal?

Fig. P1.3
The Taj Mahal.

Geometry of crystals and molecules

2

SYNOPSIS

- Crystal morphology
- Stereographic projection
- Spherical trigonometry
- Molecular geometry and its precision
- Introduction to quantum chemistry
- Crystal packing

2.1 Introduction

Crystallography, which may be called the science of structure in the widest sense, has its origins in the field of mineralogy; originally, it involved the recognition, description and classification of naturally occurring crystalline substances. Today, crystallography is a subject in its own right, of immense extent, and encompassing studies of the structures of materials ranging from those of simple elements to proteins, enzymes and polycrystalline aggregates of all types. In this chapter, crystal morphology is introduced, and methods for the determination of molecular geometry discussed.

The distinctive external feature of crystals is their bounding plane faces that intersect in straight lines and sharp angles (Fig. 2.1); nothing else in Nature exhibits such simple and sharp outlines. Figure 2.2 shows idealized quartz crystals, corresponding to the optically active left handed and right handed forms. Another important property of crystals is that the angles between corresponding faces of a crystal are the same for all different samples of the same crystalline substance. This feature was noted by the Dane, Niels Stensen [1] (1638–1686), also known as Nicolaus Steno. He demonstrated that crystals of quartz, whatever their state of development, preserved the same interfacial angles; Fig. 2.3 shows some of Steno's drawings of transverse sections of quartz; the angles between the traces of the sections are 120° in each case. This work was extended and confirmed by others, particularly by the

Fig. 2.1
Specimen of a naturally occurring quartz, SiO_2, cluster; its pale amethyst colour arises from inclusions of traces of iron and manganese. (**See Plate 3**)

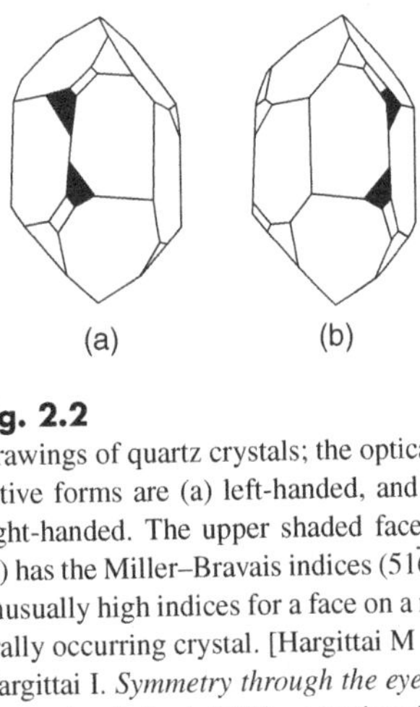

Fig. 2.2
Drawings of quartz crystals; the optically active forms are (a) left-handed, and (b) right-handed. The upper shaded face on (b) has the Miller–Bravais indices $(51\bar{6}1)$, unusually high indices for a face on a naturally occurring crystal. [Hargittai M and Hargittai I. *Symmetry through the eyes of a chemist*. 3rd ed. 2009; reproduced by courtesy of Springer Science+Business Media, NY].

Frenchman, Jean-Baptiste Romé de l'Isle [2] (1736–1790). His work led to the Law of Constant Interfacial Angles, namely, that *the angles between the faces of a given crystal species are constant and characteristic for that species*, whatever the degree of development of these faces or origin of the crystal.

A compatriot of de l'Isle, René Just (Abbé) Haüy [3] (1743–1822), published a treatise on mineralogy in which he considered crystals built up by stacking identical blocks of structure in ways that led to the shapes of naturally occurring crystals. His deductions were based initially on the observation that crystals of calcite when broken always formed rhombohedral shaped fragments, whatever the shape of the original crystal. Other crystals cleaved into different shapes, such as cubes; Fig. 2.4 illustrates some of Haüy's figures. His work laid the foundation for the law of rational intercepts, developed further by Bravais [4], which prefigured, albeit unknowingly, a shape for the unit cell of a crystal. He described also different crystal shapes, or *habits*, obtained by the packing of different shaped blocks.

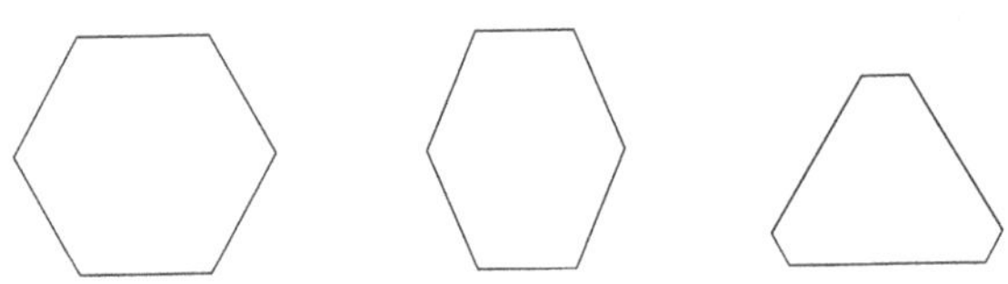

Fig. 2.3
Examples of Steno's figurers: transverse sections of quartz crystals.

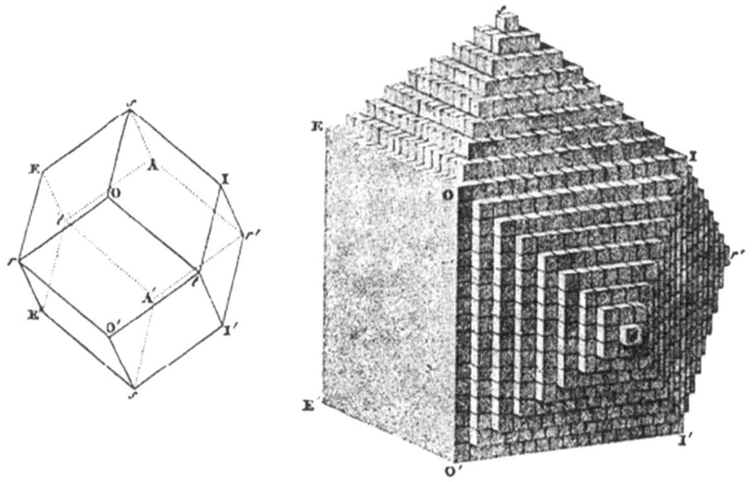

Fig. 2.4
Formation of a rhombic dodecahedron from stacked, identical cubes; alternatively expressed, as the progressive decrement of one row of cubes on each edge of lamellae added successively to a cube nucleus. The rhombic dodecahedron is the form $\{110\}$: *txsO*, (011); *rOsR*, (101); *O′tOr*, (110); a chemical example is potassium manganese sulphate, $K_2Mn_2(SO_4)_3$. [Reproduced from Häuy RJ. *Traité de mineralogy*. Paris: Louis, 1801.]

2.2 Reference axes

In order to link the faces on a crystal into a coherent spatial aggregate, they are referred to a set of reference axes, which may or may not be orthogonal, according to the requirements of the crystal and its symmetry. Consider first a straight line referred to rectangular x, y axes (Fig. 2.5). A line such as AB can be represented in a simple manner by the equation

$$Y = mX + b \tag{2.1}$$

where m is the slope, $\tan\phi$, of the line, and ϕ is the oblique angle shown on the figure. If the axes were not rectangular such that the $\angle bOa$ had a value γ, an equation of the type Eq. (2.1) could still be used, but m would now be $(\tan\phi \sin\gamma - \cos\gamma)$. Evidently, oblique axes do not lead to a convenient representation of a straight line.

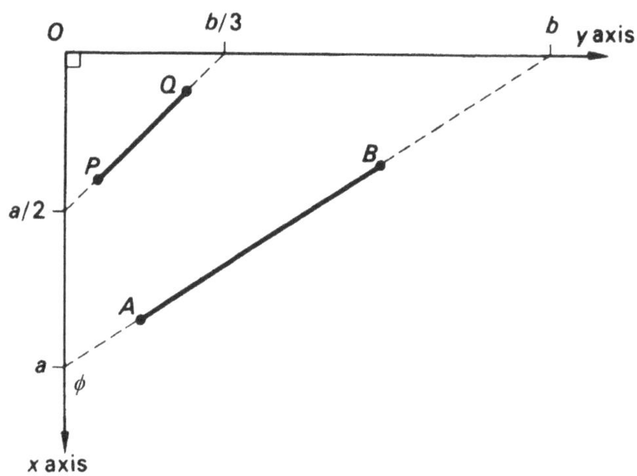

Fig. 2.5
Lines AB and PQ referred to rectangular axes.

The line AB may be described in another way. Let it intersect the x and y axes at a and b, as shown in the illustration. At $X = a$, $Y = 0$, so that from Eq. (2.1) $ma + b = 0$ whereupon $m = -b/a$. Thus, the line can be specified by the equation

$$\frac{X}{a} + \frac{Y}{b} = 1 \tag{2.2}$$

Equation (2.2) is the *intercept form* of the equation of the line AB, which will be chosen as a reference or *parametral* line; the extension to three-dimensions will be discussed shortly.

Consider next the line PQ and let its intercepts on the x and y axes be $a/2$ and $b/3$, respectively. This line can be identified uniquely by two numbers h and k such that h is the ratio of the intercept made on the x axis by a parametral line (AB) to that made by the line in question, PQ. Thus, $h = a/(a/2) = 2$. Similarly $k = 3$, and PQ may be described as the line (23); it follows that the parametral line has the indices (11). Although the magnitudes of a and b are unspecified, once the parametral line has been chosen, any other line within the same axial frame can be defined uniquely by integers h and k. It may be noted that h and k need not be integers, although they have been so chosen, but it will become clear that integral values for these indices are a feature of crystals and crystal planes.

Reference axes for a body may be chosen in an infinite number of ways (Fig. 2.6), but common sense, and convention, will be seen to dictate a choice of axes parallel to important features of the object in question. Thus, the conventional choice of axes in Fig. 2.6 is that of (c), with AB as the parametral line (11); the lines PQ, QR, RS and SP are thus indexed as (10), (01), ($\bar{1}0$) and ($0\bar{1}$). A line parallel to an axis may be said to intercept it at infinity, and

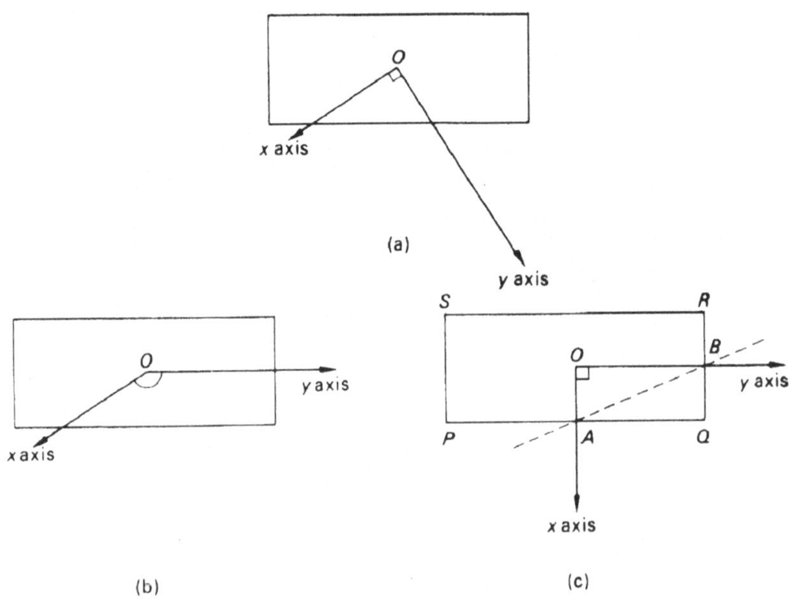

Fig. 2.6
Rectangle $PQRS$ referred to rectangular and oblique axes.

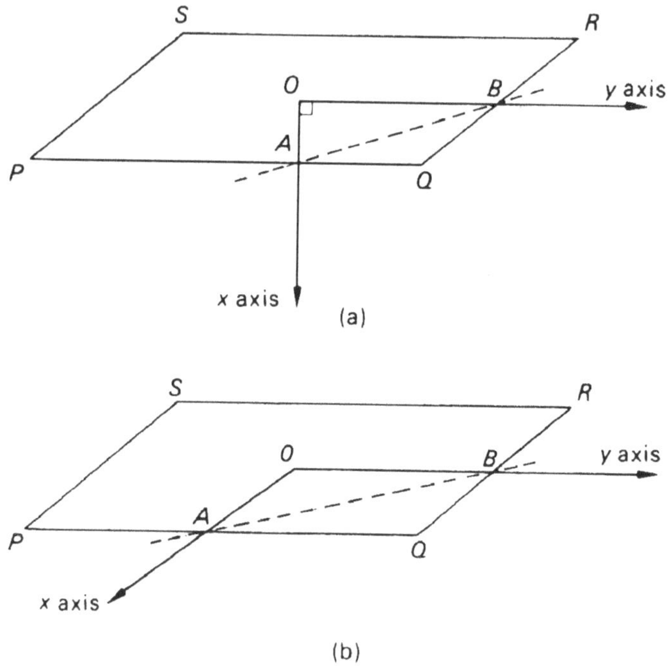

Fig. 2.7
Parallelogram referred to rectangular and oblique axes.

the corresponding index is zero; a line that intercepts an axis on the negative side of the origin, *O*, is given a negative value; hence, *RS* is $(\bar{1}0)$, read as 'bar-one zero'[1]. It is evident that the above integer indices for the perimeter lines of the rectangle would not be obtained by the orientation (a) or (b) or, indeed, any other general setting of the axes.

[1] 'one-bar zero' in the USA.

In considering a parallelogram, however, oblique axes (Fig. 2.7) are appropriate. It is left to the reader to decide that the lines *PQ*, *QR*, *RS* and *SP* are again (10), (01), $(\bar{1}0)$ and $(0\bar{1})$, respectively, provided that the reference axes are chosen parallel to the sides of the parallelogram, as in (b), with *AB* as the parametral line.

2.2.1 Crystallographic axes

Three reference axes are required in the description of a crystal. An extension of the arguments already developed leads to *x*, *y* and *z* axes being chosen parallel to important directions in the crystal. Figure 2.8 illustrates a crystal with the axes in position: the axes are not necessarily orthogonal and form a *right handed* set. Thus, if *y* and *z* lie in the plane of the paper, *x* is directed towards the reader; the succession $x \rightarrow y \rightarrow z$ simulates an *anticlockwise*, or *right handed*, screw movement. The angles between the axes are denoted $\alpha = \angle yz$, $\beta = \angle zx$, $\gamma = \angle xy$, thus providing a mnemonic connection between these six parameters.

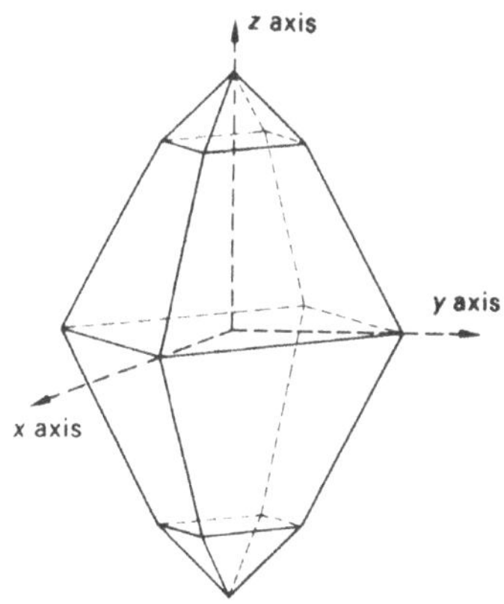

Fig. 2.8
Idealized crystal, showing the conventional, right-handed *x*, *y* and *z* axes. [Reproduced by courtesy of Springer Science+Business Media, NY, © Kluwer Academic/Plenum Publishing.]

2.3 Equation of a plane

In Fig. 2.9, the line ON is the perpendicular to the plane ABC, and its direction cosines are $\cos\chi$, $\cos\psi$ and $\cos\omega$ with respect to the x, y and z axes, respectively, as discussed in Appendix A2. The point $P(X, Y, Z)$ lies in the plane ABC; PK is parallel to OC and KM is parallel to OB. The lengths OM, MK and KP are then equal to X, Y and Z, respectively. Since ON is the projection of OP on to ON, it is equal to the sum of the projections of OM, MK and KP, all on to ON. Hence

$$d = X\cos\chi + Y\cos\psi + Z\cos\omega \tag{2.3}$$

In triangle OAN, $d = OA\cos\chi = OB\cos\psi = OC\cos\omega$. If OA, OB and OC are written as a, b and c, respectively, then Eq. (2.3) becomes

$$\frac{X}{a} + \frac{Y}{b} + \frac{Z}{c} = 1 \tag{2.4}$$

which is the *intercept equation* for the plane ABC. If the plane passes through the origin, then

$$\frac{X}{a} + \frac{Y}{b} + \frac{Z}{c} = 0 \tag{2.5}$$

Although derived for orthogonal axes, Eqs. (2.4) and (2.5) are applicable to all axial systems [5] provided that a, b and c are the intercepts of the parametral plane on the x, y, and z axes, respectively.

This equation may be viewed in another way: for this and later analyses, a short treatment of vectors and matrices is given in Appendix A3. Let OP be a vector $\mathbf{r}$, and $\mathbf{n}$ a unit vector along the normal d. Then, the scalar product $\mathbf{r}\cdot\mathbf{n}$ is given by

$$\mathbf{r}\cdot\mathbf{n} = rn\cos\angle PON = r\cos\angle PON = d \tag{2.6}$$

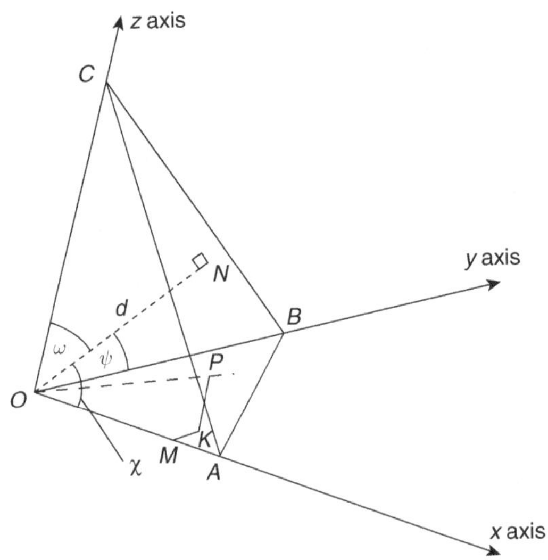

Fig. 2.9
Plane ABC intersecting the three crystallographic axes; ON is the perpendicular from the origin O to the plane.

For the case that the x, y and z axes are orthogonal, $\mathbf{r}$ and $\mathbf{n}$ are replaced by their components:

$$\mathbf{r}\cdot\mathbf{n} = (\mathbf{i}X + \mathbf{j}Y + \mathbf{k}Z)\cdot(\mathbf{i}n_X + \mathbf{j}n_Y + \mathbf{k}n_Z) = n_X X + n_Y Y + n_X Z = d \quad (2.7)$$

where $\mathbf{i}$, $\mathbf{j}$ and $\mathbf{k}$ are unit vectors parallel to the x, y, and z axes, respectively, which is an expression of the familiar general form for a plane

$$AX + BY + CZ = D \quad (2.8)$$

With oblique axes, cross terms in Eq. (2.7) must be taken into account. Since $n_X = d/a$, and similarly for n_Y and n_Z *mutatis mutandis*[2], substitution into Eq. (2.7) leads back to Eq. (2.4); in this text, the forms of Eqs. (2.4) and (2.5) are of most interest.

[2]'the necessary changes having been made'

2.4 Miller indices

Notations for specifying crystal faces have been reported over the years by Bernhardi (1808), Grassmann (1829), Frankenheim (1829), but principally by William Whewell (1825) and developed fully by his student William Hallowes Miller [6] in 1839, after whom the indices are now named. Once a set of crystallographic axes is defined for a crystal and one crystal face chosen as the parametral plane, any other face can be described in terms of three numbers, the Miller indices h, k and l. If the parametral plane is assigned suitable integral Miller indices, normally (111), then the indices of all other faces on the crystal are integers, generally *small* integers. The Miller indices h, k and l of a plane specify its orientation uniquely with respect to the reference x, y and z axes. It is conventional to write a plane as (hkl), and a set of such planes, a *form*, as $\{hkl\}$.

In an extension of the argument in Section 2.2, Fig. 2.10 shows a plane ABC with intercepts on the x, y and z axes at a, b and c respectively. This plane is chosen as the parametral plane and is labelled (111). Another plane, LMN, makes intercepts a/h, b/k and c/l along the x, y and z axes: its Miller indices are expressed as the ratios of the intercepts of the parametral plane to those of the plane LMN. The diagram has been drawn such that $OL = a/4$, $OM = b/3$ and $ON = c/2$, whereupon the plane LMN is designated (432). If fractions remain in the indices after the division, they are cleared by multiplying throughout by the lowest common denominator. If LMN had been chosen as (111), then ABC would have had the Miller indices (346). Experience has shown that a correct choice of parametral planes leads to indices of crystal faces that are rarely numerically greater than five.

The plane $ABDE$ is parallel to the z axis, and its intercept on that axis may be said to be at infinity; hence its Miller indices are (110). Similarly, $BDFG$ is the plane (010). A plane that makes a negative intercept with an axis has a corresponding negative Miller index. Thus, the plane QPB, which makes intercepts $-a/2$, b and $-c/3$, is designated $(\bar{2}1\bar{3})$. Following the argument leading to Eq. (2.4), the intercept equation of the plane (hkl) is

$$\frac{hX}{a} + \frac{kY}{b} + \frac{lZ}{c} = 1 \quad (2.9)$$

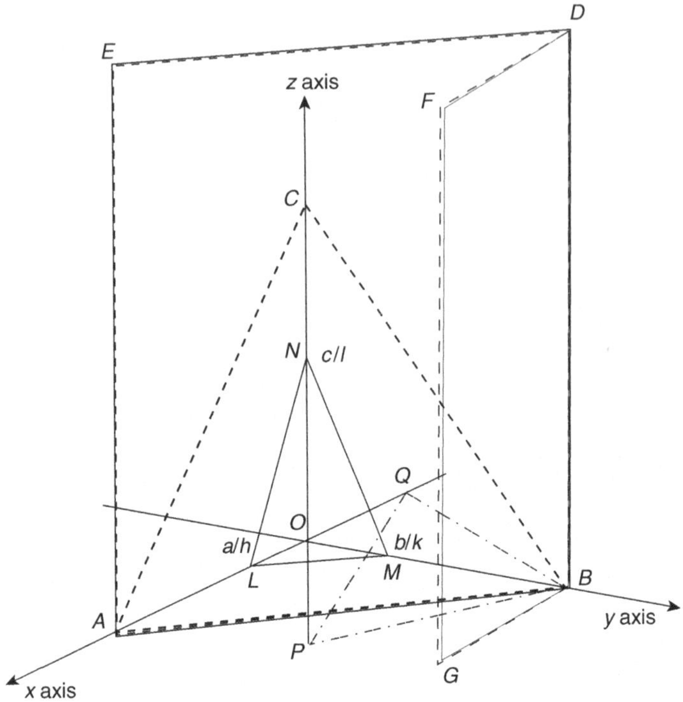

Fig. 2.10
Illustration of Miller indices of crystal planes.

and if it passes through the origin

$$\frac{hX}{a} + \frac{kY}{b} + \frac{lZ}{c} = 0 \tag{2.10}$$

since it must then satisfy the condition $X = Y = Z = 0$.

2.4.1 Miller-Bravais indices

In crystals that exhibit hexagonal symmetry, discussed in the next chapter, the conventional crystallographic axes are x, y and u, lying in a plane and at 120° one to the another, with the z axis at right angles to this plane and passing through the origin. Thus, four indices, the Miller-Bravais indices ($hkil$), define the orientation of a plane in a hexagonal crystal. Problem 2.15 will demonstrate that the index i is equal to $-(h + k)$.

2.5 Zones

An examination of a well formed crystal reveals that its faces are often symmetrically disposed in sets of two or more with respect to certain directions in the crystal. In other words, the crystal exhibits symmetry, an external manifestation of the internal ordered arrangement of the atoms and molecules that make up the crystal; Fig. 2.11 illustrates zircon, a highly symmetrical crystal.

It is clear that several faces in this crystal have a certain direction in common: these faces are said to lie in a *zone*, and the common direction is the *zone axis.* In the diagram, all vertical faces lie in a zone, and the zone axis is in the vertical direction: the normals to these faces are perpendicular to the z axis. A zone may be identified by two intersecting faces: if these faces are $(h_1\, k_1\, l_1)$ and $(h_2\, k_2\, l_2)$, then their line of intersection is given by the solution of the equations

$$\frac{h_1X}{a} + \frac{k_1Y}{b} + \frac{l_1Z}{c} = 0 \qquad (2.11)$$

$$\frac{h_2X}{a} + \frac{k_2Y}{b} + \frac{l_2Z}{c} = 0 \qquad (2.12)$$

following Eq. (2.10), since the faces can always be set to pass through the origin. The solution is the straight line

$$\frac{X}{a(k_1l_2 - k_2l_1)} = \frac{Y}{b(l_1h_2 - l_2h_1)} = \frac{Z}{c(h_1k_2 - h_2k_1)} \qquad (2.13)$$

which may be written as

$$\frac{X}{aU} = \frac{Y}{bV} = \frac{Z}{cW} \qquad (2.14)$$

where $[UVW]$ is the *zone symbol.* The result applies to all sets of reference axes, and the solution may be obtained in a simple manner.

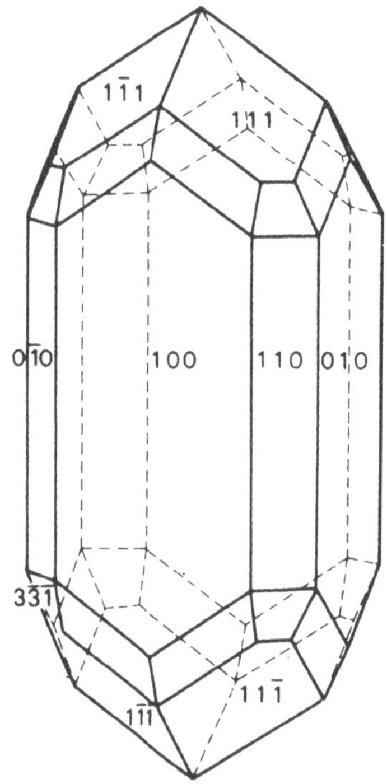

Fig. 2.11
Zircon, $ZrSiO_4$, a highly symmetrical crystal, showing Miller indices of the faces. The zone common to (111) and $(11\bar{1})$ is $[\bar{1}\,10]$. The faces $(\bar{3}31)$ and $(3\,\bar{3}\,\bar{1})$ lie in this zone: $3 \times (-1) + 3 \times 1 + 1 \times 0 = 0$. [Reproduced by courtesy of Springer Science+Business Media, New York, © Kluwer Academic/Plenum Publishing.]

Example 2.1
If two crystal faces in a zone are (123) and $(2\bar{3}1)$, what is the zone symbol?
Write the indices twice:

$$\begin{array}{c|cccccc|c} 1 & 2 & & 3 & & 1 & & 2 & 3 \\ & & \times & & \times & & \times & & \\ 2 & \bar{3} & & 1 & & 2 & & \bar{3} & 1 \end{array}$$

Then, neglecting the first and last columns, cross multiply, somewhat similarly to the solution of a determinant: $U = 2 - (-9) = 11$; $V = 6 - 1 = 5$; $W = -3 - 4 = -7$, so that the zone symbol is $[11, 5\bar{7}]$. If the two rows above in are interchanged, the zone symbol evaluates as $[\overline{11}, \bar{5}7]$; any two lines *[UVW]* and $[\bar{U}\,\bar{V}\,\bar{W}]$ are coincident but opposite in direction.

It should be noted that a first or second two-digit index in both zone symbols and Miller indices is followed by a comma.

2.5.1 Weiss zone equation

If $[UVW]$ is the zone symbol for the two faces $(h_1\, k_1\, l_1)$ and $(h_2\, k_2\, l_2)$, then any other face (hkl) lying in the same zone satisfies the Weiss zone equation [7]:

$$hU + kV + lW = 0 \qquad (2.15)$$

For the equation of the plane parallel to (*hkl*) and passing through the origin is

$$\frac{hX}{a} + \frac{kY}{b} + \frac{lZ}{c} = 0 \qquad (2.16)$$

and if $(h_1\, k_1\, l_1)$, $(h_2\, k_2\, l_2)$ and (*hkl*) are cozonal, Eq. (2.16) must contain the line Eq. (2.14). Substitution of Eq. (2.14) in Eq. (2.16) leads directly to Eq. (2.15), the *Weiss zone law*. It is normal practice to write a zone symbol as [*UVW*], and a form of such symbols as <*UVW*>.

Two zone symbols $[U_1\, V_1\, W_1]$ and $[U_2\, V_2\, W_2]$ define a face (*hkl*) that is common to both zones. From Eq. (2.15),

$$hU_1 + kV_1 + lW_1 = 0 \qquad (2.17)$$

$$hU_2 + kV_2 + lW_2 = 0 \qquad (2.18)$$

These equations may be solved for *h*, *k* and *l* in the manner of Eq. (2.13) or Eq. (2.15):

$$\left.\begin{aligned} h &= V_1W_2 - V_2W_1 \\ k &= W_1U_2 - W_2U_1 \\ l &= U_1V_2 - U_2V_1 \end{aligned}\right\} \qquad (2.19)$$

Thus, if two zone symbols are $[1\bar{3}2]$ and $[02\bar{1}]$, the face common to these zone is $(\bar{1}\,12)$. The program ZONE in the Web Program Suite solves equations of this type for zone symbols and Miller indices.

It should be noted that a study of the external features, or *morphology*, of a crystal does not permit one to distinguish between (*hkl*) and (*nh*, *nk*, *nl*), where *n* is an integer; in a morphological description, such distinction is not required. In x-ray crystallographic studies, however, general indices *nh*, *nk* and *nl* ($n = 1, 2, 3, \ldots$) for planes, or possible planes, are necessary: they represent a *family* of parallel, equidistant planes, or possible planes, parallel to (*hkl*), but $1/n$ times the perpendicular distance of (*hkl*) from the origin of the axes (Fig. 2.12).

2.5.2 Addition rule for crystal planes

If two faces $(h_1\, k_1\, l_1)$ and $(h_2\, k_2\, l_2)$ lie in the zone [*UVW*], then from Eqs. (2.18)–(2.19) it follows that

$$(ph_1 + qh_2)U + (pk_1 + qk_2)V + (pl_1 + ql_2)W = 0 \qquad (2.20)$$

where *p* and *q* are integers. This equation can be used to index crystal faces. A face *H*(*hkl*) lies at the intersection of the two zones [(210), (011)] and [(010), (101)]; using Eq. (2.20) with the first zone, $H\,(2p_1, p_1 + q_1, q_1)$ is obtained, and from the second zone $H(q_2, p_2, q_2)$. By equating the two sets of indices, it follows that

$$2p_1 = q_2,\; p_1 + q_1 = p_2,\; q_1 = q_2 \qquad (2.21)$$

whence $p_2 = \frac{3}{2}q_2$, $q_1 = q_2$. Thus, the indices for *H* are in the ratio $q_2 : p_2 : q_2 = 1 : \frac{3}{2} : 1$, namely, (232); $(\bar{2}\,\bar{3}\,\bar{2})$ is an equivalent result. Of course, the

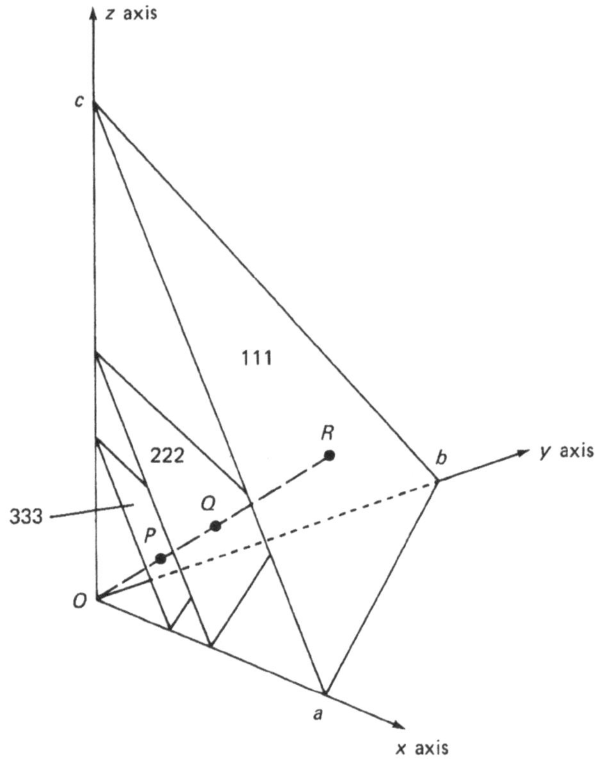

Fig. 2.12
Part of a family of parallel, equidistant planes: *OR* is perpendicular to the three planes of the family shown in the diagram; *OQ* = *OR*/2 and *OP* = *OR*/3.

same results can be obtained by first determining the two zone symbols, and then using them to determine the plane common to both zones.

2.6 Projection of three-dimensional features

'And there is a certain facility for learning all other subjects in which we know that those who have studied geometry lead the field' [8]. It is often necessary to represent three-dimensional features of an object by means of a plane projection. Architects use orthographic projections, such as plans, elevations and sections; they are forms of parallel projection in which all lines from the object project orthogonally on to the projection plane. A Mercator projection is a cylindrical map projection in which the bearings on a globe are equal to those on the projection; thus, it is used for the atlas of the world and by navigators. For crystals, however, two projections have been used:

- The *gnomonic projection* has its projection plane at the top of a spherical representation of the crystal, normal to [001], and the normals to crystal faces are extended to intersect the plane in points so as to form the projection. However, many such intersections would lie outside the edges of a projection plane of workable size.
- The *stereographic projection*, which was known to Ptolemy (90–168), maps a spherical representation of the crystal on to a plane that is normal to [001],

but which passes through the centre of the sphere. Among the properties of the stereographic projection, that which is important in crystal morphology is its property of preserving angular relationships truly, which means also that it preserves the symmetry underlying those relationships. The stereographic projection will be discussed with respect to the crystal shown in Fig. 2.13.

2.6.1 Stereographic projection

The combination of three sets of symmetry related faces is shown by the idealized crystal in Fig. 2.13: six derived from a cube, eight from an octahedron and twelve from a rhombic dodecahedron. Each of the faces of the crystal has a related parallel face on the crystal, d as shown and d' on the opposite side, across the origin of the x, y and z axes. Lines are drawn from an origin point O, placed conveniently at the centre of the crystal, normal to all faces on the crystal, and extended as necessary. A sphere of arbitrary radius ρ, centred at O, is described around the bundle of radiating normals, such that they intersect the surface of the sphere, forming the spherical projection of the crystal (Fig. 2.14). The plane of projection is shown as $ABCD$ in Fig. 2.15, and its intersection with the sphere is the *primitive plane*, or just the *primitive.*

Each point of intersection, such as R, on the upper hemisphere is now joined to P, the lowest point on the sphere, and the intersection r of the line with the primitive represents the stereographic projection, or *pole*, of the corresponding face on the crystal; it is represented as • on the stereographic projection, or *stereogram*. If the crystal is oriented such that the vertical direction dd' in the crystal corresponds to the diameter PP' of the sphere, then the poles b, f, e, g', b', f', e' and g lie on the perimeter of the primitive.

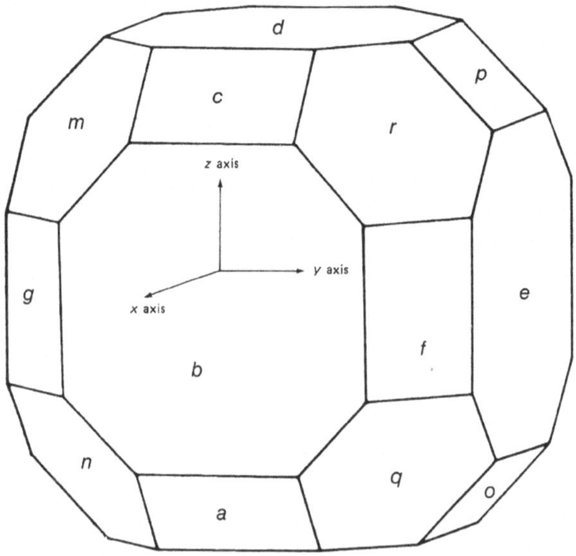

Fig. 2.13
Cubic crystal showing three forms of planes: cube—b, d, e and parallel faces; octahedron—m, n, q, r and parallel faces; rhombic dodecahedron—a, f, c, g, o, p and parallel faces. [Reproduced by courtesy of Springer Science+Business Media, New York, © Kluwer Academic/Plenum Publishing.]

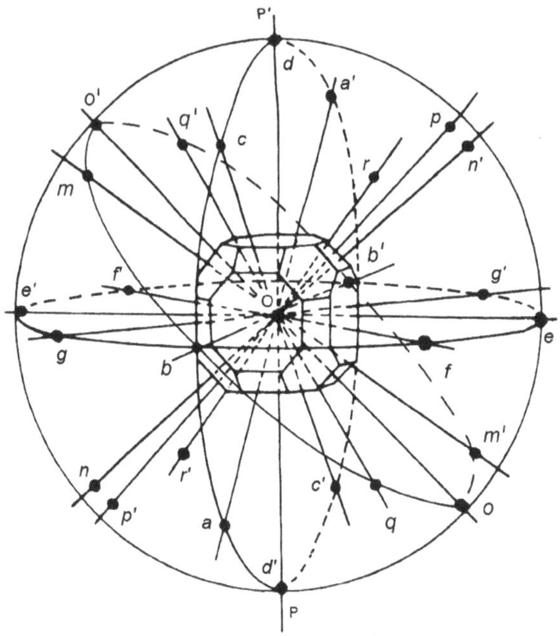

Fig. 2.14
Spherical projection of the crystal in Fig. 2.13.

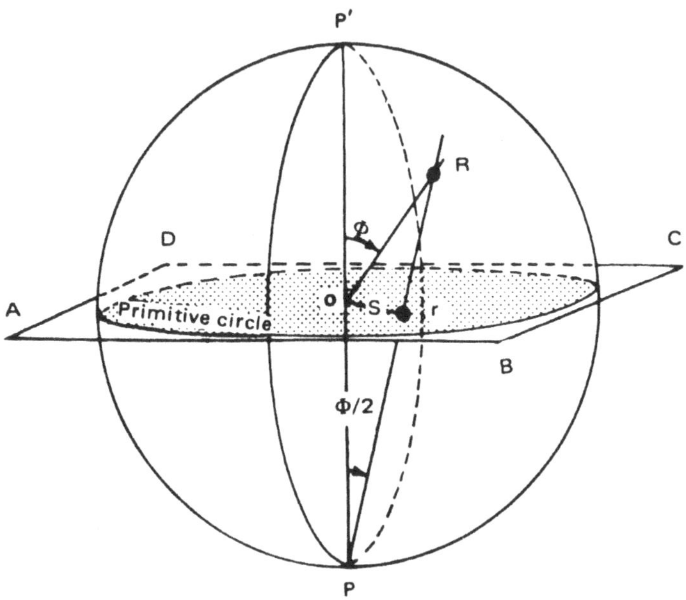

Fig 2.15
Development of the stereographic projection from the spherical projection. [Reproduced by courtesy of Springer Science+Business Media, New York, © Kluwer Academic/Plenum Publishing.]

A continuation of this process with the intersections from the lower hemisphere would lead to an unwieldy extension of the stereogram, as shown by Fig. 2.16. This situation is avoided by joining the intersections on the lower hemisphere to the upper point P' on the sphere; these poles are marked as ○ on the stereogram, as exemplified by the poles $1 1 \bar{1}$, $0 1 \bar{1}$ and $\bar{1} 1 \bar{1}$, for example, on this figure.

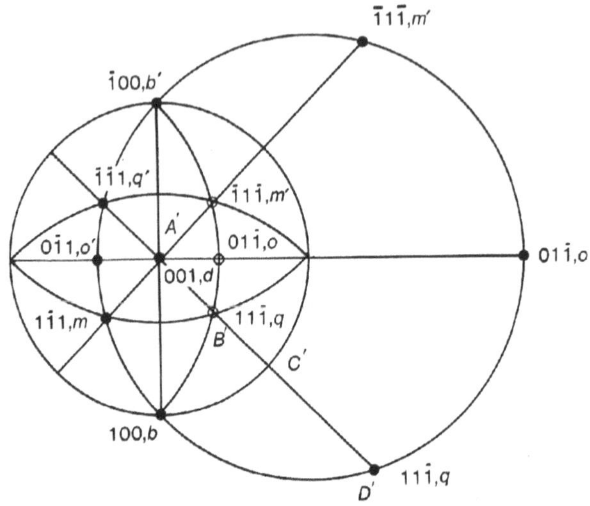

Fig. 2.16
Partial stereogram of the crystal in Fig. 2.13, showing the zone b, q, o, m', b', q', o', m projected both conventionally and from the lower point P; in the latter case, m', o and q now lie well outside the primitive. [Reproduced by courtesy of Springer Science+Business Media, New York, © Kluwer Academic/Plenum Publishing.]

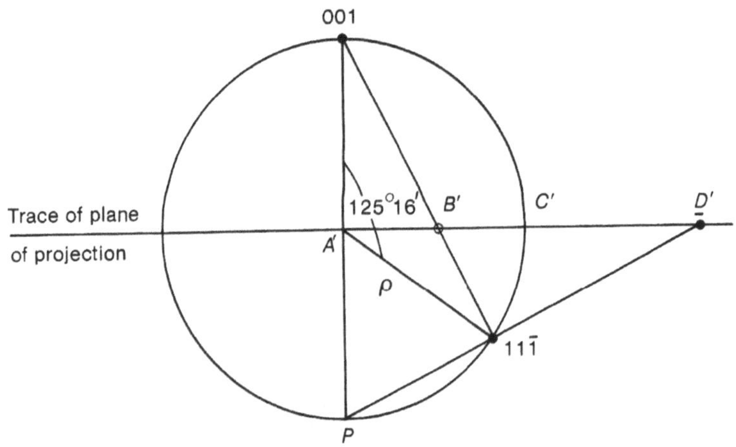

Fig. 2.17
Rationale for the conventional projection of a pole for a face lying in the lower hemisphere.

The angle $\angle 001 - 11\bar{1}$, the angle between the normals to (001) d, and ($11\bar{1}$) q, is 125.27° on the cubic crystal, and is represented by the distance $A'D'$ on the stereogram, but also by $A'C' + C'B'$ (90° + 35.27°), Fig. 2.17. The pole of face q if projected from P is the • at D', and the stereographic distance $A'D'$ is $\rho\tan(125.27°)$. The conventionally projected pole is the ○ at B', so that the angle $\angle 001 - 11\bar{1}$ is correctly represented by $A'C' + C'B'$.

The completed stereogram is shown in Fig. 2.18. Two important features have emerged so far: the distances on a stereogram are non-linear measures of the angles that they represent, and all circles drawn on the sphere (Fig. 2.14) project as circles, as proved in Appendix A4. Thus, the curve $Z_1\,Z_1'$ in Fig. 2.18 is an arc of a circle; it is the projection of the great circle b, q, o, m', b', q', o', m in Fig. 2.14. A *great circle* is the trace on the sphere made by a plane passing through the centre, O. The primitive circle and straight lines such as $Z_2\,Z_2'$ are special cases of great circles that are either the plane of projection or a plane normal to it. Great circles may be likened to meridians on a globe of the world: circles formed on the sphere by the intersections of planes that do not pass through the centre are called *small circles*; they may be likened to parallels of latitude.

The pole of a great circle is the stereographic projection of a point on the sphere that is 90° away from all points on the great circle. Thus, the pole for the faces b, f, e, g', b', f', e', g is at d (or d'); the eight faces lie on a *zone circle* for which dd' is the zone axis.

In order to construct a stereogram such as that in Fig. 2.18, the interfacial angles on the crystal must be measured. This is carried out with a goniometer, of which Fig. 2.19 represents the first such instrument, devised by Arnould Carangeot (*ca.* 1780), a student of Romé de l'Isle. Large crystals are needed with this instrument, and developments led to the reflecting goniometer (Fig. 2.20) with which precise goniometric measurements can be made.

The principle of the reflecting goniometer is shown in Fig. 2.21. A crystal is arranged to rotate about a zone axis, indicated by the point O, which is set perpendicular to the plane containing the incident light and that reflected from the crystal planes. Parallel light reflected from the face AB is received by a telescope. If the crystal is rotated in a clockwise direction from this reflecting position, a reflection from the face BC will be received when the crystal has been turned through the angle Φ; then, the interfacial angle is $(180 - \Phi)°$. However, it is the angle Φ between the normals that is actually plotted on the stereogram. Accurate goniometry brought a quantitative significance to observable angular relationships in crystals.

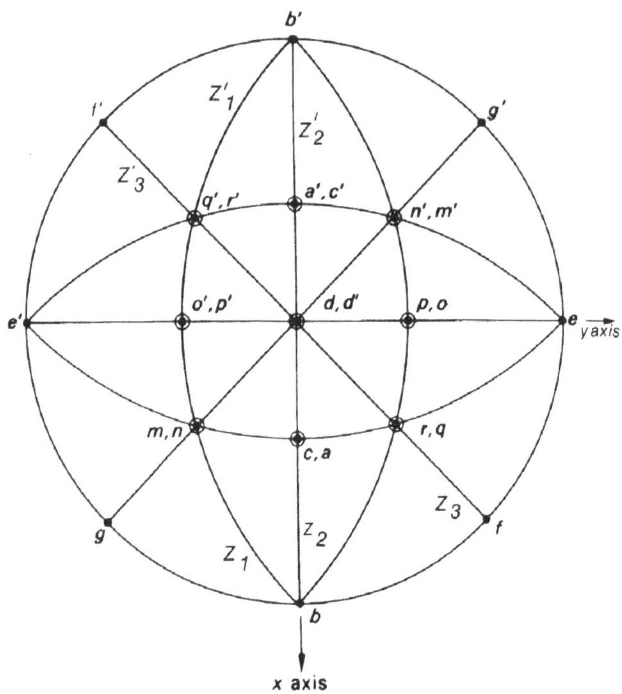

Fig. 2.18
The completed stereogram of the crystal in Fig. 2.13. The zone circles $Z_1\,Z_1'$, $Z_2\,Z_2'$ and $Z_3\,Z_3'$ have the zone symbols [011], [010] and [1$\bar{1}$0], respectively.

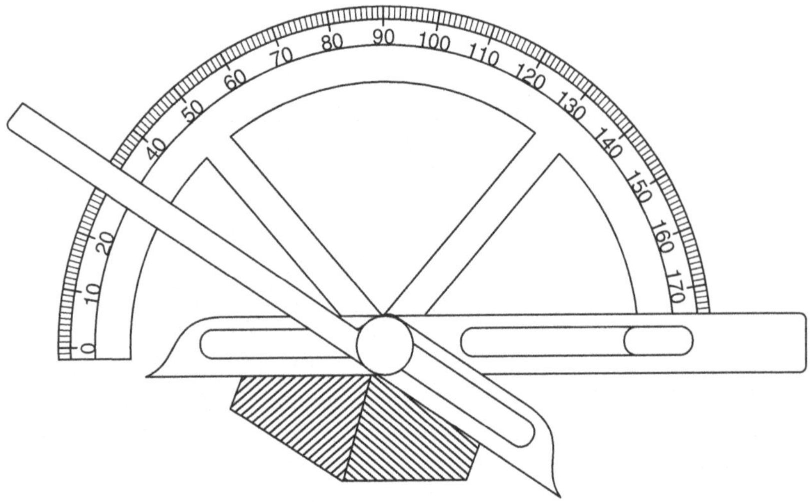

Fig. 2.19
Contact goniometer, with a crystal in the measuring position. [Reproduced by courtesy of Woodhead Publishing, UK.]

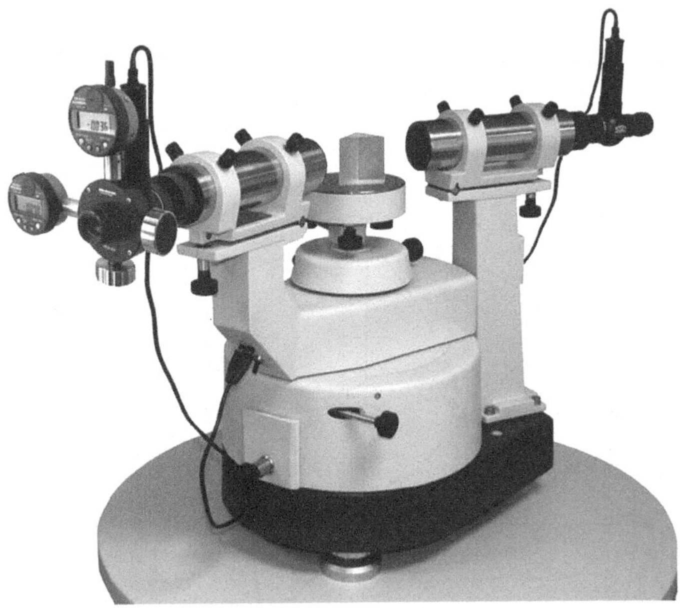

Fig. 2.20
Modern reflecting goniometer. [Reproduced by courtesy of Moeller-Wedel Optical GmbH.]

Once the necessary angular measurements of have been made, a crystal orientation with respect to the sphere must be chosen. For the crystal example of Fig. 2.13, the poles of the zone $b, f, e, g', b', f', e', g$ have been chosen to lie on the primitive circle (Fig. 2.18) and the poles of the zone normal to it,

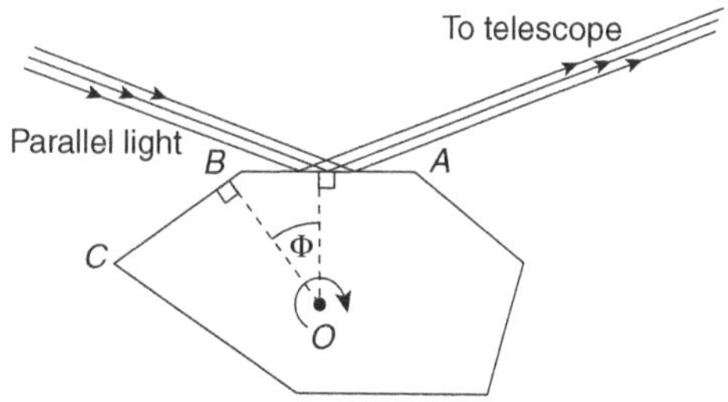

Fig. 2.21
Principle of the reflecting goniometer. Starting at the given reflecting position, light next appears in the telescope after a rotation of the crystal through the inter-normal angle ϕ. [Reproduced by courtesy of Springer Science+Business Media, New York, © Kluwer Academic/Plenum Publishing.]

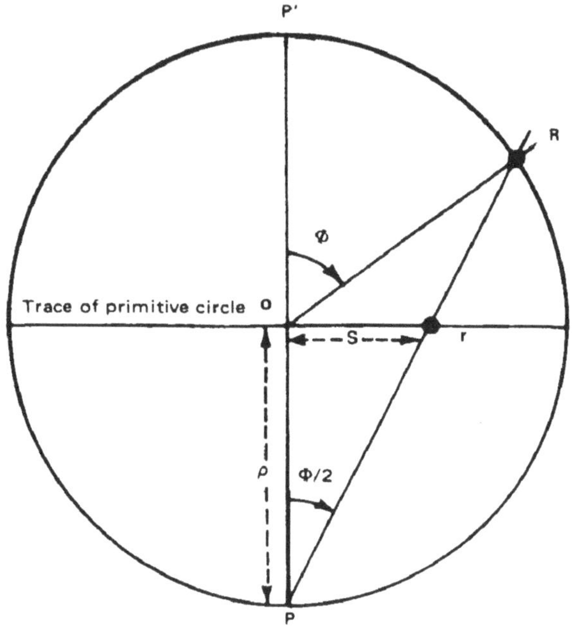

Fig. 2.22
Vertical section through R in Fig. 2.15. [Reproduced by courtesy of Woodhead Publishing, UK.]

b, c, d, a', b', c', d', project as the line $Z_2\ Z_2'$. The angle bf is found by measurement to be 45°, and the zone $f, r, d, q', f', r', d', q$, because it is normal to the primitive, projects as the line $Z_3\ Z_3'$. The distance S of the pole r from the centre of the stereogram is given, from Fig. 2.22, as

$$S = \rho \tan(\Phi/2) \tag{2.22}$$

where Φ is the angle between the normals to the faces d and r. Other faces may be plotted in like manner.

Poles may be plotted on a stereogram also by means of a Wulff net (Fig. 2.23): the curves running from top to bottom on the chart are projected great circles, and those left to right are projected small circles. In use, the centre of the net is pivoted about the centre of the stereogram, the interfacial angle measured along the appropriate great circle on the chart and the pole plotted. The pole r lies at the intersections of the b, r, p and f, r, d zones: thus, if $\angle dp$ and $\angle df$ are known, the two zone circles can be plotted and the pole r located.

It follows that a Wulff net may be used to measure the angle between two poles on a stereogram. The net is aligned as before, and rotated until the two

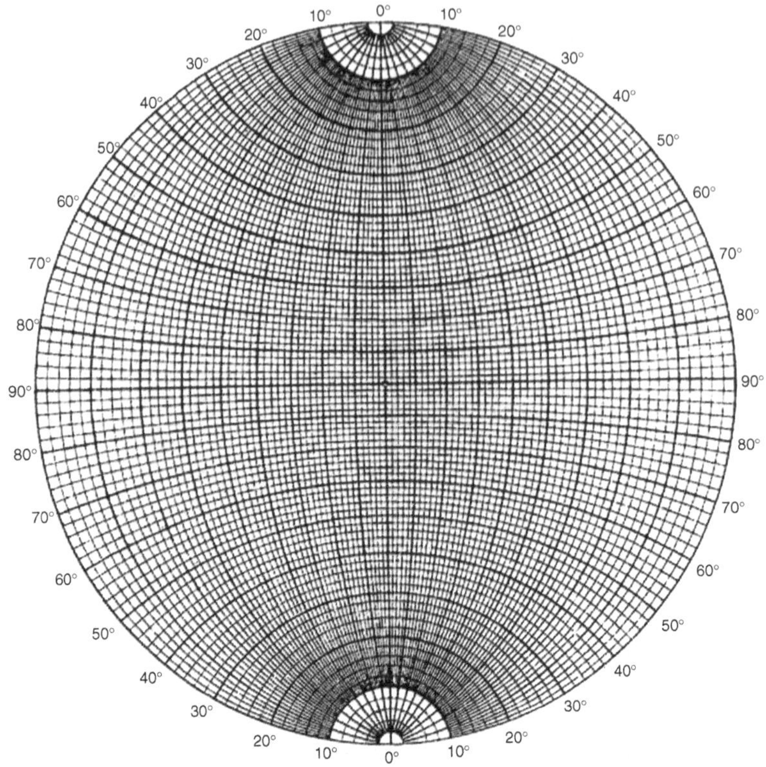

Fig. 2.23
A Wulff net. [Reproduced by courtesy of Woodhead Publishing, UK.]

poles lie on one and the same great circle; the angle is then read directly from the net, and the precision is about $^1/_2$° with a 60 mm radius net.

In order to index all poles on the stereogram, a parametral plane must be chosen. The face *r* which intersects all axes may be taken as the pole of (111), and the remaining faces indexed as shown in the diagram (Fig. 2.24), using the addition rule and the Weiss zone equation. It is sufficient to write the indices of the poles marked • on the stereogram; if *r* is 111, the symbol ○ at the same location is clearly $11\bar{1}$. Observe the notation: the *face* (111), the *pole* 111, the *zone* [111]. In general, a zone [*UVW*] does not coincide with the normal to the face (*UVW*), although it does so for the cubic example just studied.

2.6.2 Calculations in stereographic projections

In order to utilize fully the precision of goniometric measurements, the equations of spherical trigonometry must be applied. Figure 2.25 shows a spherical triangle *ABC* on a portion of the surface of a sphere, centre *O*, formed by the intersections of great circles that are represented by their arcs *AB*, *AC* and *BC*. The arcs *a*, *b* and *c* are the *sides* of the spherical triangle and *A*, *B* and *C* are its *angles*. The sides are measured by the angles they subtend at the centre; for example, *a* is determined by the ∠*BOC*.

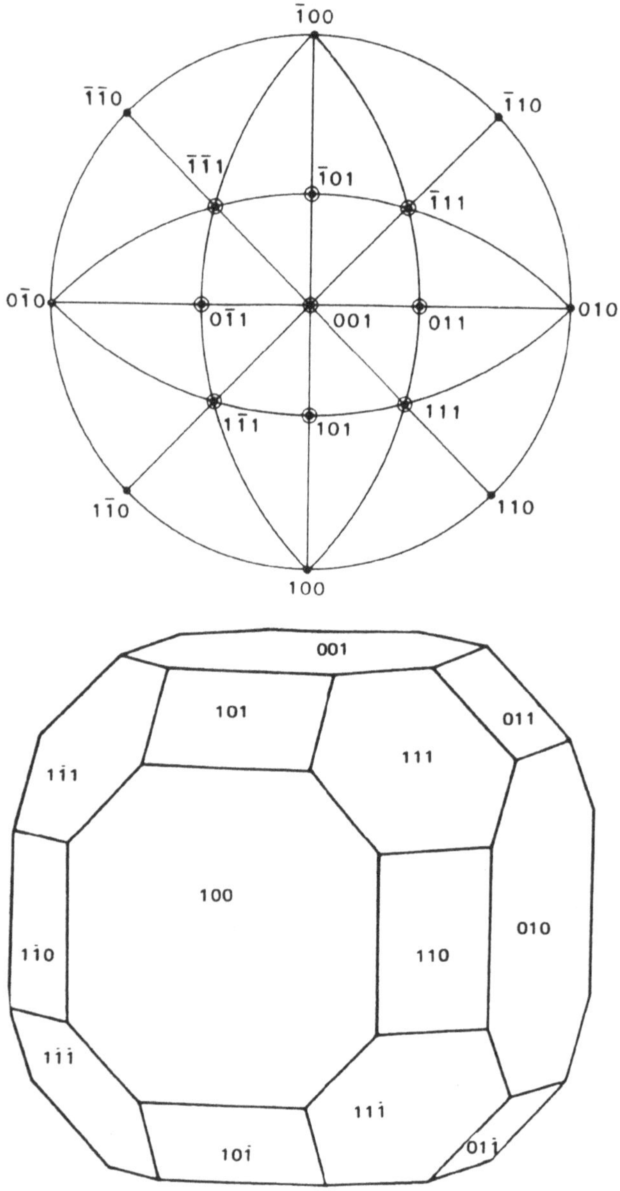

Fig. 2.24
Crystal of Fig. 2.13: (a) Indexed stereogram. (b) Miller indices attached to the crystal faces. [Reproduced by courtesy of Springer Science+Business Media, New York, © Kluwer Academic/Plenum Publishing.]

The angle *A* is measured by the angle between the tangents at *A* to the arcs *AB* and *AC*; it is the *dihedral* angle between the planes *AOB* and *AOC*. Since *AB* and *AC* are portions of zone circles, the angle at their intersection is the angle between the corresponding zone axes, that is, the angle between two edges, or possible edges, of a crystal, and is equal to the plane angle $\angle PAQ$.

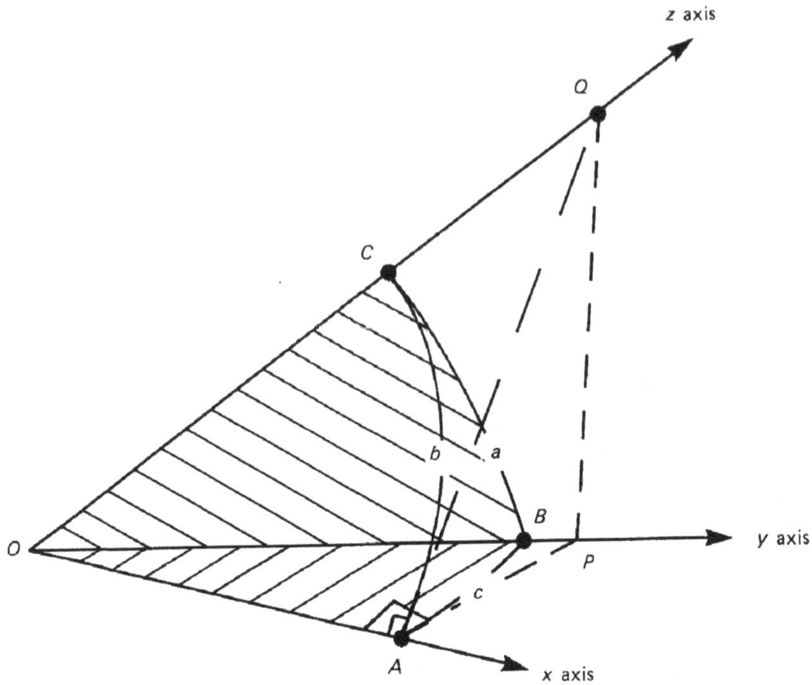

Fig. 2.25
Spherical triangle ABC on the surface of a sphere; APQ is the tangent plane at A. [Reproduced by courtesy of Woodhead Publishing, UK.]

2.6.2.1 Cosine formulae

In triangle APQ, for $A < 180°$

$$PQ^2 = AP^2 + AQ^2 - 2APAQ\cos A$$

and in triangle OPQ,

$$PQ^2 = OP^2 + OQ^2 - 2OPOQ\cos a$$

Hence,

$$0 = AP^2 - OP^2 + AQ^2 - OQ^2 - 2AP\,AQ\cos A + 2OP\,OQ\cos a$$

Since $\angle OAQ = \angle OAP = 90°$

$$AO^2 = OQ^2 - AQ^2 = OP^2 - AP^2$$

Thus

$$OP\,OQ\cos a = AO^2 + APAQ\cos A$$

and

$$\cos a = \frac{AO^2}{OP\,OQ} + \frac{APAQ\cos A}{OP\,OQ}$$

whence

$$\cos a = \cos b\cos c + \sin b\sin c\cos A \qquad (2.23)$$

with similar expressions for $\cos b$ and $\cos c$ by cyclic permutation.

2.6.2.2 Sine formulae

From Eq. (2.23)

$$\sin^2 A = (1 - \cos^2 A) = \frac{\sin^2 b \sin^2 c - \cos^2 a - \cos^2 b \cos^2 c + 2\cos a \cos b \cos c}{\sin^2 b \sin^2 c} \tag{2.24}$$

Similarly

$$\sin^2 B = (1 - \cos^2 B) = \frac{\sin^2 c \sin^2 a - \cos^2 b - \cos^2 c \cos^2 a + 2\cos a \cos b \cos c}{\sin^2 c \sin^2 a} \tag{2.25}$$

Dividing Eq. (2.24) by Eq. (2.25) and replacing $\sin^2 a$ by $(1 - \cos^2 a)$:

$$\frac{\sin^2 A \sin^2 b}{\sin^2 B \sin^2 a} = 1$$

and similarly for $\sin^2 b$ and $\sin^2 c$. Since the angles in a spherical triangle cannot be greater than 180°, the negative square root is ignored; hence,

$$\frac{\sin A}{\sin a} = \frac{\sin B}{\sin b} \tag{2.26}$$

and, by analogy, equal also to $\frac{\sin C}{\sin c}$.

2.6.2.3 Tangent formula

From Eq. (2.23),

$$\cos A = \frac{\cos a - \cos b \cos c}{\sin b \sin c}$$

then

$$1 - \cos A = 1 - \frac{\cos a - \cos b \cos c}{\sin b \sin c} = \frac{\cos(b - c) - \cos a}{\sin b \sin c}$$

Next, and using the formulae for double angle and the sum of two cosines,

$$2\sin^2\frac{A}{2} = \frac{\cos(b - c) - \cos a}{\sin b \sin c} = -2\sin\frac{(b - c) + a}{2\sin b \sin c}\sin\frac{(b - c) - a}{2\sin b \sin c}$$

$$= \frac{\tfrac{1}{2}\sin(a + b - c)\tfrac{1}{2}\sin(a - b + c)}{\sin b \sin c}$$

Let $s = \dfrac{a+b+c}{2}$; then, $a+b-c = 2s-2c = 2(s-c)$ and $a-b+c = 2s-2b = 2(s-b)$. Thus,

$$\sin\frac{A}{2} = \sqrt{\frac{\sin(s-b)\sin(s-c)}{\sin b \sin c}}$$

From a similar argument beginning with $1+\cos A = 1 + \dfrac{\cos a - \cos(b+c)}{\sin b \sin c}$,

$$\cos\frac{A}{2} = \sqrt{\frac{\sin s \sin(s-a)}{\sin b \sin c}}$$

Hence, the tangent formula:

$$\tan\frac{A}{2}\sqrt{\frac{\sin(s-b)\sin(s-c)}{\sin s\ \sin(s-a)}} \tag{2.27}$$

with similar equations for $\tan B/2$ and $\tan C/2$ by cyclic permutation. Simplifications of the formulae can be obtained for right-angled spherical triangles.

2.6.2.4 Right-angled spherical triangles

If an angle or a side of a spherical triangle is 90°, it can be solved by one or other of the rules devised by Napier [9]. However, adequate simplification derives readily from Eqs. (2.23) and (2.26), so that such rules are hardly necessary. Consider the spherical triangle *ABC* in Fig. 2.25, and let the element *A*, the angle ∠*BAC*, be 90°. Then, from Eq. (2.23),

$$\cos a = \cos b \cos c \tag{2.28}$$

and from Eq. (2.26),

$$\sin b = \sin a \sin B \tag{2.29}$$

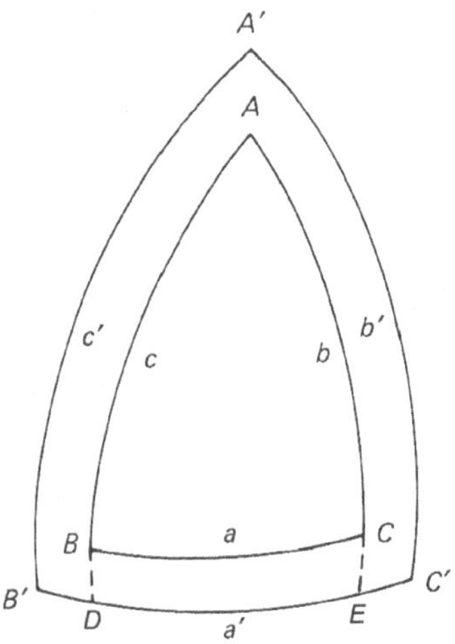

Fig. 2.26
Spherical triangle *ABC* and its polar spherical triangle *A′B′C′*. [Reproduced by courtesy of Woodhead Publishing, UK.]

2.6.2.5 Polar spherical triangles

In Fig. 2.26, *ABC* is a spherical triangle. Another spherical triangle *A′B′C′* is drawn such that all points on the arc *B′C′* are 90° from *A*, so that *A* is the pole of the great circle represented by *B′C′*; similarly *B* and *C* are poles of the arcs *C′A′* and *A′B′*, respectively. Then, triangle *A′B′C′* is defined as the *polar triangle* of *ABC*.

Since also *A* is the pole of *B′C′* and *C* is the pole of *A′B′*, it follows that *B′* is 90° from the great circle arc *CA*. Similarly, *A′* and *C′* are 90° from *BC* and *AB* respectively. Thus, *ABC* is the polar triangle of *A′B′C′*.

Let the arcs *AB* and *AC* be extended to cut *B′C′* in *D* and *E*, respectively; *A* is the pole of *DE*, so that *DE* is a measure of the angle *A*. But $B'E + C'D = B'C' + DE$ and, since *B′* and *C′* are the poles of *CE* and *BD* respectively, $B'E = C'D = 90°$. Thus, $B'C' + DE = a' + A = 180°$ or

$$a' = 180 - A \tag{2.30}$$

and because triangles ABC and $A'B'C'$ are polar

$$a = 180 - A' \tag{2.31}$$

Similarly, $b' = 180 - B$, $b = 180 - B'$, $c' = 180 - C$, $c = 180 - C'$. Substituting for a', b', c' and A', and using Eq. (2.23) for triangle $A'B'C'$,

$$\begin{aligned}\cos(180 - A) &= \cos(180 - B)\cos(180 - C)\\ &\quad - \sin(180 - B)\sin(180 - C)\cos(180 - a)\end{aligned} \tag{2.32}$$

which rearranges to

$$\cos a = \frac{\cos A + \cos B \cos C}{\sin B \sin C} \tag{2.33}$$

with similar results for $\cos b$ and $\cos c$ by cyclic permutation.

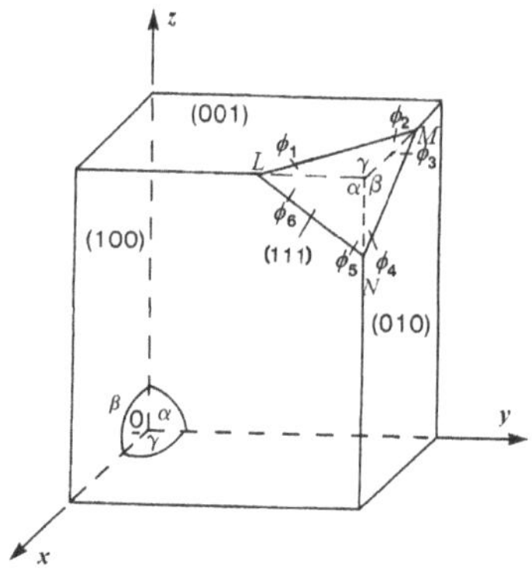

Fig. 2.27
Interaxial angles α, β, γ and parametral plane *LMN*. The crystal exhibits the {100}, {010} and {001} forms and the plane (111). The zone symbols are [100] from (010) and (001), [010] from (001) and (100), and [001] from (100) and (010). The angles ϕ are discussed in the text, and appear on Fig. 2.28.

2.6.3 Axial ratios and interaxial angles

Although the stereogram does not reveal information about the individual lengths a, b and c, it does provide a measure of the axial ratios and interaxial angles. The *axial ratios* are written normally in the form $\frac{a}{b} : 1 : \frac{c}{b}$. Consider the general case of a triclinic crystal (Fig. 2.27) so that three oblique *interaxial angles* are involved. The zones with symbols [100], [010] and [001] lie normal to the x, y and z axes, respectively. In the spherical triangle ABC (Fig. 2.28), the interaxial angles α, β and γ are the supplements of the angles at A, B and C, respectively. From the plane triangle LMN in Fig. 2.27, the axial ratios

$$\frac{a}{b} = \frac{\sin \phi_1}{\sin \phi_2} \tag{2.34}$$

$$\frac{c}{b} = \frac{\sin \phi_6}{\sin \phi_5} \tag{2.35}$$

are obtained. These angles may not always be obtainable experimentally, whereupon it is necessary to apply the equations derived above in other ways, as the following examples show.

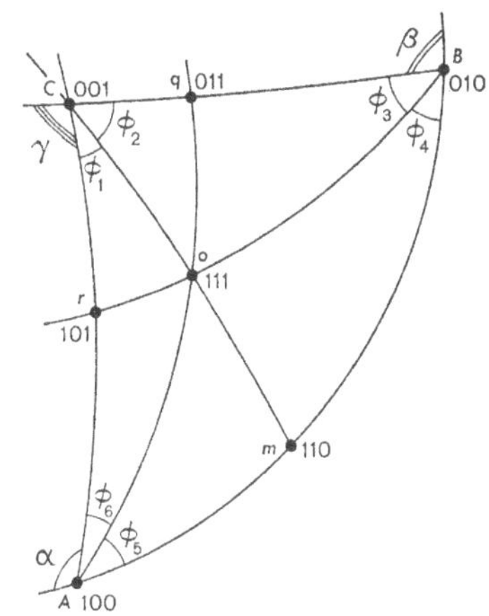

Fig. 2.28
Partial stereogram corresponding to the crystal drawing in Fig. 2.27.

Example 2.2
In potassium sulphate, K_2SO_4, which is orthorhombic, the following average goniometric results were obtained: $\angle 100 - 011 = 43.87°$ $\angle 001 - 111 = 56.18°$. What are the axial ratios for this substance?

Solve first for ψ. In the triangle 001 − 011 − 111 (Fig. 2.29)

(a) $\dfrac{\sin 90°}{\sin 56.18°} = \dfrac{\sin \psi}{\sin(90 - 43.87)°}$, so that $\psi = 60.20°$. $\dfrac{a}{b} = \tan 100 - 110 = \tan(90 - \psi)° = 0.573$

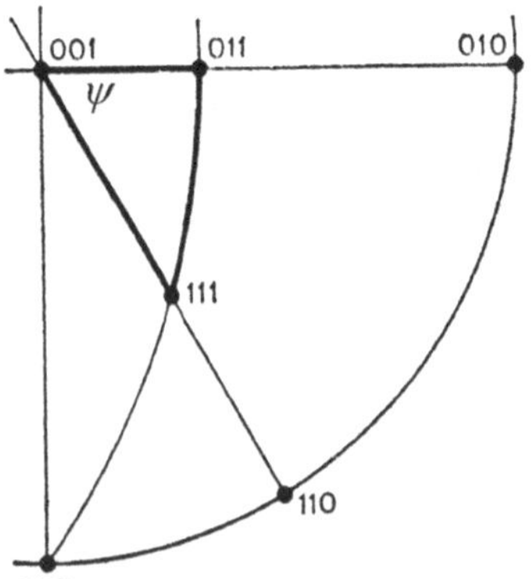

Fig. 2.29
Partial stereogram for potassium sulphate: 111–001 = 56.18°; 111–100 = 43.87°.

(b) $\cos 56.18° = \cos \angle 001 - 011 \cos 46.13°$; $\dfrac{c}{b} = \tan 001 - 001 =$

$$\tan\left(\cos^{-1}\frac{\cos 56.18°}{\cos 46.13°}\right) = 0 \cdot 7429.$$

Hence, $a : b : c = 0.573 : 1 : 0.742$.

Example 2.3
The following data were measured for the monoclinic crystal lead chromate, $PbCrO_4$: $\angle 1\bar{1}0 - 110 = 86.32°$ $\angle 010 - 111 = 80.95°$ $\angle 110 - 111 = 48.22°$. Find the axial ratios and the β angle.

Solve first for $\angle 001 - 100$ and β. See Fig. 2.30. In triangle 001 − 100 − 011 : $\cos 80.95° = \cos 43.16° \cos \angle 001 - 100$, so that $\angle 001 - 100 = 77.55$; hence, $\beta = 102{\cdot}45°$.

(a) $\sin 43.16° = \sin 80.95° \sin \psi$, so that $\psi = 43.84°$ and $\dfrac{a}{b} = \tan 43.83° = 0.960$.

(b) Again, $\dfrac{c}{b} = \dfrac{\tan \angle 001 - 011}{\sin \beta} = \dfrac{\tan(90 - 48.21)°}{\sin 102.45°} = 0.915$

Hence, $a : b : c = 0.960 : 1 : 0.916$, and $\beta = 102.45°$.

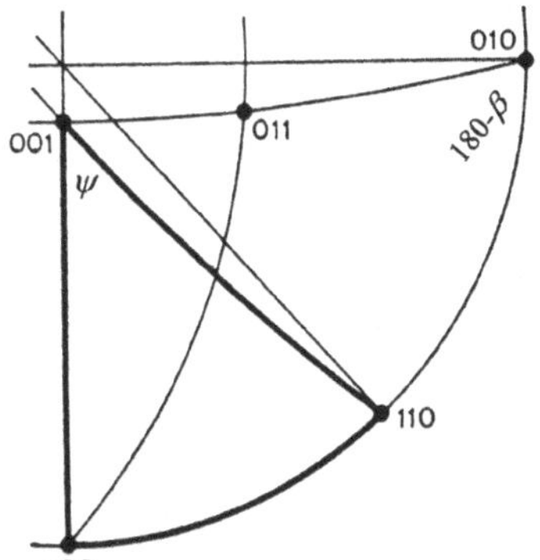

Fig. 2.30
Partial stereogram for lead chromate: 110–001 = 80.95°; 110–100 = 43.16°.

2.7 Molecular geometry: VSEPR theory

One approach to molecular geometry is achieved by the valence shell electron pair repulsion theory, generally written as VSEPR, of which descriptions are available in most modern textbooks on chemistry, and also on appropriate web sites. The theory attempts to predict the shapes of polyatomic species in terms of the repulsions between pairs of electrons, particularly lone pair electrons, on the component atoms. It begins with the Lewis bonding-pair/lone-pair model, and assumes that a molecular species adopts that shape which minimizes the repulsions between pairs of electrons; in other words, it seeks to place electron pairs as far apart as possible. The theory is applied by determining the bonding and non-bonding electrons from a model of the molecule, and then comparing the results with the standard conformations of Table 2.1, or similar compilation, which have been confirmed experimentally. It is necessary to know also the atomic connectivity: the formula C_3H_8O, for example, could represent either propanol or methoxyethane.

In a molecule of the type MX_n, each electron pair may be represented by a point on a sphere with the species M at its centre, and the positions of the n species X are then arranged so as to minimize repulsive energies. Two common arrangements for MX_4 species are

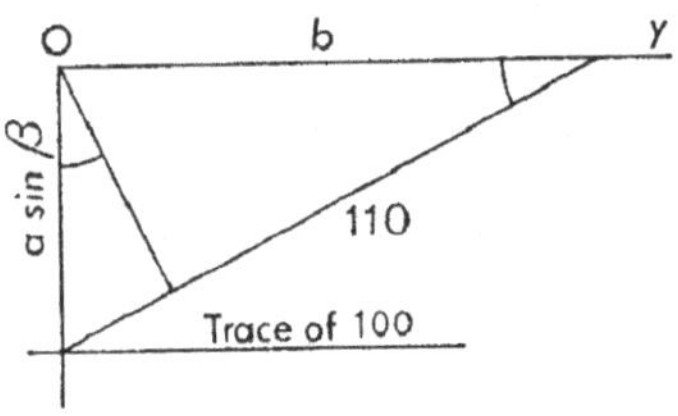

Fig. 2.31
Calculation of the b/a axial ratio for a monoclinic crystal.

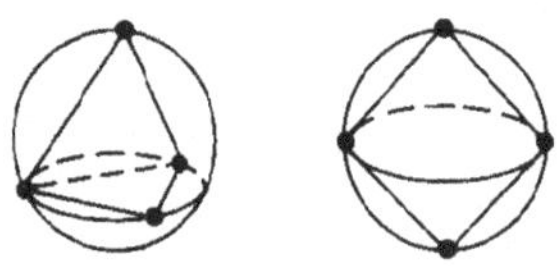

(a) (b)

Table 2.1 Molecular shapes according to VSEPR theory.

Bonding electrons on central atom	Non-bonding (lone pair) electrons	Shape and angle/ °	Example
2	0	Linear/180	BeF_2
1	1	Linear/180	CO_2
3	0	Trigonal planar/120	BF_3
2	1	Bent/< 120	SO_2
1	2	Linear/180	O_2
4	0	Tetrahedral/109.5	CH_4
3	1	Trigonal pyramidal/< 109.5	NH_3
2	2	Bent/< 109.5	H_2O
1	3	Linear/180	HF
5	0	Trigonal bipyramidal/ 90,120	PCl_5
4	1	See-saw/< 180, < 109.5	SF_4
3	2	T-shape/< 90, < 180	ClF_3
2	3	Linear	I^-
6	0	Octahedral/90	SF_6
5	1	Square pyramidal/< 90	$[SbCl_5]^{2-}$
4	2	Square planar/90	$[ICl_4]^-$

The tetrahedral arrangement (a) is the preferred conformation; square-planar is an alternative, but it would place electron pairs closer together. The model may be refined by allowing each electron pair to occupy the space of a sphere, which has the advantage of indicating domains in space that may not be overlapped by other similar domains. This *point charge* model has the following order for electron pair repulsion:

lone pair/lone pair > lone pair/bonding pair > bonding pair/bonding pair

The domain model then superimposes relative size factors on to this energy order. A species may be written as MX_nL_m, where L_m refers to m lone pairs on the species M. The following conditions should be noted:

- A bonding domain involves the M and X valence electrons, which are attracted to the nuclei; the non-bonding domain L belongs only to M. Classical double and triple bonds involve, respectively, two and three shared pairs of electrons, and the sizes of the domains increase in the following order:

  single bond < double bond < triple bond

- Electronegativities govern the extent to which electrons may be transferred from M to X.
- A lone pair domain L occupies a larger sphere than a bonding domain X and, because it belongs to M, it will tend to be closer to M than are the X species. If the species NH_3 (MX_3L type) and H_2O (MX_2L_2 type) are considered, the total of four domains would be expected to lead to tetrahedral arrangements. However, if the domain for L is larger than that for X and more strongly attracted to M, then the X–M–X angle will be less than the tetrahedral value: in fact, in NH_3 it is 107°, whereas in H_2O it is 104.4°.

The molecules SF_2 and SCl_2 (MX_2L_2 type) resemble the water molecule, so that tetrahedral arrangements would be expected. Since fluorine is more electronegative than chlorine, it will draw more electron charge from sulphur than will chlorine. This effect will, in turn lead to a smaller X–M–X angle in SF_2 than in SCl_2. From experiment, the angles are 98° in SF_2 and 102° in SCl_2 : the possible linear MX_2 conformations are perturbed by the lone pairs on the M species, which repel the X atoms to produce an angular shaped molecule.

Sulphur dioxide has multiple bonds, with the formal structure 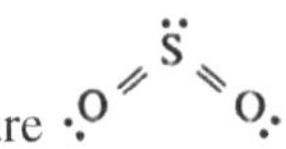 and it may be classed as an MX_2L species. The conformation tends towards trigonal-planar, but larger domains are assumed for multiple bonds and the single lone pair has only a small repulsion effect: the bond angle is just less than that for trigonal-planar at 119.1°. In SO_3, symmetry is restored and a true trigonal-planar conformation obtains.

In carbon dioxide, although the multiple bonds produce large domains, there are no lone pairs on the central atom: 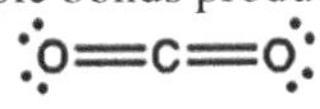 thus, the molecule is expected to be linear, which is confirmed by experiment.

The VSEPR theory is a simple and straightforward method for predicting the shapes of small molecular species. It is particularly successful where M is a main-group element, but is less satisfactory for transition-metal compounds because the central atom domain does not always have a spherical shape. However, the particular configurations d^0, d^5 and d^{10} do respond fairly well to VSEPR treatment: $TiCl_4$ (d^0) tetrahedral, $[CoF_6]^{2-}$ (d^5) octahedral and $[Ag(NH_3)_2]^+$ (d^{10}) linear are all predicted correctly. Table 2.1 summarizes the findings of the application of VSEPR theory.

2.8 Molecular geometry from experimental studies

Where good structural detail is required, recourse to spectroscopic methods may produce satisfactory results. However, if it is desired to obtain the precise geometry of a species, x-ray or neutron crystallography would be probably be the necessary procedure. From the numerical results so obtained, it is a straightforward matter to determine molecular geometry.

The data normally sought are bond lengths, intermolecular distances and torsion angles, each with its measure of precision. However, it must be borne in mind that intermolecular forces in the crystal may perturb the conformation of the species from that which would exist in the free state. A good example of this situation is given through Problem 5.7, which considers the conformations of biphenyl in the crystal and free molecule states.

In the following discussion of calculations in molecular geometry, it is assumed that the necessary results from X-ray or neutron crystallographic structure analyses are available, namely, unit cell dimensions and atomic coordinates.

2.8.1 Interatomic distances and angles

The distances between atoms in a crystal structure can be calculated from their coordinates in the crystal unit cell by a straightforward application of the scalar product. It may be desirable here to review some of the results in Appendix A3.2.1 and Appendix A.3.2.2.

2.8.1.1 Bond distances

In Fig. 2.32, the distance d_{12} between atoms 1 and 2 in the most general case of a triclinic unit cell is considered. The vectors from the origin to points 1 and 2 are $\mathbf{r}_1$ and $\mathbf{r}_2$ respectively, where

$$\mathbf{r}_1 = x_1\mathbf{a} + y_1\mathbf{b} + z_1\mathbf{c} \tag{2.36}$$

$$\mathbf{r}_2 = x_2\mathbf{a} + y_2\mathbf{b} + z_2\mathbf{c} \tag{2.37}$$

and the distance d_{12} is

$$d_{12} = |\mathbf{r}_2 - \mathbf{r}_1| \tag{2.38}$$

From Appendix A3,

$$\mathbf{d}_{12} = \mathbf{r}_2 - \mathbf{r}_1 \tag{2.39}$$

and using Eqs. (2.36)–(2.37),

$$\mathbf{d}_{12} = (x_2 - x_1)\mathbf{a} + (y_2 - y_1)\mathbf{b} + (z_2 - z_1)\mathbf{c} \tag{2.40}$$

From Appendix A3.2.1, the dot product of each side with itself is formed, leading to

$$\begin{aligned} d_{12}^2 = {} & (x_2 - x_1)^2a^2 + (y_2 - y_1)^2b^2 + (z_2 - z_1)^2c^2 + 2(y_2 - y_1)(z_2 - z_1)bc\cos\alpha \\ & +2(z_2 - z_1)(x_2 - x_1)ca\cos\beta + 2(x_2 - x_1)(y_2 - y_1)ab\cos\gamma \end{aligned} \tag{2.41}$$

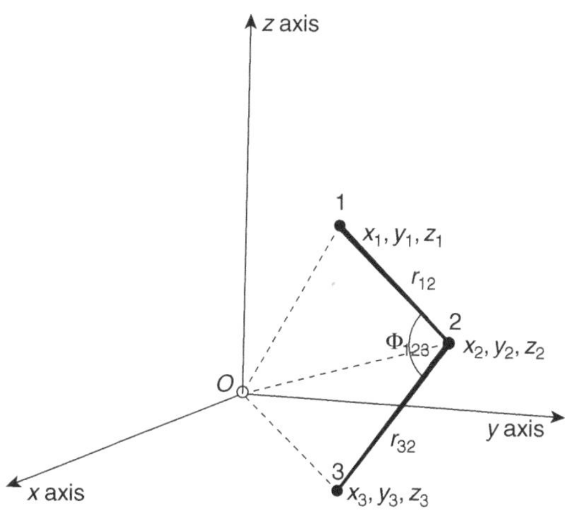

Fig. 2.32
Atoms 1, 2 and 3, showing bond lengths $|\mathbf{r}_{12}|$ and $|\mathbf{r}_{23}|$, and bond angle Φ_{123}.

Hence, any distance d between two atoms may be evaluated. In the more symmetrical crystal systems, the equation is simplified; thus, in the tetragonal system, the equation becomes:

$$d_{12}^2 = \left[(x_2 - x_1)^2 + (y_2 - y_1)^2\right] a^2 + (z_2 - z_1)^2 c^2$$

2.8.1.2 Bond angles

In a similar approach, the angle Φ_{123} between r_{12} and r_{32} may be determined:

$$\cos \Phi_{123} = \frac{\mathbf{r}_{12} \cdot \mathbf{r}_{32}}{r_{12} r_{32}} \tag{2.42}$$

The procedure evaluates the scalar product $\mathbf{r}_{12} \cdot \mathbf{r}_{32}$, and the magnitudes r_{12} and r_{32}, according to Eqs. (2.39)–(2.41), and so obtains Φ_{123}.

2.8.1.3 Torsion angles

In addition to distance and angle calculations in a molecule, it is frequently necessary to evaluate torsion angles so as to compare conformations of related molecular species. Figure 2.33 shows a pattern of four atoms, and the torsion angle τ_{1234} is defined as the dihedral angle between the planes of atoms 1, 2, 3 and 2, 3, 4, and lies in the range $-180° < \tau_{1234} \leq 180°$. The vectors $\mathbf{r}_{12}$ and $\mathbf{r}_{23}$ define the normal to the first plane, and vectors $\mathbf{r}_{23}$ and $\mathbf{r}_{34}$ the second.

The torsion angle is represented as the exterior spherical angle τ, shown below, and is given by

$$\tau_{1234} = \text{atan2}\{\ |\mathbf{r}_{23}|\mathbf{r}_{12} \cdot (\mathbf{r}_{23} \times r_{34}),\ (\mathbf{r}_{12} \times \mathbf{r}_{23}) \cdot (\mathbf{r}_{23} \times r_{34})\} \tag{2.43}$$

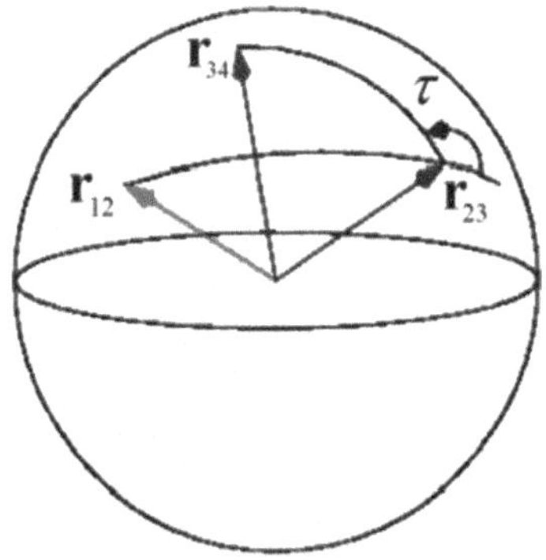

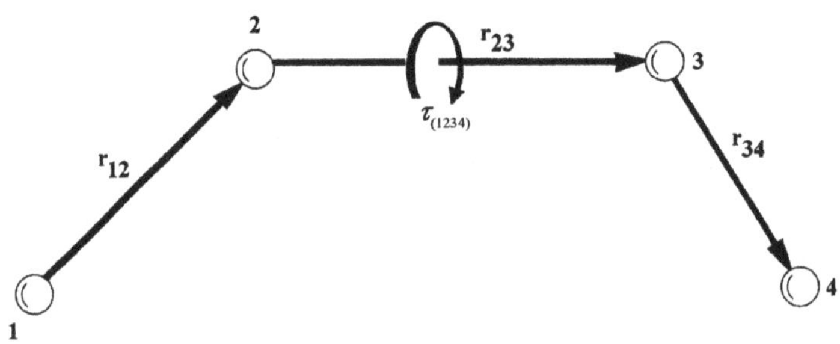

Fig. 2.33
Atoms 1, 2, 3 and 4: the torsion angle τ_{1234} is the amount of rotation of bond 1–2 about bond 2–3, as seen along the direction $2 \rightarrow 3$, that eclipses atom 4 by atom 1; if this is rotation is clockwise, the torsion angle is positive in sign.

where atan 2 is a function, given here in Fortran, that evaluates the function $\tan^{-1}$ with two arguments and leads to a signed result over the complete angular range. These three calculations are easily programmed, and are provided in the program MOLGOM in the Web Program Suite.

2.8.2 Conformational parameters

Among the terms used to describe chemical conformations, the better known are *cis*, $\tau \approx 0°$; *trans*, $\tau \approx 180°$; $\pm$ *gauche*, $\tau \approx \pm 60°$. The E/Z notation is preferred with double bonded species where several substituents are present. In this notation, the substituents are given a priority in terms of their atomic number. If the species with the higher priorities are on opposite sides of the double bond the conformation is *E* (Ger. *entgegen* = opposite) whereas if they are on the same side of the double bond the conformation is *Z* (Ger. *zusammen* = together):

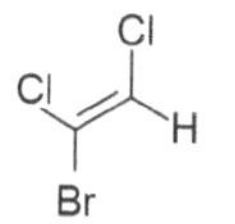

(*E*)-1-Bromo-1,2-dichloroethene

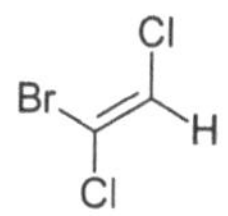

(*Z*)-1-Bromo-1,2-dichloroethene

An alternative procedure, particularly with more complex species, states the values of the appropriate torsion angles τ, as described above.

In defining the conformations of species with ring structures, two symmetry elements are generally considered: twofold rotation axes in the plane of the ring, and mirror planes perpendicular to the principal molecular plane. In six-membered rings, the symmetry elements either pass through the centres of opposite bonds or through atoms in the ring that are directly opposite each other. The maximum number of these symmetry elements in a six-membered ring is twelve: six mirror symmetry planes and six twofold symmetry axes, but this total is found only in planar hexagonal molecules such as benzene, C_6H_6, point group $\frac{6}{m}mm$.

Conformations other than a planar hexagon arise with six-membered rings, and the most common of them are modelled on cyclohexane, C_6H_{12}, in Fig. 2.34. The most stable form is the *chair*; the other forms have conformations similar to those of the staggered conformation of ethane (Fig. P1.1a), which represents its minimum energy form. In cyclohexane, the relative energies for the conversion sequence *chair* (a) → *twist-boat* (c) → *boat* (b) → *half-chair* (d) in Fig. 2.34 are in the approximate ratio 1 : 15 : 20 : 30. An interesting display of these interconversions can be found on a web site [10].

In five-membered rings, of which cyclopentane C_5H_{10} is typical, ten symmetry elements, five mirror planes and five twofold axes, can arise only in the

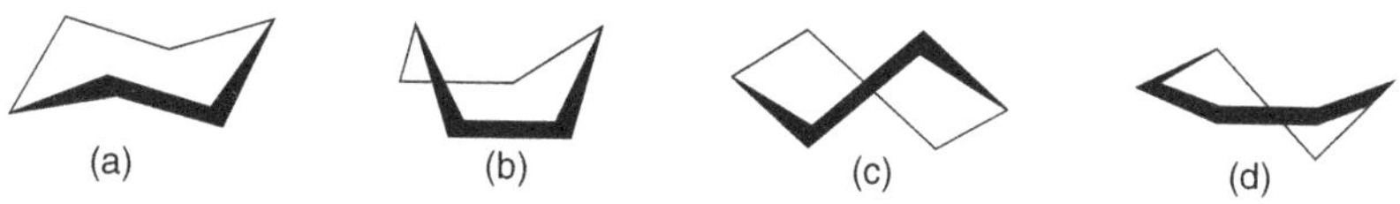

Fig. 2.34
Non-coplanar conformations of six-membered rings, exemplified by the molecule of cyclohexane, C_6H_{12}; point group symbols are listed in parentheses: (a) *chair* ($\bar{3}m$), (b) *boat* ($mm2$), (c) *twist-boat* (222), *half-chair* (2).

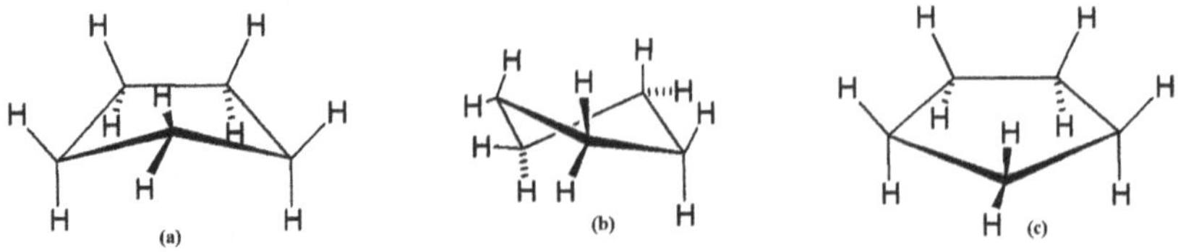

Fig. 2.35
Conformations of five-membered rings, typified by cyclopentane, C_5H_{10}: (a) *envelope*, (b) *half-chair*, (c) *planar*.

planar form. However, this form has a strain energy of *ca* 25 kJ mol^{-1} compared to the envelope and half-chair forms (Fig. 2.35), which interconvert at only one quarter of this energy. The strain energy between forms (a) and (b) is only about 8 kJ mol^{-1}, so a dynamic equilibrium exists between them. The planar form (c) is less favoured, on account of its greater strain energy.

2.8.2.1 Asymmetry parameters

Ring conformations in an experimentally determined cyclic structure frequently deviate from the ideal models just described, owing to both the nature of substituents on the ring and the interatomic and intermolecular forces in play. The amount of this departure may be expressed by the asymmetry parameters [11, 12], ΔC_s for mirror symmetry related angles, and ΔC_2 for twofold symmetry related angles, by comparing symmetry related or near symmetry related torsion angles τ_i and τ_i':

$$\Delta C_s = \left(\frac{\sum_n (\tau_i + \tau_i')^2}{n} \right)^{1/2} \tag{2.44}$$

$$\Delta C_2 = \left(\frac{\sum_n (\tau_i - \tau_i')^2}{n} \right)^{1/2} \tag{2.45}$$

where the sum is taken over n comparisons.

2.8.2.2 Ring planarity

A molecular structure, or part of it, may be checked for planarity by calculating the best-fit plane through the appropriate atoms by a linear least squares procedure, as discussed briefly in Appendix A5. The equation of a plane, often written as $AX + BY + CZ + D = 0$, may be recast conveniently as

$$AX + BY + C = Z \tag{2.46}$$

where X, Y and Z are orthogonal coordinates; the constants A, B and C may be found by a least squares procedure. The error ε_i in the fit of the ith datum is given by

$$\varepsilon_i = AX_i + BY_i + C - Z_i$$

and the sum of the squares of the errors

$$\varepsilon_i^2 = \sum_{i=1}^{m} [(AX_i + BY_i + C) - Z_i]^2$$

is minimized by a least squares procedure, where m is the number of X, Y, Z triplets. The partial differentials of this expression with respect to A, B and C are formed and set equal to zero, whereupon the least squares procedure leads to the *normal equations*

$$\underset{\mathbf{S}}{\begin{pmatrix} \sum_{i=1}^{m} X_i^2 & \sum_{i=1}^{m} X_iY_i & \sum_{i=1}^{m} X_i \\ \sum_{i=1}^{m} X_iY_i & \sum_{i=1}^{m} Y_i^2 & \sum_{i=1}^{m} Y_i \\ \sum_{i=1}^{m} X_i & \sum_{i=1}^{m} Y_i & \sum_{i=1}^{m} 1 \end{pmatrix}} \underset{\mathbf{A}}{\begin{pmatrix} A \\ B \\ C \end{pmatrix}} = \underset{\mathbf{Z}}{\begin{pmatrix} \sum_{i=1}^{m} X_iZ_i \\ \sum_{i=1}^{m} Y_iZ_i \\ \sum_{i=1}^{m} Z_i \end{pmatrix}}$$

which may be written concisely as

$$\mathbf{S\,A} = \mathbf{Z} \tag{2.47}$$

The solution, following Appendix A3.4.10 and Appendix A3.6, is

$$\mathbf{S}^{-1}\mathbf{S\,A} = \mathbf{A} = \mathbf{S}^{-1}\mathbf{Z} \tag{2.48}$$

Hence, the values of A, B and C; the procedure may be examined through Problem 2.12.

2.8.2.3 Least distance of atoms to a best-fit plane

In conjunction with the determination of the best-fit plane, knowledge of the distances of atoms from that plane is desirable. Consider Fig. 2.36. The plane, Eq. (2.46), can be set in the form

$$AX + BY + C'Z = D' \tag{2.49}$$

where $C' = -1$; and $D' = -C$ from the above analysis. The point P is any point not on the plane while Q lies in the plane. The shortest distance L from P to the plane is the perpendicular PO and is equal to $|\mathbf{r}|\cos\phi$, which is the projection of $\mathbf{r}$ on to the vector normal of which $\mathbf{n}$ is the unit vector. Thus,

$$L = |\mathbf{r}|\cos\phi = \frac{\mathbf{n}}{|\mathbf{n}|}\cdot\mathbf{r} \tag{2.50}$$

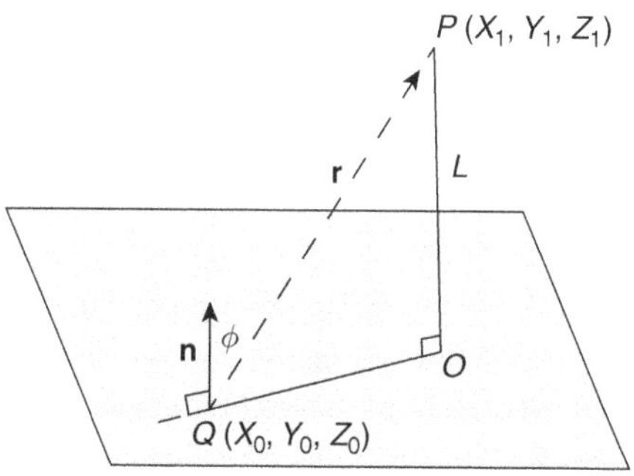

Fig. 2.36
Calculation of the distance from a point to a plane: the point P is distant L from the plane, and $\mathbf{r}$ is the vector **OP–OQ** and $\mathbf{n}$ is the unit vector normal to the plane.

The vector **n** has components (A, B, C'), so that $|\mathbf{n}| = \sqrt{A^2 + B^2 + C'^2}$. Writing Eq. (2.50) *in extenso*,

$$L = \frac{1}{\sqrt{A^2 + B^2 + C'^2}} \begin{pmatrix} A & B & C' \end{pmatrix} \cdot (X_1 - X_0)(Y_1 - Y_0)(Z_1 - Z_0)$$

Evaluating the dot product:

$$L = \frac{AX_1 + BY_1 - C'Z_1 - \left(AX_0 + \ BY_0 + \ C'Z_0\right)}{\sqrt{A^2 + B^2 + C'^2}}$$

But since Q lies in the plane,

$$\left(AX_0 + BY_0 + C'Z_0\right) = -D'$$

Hence,

$$L = \frac{AX_1 + BY_1 - C'Z_1 + D'}{\sqrt{A^2 + B^2 + C'^2}} \tag{2.51}$$

The calculations of best-fit plane and least distances have been programmed in PLANE in the Web Program Suite.

2.8.3 Internal coordinates

In certain investigations in the determination of crystal and molecular structures, such as Patterson search methods, crystal structure prediction [13] and *ab initio* theoretical calculations of molecular geometry, it is often necessary to generate a model chemical structure. For this application, use is made of standard interatomic distances, bond angles and atomic fragments of known geometry, such as a phenyl ring. Torsion angles may be derived by inspection of a physical model of the chemical entity, adopting the most likely configuration for non-rigid moieties in the model. The calculation requires a conversion of the internal coordinates, that is, bond lengths, bond angles and torsion angles of the model, to a set of coordinates in absolute measure referred to a system of axes, often orthogonal.

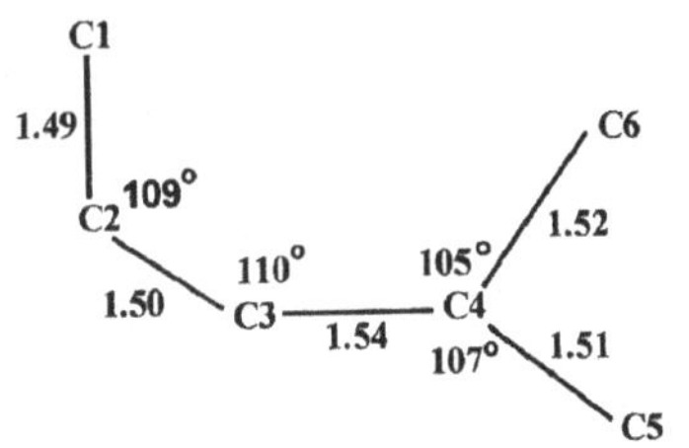

Fig. 2.37
A six-carbon aliphatic skeleton, omitting hydrogen atoms, showing bond lengths and bond angles; the torsion angle τ_{2346} is –30°.

Consider the aliphatic fragment 2-methylpentane, shown in Fig. 2.37 (omitting the hydrogen atoms) in which C1–C5 form a planar moiety. The torsion angle $\tau_{1234} = 0°$, $\tau_{2345} = 180°$ and τ_{2346}, obtained by measurement on a structure model is *ca.* –30°. Is the C4–C6 bond above or below the plane of the diagram? The calculation of Cartesian coordinates for the six atoms can be carried out with the program INTXYZ in the Web Program Suite. The data file is prepared either from standard geometry, Tables 2.2–2.4 and Table 2.9, or from a similar fragment of know geometry, and set out as the numerical data below. The program is written in FORTRAN 90, and the format for the data, (A4, I6, 3F10.5), as in the example, must be followed exactly:

Atom name	Atom code	Bond angle/deg	Torsion angle/deg	Bond length/Å
C1	0	0 · 0	0 · 0	0 · 0
C2	1	0 · 0	0 · 0	1 · 49
C3	2	109 · 0	0 · 0	1 · 50
C4	3	110 · 0	0 · 0	1 · 54
C5	4	107 · 0	180 · 0	1 · 51
C6	3	105 · 0	−30 · 0	1 · 52

The interpretation of this data set is as follows:

- Column 1 gives the *atom name*, which does *not* enter into the calculation.
- Column 2 gives the atom *code number*, and indicates the bond to the atom coded above it, with the bond length given in column 5.
- Column 3 gives the *bond angle* formed with the atoms coded above and below it: thus atom code 2(C3) is linked with atom codes 0(C1) and 2 (C3), forming the angle C1–C2–C3 of 109°.
- Column 4 gives the *torsion angle* formed by a coded atom and the three coded above it, Thus, τ is always zero for the first three entries, because they lie in one and the same plane; $\tau_{1234} = 0°$ for a planar four atom fragment in *cis* conformation, and $\tau_{2345} = 180°$ for a four atom planar fragment in *trans* conformation; the torsion angle τ_{2346} was estimated as –30.0° from a molecular model. The geometry of the calculation is straightforward, and its use is examined in Problem 2.13.

Table 2.2 Radii/Å of common ionic species, referred to coordination number 6[a].

Ion	Ladd	Shannon & Prewitt	Pauling
Li^+	0.86	0.86	0.90
Na^+	1.12	1.02	1.16
K^+	1.44	1.38	1.52
Rb^+	1.58	1.52	1.66
Cs^+	1.84	1.67	1.81
$NH_4{}^+$	1.66	–	–
Ag^+	1.27	1.15	1.27
Tl^+	1.54	1.50	1.64
Be^{2+}	0.48	0.45	0.59
Mg^2	0.87	0.89	0.86
Ca^{2+}	1.18	1.00	1.14
Sr^{2+}	1.32	1.18	1.32
Ba^{2+}	1.49	1.35	1.49
H^-	1.39	–	–
F^-	1.19	1.33	1.19
Cl^-	1.70	1.82	1.67
Br^-	1.87	1.96	1.82
I^-	2.12	2.20	2.06
O^{2-}	1.25	1.40	1.26
S^{2-}	1.70	1.84	1.70
Se^{2-}	1.81	1.98	1.84
Te^{2-}	1.97	2.23	2.07

[a] The changes in ionic radius from coordination number 6 to coordination numbers 8, 4, 3, 2 are approximately +1.5, –1.5, –3.0 and –3.5 %, respectively.

Table 2.3 Standard bond lengths/Å.

Single bonds

C4–H	1.09	C3–C2	1.45
C3–H	1.08	C3–N3	1.40
C2–H	1.06	C3–N2	1.40
N3–H	1.01	C3–O2	1.36
N2–H	0.99	C2–C2	1.38
O2–H	0.96	C2–N3	1.33
C4–C4	1.54	C2–N2	1.33
C4–C3	1.52	C2–O2	1.36
C4–C2	1.46	N3–N3	1.45
C4–N3	1.47	N3–N2	1.45
C4–N2	1.47	N3–O2	1.36
C4–O2	1.43	N2–N2	1.45
C3–C3	1.46	N2–O2	1.41

Double bonds

C3–C3	1.34	C2–O1	1.16
C3–C2	1.31	N3–O1	1.24
C3–N2	1.32	N2–N2	1.25
C3–O1	1.22	N2–O1	1.22
C2–C2	1.28	O1–O1	1.21
C2–N2	1.32		

Triple bonds

C2–C2	1.20	N1–N1	1.10
C2–N1	1.16		

Aromatic bonds

C2–C3	1.40	N2–N2	1.35
C2–N2	1.34		

2.8.4 Errors and precision

'*In this world nothing can be said to be certain, except death and taxes*' [14]; had this writer been alive today, he would certainly have added '*uncertainty*'. Consider measuring a given distance with a ruler and quoting the result together with an estimate of its uncertainty. The measurement may be repeated many times in order to provide a better measure of the uncertainty, and the final result expresses the measured value and its *precision*. However, it has been assumed that the ruler is providing exact results. If it transpires that the ruler itself was 1% shorter than the stated value then, although the results may be very precise, they are *inaccurate* by at least 1%. The ruler contains a *systematic* error which is transmitted to the result, as well as the *random error* inherent in any physical measurement. Errors may also be observational; they are unlikely to be systematic, and can be corrected by careful measurement.

2.8.4.1 Random errors

Random errors are a feature of experimentally determined results, and arise from unpredictable and unknown effects during the measurement process. An important distribution of experimental results is a *Gaussian*, or *normal*, distribution, $f(x)$. It follows the equation

$$f(x) = \frac{1}{\sigma\sqrt{2\pi}} \exp\left[(x - \overline{x})^2\right] \qquad (2.52)$$

where $\overline{x}$ here represents the mean of a series of observations x and σ its standard deviation, or square root of its variance σ^2. The distribution is symmetrical about its mean, and an example involving the heights of German Shepherd dogs is shown by Fig. 2.38. According to the central limit theorem, if in a population there is a sufficiently large number of independent random variables

Table 2.4 Standard bond angles/°.

Atom	Geometry	Angle
C4	Tetrahedral	109.5
C3	Planar	120
C2	Bent	109.5
C2	Linear	180
N4	Tetrahedral	109.5
N3	Pyramidal	109.5
N3	Planar	120
N2	Bent	109.5
N2	Linear	180
O3	Pyramidal	109.5
O2	Bent	109.5

x_i each of which has its own mean m_i and variance σ_i^2, then the population has a normal probability distribution, with a mean and variance given by

$$\left.\begin{aligned} m &= \sum_i m_i \\ \sigma^2 &= \sum_i \sigma_i^2 \end{aligned}\right\} \qquad (2.53)$$

In the experiment on the heights of German Shepherd dogs, the central limit theorem implies that the mean m of the normal distribution of the sample is the value that would obtain for the whole population, and the variance of the mean, equal to σ^2/n, also that of the whole population. For a sufficiently large sample, the heights shown follow the normal, or Gaussian, distribution, and the standard deviation determines the percentages of observations between chosen limits, as shown in Table 2.5. In this one sample of measurements, the best value for the mean of the population is m and the variance σ^2.

In many scientific experiments, however, it is not feasible to make repetitious measurements, and the usual measure of precision is the *estimated standard deviation*, generally abbreviated as esd or σ, the

Table 2.5 Percentage of observations (frequencies) within limits of σ.

$m \pm \sigma$	68.3%
$m \pm 1.96\sigma$	95.0%
$m \pm 2\sigma$	95.5%
$m \pm 2.58\sigma$	99.0%
$m \pm 3\sigma$	99.7%

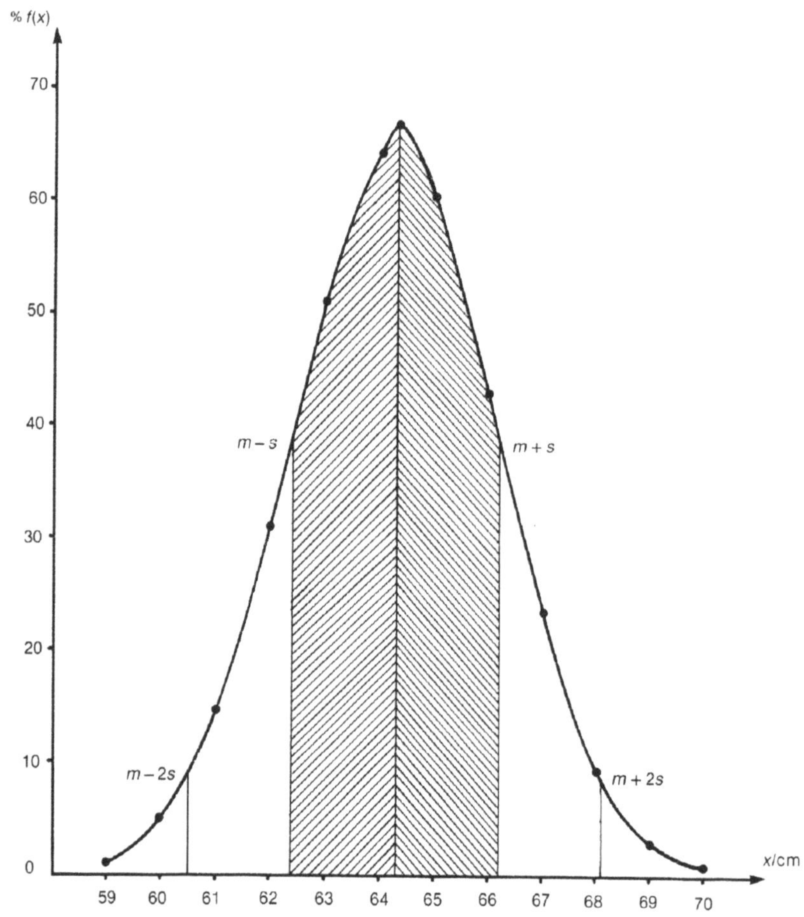

Fig. 2.38
Percentage frequency $f(x)$ of male German Shepherd dogs as a function on their height. The distribution is normal, with a mean of 62.5 cm and a standard deviation of 2.5 cm. The shaded area represents 1σ about the mean, and encloses 68.3% of the sample of 308 dogs.

positive square root of the variance σ^2; for a large number of representative measurements, σ^2 may be taken to apply to the whole population. While the analysis of German Shepherd dogs is not of particular concern here, similar principles come into play in all statistical procedures.

In a crystal structure analysis, for example, a subject very closely related to the theme of this book, estimated standard deviations are obtained for the atomic coordinates and other parameters through a least squares refinement procedure, since the number of experimental data normally greatly outweighs the number of parameters to be determined. In this application, the quantity minimized may be expressed as

$$\sum_i w_i \Delta_i^2$$

where Δ_i depends upon the difference between experimentally observed and calculated parameters. The quality of refinement, as expressed in the esds of the parameters, evolves numerically from the diagonal elements of the inverse of a least squares matrix **A**. Thus, for any parameter p_j

$$\sigma_{p_j} = \left((a^{-1})_{jj} \frac{\sum_i w_i \Delta_i^2}{n - P} \right) \tag{2.54}$$

where p_j is the jth parameter among the total of P parameters and n is the total number of observations [13].

2.8.4.2 Transmission of errors

In molecular geometry, a measured property may depend upon a number of parameters, each of which has its own uncertainty. The uncertainties in these component parameters are propagated in any quantity that is dependent upon them.

Let q be a function of several variables p_i ($i = 1, 2, 3, \ldots, n$), each with its own esd, $\sigma(p_i)$. Then, if the variables are independent, the esd σ of q is given by

$$\sigma^2(q) = \sum_i \left(\frac{\partial q}{\partial p_i} \sigma(p_i) \right)^2 \tag{2.55}$$

A simple example may be given, using the discussion above on bond lengths.

Consider a bond between two atoms lying along the vertical edge of a tetragonal unit cell, with $c = 10.06\,(1)$ Å. Two atoms A and B have z fractional coordinates 0.3712 (3) and 0.5418 (2), respectively. The fractional coordinate z is Z/c, where both quantities are expressed in one and the same length unit, and the figures in parentheses are esds, to be applied to the final digits of the coordinates. From Eq. (2.41), inserting the particular values of the parameters,

$$r_{AB} = (z_2 - z_1)c = (0.5418 - 0.3712)10.06 = 1.7162 \text{ Å}$$

Then, applying Eq. (2.55),

$$\sigma^2(r_{AB}) = (0.5418 - 0.3712)^2(0.01)^2 + (10.06)^2(0.0002)^2 + (10.06)^2(0.0003)^2$$
$$= 1.597 \times 10^{-5}$$

Hence, $\sigma\ (r_{AB}) = 0.0040$ Å, and r_{AB} is written as 1.716 (4) Å. Similar calculations may be used for all distance and angle calculations in all crystal systems; the general equations are quite involved numerically and best handled by computer methods.

Relatively recently the concept of combined standard uncertainty (csu) has been used as a better measure than the esd in statements of uncertainty in data and results, and quoted together with a description of the associated experimental and computational procedures. The csu is the positive square root of the estimated variance u_c^2, given by

$$u_c^2(q) = \sum_{i=1}^{N} \left(\frac{\partial q}{\partial p_i}\right)^2 u^2(p_i) + 2\sum_{i=1}^{N-1}\sum_{j=i+1}^{N}\left(\frac{\partial q}{\partial p_i}\frac{\partial q}{\partial p_j}\right) u(p_i, p_j) \qquad (2.56)$$

based on a first order Taylor approximation for a function $Q = f(P_1, P_2, \ldots P_N)$, where q and $p_i\ (i = 1,\ N)$ are the experimental measures of the true values Q and $P_i\ (i = 1,\ N)$. Where the measured quantities p_i are uncorrelated, the second term is zero and the expression reverts to Eq. (2.55). For further details on this procedure the reader is referred to the literature [15–17].

2.9 Molecular geometry from theoretical studies

A third method of determining the geometry and symmetry of a chemical species employs theoretical calculations; in this process, the energetically most stable conformation of the species is sought. *A priori* knowledge of the chemical system is helpful, in the form of standard bond lengths and bond angles, from which an approximate set of atomic coordinates can be determined (see Section 2.8.3 and Section 9.4). There is space here for but a brief outline of some the principles involved in the calculations of quantum chemistry.

2.9.1 The Schrödinger equation

The bonded state of a chemical species can be represented by the Schrödinger equation

$$\mathscr{H}\psi = E\psi \qquad (2.57)$$

where $\mathscr{H}$ is here the time independent Hamiltonian operator that describes the kinetic energy and the potential energy of the chemical system in terms of the positions and masses of the particles in the system: the wave function ψ is a solution of Eq. (2.57) and E is the total energy of the system. The formidable computational task involved in solving this equation for any chemical species larger than the hydrogen molecule ion is alleviated by the Born–Oppenheimer approximation, that the electronic and nuclear motions are separable. The success of this procedure depends upon the fact that the nuclear masses in an atom are so much greater than those of the electrons that the nuclei can be treated effectively as stationary. Thus, the wave function is then written as

$$\psi_{\text{molecule}} = \psi_{\text{electrons}}\,\psi_{\text{nuclei}} \tag{2.58}$$

Of the two parts, the concern here is for the contributions from the electrons.

If the system consists of n electrons under a potential energy V, which is itself a function of position, then the time independent electronic Hamiltonian $\mathscr{H}_e$ takes the form

$$\mathscr{H}_e = -\sum_{j=1}^{n} [\hbar^2/(2m_e)] \left(\frac{\partial^2}{\partial x_j^2} + \frac{\partial^2}{\partial y_j^2} + \frac{\partial^2}{\partial z_j^2} \right) + V(x_j, y_j, z_j) \tag{2.59}$$

where $\hbar = h/2\pi$ and m_e is the mass of an electron; strictly, the reduced mass of an electron and nucleus should be used here, but the error in using m_e is no more than 0.05%.

2.9.2 Atomic orbitals

Consider the situation for a single electron, $n = 1$ in Eq. (2.59), which could be taken as the electron on a hydrogen atom. The term $\left(\frac{\partial^2}{\partial x^2} + \frac{\partial^2}{\partial y^2} + \frac{\partial^2}{\partial z^2}\right)$ is known as the *Laplacian*, ∇^2, and in spherical polar coordinates (Appendix A8) it takes the form $\frac{\partial^2}{\partial r^2} + \frac{2}{r}\frac{\partial}{\partial r} + \frac{1}{r^2}\left(\frac{1}{\sin^2\theta}\frac{\partial^2}{\partial\phi^2} + \frac{1}{\sin\theta}\frac{\partial}{\partial\theta}\sin\theta\frac{\partial}{\partial\theta}\right)$. The solution of the wave equation will not be considered here, but it is notable that it is separable into a radial term that depends on the radial coordinate r, and an angular term, or surface harmonic[3], that depends on the angular coordinates θ and ϕ. Acceptable wave functions for each energy level characterized by the principal quantum number n, are further specified by an orbital angular momentum quantum number l, $(l = 0,\ 1,\ 2\ldots n-1)$, and a magnetic quantum number m_l, $(m_l = l,\ l-1\ldots 0\ldots -l+1,\ -l)$, arising from the angular momentum of the electron around the nucleus; in addition, there is a spin quantum number m_s with a value of either $+\frac{1}{2}$ or $-\frac{1}{2}$, which evolves from the wave mechanical treatment of the atom.

[3] Also referred to as 'spherical' harmonic.

The wave function, because of its separable nature, may be written as $\psi = R_n(r)\,Y_{l,m_l}(\theta,\phi)$. The radial function $R_n(r)$ governs the size of the function, and the surface harmonic[a] $Y_{l,m_l}(\theta,\phi)$, or angular function, governs its shape. The terms 'size' and 'shape' as used here refer to commonly used graphical illustrations of the wave functions. Values of radial functions and surface harmonics are listed in Table 2.6 and Table 2.7, respectively.

An atomic orbital may be thought of qualitatively as a region in space around an atomic nucleus where an electron is most likely to be found, that is, where the electron density is greatest. Specifically, it is a one-electron, hydrogenic wave function for an electron in an atom, denoted by the quantum numbers n, l and m_l. The lowest energy orbital corresponds to $n = 1$; hence, $l = 0$ and $m_l = 0$. The corresponding wave function may be written as

$$\psi_{1,0,0} = N\exp(-Zr/a_0) \tag{2.60}$$

Table 2.6 Normalized hydrogenic radial wave functions $R_{n,l}$ for $n = 1-3$: $\rho = 2Zr/na_0$.

Orbital	n	l	$R_{n,l}$
1s	1	0	$(Z/a_0)^{3/2}\, 2\exp(-\rho/2)$
2s	2	0	$(Z/a_0)^{3/2}\,(1/2\sqrt{2})\,(2-\rho)\exp(-\rho/2)$
2p	2	1	$(Z/a_0)^{3/2}\,(1/2\sqrt{6})\,\rho\exp(-\rho/2)$
3s	3	0	$(Z/a_0)^{3/2}(1/9\sqrt{3})\,(6-6\rho+\rho^2)\exp(-\rho/2)$
3p	3	1	$(Z/a_0)^{3/2}\,(1/9\sqrt{6})\,(4\rho-\rho^2)\exp(-\rho/2)$
3d	3	2	$(Z/a_0)^{3/2}\,(1/9\sqrt{30})\,\rho^2\exp(-\rho/2)$

Table 2.7 Normalized surface harmonics Y_{l,m_l} for $l = 0-2$.

l	m_l	Y_{l,m_l}	Orbital
0	0	$\tfrac{1}{2}\sqrt{1/\pi}$	s
1	0	$\tfrac{1}{2}\sqrt{3/\pi}\cos\theta$	p_0
1	+1	$-\tfrac{1}{2}\sqrt{3/2\pi}\sin\theta\exp(\mathrm{i}\phi)$	p_1
1	−1	$\tfrac{1}{2}\sqrt{3/2\pi}\sin\theta\exp(-\mathrm{i}\phi)$	p_{-1}
2	0	$\tfrac{1}{4}\sqrt{5/\pi}(3\cos^2\theta-1)$	d_0
2	1	$-\tfrac{1}{2}\sqrt{15/2\pi}\cos\theta\sin\theta\exp(\mathrm{i}\phi)$	d_1
2	−1	$\tfrac{1}{2}\sqrt{15/2\pi}\cos\theta\sin\theta\exp(-\mathrm{i}\phi)$	d_{-1}
2	2	$\tfrac{1}{4}\sqrt{15/2\pi}\sin^2\theta\exp(\mathrm{i}2\phi)$	d_2
2	−2	$\tfrac{1}{4}\sqrt{15/2\pi}\sin^2\theta\exp(-\mathrm{i}2\phi)$	d_{-2}

where N is a normalization constant. Any occupied orbital contains one or two electrons; in the latter case the two electrons have opposite spins, in accordance with the Pauli exclusion principle, which states that *no two electrons in a given atom can have the same four quantum numbers*. Thus, if there are two electrons represented in any orbital, their spin quantum numbers m_s are $+1/2$ and $-1/2$.

2.9.3 Normalization

The probability of finding a given electron over all points in space must be unity; hence the *normalization constant* N for a wave function is given by

$$N^2 \int_V \psi\psi^* \mathrm{d}\tau = 1 \tag{2.61}$$

where the integral is over all space covered by τ, and ψ^* is the complex conjugate of the function ψ. The expression (2.61) is best set in spherical coordinates (Appendix A8):

$$N^2 \int_0^\infty \int_0^\pi \int_0^{2\pi} \psi\psi^* r^2 \sin\theta \,\mathrm{d}r\,\mathrm{d}\theta\,\mathrm{d}\phi = 1 \tag{2.62}$$

The radial $R_{n,l}(r)$ and angular $Y_{l,m_l}(\theta,\phi)$ components of a wave function may be normalized separately, according to the expression

$$\left.\begin{aligned} &N_r^2 \int_0^\infty |R_{n,l}(r)|^2 r^2 \mathrm{d}r = 1 \\ &N_{\theta,\phi}^2 \int_0^{2\pi}\int_0^{\pi} |Y_{l,m_l}(\theta,\phi)|^2 \sin\theta\,\mathrm{d}\theta\,\mathrm{d}\phi = 1 \end{aligned}\right\} \tag{2.63}$$

The $\psi_{1,0,0}$ wave function may be written for normalization, following Eq. (2.62), as

$$N^2 \int_0^\infty \exp(-2Zr/a_0)\,r^2\,\mathrm{d}r \int_0^\pi \sin\theta\,\mathrm{d}\theta \int_0^{2\pi} \mathrm{d}\phi = 1$$

The integrals over ϕ and θ are readily seen to be 2π and 2, respectively. The integral $\mathscr{I}$ over r can be carried out readily by making the substitution $t = 2Zr/a_0$, so that $\mathrm{d}r = (a_0/2Z)\,\mathrm{d}t$. Then,

$$\mathscr{I} = \left(\frac{a_0}{2Z}\right)^3 \int_0^\infty t^2 \exp(-t)\,\mathrm{d}t = \Gamma(3) = \left(\frac{a_0}{2Z}\right)^3 2! \tag{2.64}$$

where Γ is the gamma function (Appendix A9). Hence, $N^2 = 4\pi \times 2\left(\frac{a_0}{2Z}\right)^3 = 1$, so that $N = \left(\frac{Z}{\pi a_0}\right)^{3/2}$, and the normalized $\psi_{1,0,0}$ wave function, or atomic orbital, is

$$\psi_{1,0,0} = (Z/\pi a_0)^{3/2} \exp(-Zr/a_0) \tag{2.65}$$

Normalizing the radial component alone leads to

$$R_{1,0}(r) = (Z/a_0)^{3/2} 2 \exp(-Zr/a_0) \tag{2.66}$$

2.9.4 Probability distributions

The physical interpretation for the wave function requires that $\psi\psi^*$ be a probability distribution of the electron, or electron density, relative to the nucleus, and its form can be approached on the basis of Eq. (2.65) which, for hydrogen ($Z = 1$), may be simplified to $\psi_{1,0,0} = (1/\pi a_0^3)^{1/2} \exp(-r/a_0)$. Thus,

$$\psi_{1,0,0}\,\psi_{1,0,0}^* = (1/\pi a_0^3) \exp(-2r/a_0) \tag{2.67}$$

As this equation is independent of θ and ϕ, it represents a function with spherical symmetry. The likelihood that the electron be in a volume element[4] $\mathrm{d}\tau$ and lying between distances r and $r + \mathrm{d}r$ from the nucleus is given by $\psi_{1,0,0}\psi_{1,0,0}^*\mathrm{d}\tau = (1/\pi a_0^3)\exp(-2r/a_0)r^2\mathrm{d}r \sin\theta\,\mathrm{d}\theta\,\mathrm{d}\phi$. Integrating over θ and φ, that is, the surface of a sphere, leads to the expression

$$P(r)\,\mathrm{d}r = (4/a_0^3)\,r^2 \exp(-2r/a_0)\,\mathrm{d}r \tag{2.68}$$

which represents the probability that the electron lies between r and $r + dr$ from the nucleus.

From Eq. (2.67), it evolves that $\psi\psi^*$ has its maximum value at $r = 0$, implying that the most probable position for the electron is in the immediate neighbourhood of the nucleus, very much more probable than it would be

[4]The volume element $\mathrm{d}\tau = r^2\mathrm{d}r \sin\theta\,\mathrm{d}\theta\,\mathrm{d}\phi$ is equivalent to $\mathrm{d}x\,\mathrm{d}y\,\mathrm{d}z$ in Cartesian space.

in an element of the same volume further away from the nucleus. In fact, $\psi\,\psi^*$ tends to a maximum as r tends to zero for all spherically symmetrical functions ($l = 0$). More interesting than the probability at a specific location is the probability of finding the electron at a given distance irrespective of direction. This probability is the *radial distribution function* for the wave function given by $4\pi r^2\psi\,\psi^*$, which is effectively $P(r)$ in Eq. (2.68):

$$P_{1,0,0}(r) = 4\pi r^2\psi\,\psi^* = (4/a_0^3)r^2\exp(-2r/a_0) \tag{2.69}$$

Example 2.4

What are the probabilities of finding the electron of the hydrogen $\psi_{1,0,0}$ wave function in spherical shell of radii between r and $r + dr$ from the nucleus, for $r = 0$–100 pm, in steps of 20 pm? How many nodes exist in the probability distribution? ($a_0 = 5.2918 \times 10^{-11}$ m.)

The calculation is straightforward, and the following results ($Z = 1$) obtain:

r/pm	0	20	40	60	80	100
$R_{1,0,0}(r)$/pm^{-1}	0	5.1×10^{-3}	9.5×10^{-3}	1.0×10^{-3}	8.4×10^{-3}	6.6×10^{-3}

The number of (radial) nodes is zero. This can be checked by calculating (easily programmed) and plotting Eq. (2.68) from $r = 0$ to any chosen value. (The probability is zero at the nucleus because the spherical shell has zero volume at $r = 0$.)

It is clear from the results that a maximum in $R_{1,0,0}(r)$ exists at approximately 60 pm, and decreases rapidly with increasing distance r. By differentiation, it can be shown that the maximum occurs at a_0, the Bohr radius for hydrogen, or *ca.* 59.2 pm. Where would the maximum lie for the species He^+?

A position of zero probability is termed a *node*; the number of nodes depends on n and l: the number of radial nodes being $n - l - 1$, and the angular nodes number l.

2.9.5 Atomic *s* and *p* orbitals

All *s* type orbitals are characterized by the quantum number l of zero: whatever the value of the principal quantum number n they have no angular dependence, and their spherical symmetry gives them the shape of a sphere and imposes no restrictions on them. The first surface harmonic may be written as $Y_{0,0}$, which from Eq. (2.70) is $1/2\sqrt{\pi}$, which is effectively the normalizing constant in the complete 1s wave function:

$$Y_{l,m_l}(\theta,\phi) = (-1)^{(m_l+|m_l|)/2}\left[\left(\frac{2l+1}{4\pi}\right)\frac{(l-|m_l|)!}{(l+|m_l|)!}\right]^{1/2} P_l^{|m_l|}(\cos\theta)\exp(im_l\phi) \tag{2.70}$$

where $P_l^{|m|}(\cos\theta)$ are associated Legendre polynomials in angular form; their development is not a concern here, but the first few results are listed below:

$$P_0^0 = 1 \quad P_1^0 = \cos\theta \quad P_1^1 = -\sin\theta \quad P_1^{-1} = \sin\theta \quad P_2^0 = \tfrac{1}{2}(3\cos^2\theta - 1)$$

For a principal quantum number n of 2, $l = 1$, so that $m_l = 0, \ \pm 1$. The corresponding wave functions are designated p; there are three such functions, since $m_l = 0, \ldots, \pm l$, named p_0, p_1 and p_{-1}; they have the same energy, indicating a *degeneracy* of three.

The surface harmonics $Y_{l,m_l}(\theta, \phi)$ are of importance in many applications in physical science, and are complex functions for m not equal to zero. Further details on this topic can be gleaned from the literature [18], and the surface harmonics are listed in Table 2.7.

The p_0 orbital has a zero value of m_l and, hence, no component of angular momentum about its axis, which is normally chosen as the z reference axis for a molecular species. The p_1 and p_{-1} orbitals are complex and their amplitudes have maximum values in the x, y plane. The p orbitals are formulated within Table 2.6, normalized radial functions, and Table 2.7, normalized surface harmonics. The p_0, which is generally referred to as p_z is a real function, whereas the $p_{\pm 1}$ functions are complex. It is usual to set up real and imaginary functions, using Euler's formula [19], and to designate them p_x and p_y orbital functions respectively:

$$\left.\begin{aligned} p_z &= \ (3/4\pi)^{1/2} R_{n,l}(r) \cos\theta \\ p_1 &= -(3/8\pi)^{1/2} R_{n,l}(r) \sin\theta \exp(\mathrm{i}\phi) \\ p_{-1} &= \ (3/8\pi)^{1/2} R_{n,l}(r) \sin\theta \exp(-\mathrm{i}\phi) \end{aligned}\right\} \tag{2.71}$$

Linear combinations are formed, normally termed p_x and p_y:

$$\left.\begin{aligned} p_x &= (1/\sqrt{2})(p_- - p_+) = (3/4\pi)^{1/2} R_{n,l}(r) \sin\theta \cos\phi \\ p_y &= (\mathrm{i}/\sqrt{2})(p_- + p_+) = (3/4\pi)^{1/2} R_{n,l}(r) \sin\theta \sin\phi \end{aligned}\right\} \tag{2.72}$$

In these two equations, $1/\sqrt{2}$ is an additional normalization factor arising from the linear combination. Species without defined axes, such as atoms and linear molecules may be examined by the complex functions; with other species, where x, y and z axes have been assigned, real functions are more appropriate. The angular components are of particular interest here, since the associated radial functions do not change parity under symmetry transformations.

The p_x, p_y and p_z atomic functions are proportional to their parent wave functions, and carry a representation of the full rotation group, that is, the group that has the symmetry operators of a sphere [20], and form correct combinations to carry irreducible representations of its subgroups, namely the point groups; the latter topics will be discussed in subsequent chapters. The d wave functions, and others with $n > 3$, can be treated in a manner similar to that applied to the p wave functions; Table 2.8 lists the normal angular functions for $n = 1 - 3$; the shapes of these functions are illustrated in Fig. 2.39.

In applying group theory to these so-called orbitals in order to study covalent bonding, which is standard chemical practice, it should not be forgotten that they are not all true wave functions: the functions derived by linear combinations, such as Eq. (2.72), cannot be described by quantum numbers and cannot be used to describe electronic orbital angular momentum or spin [21–22]; the term orbital is applied generally with this understanding. Fuller

Table 2.8 Angular components of the *s*, *p* and *d* atomic orbitals.

Orbital	Function	Normalization constant	$f(x,y,z)$
s	—	$\frac{1}{2}\sqrt{\pi}$[a]	—
p_z	$\cos\theta$	$\frac{1}{2}\sqrt{3/\pi}$	z
p_x	$\sin\theta\cos\phi$	$\frac{1}{2}\sqrt{3/\pi}$	x
p_y	$\sin\theta\sin\phi$	$\frac{1}{2}\sqrt{3/\pi}$	y
d_{z^2}	$3\cos^2\theta - 1$	$\frac{1}{4}\sqrt{5/\pi}$	$2z^2 - (x^2+y^2+z^2) \equiv z^2$
$d_{x^2-y^2}$	$\sin^2\theta\cos 2\phi$	$\frac{1}{4}\sqrt{15/\pi}$	$x^2 - y^2$
d_{xz}	$\sin\theta\cos\theta\cos\phi$	$\frac{1}{2}\sqrt{15/\pi}$	xz
d_{yz}	$\sin\theta\cos\theta\sin\phi$	$\frac{1}{2}\sqrt{15/\pi}$	yz
d_{xy}	$\sin^2\theta\sin 2\phi$	$\frac{1}{4}\sqrt{15/\pi}$	xy

[a] This constant multiplies the radial term in deriving the complete 1*s* wave function.

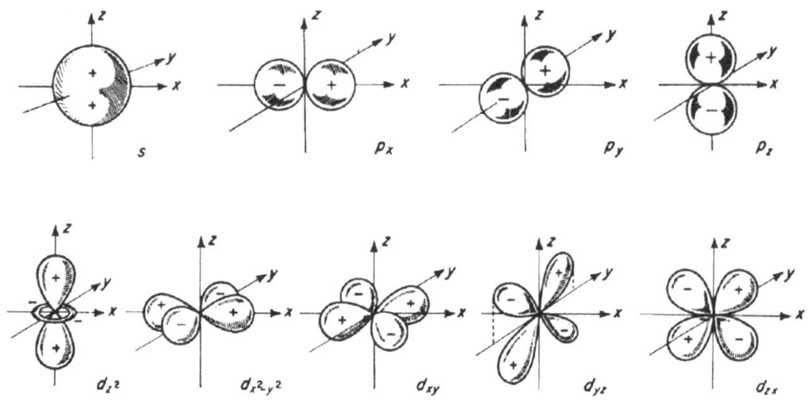

Fig. 2.39
Shapes of *s*, *p* and *d* atomic orbitals, generally considered as enclosing *ca.* 90% of their nominal electronic charge. [Reproduced by courtesy of Woodhead Publishing, UK.]

tables of surface harmonics, radial functions and complete wave functions may be found in the literature [23–25].

2.9.6 Chemical species and molecular orbitals

A complete hydrogenic wave function may be obtained by combining the appropriate terms from Table 2.6 and Table 2.7, and ensuring that the combination is normalized. Thus, for $n = 2$, $l = 1$ and $m_l = 1$,

$$\psi_{2,1,1} = (1/\sqrt{64\pi})(Z/a_0)^{3/2}\rho\ \exp(-\rho/2)\exp(\mathrm{i}\phi) \qquad (2.73)$$

or in real terms, as

$$\psi_{p_x}(= p_x) = (1/\sqrt{32\pi})(Z/a_0)^{3/2}\rho\exp(-\rho)\sin\theta\cos\phi \qquad (2.74)$$

where $\rho = 2Zr/na_0$.

The solution of the wave equation for a many-electron system in chemical application is complex and approximations must be used. One way in which the hydrogenic wave function can be modified for other species is by Slater's method, which can be used successfully with atoms of principal quantum number up to 3. The atomic number Z is modified by a quantum mechanical screening constant σ to form an effective atomic number Z_{eff} to replace the term Z in the wave function:

$$Z_{\text{eff}} = Z - \sigma \tag{2.75}$$

where σ is the screening constant calculated according to Slater's rules [26, 27]; a few examples are listed hereunder:

Atom	Z	Electron	σ	Z_{eff}	Z'_{eff}
He	2	1s	0 · 30	1 · 70	1 · 6875
Be	4	2s	2 · 05	1 · 95	1 · 9120
C	6	1s	0 · 30	5 · 70	5 · 6727
C	6	2s, 2p	2 · 75	3 · 25	3 · 2166 (2s) 3 · 1358 (2p)

Refinements have been applied to the original Slater calculation of Z_{eff}, listed here as Z'_{eff} in the table [28]. Theoretical calculations may be carried out with well-established computer programs, such as *Gaussian* [29], which computes *inter alia* energies, geometries (symmetries) and vibrational frequencies of chemical species.

Molecules may be visualized in terms of the overlap of atomic orbitals to form molecular orbitals or, more precisely, by combinations of wave functions. Molecular orbitals formed in this manner are normalized linear combinations of atomic orbitals of the species concerned; the number of molecular orbitals is equal to the number of atomic orbitals included in the linear combination. The ith molecular orbital of a species would be written as the linear combination

$$\Phi_i = \sum_j c_{i,j}\,\psi_j \tag{2.76}$$

where the coefficients $c_{i,j}$ are adjusted for a minimum energy conformation. The molecular configuration thus determined reveals geometric and symmetric properties of the species under examination, such as the C_{2v} symmetry of the water molecule, the equivalence of three coplanar B–F bonds in boron trifluoride, and the tetrahedral disposition of the four equivalent C–H bonds in methane.

An exposition of quantum chemistry lies outside the scope of this book, although a little more will be said of it in Chapter 8. Most modern books on quantum mechanics and quantum chemistry treat molecular orbitals and the linear combination of atomic orbitals (LCAO) technique in detail [24,25,30,31].

2.10 Crystal packing

Crystals form because interatomic forces of attraction and repulsion between their components, when brought sufficiently close to one another, overcome the thermal energy of the separate species; in a crystal structure, the result of

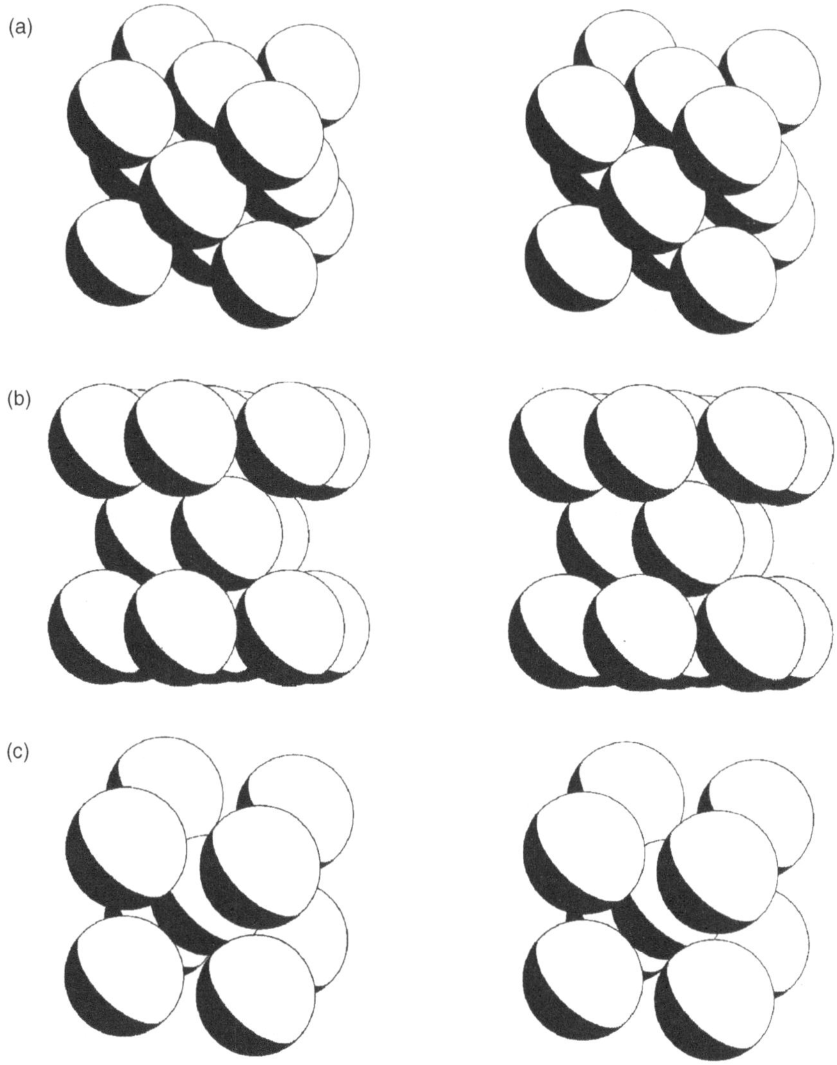

Fig. 2.40
Stereoscopic views of space-filling structures of metals: (a) close packed cubic, (b) close packed hexagonal, (c) body-centred cubic. The fraction of space occupied by the atoms, or packing efficiency, is 0.74 for (a) and (b), and 0.68 for (c). [Reproduced by courtesy of Woodhead Publishing, UK.]

the interplay between these forces is realized. Most metallic and other simple structures have atomic packing modes that are determined largely by geometry. Thus, the close packed cubic, close packed hexagonal and body-centred cubic structures (Fig. 2.40) represent a large majority of elemental metals; Fig. 1.8a is the body-centred cubic unit cell of chromium.

Ionic crystal structures, too, are influenced mainly by the sizes of their component ions. Among simple structures high symmetry is exhibited, for example, by the sodium chloride structure (Fig. 2.41), space group $Fm\bar{3}m$, and the caesium chloride structure (Fig. 2.40c) if Cs^+ is the central species in the unit cell and Cl^- at its eight corners, space group $Pm\bar{3}m$. In these structures, ions are packed as closely as geometry allows, and where the ions are of significantly disparate in size, as in lithium iodide or magnesium selenide, the anions are in contact with one another across the diagonals of the face of

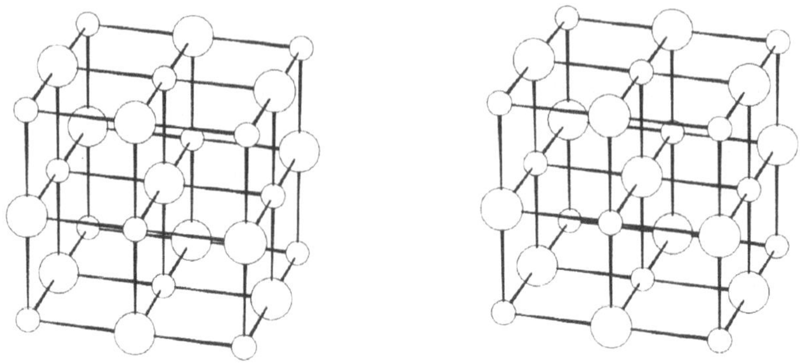

Fig. 2.41
Stereoview of the face-centred cubic unit cell of sodium chloride; circles in decreasing order of size represent Cl^- and Na^+. [Reproduced by courtesy of Woodhead Publishing, UK.]

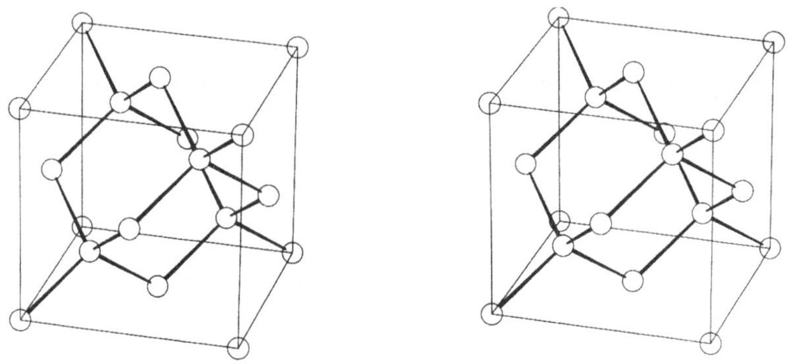

Fig. 2.42
Stereoview of the face centred cubic unit cell of diamond, C; three-dimensional covalent bonding is present in this crystal. [Reproduced by courtesy of Woodhead Publishing, UK.]

a cube, with the cations in the interstices. Figure 2.41 is a stereoview of the sodium chloride structure type, which is found for many halides and chalcogenides. If the anions are in contact across the cube face diagonals, the radius of the anion is given by $4r_- = a\sqrt{2}$, so that $r_- = \frac{a}{2\sqrt{2}}$ where a is the length of the cube side determined by experiment; thus, if $a_{\text{LiI}} = 6.00$ Å, $r_{\text{I}^-} = 2.12$ Å, whereupon, $r_{Li^+} = 0.88$ Å, assuming the additivity of ionic radii. The radii in Table 2.2 , column 2 were deduced in this way [32]. For comparison, two other sets of ionic radii have been included [33,34]. It seems that there remains still some discrepancy in the true values ionic radii.

Of truly covalent crystals there are but few, though many organic substances are often incorrectly named as covalent crystals. Cohesion in organic crystals occurs principally by van der Waals forces, often enhanced by dipolar attraction and by hydrogen bonding. A truly covalent crystal requires three-dimensional covalent bonding, and an excellent example is the structure of the diamond crystal (Fig. 2.42). Other highly covalent structures include silicon carbide, boron nitride, quartz (Fig. 2.43) and, to some extent, zinc sulphide.

Van der Waals forces exist between the molecules of all substances. In many structures, hydrogen bonding is the main source of cohesive attraction. Extensive hydrogen bonding exists in the structure of pentaerythritol (Fig. 2.44). In this equant, dipolar crystal, the H$-$O $\cdots$ H distance is 3.70 Å, whereas the sum of the corresponding van der Waals radii (Table 2.9) is 3.92 Å, which difference is evidence of an increased attraction between non-bonded atoms in this structure. In the structure

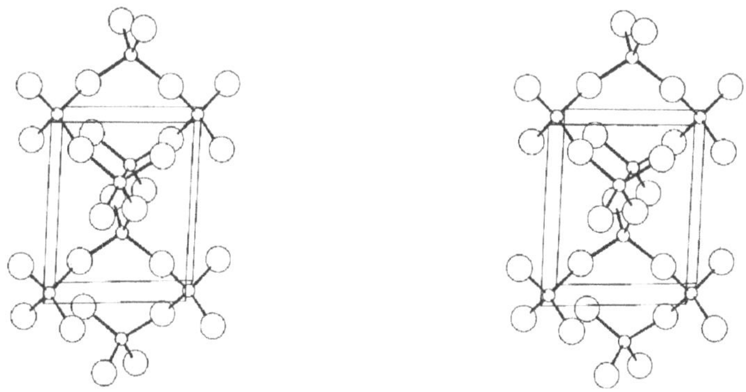

Fig. 2.43
Stereoview of the crystal structure of β-quartz, SiO_2: circles in decreasing order of size represent O and Si, respectively. [Reproduced by courtesy of Woodhead Publishing, UK.]

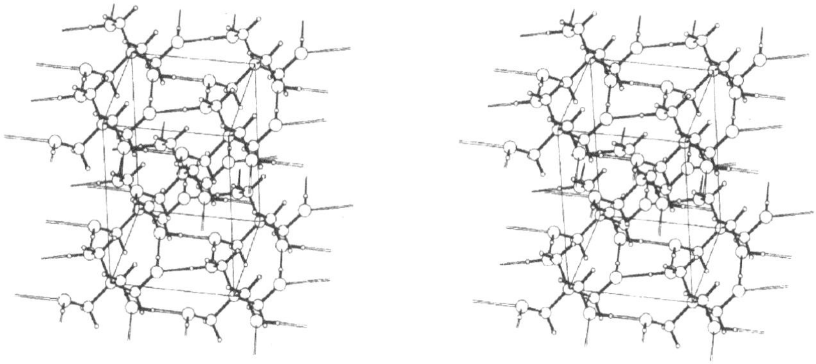

Fig. 2.44
Stereoview of the crystal structure of pentaerythritol [2,2-bis(hydroxymethyl) propan-1,2-diol, $C(CH_2OH)_4$]: circles in decreasing order of size represent O, C and H; hydrogen bonds are shown by doubled lines. [Reproduced by courtesy of Woodhead Publishing, UK.]

of gypsum, calcium sulphate dihydrate (Fig. 2.45), hydrogen bonding is wholly responsible for cohesion in one direction in the crystal.

Graphite (Fig. 2.46) exhibits a layered structure: the carbon atoms within each layer are strongly bonded fused hexagonal rings of aromatic character, but the cohesion between the layers is weak, giving rise to glide deformation, a property that invests graphite with its lubricant character. The linkage between the layers is through van der Waals forces, which give rise to the characteristic intermolecular contact distances; the layer separation distance in graphite is 3.7 Å, which is also the sum of the van der Waals radii. Van der Waals radii are of great significance in crystal packing, particularly with organic molecules.

Table 2.9 Van der Waals radii/Å.

Atom	Radius	Atom	Radius
H	1.20	As	2.20
C	1.85	Se	2.00
N	1.50	Br	1.95
O	1.52	Sb	2.20
F	1.35	Te	2.20
Si	2.10	I	2.15
P	1.90	$-CH_3$	2.00
Cl	1.80	$>CH_2$	2.00
S	1.83	$-C_6H_5$	1.85[a]

[a] Half-thickness of a phenyl ring.

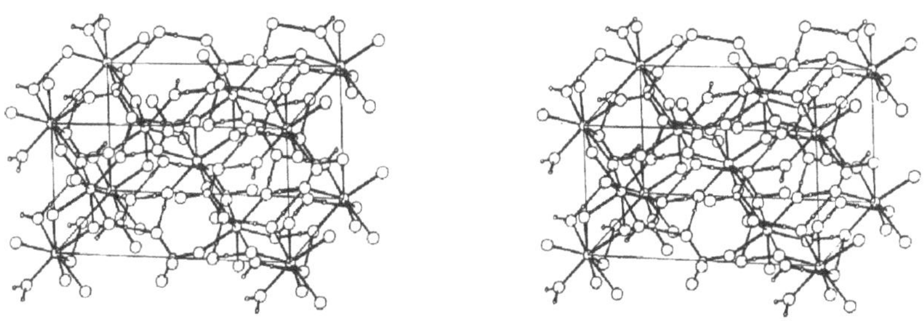

Fig. 2.45
Stereoview of the crystal structure of gypsum, calcium sulphate dihydrate, $CaSO_4.2H_2O$: circles in decreasing order of size represent O, S, Ca and H; hydrogen bonds, shown by double lines, hold the structure together in one direction. [Reproduced by courtesy of Woodhead Publishing, UK.]

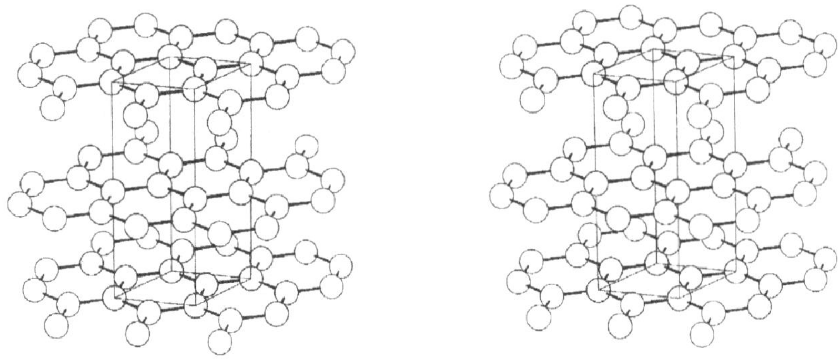

Fig. 2.46
Stereoview of the unit cell of graphite; the layered structure parallel to the (0001) planes is the source of its mechanical weakness, leading to its use as a lubricant material. [Reproduced by courtesy of Woodhead Publishing, UK.]

Crystal packing is a vast subject, and only a small section of it has been treated here. It is well documented in the literature [35,37] to which the reader should turn for greater detail on this topic.

References 2

[1] Great creation scientists: Nicolas Steno (1638–1686). <http://www.answersingenesis.org/articles/cm/v23/n4/steno>.
[2] Romè de I'Isle. <http://www.yourmineralcollection.com/Rome.htm>.
[3] Rene-Just Haüy and his influence. <http://www.minsocam.org/msa/collectors_corner/arc/hauyiv.htm>.
[4] Bravais A. *Études cristallographiques*. Paris: Académie des Sciences, 1866.
[5] Jeffreys H and Jeffreys BS. *Methods of mathematical physics*. Cambridge, UK: Cambridge Mathematical Library, 1999.
[6] Miller WH. *Treatise on crystallography*. Cambridge: 1839.
[7] Weiss CS. *Abhandlungen, der Königlichen Akademie der Wissenschaften,* Berlin: 1815.
[8] Plato. *The republic*, Book VII. *ca*. 380 BC.
[9] Phillips FC. *Introduction to crystallography*. Longmans, 1986.
[10] Learners TV. <http://www.learnerstv.com/animation/animation.php?ani=55&cat=Chemistry>.
[11] Duax W, et al. *Lipids*. New York: Springer, 1980.
[12] Nardelli M. *Acta Crystallogr. C* 1983; 39: 114.
[13] Ladd M and Palmer R. *Structure determination by X-ray crystallography*. 5th ed. Springer Science+Business Media, 2013.
[14] Franklin B. *Letter to Jean-Baptiste Leroy*. 1789.
[15] Leligny H and Grebille D. *Acta Crystallogr. A*1997; 53: 676.
[16] Schwarzenbach G, et al. *Acta Crystallogr. A* 1995; 51: 565.
[17] Taylor JR. *An introduction to error analysis*. University Science Books, 1997.
[18] McWeeny R. *Symmetry*. Pergamon, 1963.
[19] Euler L. *Introductio in Analysin Infinitorum*. Lausanne: Bosquet, 1748; (Eng. trans: Blanton J. *Introduction to analysis of the infinite,* Book I. Springer Verlag, 1988).
[20] Burns G. *Introduction to group theory with applications*. Academic Press, 1977.
[21] Boeyens JCA. *Chemistry from first principles*. Springer, 2008.
[22] Shpenkov GP. *Hadronic J*. 2006; 29: 455.
[23] Pauling L and Wilson EB. *Introduction to quantum mechanics with applications to chemistry*. [First published, 1935] Dove Books, 1986.
[24] House JE. *Fundamentals of quantum chemistry*. Academic Press, 2003.
[25] McQuarrie DA. *Quantum chemistry*. University Science Books, 2007.
[26] Slater JC. *Phys. Rev*. 1930; 36: 57.

[27] Atkins PW and Friedman RL. *Molecular quantum mechanics*. Oxford University Press, 2005.
[28] Clementi E and Raimondi DL. *J. Chem. Phys.* 1963; 38: 2686.
[29] Gaussian, Inc. <http://www.gaussian.com/>.
[30] Murrell JM, et al. *Valence theory*. John Wiley and Sons, 1979.
[31] Atkins PW and de Paula J. *Physical chemistry*. 8th ed. Oxford University Press, 2006.
[32] Ladd MFC. *Theor. Chim. Acta* 1968; 12: 333.
[33] Shannon RD and Prewitt CT. *Acta Crystallogr. B* 1968; 25: 925.
[34] Pauling L. *The nature of the chemical bond*. Cornell University Press, 1960.
[35] Kitaigorodskii AI. *Organic chemical crystallography*. New York: Consultants Bureau, 1961.
[36] Brock CP and Dunitz JD. *Chem. Mater.* 1994; 6: 1118.
[37] Gavezzotti A and Flack H. Crystal packing. <http://www.iucr.org/education/pamphlets/21/full-text>.

Problems 2

2.1 Write the Miller indices of planes that make the following intercepts on the x, y and z axes:

	On the x axis	On the y axis	On the z axis
(a)	$a/3$	$-b$	parallel
(b)	$-a$	$3b/2$	$c/4$
(c)	$a/4$	parallel	$-c/3$
(d)	$2a$	$-b/2$	c
(e)	parallel	$b/4$	$-2c$
(f)	$3a$	$-3b$	$c/2$

2.2 Evaluate the zone symbol for each of the pairs of planes:
(a) (001), (111) (b) $(0\bar{1}3)$, (100) (c) $(1\bar{2}1)$, $(\bar{3}\bar{2}1)$
(d) $(0\bar{1}0)$, $(\bar{1}\bar{1}3)$ (e) $(\bar{1}02)$, $(0\bar{1}3)$ (f) (123), $(\bar{2}31)$

2.3 In which of the zones evaluated in 2.2 does each of the following faces lie? Some faces may lie in more than one zone.
(a) $(\bar{1}\bar{1}2)$ (b) $(11\bar{3})$ (c) $(0\bar{2}1)$
(d) $(01\bar{3})$ (e) $(\bar{1}1\bar{1})$ (f) (101)

2.4 What faces lie at the intersections of each of the following pairs of zones?
(a) [001], [111] (b) [120], [021] (c) [113], [214]
(d) [010], $[1\bar{1}\bar{1}]$ (e) [103], [100] (f) $[\bar{1}10]$, $[\bar{1}3\bar{2}]$

2.5 Show how the Weiss zone equation may be used to specify two possible planes in the zone $[1\bar{2}3]$ There is more than one answer.

2.6 If a zone is defined by the planes $(h_1\,k_1\,l_1)$ and $(h_2\,k_2\,l_2)$, show that the plane $(mh_1 + nh_2,\ mk_1 + nk_2,\ ml_1 + nl_2)$ is cozonal; m and n are integers.

2.7 If the normals to two planes $(h_1\,k_1\,l_1)$ and $(h_2\,k_2\,l_2)$ in a crystal with axial ratios 1:1:1 have the direction cosines $\cos\chi_1, \cos\psi_1, \cos\omega_1$ and $\cos\chi_2, \cos\psi_2, \cos\omega_2$, respectively, show that the angle ϕ between the normals is given by

$$\cos\phi = \frac{\frac{h_1h_2}{a^2} + \frac{k_1k_2}{b^2} + \frac{l_1l_2}{c^2}}{\left(\frac{h_1^2}{a^2} + \frac{k_1^2}{b^2} + \frac{l_1^2}{c^2}\right)^{1/2}\left(\frac{h_2^2}{a^2} + \frac{k_2^2}{b^2} + \frac{l_2^2}{c^2}\right)^{1/2}}$$

2.8 A partial stereogram of an orthogonal[5] crystal is shown in Fig. P2.1. The axial ratios are 0.5287 : 1 : 0.9539 and $\angle 001 - 111 = 63.90°$. Find the value of $\angle 111 - 102$.

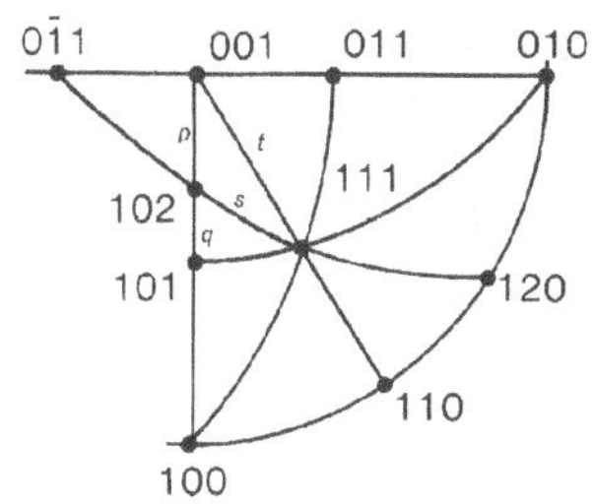

Fig. P2.1
Partial stereogram of an orthogonal crystal.

[5]Here, orthogonal means referred to mutually perpendicular axes.

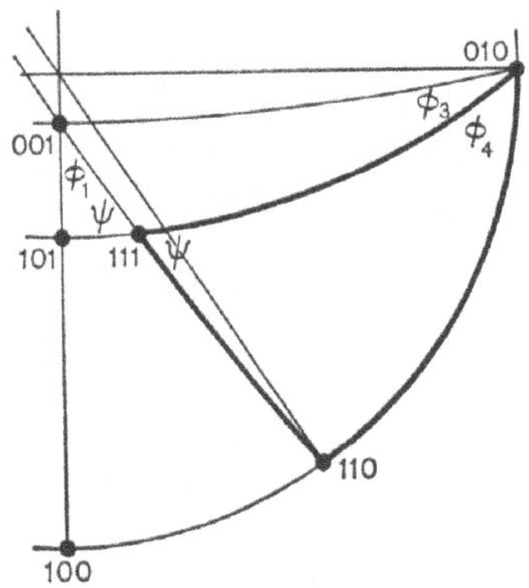

Fig. P2.2
Portion of the stereogram of gypsum, $CaSO_4.2H_2O$.

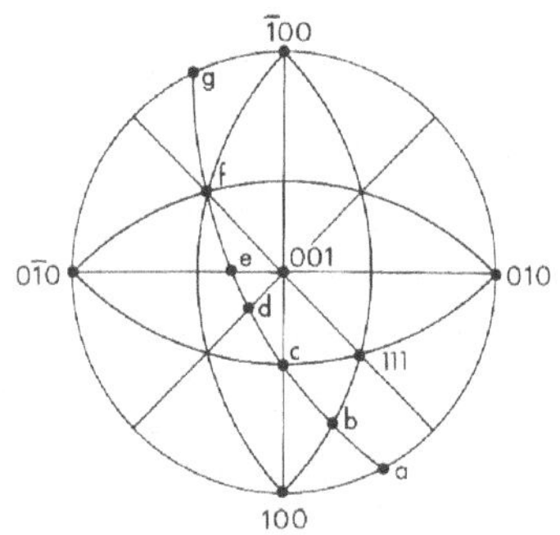

Fig. P2.3
Partial stereogram for an orthorhombic crystal. [McKie D and McKie C. *Essentials of crystallography.* 1986; reproduced by courtesy of Blackwell Scientific Publications.]

2.9 A portion of the stereogram for gypsum, $CaSO_4.2H_2O$, is shown in Fig. P2.2. The following angles between normals were recorded:

$$\angle 010-110=55.75^\circ,\ \angle 010-111=71.90^\circ,\ \angle 110-111=49.15^\circ.$$

Determine the axial ratios and the value for the β angle.

2.10 Find the zone axes for the pairs of planes (121) and (231). Show that the plane (011) lies in the zone. Find the values of p and q that enable (011) to be written as $(ph_1+qh_2,\ pk_1+qk_2,\ pl_1+ql_2)$, where the subscripts refer to (121) and (231), respectively.

2.11 Index the faces a–g on the stereogram of Fig. P2.3. Then, find the zone symbol for the zone containing these faces.

2.12 Use the program PLANE to obtain the best fit to the following six points,

X	1	2	−1	1	3	−2
Y	2	3	2	1	1	2
Z	3	−4	3	2	−2	1

and list the root mean square deviation of the points from the plane and its esd.

2.13 The data listed in Section 2.6.5 are provided in the file CART.TXT. Use the program INTXYZ to obtain the atomic coordinates. Then, use these coordinates in program MOLGOM to check the bond length and angles, and torsion angles. Does atom C6 lie above or below the plane of C1–C5?

2.14 A crystal contains four molecules of a compound in a unit cell of dimensions $a=7.210\,(4)$, $b=10.43\,(1)$, $c=15.22\,(2)$ Å. The fractional coordinates of the chlorine atoms in orthogonal unit cell unit are $\pm\left[{}^1\!/_4,\ 0.140\,(2), 0.000\,(2)\right]$ and $\pm\left[{}^1\!/_4,\ 0.640\,(2)\ 0.500\,(2)\right]$. Calculate the shortest Cl ... Cl contact distance and its estimated standard deviation.

2.15 Show that for the Miller–Bravais indices $hkil$, $i=-(h+k)$.

2.16 Show that the wave function $\psi_{2,1,1}$ in Eq. (2.67) is normalized.

Point group symmetry

3

SYNOPSIS

- Symmetry elements and operations
- Point groups and their representation
- Combinations of symmetry operations
- Crystal systems and crystal classes
- Derivation and recognition of point groups
- Point groups and physical properties
- Point groups of chemical species in stereoview
- Non-crystallographic point groups
- Hermann–Mauguin and Schönflies symmetry notations
- Matrix representation of point group symmetry
- Periodic and aperiodic crystals

3.1 Introduction

In the first chapter, the existence of reflection symmetry lines and planes in certain well known objects was introduced, the meaning of symmetry discussed and the manifestation of symmetry over a wide spectrum of life reviewed. While studying zones in the previous chapter, it became evident that crystals usually exhibit symmetry to some degree. Now, the study of crystals and molecules is deepened by examining their symmetry as finite bodies, introducing point groups and, subsequently, matrix methods of handling symmetry operations.

3.2 Symmetry elements, symmetry operations and symmetry operators

A *symmetry element* is a geometrical entity, a point, line or plane, in a body or assemblage, with which a symmetry operation is associated. A symmetry element is strictly conceptual, but it is advantageous to endow it with a sense of reality. The symmetry element associates all parts of the assemblage in which it is present into a number of symmetrically related sets. The term assemblage was introduced in Chapter 1 and describes usefully sets of faces on a crystal and a bundle of radiating normals from a central atom in a chemical entity: compare the stereoviews of Fig. 3.1, a hypothetical tetrahedral species C_4, and Fig. 3.2, a molecule of methane, in which the four bonds from the central carbon atom radiate to the hydrogen atoms, thus forming a similar tetrahedral arrangement.

A *symmetry operation* when applied to a body transforms it to a situation that is indistinguishable from its initial situation, and thus *reveals* the symmetry inherent in the body according to the nature of the operation. In many examples, a single symmetry element can give rise to more than one symmetry operation. Thus, the symmetry operations $\boldsymbol{4}^2$ and $\boldsymbol{4}^3$ may be regarded as either multiple steps of symmetry operation ***4*** itself, which could be written as $\boldsymbol{4}^1$, or as single-step operations in their own right: but all are contained within the symmetry of the assemblage to which they refer, and are associated with the single symmetry element, 4. This aspect of symmetry will be important in the study of group theory in later chapters. It is in the application of symmetry operations that the essence of symmetry, as described in Section 1.4, is made manifest; it is the property of the invariance of an assemblage under a transformation. Symmetry operators and symmetry operations are written here in bold italic font.

A *symmetry operator* is perhaps best thought of as a mathematical function that carries out a symmetry operation in a definite manner and orientation, and

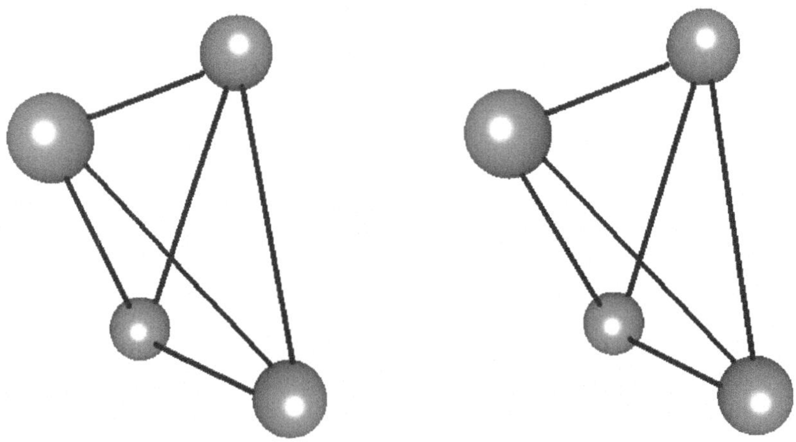

Fig. 3.1
Tetrahedral structure of a hypothetical C_4 species.

Fig. 3.2
Molecule of methane, showing the tetrahedral arrangement of C–H bonds radiating from the central carbon atom. [Reproduced by courtesy of Woodhead Publishing, UK.]

is most usefully represented in matrix form. It is that entity by virtue of which its corresponding symmetry operation is executed.

3.3 Point groups

Symmetry elements may occur singly in a finite body, as shown by Fig. 1.5a, or in combinations, (Fig. 1.5b–d). The set of associated symmetry operators in the body, or just one such operator, is referred to as a *point group*. A point group may be defined as *a set of symmetry operations all of which pass through a single fixed point*: this point is also the origin of the reference axes for the body. It follows that the symmetry operations of a point group must leave at least one point unmoved: in some cases, it is a line or plane that is unmoved under the action of the point group.

Real objects, even crystals, have imperfections, albeit on a microscopic scale, and an exact indistinguishable aspect may be obtained only by a 360° rotation; this operation is *identity*, effectively applying no action to the body. For practical purposes, however, the effects of minute imperfections are small and statistically insignificant, as discussed in Section 1.2.

In Section 1.4, the dependence of symmetry upon the type of examination applied to it was noted. Here, the concern is with the symmetry shown by an assemblage of directions in space. These directions are conveniently represented by stereograms, Section 2.6.1, and this method of representation will be used in the discussions that follow.

Several concepts in symmetry can be introduced with two-dimensional objects; subsequently, the third dimension can be added, mainly as a geometrical extension of the two-dimensional arguments. In one dimension, two point groups are recognized, 1 and m, wherein the latter, reflection occurs across a point; little use will be made here of one-dimensional point groups.

3.4 Symmetry in two dimensions

'*We proceeded straight from plane geometry to solid bodies in motion without considering solid bodies first on their own. The right thing is to proceed from second dimension to third, which brings us to cubes and other three-dimensional figures*'[1]. Symmetry in two dimensions implies operations on an object strictly in two-dimensional space, the plane of the paper with the illustrations herein. Any operation that would move the object wholly or partly out of its plane, even though indistinguishability may be achieved after the operation, is inconsistent with two-dimensional symmetry. There are two types of symmetry operations in two dimensions, those of rotation about a point and reflection across a line.

3.4.1 Rotation symmetry

If an object can be brought from an initial state to an indistinguishable state R times during a complete rotation of 360° about a *point* in it, it is said to possess R-fold rotational symmetry about that point. In principle, R may take any value from 1 to infinity: for the moment, only the values of 1, 2, 3, 4 and 6, will be considered because they are the values permitted in crystals, although other degrees of rotation are found in certain solids and in molecules, as will be discussed later; the reasons for the restriction will evolve in a later discussion.

3.4.2 Reflection symmetry

In two dimensions, reflection symmetry takes place across a *line*. The line may be imagined to divide the object into halves such that one half is the mirror image of the other. The symmetry operation again leads to indistinguishability of the initial and final states of the object. The operation is physically non-performable on the object, unlike rotation, but its existence can be appreciated readily from a study of the object or from its stereogram.

3.4.2.1 Representation of two-dimensional symmetry

Reflection symmetry involves enantiomorphous, or change-of-hand, aspects, whereas rotational symmetry involves congruent aspects. Figure 3.3 illustrates six simple two-dimensional objects and their stereograms. Each object comprises the motif in (a) or an assemblage of more than one of them; diagrams (a)–(e) depict rotational symmetry and diagram (f) reflection symmetry. In the stereograms of these two-dimensional objects, all poles must of necessity lie on the perimeter of the circle. The standard graphic symbols for the rotation operations are shown at the centres of the stereograms, and that for reflection symmetry is a line of thicker extent than those on the rest of the diagram. There is a clear mnemonic connection between the shape of the symbol and the degree R of rotation that it represents. The poles representative of the motif are shown by the symbol • on the stereogram: the number of poles in general (asymmetric) positions on a stereogram is equal to the number of symmetry operations in the corresponding group. Thus, the operations $\boldsymbol{4}$, $\boldsymbol{4}^2\ (\equiv \boldsymbol{2}), \boldsymbol{4}^3\ (\equiv \boldsymbol{4}^{-1})$ and identity ($\boldsymbol{1}$) mentioned in Section 3.2 relate the four poles on Fig. 3.3d.

3.4.3 Combinations of symmetry operations in two dimensions

The objects in Fig. 3.3 each contain a single symmetry element and illustrate six two-dimensional point groups, the *plane* point groups 1, 2, 3, 4, 6 and m. Other plane point groups arise from the combinations or R and m. Formally, $1m$, $2m$, $3m$, $4m$, $6m$ and mm can be written as symbols for these combinations. Point group $1m$ is equivalent to m, and mm in standard notation is $2mm$. Remembering that the symmetry elements of a point group intersect in a point, Fig. 3.4 can be drawn to represent the possible combinations of the symmetry operators $\boldsymbol{R}$ and $\boldsymbol{m}$ in two dimensions. Notice that the combination of two $\boldsymbol{m}$ line operators at right angles to each other introduces a twofold rotation operator at their intersection: operation $\boldsymbol{3}$ and $\boldsymbol{m}$ generate two more m lines, because the rotational operation acts on all symmetry elements present; the operations $\boldsymbol{4m}$ also introduces two additional m lines, and $\boldsymbol{6m}$, six more m lines; the symbolism $\boldsymbol{Rm}$ implies a combination of the operations $\boldsymbol{R}$ and $\boldsymbol{m}$.

In Figs. 3.3 and 3.4, the poles have not been placed on any symmetry element. Each stereogram thus represents the full set of symmetry related poles or *general form*. Poles lying on a symmetry element constitute a *special form*.

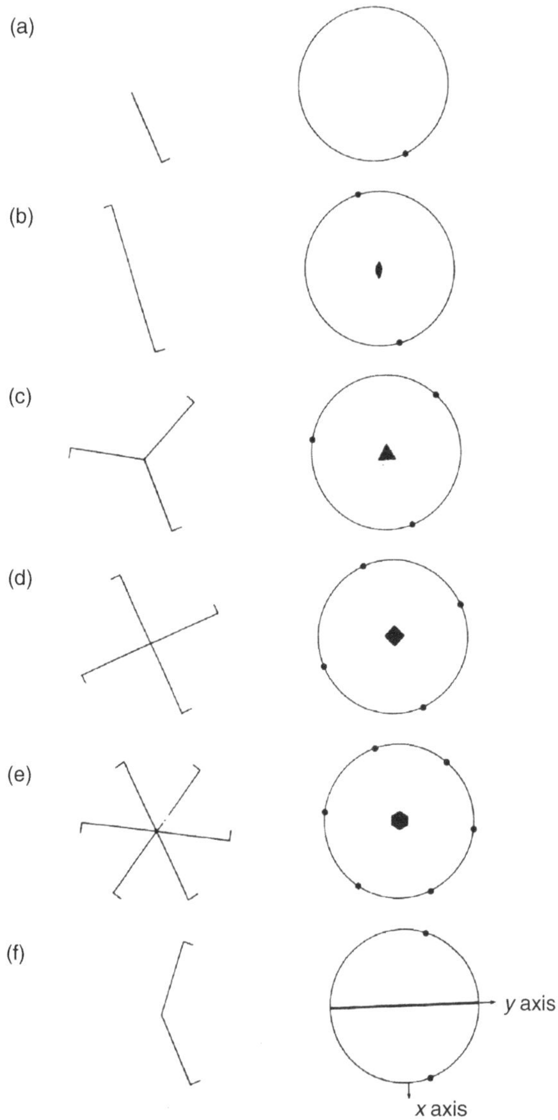

Fig. 3.3
Two-dimensional objects and their stereograms, I. The plane point group symbols are: (a) 1, (b) 2, (c) 3, (d) 4, (e) 6, (f) *m*.

Figure 3.5 shows a stereogram showing only two poles: they could represent the general form of plane point group 2 or a special form in 2*mm*. The point group can be determined with certainty from its stereogram *only* if the general form is present. An entity lying on a symmetry element must normally have a symmetry that is consistent with that element, that is, it must have the symmetry of that operator or a subgroup of it (Section 3.4.5).

3.4.4 Two-dimensional systems and point group notation

It is convenient to refer two-dimensional objects and their point groups to a two-dimensional axial system, with both the system name related to the maximum rotational symmetry of the object, and reference axes chosen,

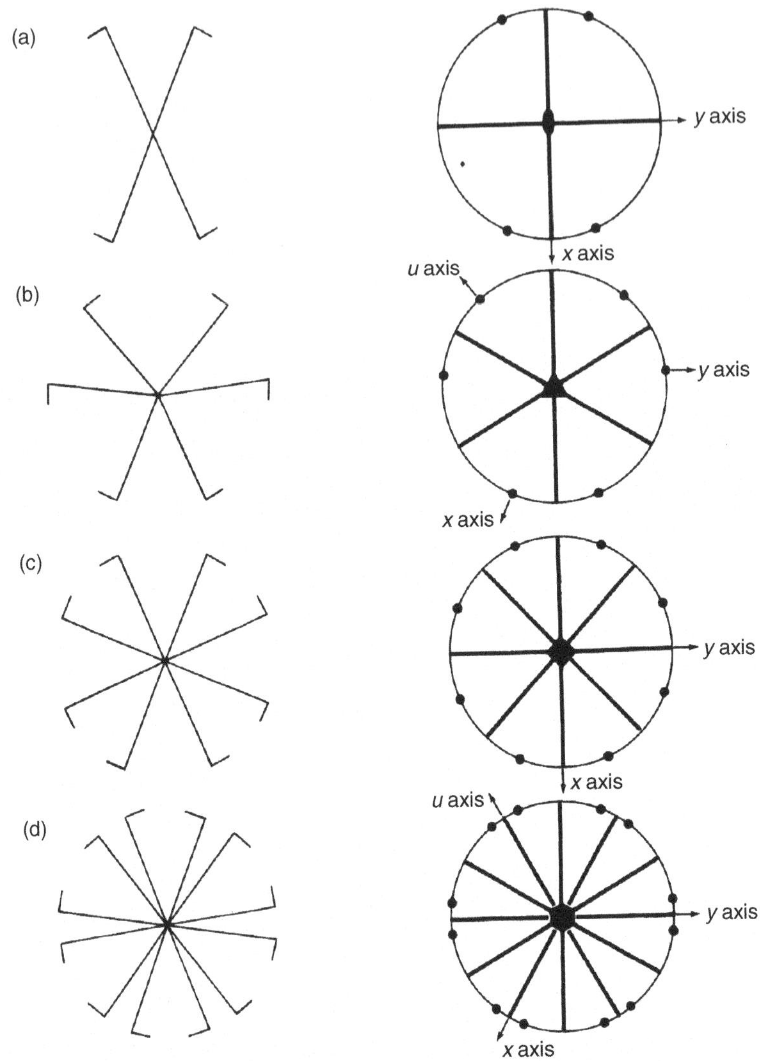

Fig. 3.4
Two-dimensional objects and their stereograms, II. The plane point group symbols are: (a) $2mm$, (b) $3m$, (c) $4mm$, (d) $6mm$.

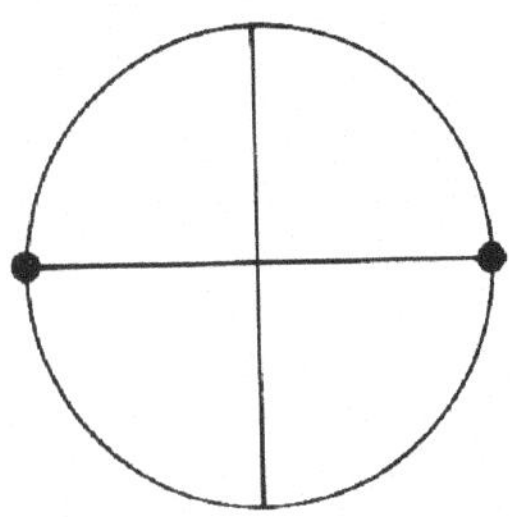

Fig. 3.5
Stereogram showing two poles: with no other information, it could represent a general form in plane point group 2 or a special form in $2mm$.

as already described in Section 2.2. There ensues a conventional, precise orientational significance to the positions in the point group symbol, which is a feature of the Hermann–Mauguin notation that will be described fully later in this chapter. Table 3.1 lists the two-dimensional system names, the associated point groups and the meanings of the point group symbols.

It is *very necessary* to understand the reason for the differing meanings of the positions in the point group symbol in different systems. Figure 3.4c, for example, shows that the mirror lines occur in two sets, each related by fourfold rotation symmetry. So the second position in the symbol refers to both the x and y directions. Compare this with Fig. 3.4a, where the x and y directions are independent, because the rotation symmetry here is only twofold; this matter will be elaborated in the following section.

Table 3.1 Two-dimensional systems and point groups.

System	Point group	Symbol meaning for each position: 1st position	2nd position	3rd position
Oblique	1, 2[a]	Rotation about a point[b]	—	—
Rectangular	1m[c]	Rotation about a point	$m \perp x$	—
	2mm	Rotation about a point	$m \perp x$	$m \perp y$
Square	4	Rotation about a point	$m \perp x, y$	—
	4mm	Rotation about a point	$m \perp x, y$	m at 45° to x, y
Hexagonal	6	Rotation about a point	$m \perp x, y, u$	m along x, y, u

[a] 2 is the two-dimensional analogue of a centre of symmetry in three dimensions.
[b] This point will always correspond to the origin of the reference axes.
[c] Normally m; the initial 1 is given here in order to indicate the position (1m1) of m with respect to the reference axes.

3.4.5 Subgroups

The point groups in Fig. 3.3 and Fig. 3.4 other that Fig. 3.3a contain point groups of lower symmetry, that is, fewer symmetry elements. If it is possible to remove certain symmetry elements from a point group and still leave a true point group, then the latter is a *subgroup* of the former. Table 3.2 lists the subgroups of the plane point groups (see also Section 7.4.3).

Table 3.2 Subgroups of the ten plane point groups.

Point group	Subgroups
1	None
2	1
3	1
4	1, 2
6	1, 2, 3
m	1
2mm	1, 2, m
3m	1, 3, m
4mm	1, 2, 4, m, 2mm
6mm	1, 2, 3, 6, m, 2mm, 3m

3.5 Three-dimensional point groups

The symmetry operations encountered in three-dimensional objects are rotation, reflection, and roto-inversion with respect to the origin. The operations associated with the first two of these symmetry operations are similar to their counterparts in two dimensions, but with geometrical extension: rotation operates now about a *line*, or *axis*, and reflection operates across a *plane*. The roto-inversion axis has no counterpart in two dimensions.

3.5.1 Rotation symmetry in three dimensions

If an R-fold rotational operation about a line, a symmetry axis, brings an object into an indistinguishable state for every successive rotation of $(360/R)°$ about that line, then that line is designated a rotation axis of degree R. Rotation is also called *proper rotation*, or an *operation of the first kind*, and is concerned with *congruent* aspects of an object.

3.5.2 Reflection symmetry in three dimensions

Reflection symmetry takes place across a *plane*, usually called a *mirror* plane, symbol m, such that the plane divides the object into mirror image halves, as in the two-dimensional case. It involves *enantiomorphous* aspects of the object, and is termed an *improper rotation* or an *operation of the second kind.*

3.5.3 Roto-inversion symmetry

An object possesses a roto-inversion symmetry axis $\overline{R}$, often called just inversion axis, if it can be brought from an initial state to an indistinguishable state by the combined actions of rotation by the degree R and inversion in the origin point. Roto-inversion also involves enantiomorphous aspects of the object and so it, too, is an operation of the second kind, or improper rotation.

The inversion operation is a little more complicated than either reflection or rotation. Figures 3.1 and 3.2 exhibit *inter alia* $\overline{4}$ inversion symmetry, possibly easier to appreciate from the second of these illustrations: a hydrogen atom is rotated about the vertical axis by 90° and then inverted in the origin, the carbon atom in this case; the two actions constitute a *single* symmetry operation. Appendix A1 shows how simple models containing a $\overline{4}$ axis may be constructed.

Another symmetry element that can be used instead of the inversion axis is the *roto-reflection*, or *alternating*, symmetry axis; it arises in the Schönflies symmetry notation and is discussed in Section 3.10.1.

3.5.4 Stereogram representations of three-dimensional point groups

It has been stated that rotation is concerned with congruent aspects of an object whereas reflection and inversion involve enantiomorphism, but the conventional stereogram notation (Fig. 2.18) does not distinguish between these two aspects of symmetry operations (see also Section 3.7.1).

Figure 3.6a shows the operation of twofold rotation along the y axis, drawn conventionally left to right in the plane of the paper, whereas (b) illustrates a fourfold inversion axis along z, normal to the plane of the paper. Points (1) and (2) in (a) are congruent aspects, and points (1) and (4) in (b) contain an enantiomorphic relationship, but their graphic symbols are identical. In Fig. 3.7, employing the same two symmetry operations, the enantiomorphic relationship is shown by the comma in the centre of the point, a notation adopted by the author [2] from space group diagrams and which will be employed here.

It is conventional that the reference axes on stereograms are set with the z axis normal to the primitive, the y axis lies in the primitive, left to right, and the

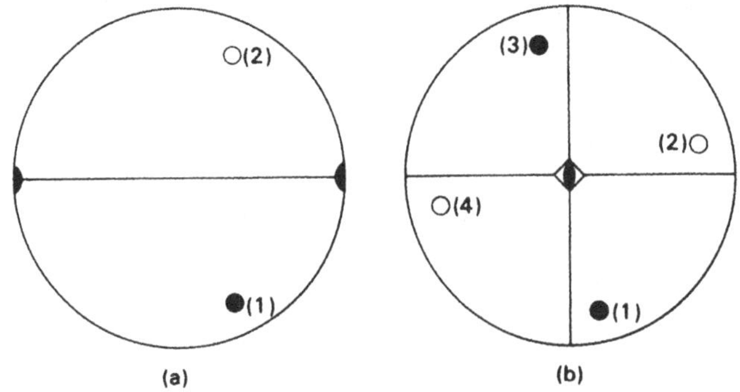

Fig. 3.6
Stereograms of general forms in the traditional notation. (a) Point group 2. (b) Point group $\overline{4}$. In (a) points (1) and (2) are in a congruent aspect, and in (b) points (1) and (4) are in an enantiomorphous aspect, but the notation gives no indication of this fact. [Reproduced by courtesy of Springer Science+Business Media, New York, © Kluwer Academic/Plenum Publishing.]

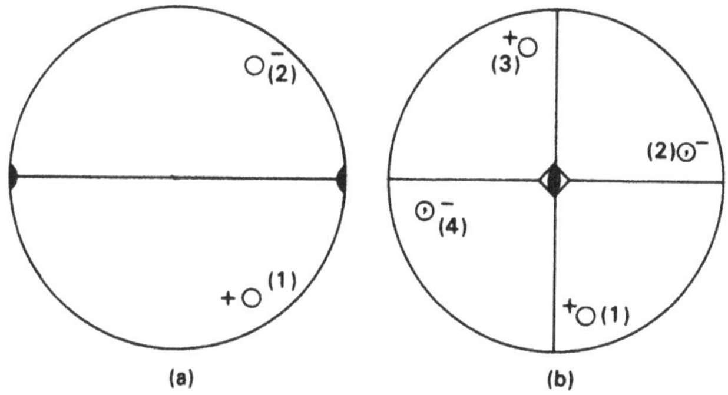

Fig. 3.7
Stereograms of Fig. 3.6 in the preferred notation. The congruent aspects are shown as ○+ and ○− whereas the enantiomorphous aspects are now clearly distinguished as ○+ and ⊙−. The + and − signs indicate the heights in z. [Reproduced by courtesy of Springer Science + Business Media, New York, © Kluwer Academic/Plenum Publishing.]

x axis lies in the primitive from the origin downwards to the primitive. A departure from the latter two orientations in the non-orthogonal crystal systems is indicated by the axial terminations in Fig. 3.12, within the primitive.

3.5.5 Crystallographic point groups

When rotation operators are combined with one another or with other symmetry operators, their relative orientations are limited but the number of possible point groups is still infinite. One reason for the symmetry restriction in crystals, that R be equal to 1, 2, 3, 4 and 6, is because crystals are a periodic, three-dimensional stack of identical parallelepipeds, and only such figures that are based on the above rotational symmetries can be packed to fill space completely, as Fig. 3.8 shows; voids are present in periodic structures based on rotation degrees of five, seven or greater (see also Section 3.13.1).

A standard notation is adopted for the symmetry elements in the crystallographic point groups, and is detailed in Table 3.3. Not surprisingly, some of the entries in the table are similar to those used for the two-dimensional point groups. The same symbolism will be used also in studying space groups.

3.5.6 Crystal classes

There are 32 *crystal classes*, derived from observations of the external shapes of crystals; each class is characterized by the name of its general form, which implies the full symmetry of the crystal. Native crystals frequently do not display their general forms, so that the true symmetry may not be immediately apparent. For example, quartz, shown in Fig. 2.1 and Fig. 2.2, has the class name trigonal trapezohedral $\{hkl\}$[1], and zircon (Fig. 2.11) has the class name ditetragonal dipyramidal $\{hkl\}$, but these general forms are rarely displayed by naturally occurring crystals.

The morphological names are well documented in the literature, together with interesting history of early crystallography [3–5], and it is not the purpose of this book to dwell on them. More importantly, it should be noted that each crystal class is characterized by a point group: quartz is 32, and zircon $\frac{4}{m}mm$. The term crystal class is not synonymous with crystallographic point group: crystals of a given class all exhibit the same point group; it is a classificatory pigeon-hole term.

[1] $\{hkil\}$ on hexagonal axes.

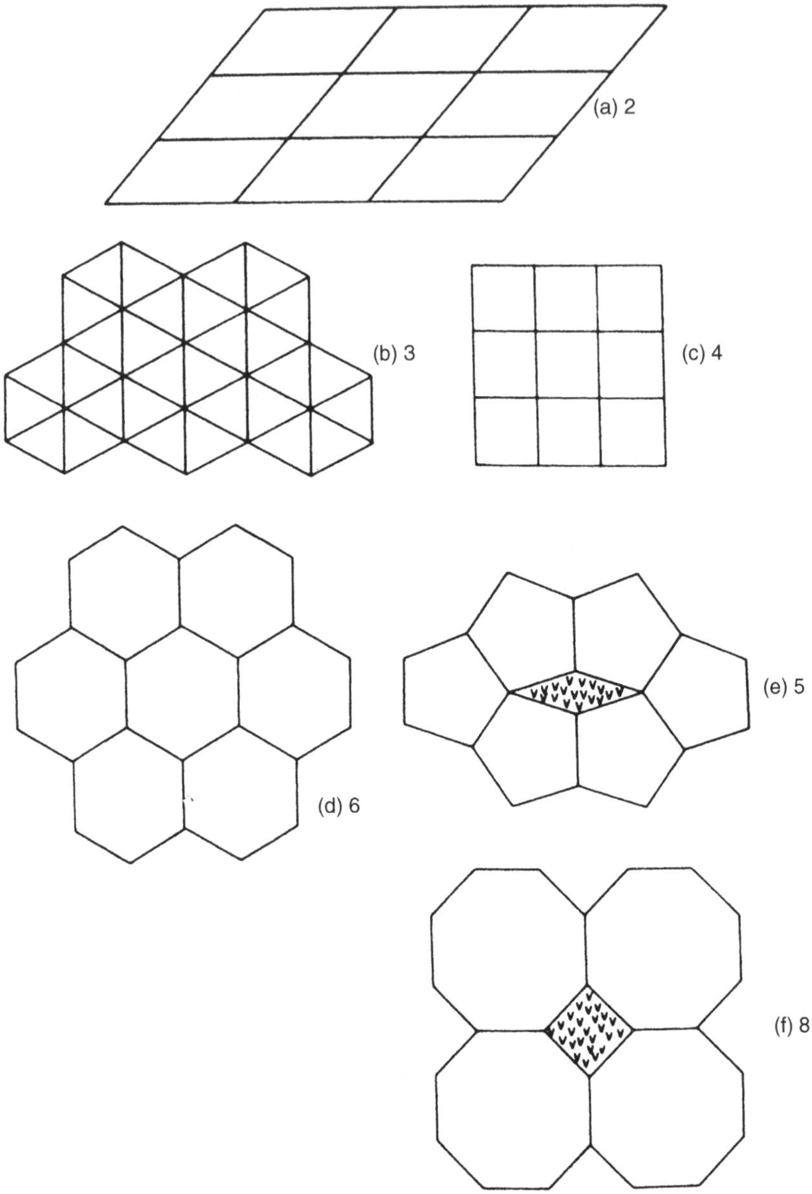

Fig. 3.8
Sections of stacked parallelepipeds normal to principal rotation axes, and the symmetries of their structural units: (a)–(d) are space-filling; (e)–(f) with rotational symmetries 5 and 8 show voids v in the structures. [Reproduced by courtesy of Springer Science + Business Media, New York, © Kluwer Academic/Plenum Publishing.]

3.5.7 Crystal systems

A broad but convenient classification of crystals is given in terms of *crystal systems*. A crystal is allocated to one of seven crystal systems dependent upon its *characteristic* symmetry, that is, the symmetry that is essential to categorize the crystal. A given crystal will frequently display more than its characteristic symmetry. The crystal symmetry determines special relationships among both

Table 3.3 Three-dimensional symmetry symbols.

Symbol	Name	Action for indistinguishability	Graphic symbols
1	Monad	360°(0°) rotation; identity	None
2	Diad	180° rotation	⬮ ⊥ projection ◖ ‖ projection
3	Triad	120° rotation	▲ ⊥ or inclined to projection
4	Tetrad	90° rotation	◆ ⊥ projection ■ ‖ projection
6	Hexad	60° rotation	⬢ ⊥ projection
$\bar{1}$	Inverse monad	Inversion[a]	○
$\bar{3}$	Inverse triad	120° rotation + inversion	⊥ or inclined to projection
$\bar{4}$	Inverse tetrad	90° rotation + inversion	⊥ projection ‖ projection
$\bar{6}$	Inverse hexad	60° rotation + inversion	⊥ projection
m	Mirror plane[b]	Reflection across plane	▬ ⊥ projection (‖ projection

[a] Point group $\bar{R}$ is equivalent to point group $R\bar{1}$ when R is an odd number; $\bar{1}$ represents the centre of symmetry, but $\bar{2}$, $\bar{4}$ and $\bar{6}$ are not centrosymmetric point groups. For R even, point group $R\bar{1} \equiv$ point group R/m.
[b] The symmetry operations m and $\bar{2}$ produce equivalent actions where $\bar{2}$ is normal to m.

Table 3.4 Crystal systems and their characteristics.

System	Characteristic symmetry axes with their orientation	Parametral plane intercepts and interaxial angles, assuming the simplest indexing of faces[a,b]
Triclinic	None	*a* ¢ *b* ¢ *c*; α ¢ β ¢ γ ¢ 90°, 120°
Monoclinic	One 2 or $\bar{2}$ axis[c] along *y*	*a* ¢ *b* ¢ *c*; α = γ = 90°; β ¢ 90°, 120°
Orthorhombic	Three mutually perpendicular 2 or $\bar{2}$ axes along *x*, *y* and *z*	*a* ¢ *b* ¢ *c*; α = β = γ = 90°
Tetragonal	One 4 or $\bar{4}$ axis along *z*	*a* = *b* ¢ *c*; α = β = γ = 90°
Trigonal[d]	One 3 axis along *z*	*a* = *b* ¢ *c*; α = β = 90°; γ = 120°
Trigonal[e]	One 3 axis along **a** + **b** + **c**	*a* = *b* = *c*; α = β = γ ¢ 90°, < 120°
Hexagonal	One 6 or $\bar{6}$ along *z*	*a* = *b* ¢ *c*; α = β = 90°; γ = 120°
Cubic	Four 3 axes inclined at 54.74°($\cos^{-1} 1/\sqrt{3}$) to *x*, *y* and *z*	*a* = *b* = *c*; α = β = γ = 90°

[a] The same relationships apply to the conventional crystallographic unit cells in lattices.
[b] The special symbol ¢ should be read as 'not constrained by symmetry to equal'.
[c] $\bar{2}$ is equivalent to m where m is perpendicular to $\bar{2}$; *see* Fig 3.9; 2 is set along *y* by crystallographic convention.
[d] Referred to hexagonal axes.
[e] Referred to rhombohedral axes; **a** + **b** + **c** is here the direction [111] in the unit cell.

the intercepts, a, b and c, of the parametral plane and the interaxial angles, α, β and γ. Table 3.4 lists the seven crystal systems and their characteristics.

3.6 Derivation of point groups

Before embarking on a derivation of point groups, a simple scheme whereby they can be assembled is given. Table 3.5 lists these point groups under the seven crystal systems as headings. The main difficulty in understanding the full meaning of point group symbols lies not in knowing the action of their symmetry operations, but rather in appreciating both the relative orientations of the different elements in the point group and the fact that these orientations change according to the nature of the principal symmetry axis, that is, the rotation axis R of highest degree. These orientations *must be assimilated*: they are a key factor in point group and space group studies, and are shown in Table 3.6; the equivalence of $\bar{2}$ and m is has been illustrated by Fig. 3.9.

The meanings of the positions in the three-dimensional point group symbols are set out in Table 3.6. This table and Table 3.5 should be studied in conjunction with the stereogram representations of the crystallographic point groups shown in Fig. 3.12. For example, consider point groups 222 and 422. In 222, the three symmetry axes are along x, y and z, respectively, and any two of the symmetry operations acting at right angles to each other introduce an action corresponding a third mutually perpendicular twofold axis. In 422, 4 is taken, by convention, along z, then the second position 2 refers to both x and y directions because they are equivalent under fourfold rotation. The combination of operations ***4*** and ***2*** introduces a second form of twofold axes along $[1\bar{1}0]$ and [110]. Similar situations occur among other point groups where the principal axis is of greater degree than 2.

In the case of point group $mm2$, Fig. 3.10 shows that the existence of m planes normal to x and y introduces a twofold axis along z. In Fig. 3.11, the symmetry operations in point group $4mm$ are clarified.

The reader should not be discouraged by the wealth of convention in crystal symmetry: it is essential for unambiguous description and communication in this subject. With practice, the conventions cease to present problems.

Table 3.5 Crystallographic point group scheme[a].

Point group type	Triclinic	Monoclinic	Trigonal	Tetragonal	Hexagonal	Cubic[b,c]
R	1	2	3	4	6	23
$\bar{R}$	$\bar{1}$	m	$\bar{3}$	$\bar{4}$	$\bar{6}$	$m\bar{3}$
$R\bar{1}$	—	$\frac{2}{m}$	—	$\frac{4}{m}$	$\frac{6}{m}$	—
		Orthorhombic				
R2		222	32	422	622	432
Rm		$mm2$	$3m$	$4mm$	$6mm$	—
$\bar{R}m$		—	$\bar{3}m$	$\bar{4}2m$	$\bar{6}m2$	$\bar{4}3m$
$R2$ and $\bar{1}$		mmm	—	$\frac{4}{m}mm$	$\frac{6}{m}mm$	$m\bar{3}m$

[a] The reader should consider the implication of the spaces marked — in the table.
[b] The cubic system is characterized by its four threefold axes: here, R refers to 2, 4 or $\bar{4}$; threefold axes are always present along < 111 >.
[c] Point groups $m\bar{3}$ and $m\bar{3}m$ will often be found written as $m3$ and $m3m$, respectively, in earlier texts.

Table 3.6 Three-dimensional point groups in the Hermann–Mauguin notation.

System	Point groups[a]	1st position	2nd position	3rd position
Triclinic	$1, \bar{1}$	All directions in crystal	—	—
Monoclinic[b]	$2, m, 2/m$	2 and/or $\bar{2}$ along y	—	—
Orthorhombic	$222, mm2, mmm$	2 and/or $\bar{2}$ along x	2 and/or $\bar{2}$ along y	2 and/or $\bar{2}$ along z
Tetragonal	$4, \bar{4}, 4/m$	4 and/or $\bar{4}$ along z	—	—
	$422, 4mm, \bar{4}2m, \frac{4}{m}mm$	4 and/or $\bar{4}$ along z	2 and/or $\bar{2}$ along x, y	2 and/or $\bar{2}$ at 45° to x, y and in the x, y plane, i. e. along <110>
Cubic[d]	$23, m\bar{3}$	2 and/or $\bar{2}$ along x, y, z	3 and/or $\bar{3}$ at[c] 54.73° to x, y, z, i. e. along <111>	—
	$432, \bar{4}3m, m\bar{3}\,m$	4 and/or $\bar{4}$ along x, y, z	3 and/or $\bar{3}$ at[c] 54.73° to x, y, z, i. e. along <111>	2 and/or $\bar{2}$ at 45° to x, y, z, i. e. along <110>
Hexagonal	$6, \bar{6}, \frac{6}{m}$	6 and/or $\bar{6}$ along z	—	—
	$622, 6mm, \bar{6}\,m2, \frac{6}{m}mm$	6 and/or $\bar{6}$ along z	2 and/or $\bar{2}$ along x, y, u	2 and/or $\bar{2}$ $\perp$ x, y, u and in the x, y, u plane
Trigonal[e]	$3, \bar{3}$	3 and/or $\bar{3}$ along z	—	—
	$32, 3m, \bar{3}\,m$	3 and/or $\bar{3}$ along z	2 and/or $\bar{2}$ along x, y, u	—

[a] $\frac{R}{m}$ occupies a *single* position in a point group symbol because only one direction is involved (m is perpendicular to R).
[b] In the monoclinic system, the y axis is taken here as the unique 2 or $\bar{2}$ axis; hence an m plane, if present, is normal to it, and *mutatis mutandis* in other crystal systems.
[c] Actually $\cos^{-1}(1/\sqrt{3})$.
[d] Earlier notation uses $m3$ and $m3m$ for $m\bar{3}$ and $m\bar{3}m$, respectively.
[e] For convenience, the trigonal system is referred to hexagonal axes: on the axes of a rhombohedral unit cell, the orientations of the three positions are, in order, [111], [$1\bar{1}0$] and [$11\bar{2}$].

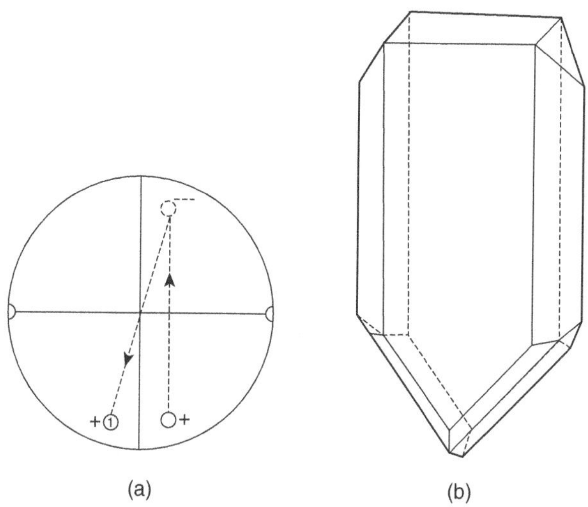

Fig. 3.9
Point group m. (a) Stereogram showing the equivalence of $\bar{2}$ and m; the symbol used here for $\bar{2}$ in the plane of the paper is not standard. (b) Crystal of potassium tetrathionate, $K_2S_4O_8$, point group m. [Reproduced by courtesy of Springer Science + Business Media, New York, © Kluwer Academic/Plenum Publishing.]

3.6.1 Ten simple point groups

The derivation of the three-dimensional point groups which are consistent with crystal rotation symmetry will now be considered.

With no further ado, the ten point groups that contain a single symmetry element can be written: $1, 2, 3, 4, 6, \bar{1}, \bar{2} \equiv m, \bar{3}, \bar{4}, \bar{6}$. In Fig. 3.12, stereograms are presented that show the general form in each of the thirty-two crystallographic

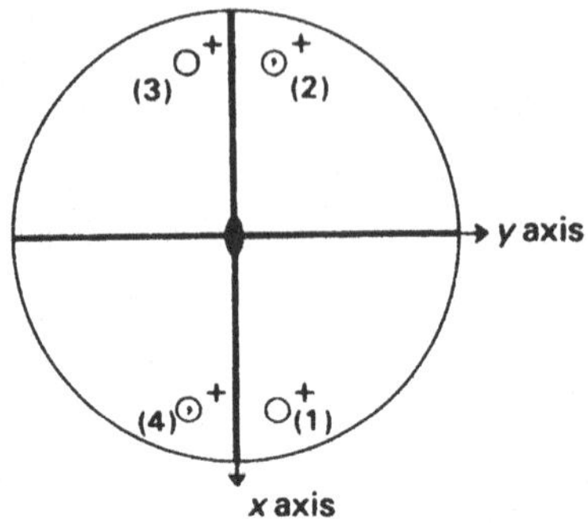

Fig. 3.10
Point group $mm2$. Point (1) in a general position is reflected across the m plane normal to x to point (2). Point (2) reflected across the m plane normal to y produces (3). Then either (3) across $m_{\perp x}$ or (1) across $m_{\perp y}$ produces point (4). The pairs of points (1), (3) and (2), (4) are each related by a twofold axis along z. [Reproduced by courtesy of Springer Science+Business Media, New York, © Kluwer Academic/Plenum Publishing.]

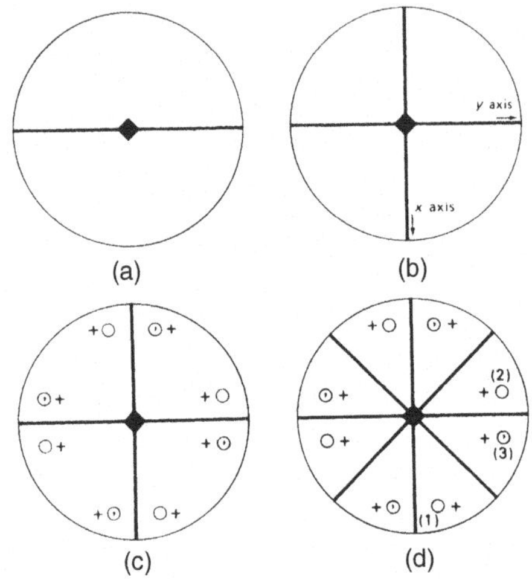

Fig. 3.11
Point group $4mm$. (a) One m plane intersecting the fourfold axis is inconsistent with fourfold symmetry. (b) The second m plane at right angles to the first satisfies the fourfold symmetry. (c) General form of points generated by the symmetry $4m$. (d) Second form of vertical m planes now shown, at 45° to the first form; they do not introduce any more points. [Reproduced by courtesy of Springer Science+Business Media, New York, © Kluwer Academic/Plenum Publishing.]

point groups, arranged under the headings of crystal systems; the positions of the crystallographic reference axes are indicated once for each system. The point group containing a single symmetry plane is generally symbolized by m although in some circumstances, such as indicated in Table 3.6, it can be helpful to use the equivalent symbol $\bar{2}$.

3.6.2 Combinations of symmetry operations in three dimensions

After considering the above simple point groups, it becomes necessary to enquire into combinations of symmetry operations. The combinations of $\boldsymbol{R1}$ and of $\boldsymbol{\bar{R}1}$ are equivalent to $\boldsymbol{R}$ and $\boldsymbol{\bar{R}}$ respectively, since 1 represents the trivial 360° rotation. The combinations $\boldsymbol{R\bar{1}}$ may be written formally as follow:

$$\boldsymbol{1\bar{1}, 2\bar{1}, 3\bar{1}, 4\bar{1}, 6\bar{1}}$$

The stereograms for point groups $\bar{1}$ and $\bar{3}$ show that the combination $\boldsymbol{R\bar{1}} \equiv \boldsymbol{\bar{R}}$ in point groups where $\boldsymbol{R}$ is an odd number, but equivalent to $\boldsymbol{R/m}$ where $\boldsymbol{R}$ is even. These two point group combinations may be written symbolically as

$$\boldsymbol{\bar{1}R}_{\text{odd}} \equiv \boldsymbol{\bar{R}}_{\text{odd}} \qquad \boldsymbol{\bar{1}R}_{\text{even}} \equiv \boldsymbol{R}_{\text{even}}/\boldsymbol{m} \tag{3.1}$$

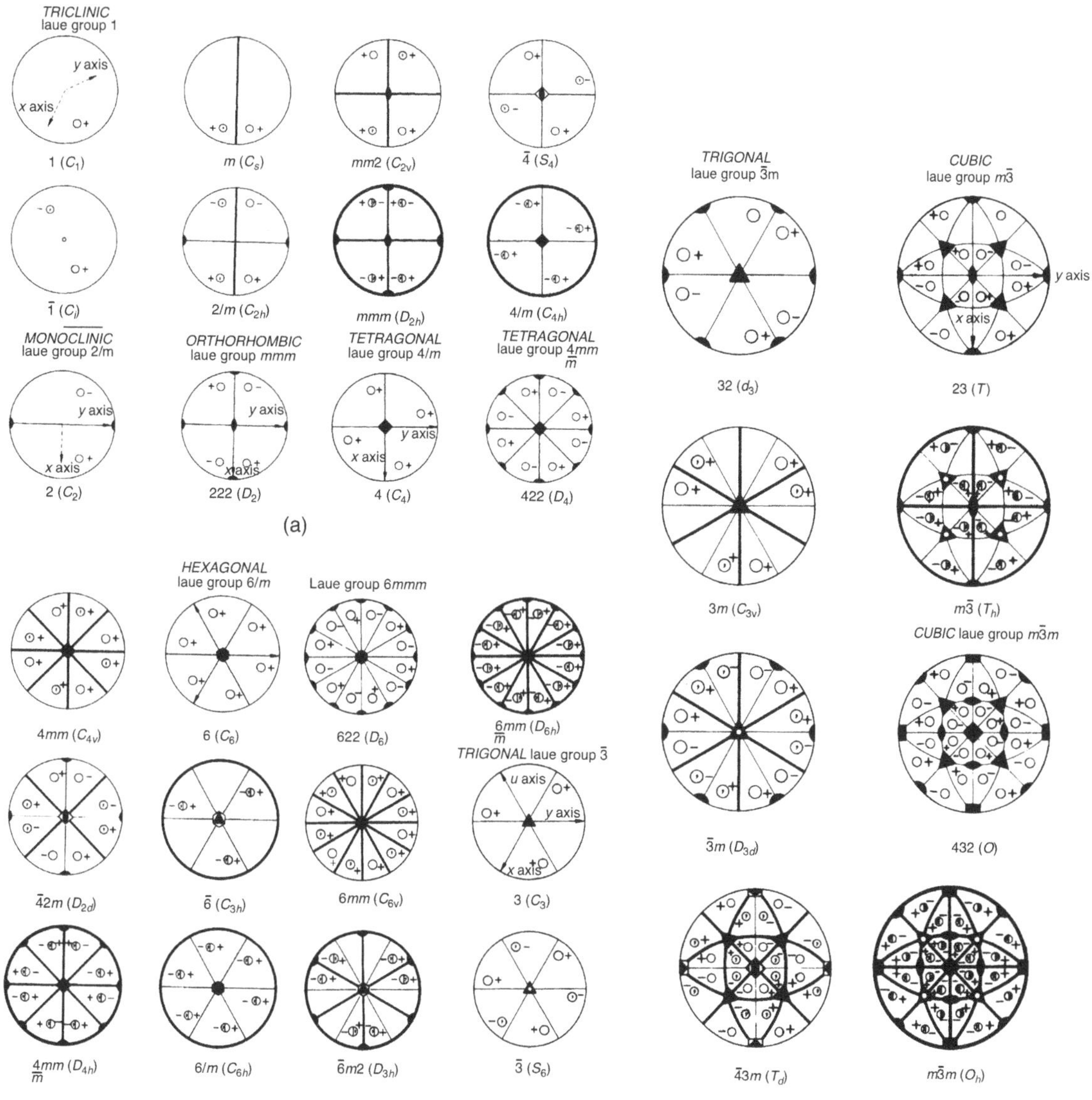

Fig. 3.12

Stereograms of the 32 crystallographic point groups, each showing the symmetry elements and the general form. They are arranged by crystal system and Laue class, and the reference axes are shown once for each system. An axial termination within the primitive indicates an oblique angle between that axis and the z axis, the latter always being normal to the plane of projection. The symbols in parentheses are the Schönflies point group symbols. [Reproduced by courtesy of Springer Science+Business Media, New York, © Kluwer Academic/Plenum Publishing.]

and may be thought of as incorporating a centre of symmetry into the stereogram of point group R.

In the stereograms for point groups such as R/m, which possess a mirror plane normal to the z axis, the notation -⦶+ is used: the comma implies an enantiomorphic relationship across the mirror plane in the diagram; a similar notation exists in point groups with inversion symmetry. The + and – signs refer to the z direction, normal to the plane of the stereogram.

It should be now clear that the combination of $\overline{\boldsymbol{R}}$ with $\overline{\boldsymbol{1}}$ will not provide any new point groups, and it will be necessary to turn to other combinations.

It is important to distinguish between the combination of symmetry operations as in a point group and the same combination as symmetry operators *per se*. For example, if a centre of symmetry is added to point group 4, the result is point group $4/m$, as is shown readily by a stereogram, whereas the operator combination of $\overline{\boldsymbol{1}}$ and $\boldsymbol{4}$ leads to the operator $\overline{\boldsymbol{4}}$; this sort of result will arise again in later chapters.

The deductions so far are summarized in Table 3.7; all point groups in the triclinic and monoclinic systems have been derived.

3.6.3 Euler's construction

The first 13 point groups, denoted here in types (*a*)–(*f*), in Table 3.7 were derived without difficulty. In the few combinations of symmetry operations used so far, there was no question of orientation because $\overline{\boldsymbol{1}}$ is a point. If $\boldsymbol{R}$ operators are now combined with another operator, say $\boldsymbol{2}$, the question of orientation arises immediately. It seems most likely that the symmetry elements R and 2 would be either coincident or perpendicular: but are these orientations correct and are there any other possibilities to consider?

Symmetry axes are zone axes, and a crystal can exhibit many zones; each of the corresponding zone axes will be of symmetry higher than identity. It might seem at first that a crystal could be symmetrical with respect to many sets of intersecting axes, but constraints exist that reduce markedly the number of possible sets. The combinations of symmetry operations, such as Eq. (3.1), may be deduced by using a principle attributed to Euler. In general, two intersecting symmetry operators give rise to a third mutually intersecting symmetry operator, and in this discussion it will be found convenient to use the symbol $\overline{2}$ instead of m.

Zone axes and, thus, symmetry axes lie normal to their respective zone circles. In Fig. 3.13, OA and OB are two symmetry axes intersecting at O, the centre of a spherical projection of a crystal (Section 2.6.1). Let $\angle BAC = \angle BAC' = \alpha/2$ and $\angle ABC = \angle ABC' = \beta/2$, where α and β are right-handed angular rotations about OA and OB, respectively. Consider the point C: a

Table 3.7 Thirteen of the 32 crystallographic point groups.

Type	Operator/s	Number of new groups	Point groups	Cumulative total
a	R	5	1, 2, 3, 4, 6	5
b	$\overline{R}$	5	$\overline{1}, \overline{2} \equiv m, \overline{3}, \overline{4}, \overline{6}$	10
c	$R1$	0	No additions	10
d	$\overline{R}1$	0	No additions	10
e	$R\overline{1}$	3	$2/m, 4/m, 6/m$	13
f	$\overline{R}\,\overline{1}$	0	No additions	13

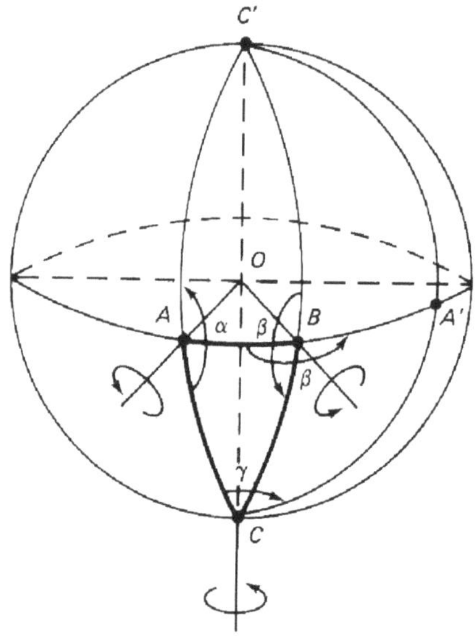

Fig. 3.13
Partial spherical projection of a crystal: OA and OB are symmetry axes. [Reproduced by courtesy of Woodhead Publishing, UK.]

right-handed (anticlockwise) rotation $\boldsymbol{\alpha}$ about OA maps C on to C', and an anticlockwise rotation $\boldsymbol{\beta}$ about OB returns C' to its original position; thus, C' behaves as the image of C in the plane OAB. The combination of rotations $\boldsymbol{\alpha}(A)$ and $\boldsymbol{\beta}(B)$ leaves the point C effectively unmoved. So, if there is to be motion of any point on the sphere arising from the operations $\boldsymbol{\alpha}(A)$ followed by $\boldsymbol{\beta}(B)$, the resultant, third operation must correspond to a symmetry element lying along the line $C'OC$.

Next, examine the motion of point A under the same rotations: $\boldsymbol{\alpha}(A)$ leaves point A unmoved and $\boldsymbol{\beta}(B)$ maps A on to A', where $\angle A'BC = \beta/2$; thus, A' is the image of A in the plane OBC. In the spherical triangles ABC and $A'BC, \angle ABC = \angle A'BC = \beta/2$, $AB = A'B$ and BC is common to both triangles. The triangles are, therefore, congruent, so that $\angle ACB = \angle A'CB$.

Let these angles be $\gamma/2$; then, the anticlockwise rotation γ about OC' maps A on to A'. Symbolically, this result may be written as

$$\boldsymbol{\beta}(B)\boldsymbol{\alpha}(A) = \boldsymbol{\gamma}(C) \tag{3.2}$$

or

$$\boldsymbol{\gamma}(C)\boldsymbol{\beta}(B)\boldsymbol{\alpha}(A) = \boldsymbol{E} \tag{3.3}$$

where $\boldsymbol{E}$ represents the operation of identity; Eq. (3.3) means that the successive symmetry operations $\boldsymbol{\alpha}$ about OA, $\boldsymbol{\beta}$ about OB and $\boldsymbol{\gamma}$ about OC, all in the same sense, are equivalent to the operation of identity.

It is necessary now to solve for the angles $\angle AB, \angle BC$ and $\angle CA$ between the axes OA, OB and OC respectively. The relevant portion of Fig. 3.13 is the spherical triangle ABC (Fig. 3.14) and it can be solved for a, b and c using Eq. (2.33). In crystals, the angles α, β and γ take only the values 360°, 180°, 120°, 90° and 60°, corresponding to R values of 1, 2, 3, 4 and 6, respectively.

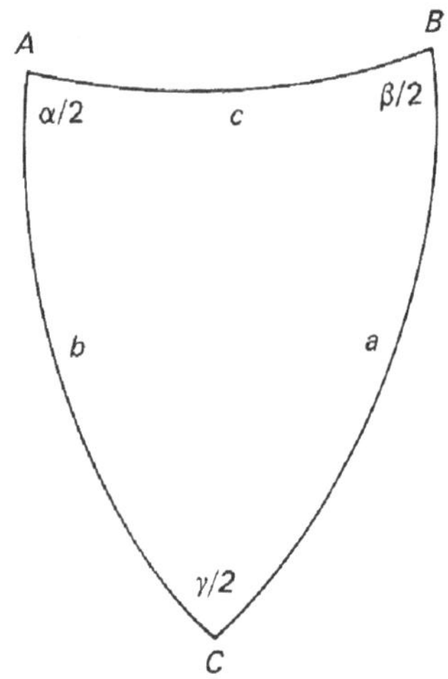

Fig. 3.14
Spherical triangle ABC from Fig. 3.13. [Reproduced by courtesy of Woodhead Publishing, UK.]

For the three rotation angles α, β and γ there are formally 5^3 solutions to Eq. (2.33), but the following four constraints reduce the amount of calculation considerably:

(a) The value of 360° for α, β and γ is ignored because it corresponds to the identity operation;
(b) $|\cos a| \leq 1, |\cos b| \leq 1$ and $|\cos c| \leq 1$;
(c) Since only the number of combinations of symmetry elements is of interest, it is necessary only to solve for α, β and γ values of 180°, 120°, 90° and 60°, subject to the conditions $\beta \leq \alpha$ and $\gamma \leq \beta$;
(d) Solutions in which one or more of α, β and γ equal zero are ignored because they correspond to a dimensionality of less than three.

Solving the triangle now leads to six non-trivial solutions, shown graphically in Fig. 3.15 and listed in Table 3.8.

Six point groups, namely, 222, 422, 32, 622, 23 and 432, are indicated by Fig. 3.15, 19 in all so far. Consider now situations where one twofold rotation axis is replaced by $\bar{2}$. Thus, for the orthorhombic system, type *g* in the table, the following eight symbols can be written:

$$\begin{array}{cccc} 222 & \bar{2}22 & \bar{2}\bar{2}2 & \bar{2}\bar{2}\bar{2} \\ & 2\bar{2}2 & \bar{2}2\bar{2} & \\ & 22\bar{2} & 2\bar{2}\bar{2} & \end{array}$$

The symbols in the second and fourth columns are wholly inadmissible: an improper rotation operation, such as $\bar{\mathbf{2}}$, combined with a proper rotation operation, such as **2**, must lead to another improper rotation. The symbols in the third column are equivalent under change of axes, and only one of them need be chosen, $\bar{2}\bar{2}2$. There remain point groups 222 and $\bar{2}\bar{2}2$, the latter being written normally as *mm*2. Neither of these point groups is centrosymmetric, but if a

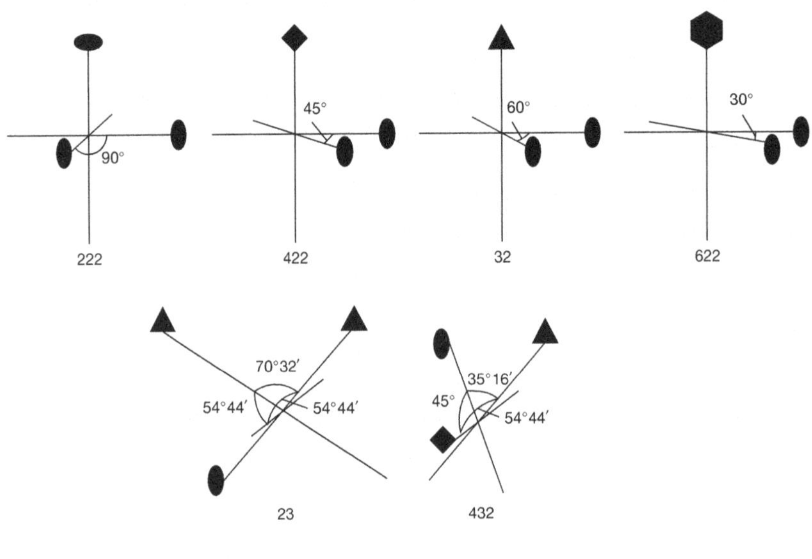

Fig. 3.15
Angles between rotation axes, derived from Euler's construction; the point groups of these combinations are shown under each set of axes.

Table 3.8 Solutions for symmetry axes and interaxial angles from Euler's construction.

Type	System	$\alpha/°$	$\beta/°$	$\gamma/°$	$a/°$	$b/°$	$c/°$
g	Orthorhombic	180	180	180	90	90	90
h	Tetragonal	180	180	90	90	90	45
i	Trigonal	180	180	120	90	90	60
j	Hexagonal	180	180	60	90	90	30
k	Cubic	180	120	120	$\cos^{-1} 1/\sqrt{3}$	$\cos^{-1} 1/\sqrt{3}$	$\cos^{-1} 1/\sqrt{3}$
l	Cubic	90	180	120	$\cos^{-1} 1/\sqrt{3}$	45	$\cos^{-1}\left(2/\sqrt{6}\right)$

centre of symmetry ($\bar{1}$) is introduced to either 222 or $mm2$, point group $\frac{2}{m}\frac{2}{m}\frac{2}{m}$ is obtained, usually written sufficiently as mmm.

Continuing the derivation, type h, Table 3.8, leads to the unique arrangements $422, 4\bar{2}\,\bar{2}$ and $\bar{4}2\bar{2}$, written conventionally as 422, $4mm$ and $\bar{4}2m$. Adding a centre of symmetry to any of these three groups leads to point group $\frac{4}{m}mm$.

Types i and j lead, by similar arguments, to the remaining trigonal and hexagonal point groups, while types k and l determine the five cubic point groups. It is left as an exercise to the reader to determine these final twelve point groups. It may be noted that there exists no third position in the symbols for the trigonal point groups (see Problem 3.14): also, be careful not to confuse 32 and $3m$ (trigonal) with 23 and $m\bar{3}$ (cubic), especially where the latter is written in the earlier form of $m3$.

Example 3.1

What unique point groups can be determined from the symbol $6RR$, where R can be 2 and/or $\bar{2}$? What are the standard symbols for the unique groups?

By permutations, the symbols $622, 6\bar{2}2, 62\bar{2}, 6\bar{2}\,\bar{2}, \bar{6}22, \bar{6}\,\bar{2}2, \bar{6}2\bar{2}$ and $\bar{6}\,\bar{2}\,\bar{2}$ can be written. Symbols $6\bar{2}2, 62\bar{2}, \bar{6}22$ and $\bar{6}\,\bar{2}\,\bar{2}$ may be discarded for reasons given above. Symbols $\bar{6}\,\bar{2}2$ and $\bar{6}2\bar{2}$ are equivalent, corresponding only to a rotation of the x, y, u axes. There remain $622, 6\bar{2}\,\bar{2}$ and $\bar{6}\,\bar{2}2$ as unique, and their usual descriptions are $622, 6mm$ and $\bar{6}m2$.

The stereograms of Fig. 3.12 should be studied alongside the derivation of the point groups, so that the meanings of the point group symbols become perfectly clear. The disposition of poles related by the fourfold axes along x and y on the stereograms of the cubic point groups 432, $\bar{4}3m$ and $m\bar{3}m$, which are not easily visualized, is clarified by Fig. 3.16.

The program EULR in the Web Program Suite has been devised in order to follow through the steps of the derivation of point groups described above. The program is not interactive, but it shows how Euler's theorem is used in the combinations of twofold symmetry rotations and the permitted values of R to develop the six sets of orientations of rotation axes upon which the combination point groups are based. Additionally, it considers the non-crystallographic rotations of 5, 7 and 8, which are encountered in species such as ferrocene, iodine heptafluoride and sulphur.

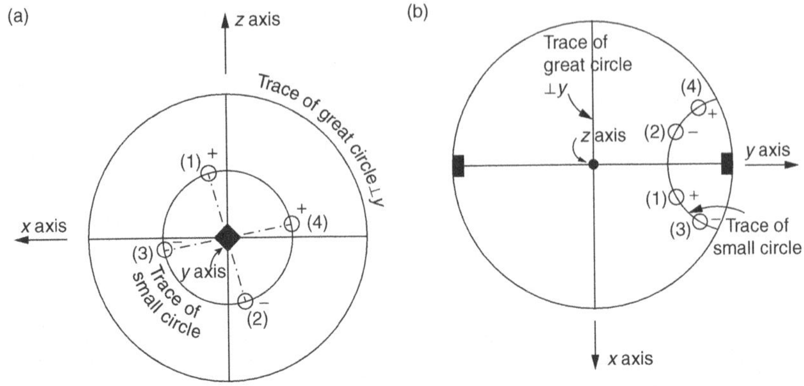

Fig. 3.16
Fourfold axes lying the primitive plane. (a) Four general intersections on the sphere, related by a fourfold rotation axis along *y* and lying on a small circle. (b) The same four positions as poles on a stereogram, shown in the preferred notation, Fig. 3.12c. Note the points on the stereogram that are related by the fourfold axis, and recall that point group 2 is a subgroup of point group 4.

3.6.4 Centrosymmetric point groups (Laue groups) and Laue classes

Eleven crystallographic point groups are centrosymmetric: they are indicated in bold type in Table 3.9. In three dimensions, they are the Laue groups, and have a special significance in X-ray crystallography (*see* Chapter 6). The thirty-two point groups in this table therefore form eleven columns of Laue classes, at the base of each of which is its Laue group. Thus, the orthorhombic Laue class members 222 and *mm*2 both react under X-ray diffraction (in the absence of anomalous scattering [2]) as though they were of the point group symmetry of their Laue group, *mmm*.

In two dimensions, a true centre of symmetry cannot exist, but a similar centrosymmetric type property exists in the groups 2, 2*mm*, 4, 4*mm*, 6 and 6*mm*, by virtue of the plane point group 2, either as point group 2 itself or as a subgroup of the other five of these groups.

3.6.5 Projected symmetry

A three-dimensional crystal form, as illustrated by a stereogram, can be projected on to a plane, and the resulting *projected symmetry* then described by one

Table 3.9 The 32 crystallographic point groups: Laue classes are in bold type.

General notation	Triclinic	Monoclinic	Trigonal	Tetragonal	Hexagonal	Cubic
R	1	2	3	4	6	23
$\bar{R}$	$\mathbf{\bar{1}}$	*m*	$\mathbf{\bar{3}}$	$\bar{4}$	$\bar{6}$	**m3**
R + centre		$\mathbf{\frac{2}{m}}$		$\mathbf{\frac{4}{m}}$	$\mathbf{\frac{6}{m}}$	
		Orthorhombic				
*R*2		222	32	422	622	432
Rm		*mm*2	3*m*	4*mm*	6*mm*	$\bar{4}3m$
$\bar{R}m$			$\mathbf{\bar{3}m}$	$\bar{4}2m$	$\bar{6}m2$	**m3m**
*R*2 + centre		**mmm**		$\mathbf{\frac{4}{m}mm}$	$\mathbf{\frac{6}{m}mm}$	

[a] In the cubic point groups 23 and $m\bar{3}$, *R* refers to 2 or $\bar{2}$ (*m*), and to 4 or $\bar{4}$ in the 432, $\bar{4}3m$ and $m\bar{3}m$ point groups.

of the plane point groups. This projected symmetry should be distinguished from Laue *projection symmetry*, a discussion of which will be deferred until a treatment of X-ray diffraction from crystals in a later chapter.

The projected symmetry pattern of a point group depends on the plane of projection: Table 3.10 lists the plane point groups of the more important projections in the seven crystal systems, and may be derived from the corresponding stereograms. For example, if in point group 2, with the y axis unique, the poles of the general form are projected on to (001) the plane group is m, whereas if they are projected on to (010) plane point group 2 results.

One may be tempted to relate the poles in the (001) projection by a twofold axis lying in the plane, but its action would move the points out of the plane, which is inadmissible in two dimensions. Another thought might be that the *congruent* aspect of point group 2 has projected on to (001) in an *enantiomorphic* aspect. But consider two infinitely thin left hands related by a twofold axis along y, and at equal distances above and below the x, y plane. Let them be projected on to the plane: they now assume an enantiomorphic relationship to each other. If you do not believe it, make two tracings of your left (or right) hand on a card, cut them out and apply a twofold rotation about an imagined y axis followed by projection on to the (imagined) x, y plane.

In studying Table 3.10, it will be seen that the important projection planes change $\{hkl\}$ indices with a change in crystal system, and Table 3.6 may be reviewed in this context.

3.7 Point groups and physical properties of crystals and molecules

The point group of a crystal expresses the symmetry common to its macroscopic properties. Consequently, the symmetry of the physical properties of a crystal must exhibit the symmetry operations of its point group, a principle due to Neumann. Conversely, the observation of a physical property of a crystal may aid the determination of its point group. Crystals generally are anisotropic, that is, the magnitude of their physical properties varies with the direction of observation or measurement. Some crystals, because of their symmetry, are isotropic for certain properties or along certain directions but anisotropic along all others.

3.7.1 Enantiomorphism and chirality

These two terms indicate the absence of any form of inversion symmetry in a crystal or molecule. Such species are termed *chiral*, or *dissymmetric* in earlier literature, and occur as non-superimposable, mirror image forms or *enantiomorphs*. Right and left hands are chiral objects; they are non-superimposable, and the only way a left hand could appear as a right hand is as a mirror reflection.

Only proper rotations are allowed for chiral species; objects with no symmetry other than identity are termed *asymmetric*. Enantiomorphous crystals

Table 3.10 Projected symmetry for the 32 crystallographic point groups.[a] The symmetries listed apply as indicated by the forms $\{hkl\}$.

Point group	$\{100\}$	$\{010\}$	$\{001\}$
1	1	1	1
$\bar{1}$	2	2	2
2	m	2	m
m	m	1	m
$2/m$	$2mm$	2	$2mm$
222	$2mm$	$2mm$	$2mm$
$mm2$	m	m	$2mm$
mmm	$2mm$	$2mm$	$2mm$
	$\{001\}$	$\{100\}$	$\{110\}$
4	4	m	m
$\bar{4}$	4	m	m
$4/m$	4	$2mm$	$2mm$
422	$4mm$	$2mm$	$2mm$
$4mm$	$4mm$	m	m
$\bar{4}2m$	$4mm$	$2mm$	m
$\frac{4}{m}mm$	$4mm$	$2mm$	$2mm$
	$\{0001\}$	$\{10\bar{1}0\}$	$\{11\bar{2}0\}$
6	6	m	m
$\bar{6}$	3	m	m
$6/m$	6	$2mm$	$2mm$
622	$6mm$	$2mm$	$2mm$
$6mm$	$6mm$	m	m
$\bar{6}m2$	$3m1$	$2mm$	m
$\frac{6}{m}mm$	$6mm$	$2mm$	$2mm$
3	3	1	1
$\bar{3}$	6	2	2
32	$31m$	m	2
$3m$	$3m1$	m	1
$\bar{3}m$	$6mm$	$2mm$	2
	$\{100\}$	$\{111\}$	$\{110\}$
23	$2mm$	3	m
$m\bar{3}$	$2mm$	6	$2mm$
432	$4mm$	$3m$	$2mm$
$\bar{4}3m$	$4mm$	$3m$	m
$m\bar{3}m$	$4mm$	$6mm$	$2mm$

[a] The trigonal point groups are here referred to hexagonal axes.

exhibit optical activity: the plane of polarized light is rotated in opposite directions by the two enantiomers. If the degree of rotation is to be measured, the light source must be monochromatic, usually the sodium D spectral line of wavelength approximately 589 nm, and the rotation expressed as an $[\alpha]_{\mathrm{D}}^{25}$ parameter.

Racemic mixtures are physical mixtures of equal amounts of the two enantiomers of a substance, and may crystallize in non-enantiomorphic, or centrosymmetric forms as do *racemates*, which are crystals containing both enantiomers in the crystal unit cell. Enantiomorphic crystals can be derived from achiral species to form chiral crystals, such as sodium chlorate, $NaClO_3$ and quartz, SiO_2.

A tragic situation, dependent upon chirality, was that involving thalidomide (**I**), (*R*/*S*)-2-(2, 6- dioxopiperidin -3-yl)-1*H*-isoindole-1, 3- (2*H*) - dione: in its *R*-form, where the hydrogen atom at *C** lies above the plane, it acts as a sedative and antiemetic drug, but in the *S*-form it is a teratogen, which interferes with foetal development. Unfortunately, the synthetic material was a racemate, and the properties of the harmful enantiomer were not known at the time.

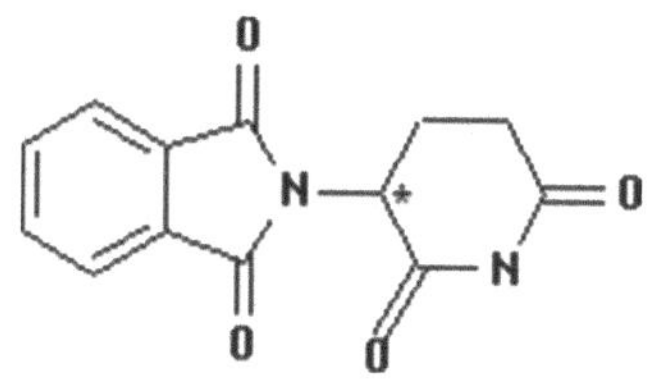

(**I**)

3.7.2 Optical properties

The extent of refraction of light by a crystal depends on the vibration direction of the electric field of the plane polarized light passing through it. In the most general case, it may be represented by the surface of a triaxial ellipsoid, or optical indicatrix, which is defined by the refractive indices of the crystal. With increase in the crystal symmetry, the indicatrix degenerates first to an ellipsoid of revolution and finally to a sphere. Hence, crystals may be divided into three categories, biaxial, uniaxial and isotropic, corresponding to the form of the ellipsoid.

If plane polarized light is passed through an optically active crystal along a direction of single refraction, the plane of polarization of the emergent light is rotated by an amount that depends upon the material, the wavelength of the light and the thickness of the optical path. A good example is sodium chlorate: from aqueous solution it gives large crystals readily, and the right-handed and left-handed forms can be separated by hand; Fig. 3.17 illustrates the enantiomorphs of sodium chlorate. Optical activity is lost on dissolution unless the species in solution is itself optically active; sucrose, for example, is optically active in both solid and solution phases.

Racemic acid is an early name for the optically inactive form of tartaric acid. It consists of a mixture of two crystal forms of tartaric acid that are mirror images of each other. Pasteur (1848) found that crystals of sodium ammonium tartrate could be grown sufficiently large for the two enantiomers (shown below) to be separated by hand.

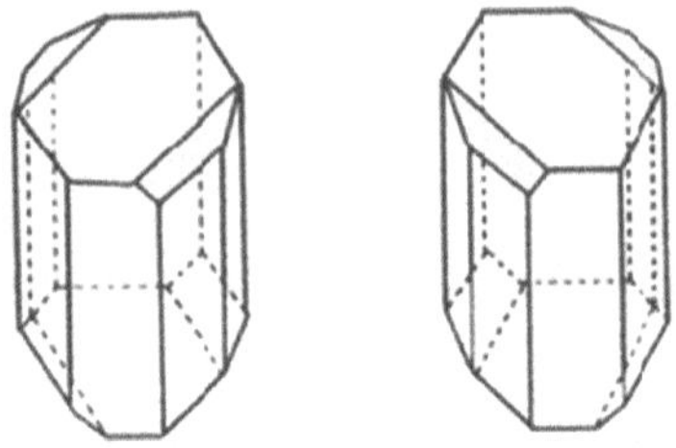

The (+) and (−) enantiomers crystallize in the non-centrosymmetric space group $P2_12_12_1$, whereas the racemic (±) mixture crystallizes in the centrosymmetric space group $P2_1/a$. Pasteur also suggested that because certain biological solutions rotated polarized light one way, the essence of living matter must be dissymmetric (chiral): a century later the elucidation of the structure of DNA marked the beginning of molecular biology.

Crystals, when viewed under the polarizing microscope in white light, may often be classified according to the scheme in Table 3.11. Birefringent crystals may be further divided in terms of their *extinction* properties: orthorhombic crystals show straight extinction on {100}, {010} and {001} and monoclinic crystals on {100} and {001}; but triclinic crystals generally show oblique extinction. Extinction here refers to the angle between an edge of the crystal under examination and the cross wires of the microscope eyepiece.

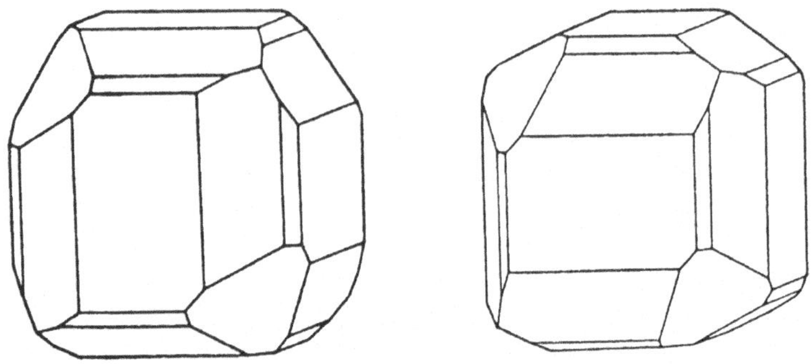

Fig. 3.17
Enantiomorphs of sodium chlorate, point group 23 (space group $P2_13$).

Table 3.11 Optical classification of crystals.

Optical properties	Indicatrix	Optical class	Crystal system
Refractive index invariant with direction; no polarization colours shown	Sphere	Isotropic	Cubic
Refractive index varies with direction; polarization colours shown	Ellipsoid of revolution	Anisotropic, uniaxial; isotropic along principal symmetry axes	Tetragonal Trigonal Hexagonal
	Triaxial ellipsoid	Anisotropic, biaxial; two general directions of single refraction	Orthorhombic Monoclinic Triclinic

Birefringent crystals also exhibit characteristic polarization figures in convergent light under crossed polaroids; a more detailed optical study may occasionally reveal useful symmetry information, and the reader is referred to the literature [2, 6].

Optical activity in the solid state is confined generally to the eleven enantiomorphous point groups, 1, 2, 222, 3, 32, 4, 422, 6, 622, 23 and 432; theoretically, it can occur in the non-enantiomorphous point groups $m, mm2, \bar{4}$ *and* $\bar{4}2m$, but both right-hand and left-hand forms are present in one and the same crystal. In some crystals, it has been measured by a specialized polarimetric technique that removes the effect of linear birefringence, which imposes constraints on the propagation of light through the crystal [7]. Point group 1 is *achiral* (asymmetric); all other groups with only proper rotation axes are chiral. Notwithstanding the above restriction, optical activity has been reported [8] for $AgGaS_2$, point group $\bar{4}2m$. This occurs because the linear birefringence is zero at certain wavelengths, thus enabling the underlying optical rotation (circular birefringence) to be observed.

Optical activity does not necessarily arise for crystals in enantiomorphous point groups; a screw structure is also necessary: sodium chlorate has such a structure; its space group is $P2_13$. Crystals of a bismuth titanate, $Bi_{12}TiO_{20}$, space group $I23$, have been found to be laevorotatory [9]; the screw nature of this structure is inherent in its space group (Section 5.4.14ff). Many examples of optical rotation in inorganic crystals have been discussed and the relationship between helical structure and optical activity demonstrated [10].

Examples of optically active crystals are listed in Table 3.12. Because optical activity can occur only in non-centrosymmetric crystals, it is probably a better test for acentricity than the pyroelectric or piezoelectric effect.

Optical activity in molecules obeys the same symmetry rules, and numerous examples exist, particularly in organic chemical species. Examples are lactic acid, $CH_3CH(OH)CO_2H$, point group C_1, and the tris(ethylenediamine)cobalt(III) cation, symmetry D_3:

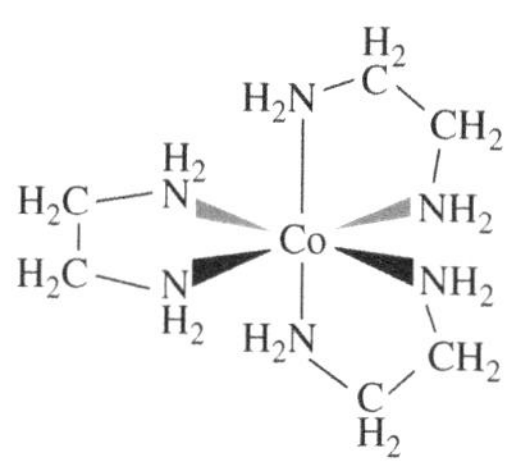

Table 3.12 Some optically active crystals.

Crystal	Indicatrix	Point group	Space group	$[\alpha]_D^{25}$
Tartaric acid	Triaxial ellipsoid	2	$P2_1$	10.8°
Magnesium sulphate heptahydrate	Triaxial ellipsoid	222	$P2_12_12_1$	2.0°
α-Quartz[a]	Ellipsoid of revolution	32	$P3_121/P3_221$	21.7°
Sodium chlorate	Sphere	23	$P2_13$	3.1°

[a] The two space groups quoted for α -quartz refer to their enantiomorphs.

3.7.3 Pyroelectricity and piezoelectricity

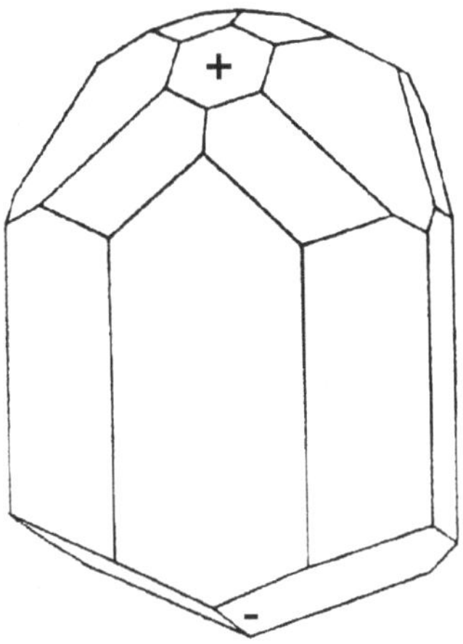

Fig. 3.18
Crystal of tourmaline, point group $3m$. Under applied heat, the unique threefold axis develops + and – charges as shown.

Certain crystals such as quartz and tourmaline when heated develop an electric charge across their polar ends. A typical crystal of tourmaline is shown in Fig. 3.18. When heat is applied, the upper end becomes positive and the lower end negative; cooling has the opposite effect. Pyroelectric crystals possess a spontaneous polarization, but the *pyroelectric effect* can be observed only in response to a perturbation, such as the application of heat; the balance of charge is upset and polarity is exhibited. This effect can arise in those crystals in which a given direction cannot be transformed to the opposite direction by any operation of its point group. Such crystals are termed *polar*, and they exist only in the polar point groups 1, 2, m, $mm2$, 3, $3m$, 4, $4mm$, 6 and $6mm$. If the process can be reversed by the application of an electric field across the polar direction, the crystal is also *ferroelectric*.

The application of heat to a crystal also sets up a strain in the material, which can produce a piezoelectric effect, so that true pyroelectricity is not easily observed. The absence of a pyroelectric effect is not necessarily a proof of the absence of centrosymmetry as the effect may be very small. A quite striking demonstration of pyroelectricity can be obtained by dipping a crystal of quartz into liquid air and then shaking on it, from a muslin bag, a mixture of red lead, Pb_3O_4, and sulphur. The muslin has the effect of charging the particles electrostatically, red lead negative and sulphur positive, and the vertical, prism faces, the normals to $\{10\bar{1}0\}$, are coated alternately red and yellow.

Another type of crystal when subjected to an applied mechanical stress along certain directions develops electric charges; this phenomenon is the *piezoelectric effect*. The stress must be applied along a polar direction in the crystal and is, in principle, possible in any of the twenty-one non-centrosymmetric point groups except 432. In this group, the high symmetry causes all the elements, or piezoelectric moduli, of the piezoelectric tensor (Section 8.2.6) to be zero. Piezoelectric effects have been observed *inter alia* in tartaric acid (2), quartz (32), tourmaline ($3m$), sphalerite, ZnS ($\bar{4}2m$), barium titanate ($4mm$) and particularly in a lead zirconium titanate that approximates to the composition $PbZr_{0.52}Ti_{0.48}O_3$ (PZT) and develops considerable electrical energy with even as little deformation as 0.1% in a linear dimension.

When pressure is applied along the polar axis of a crystal, the centres of positive and negative charge are separated, because the dipole or induced dipole lies along the unique axis. The effect is reversed in sign if tension is applied.

$$\text{Stress} \rightleftarrows \text{Electrical energy}$$

It follows that the application of an alternating electric field can be used to set up an oscillatory effect in the material. Piezoelectric effects have been used in numerous ways, such as air-bag sensors, inkjet printers, ultrasonics and medical imaging. Perhaps the best known application of the piezoelectric effect is in the high-precision quartz clock. A quartz crystal plate, parallel to (0001), is prepared, and laser-trimmed to resonate at precisely 2^{15} Hz. The power of 2 is chosen because a digital divide-by-two will produce a 1 Hz signal for the second hand, and the change of frequency with the temperature of the quartz resonator is negligibly small. A more recent application of the piezoelectric effect is 'energy harvesting', a process that uses the mechanical strain induced by traffic on road and rail to activate embedded piezoelectric materials and so provide energy for public lighting or for feed back to the electric grid.

3.7.4 Dipole moments

A molecule has a dipole moment if its structure prevents the individual bond moments from cancelling. The dipole moment μ is measured by the product of the magnitude of the charge at either end of the representative dipole vector and the distance between its positive and negative ends. Thus, a linear, symmetrical molecule such as CO_2 has zero dipole moment, as do all centrosymmetric species. Polar molecules are confined to the classes C_n, C_s and C_{nv}. A molecule with more than one C_n axis ($n > 1$) is non-polar because its bond moments cancel, as with phosphorus pentafluoride, for example. In this molecule, which has polar bonds, the two apical P–F bond moments cancel, and in the trigonal-planar arrangement of the other three bonds any two resolved along the line of the third exactly cancel it.

Theoretical calculations on the water molecule have determined the charges on the species as approximately −0.32 on oxygen and +0.16 on each hydrogen atom. The H–O–H bond angle is 104.4° and the O–H distance is *ca.* 0.98 Å. Hence, $\mu/\text{C m} = qd/\text{C m} = (0.32 \times 1.6022 \times 10^{-19}) \times (2 \times 0.98 \times \cos 52.2°) = 6.159 \times 10^{-30}$. Dipole moments are often quoted in Debye units, D; $1\text{D} = 3.3356 \times 10^{-30}$ C m, so that $\mu = 1.85$ D.

3.7.5 Infrared and Raman activity

A molecule consisting of N atoms has $3N - 6$ normal modes of vibration. The subtractive term represents three components of translational motion and three components of rotational motion. For a linear species, the number is $3N - 5$ because a component of rotation about the molecular axis cannot be observed. Thus, carbon dioxide. CO_2, has four such modes:

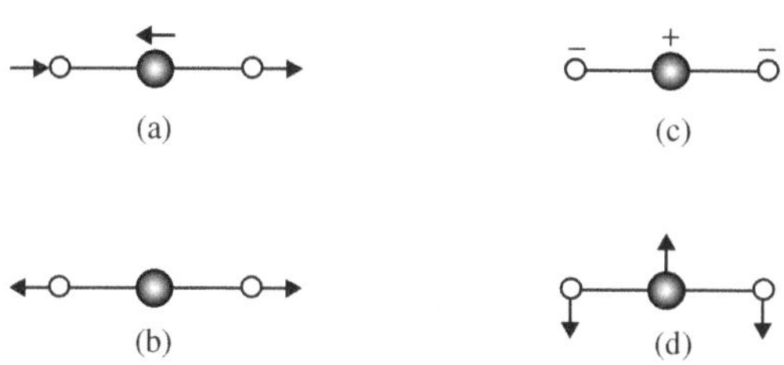

The vibrational energy can be active in the infrared region of the spectrum or Raman active or both. The internal vibrational motions of a molecule result from a superposition of a number of *normal mode* vibrations; the number of such modes may be termed the number of degrees of vibrational freedom of the species. The theoretical total number of normal modes will not always be obtained experimentally. In formaldehyde, for example, HCHO, the expected ($3N - 6$) modes are observed, whereas in chloromethane, CH_3Cl, only six modes are observed because three are doubly degenerate. Molecular vibration is discussed further in Chapter 8, and in detail in the literature [11–13].

3.8 Point groups and chemical species

In this section, chemical representatives of the crystallographic point groups will be illustrated. In some point groups, the representative species are very large molecules, and hypothetical species have been used here in order to illustrate these point groups. Thus, the molecule of α-cyclodextrin, which contains 126 atoms, has been replaced by a hypothetical species $M(AB)_6$ of the same point group. In describing the point groups of chemical species, an order similar to that used in Table 3.9 will be followed, which is a little different from the order of the derivation in Section 3.6, and in the next section some non-crystallographic point groups that arise with well-known molecular species will be reviewed; for example, elemental sulphur, S_8, which crystallizes in point group *mmm*, but exhibits the molecular symmetry $\overline{8}2m\,(D_{4d})$.

The spatial nature of the various species has been highlighted herein by the use of stereoviews, and the reader may wish to recall Appendix A1. Every atom in the molecule may not be shown, for example, hydrogen atoms in some structures based on phenyl rings or in cycloocta-1,5-diene, but sufficient detail is given to demonstrate clearly the point group symmetry. The information for the species in each symmetry group is the *compound name and formula*, where they are known with certainty, the *point group symbol*, in the Hermann–Mauguin and Schönflies (in parentheses) notations, the point group, and a *number in **bold** type* which links with the program SYMM, to be described shortly, and any other information that may be of help in viewing the diagrams correctly.

In the point groups without confirmed chemical representatives, the conformations of the $A-B$ bonds with respect to the M species in formulae such as $M(AB)_4$ should be considered carefully. The illustrations are listed in (a), (b), (c), ... order, as indicated by the legends. The circles representing atoms or groups of atoms have been sized approximately relatively to the species that they represent.

3.8.1 Point groups *R*

Molecules with point groups of the general symbol *R* are shown in Fig. 3.19. Any non-planar molecule of the type $M(ABCD)$ has point group 1. Point group 4 has been reported for tetra-azacopper(II), $C_{28}H_{12}N_{12}Cu$, and point group 6 for one form of α-cyclodextrin, $C_{36}H_{60}O_{30}$. As these molecules are large, simpler, hypothetical species have used here to illustrate these point groups.

3.8.2 Point groups $\overline{R}$

Molecules with point groups of the general symbol $\overline{R}$ are shown in Fig. 3.20. All these molecules have chemical representatives. In the dihydrogen phosphate anion, the four possible hydrogen atom positions are occupied in a statistical manner in the crystal.

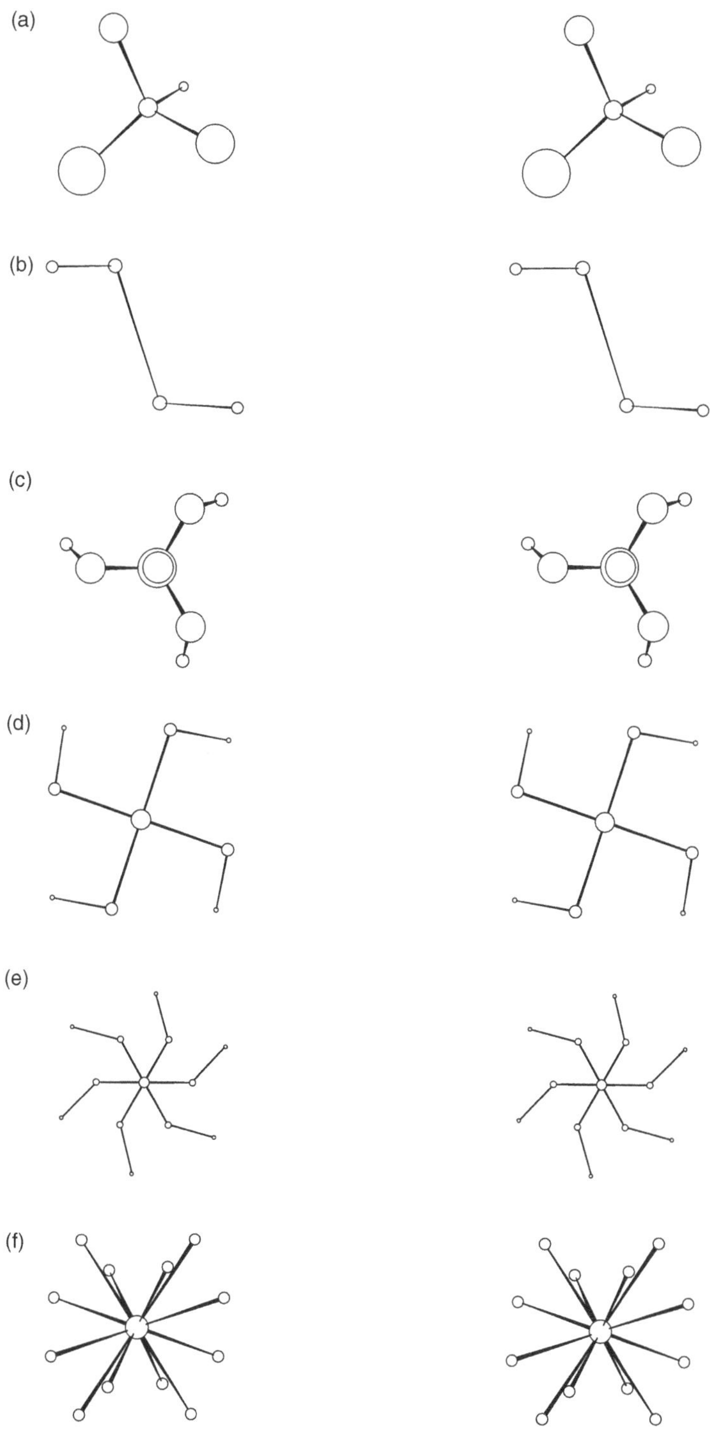

Fig. 3.19
Molecules of point group *R*. (a) Bromo (chloro)fluoromethane, CHBr(Cl)F, 1, (C_1), **91**. (b) Hydrogen peroxide, H_2O_2, 2, (C_2), **77**. (c) Phosphoric acid, H_3PO_4, 3, (C_3), **84**. (d) $M(AB)_4$, 4, (C_4), **85**; all $A-B$ bonds are directed below the MA_4 plane. (e) $M(AB)_6$, 6, (C_6), **88**; all $A-B$ bonds are directed below the MA_6 plane. (f) MA_{12}, 23, (T), **27**. [Reproduced by courtesy of Woodhead Publishing, UK.] In Figs. 3.19–3.26, the numbers in bold type are for use with the program SYMM.

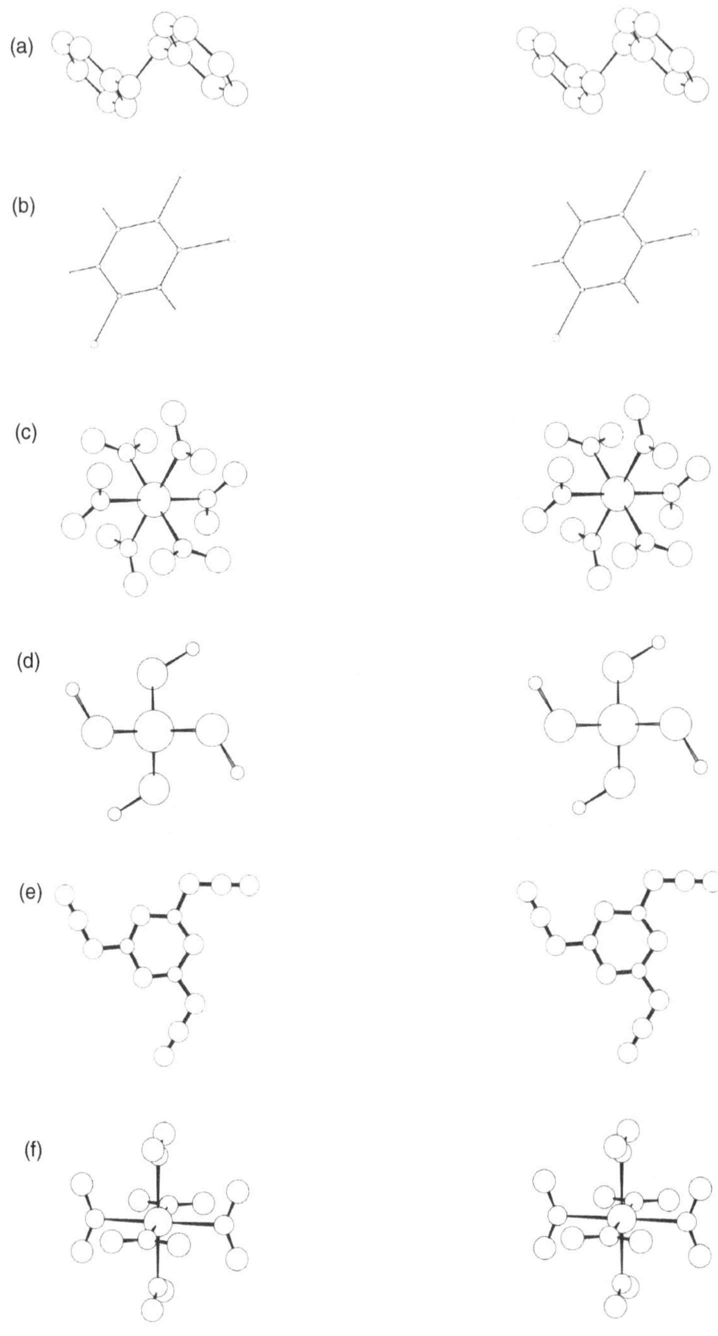

Fig. 3.20
Molecules of point group $\overline{R}$. (a) Dibenzyl, $(C_6H_5)_2$, $\overline{1}$, (C_i, or S_2), **78**. (b) 1,2,4-Trichlorobenzene, $C_6H_3Cl_3$, m or $\overline{2}$, (C_s, or S_1), **83**. (c) Hexanitronickelate(II) ion, $Ni[(NO_2)_6]^{4-}$, $\overline{3}$, (S_6), **48**. (d) Dihydrogen phosphate ion, $[H_2PO_4]^-$, $\overline{4}$, (S_4), **86**. (e) 2,4,6-Triazidotriazine, $C_3N_3(N_3)_3$, $\overline{6}$, (C_{3h}), **89**. (f) Hexanitrocobaltate(III) ion, $Co[(NO_2)_6]^{3-}$, $m\overline{3}$, (T_h), **22**. [Reproduced by courtesy of Woodhead Publishing, UK.]

3.8.3 Point groups $R\overline{1}$

Molecules with the point group combination $R\overline{1}$ are shown in Fig. 3.21. The 'pinwheel' molecule cyclophane, $C_{18}H_{24}$, has been reported [14] in point group $6/m$, but the simpler, hypothetical species $M(AB)_6$ has been illustrated here.

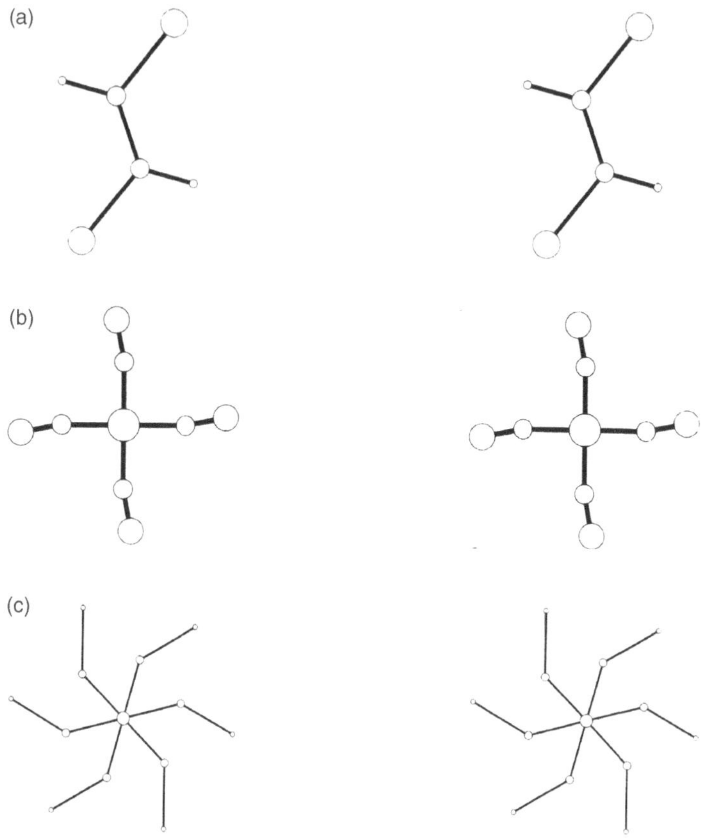

Fig. 3.21
Molecules of point group $R\bar{1}$. (a) *trans*-1, 2-Dichloroethene, $C_2H_2Cl_2, 2/m, (C_{2h})$, **68**. (b) Tetracyanonickelate(II) ion, $Ni\,[(CN)_4]^{2-}, (C_{4h})$, **56**. (c) $M(AB)_6$, $C_{16}H_{24}, (C_{6h}), 6/m$, **37**. [Reproduced by courtesy of Woodhead Publishing, UK.]

3.8.4 Point groups *R*2

Molecules with point groups of the type *R* combined with 2 are shown in Fig. 3.22. Cycloocta-1,5-diene is not a rigid molecule, and can take on conformations other than that shown here in point group 222. Point group 432 has been reported [14] for dodeca(ethylene)octamine, but a simpler MA_{12} hypothetical species has been drawn here.

3.8.5 Point groups *Rm*

Molecules with the point group combination *Rm* are shown in Fig. 3.23. Point group 6*mm* has been reported [15] for the (η^6-hexamethylbenzene)gallium(I) ion, but a simpler species is used here to illustrate this point group.

3.8.6 Point groups $\bar{R}m$

Molecules with point groups of this combination are shown in Fig. 3.24. The *chair* conformation for cyclohexane, shown here, is its most stable form.

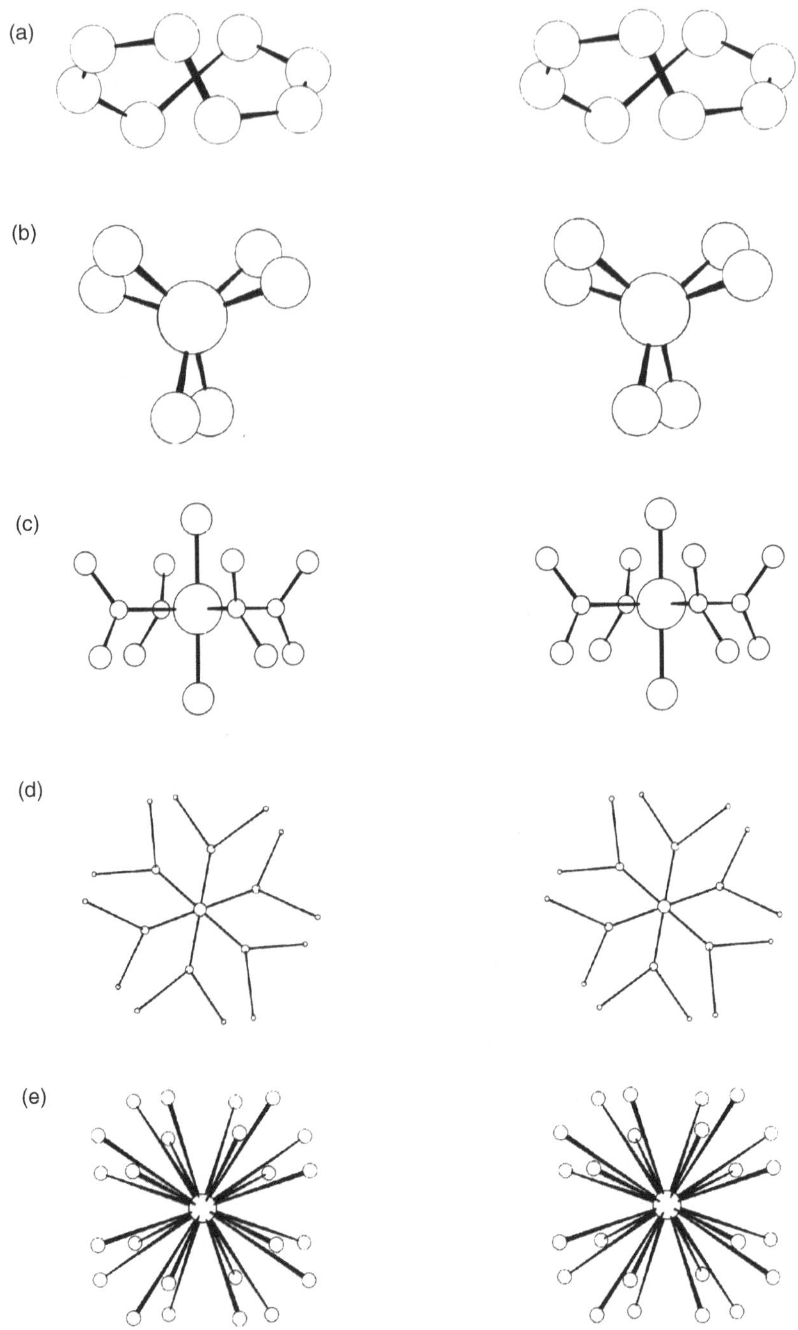

Fig. 3.22
Molecules of point group $R2$. (a) Cycloocta-1,5-diene, C_8H_{12}, 222, (D_2), **67**. (b) Dithionate ion, $S_2O_6^{2-}$, 32, (D_3), **43**; shown along the line of the S–S bond. (c) Tetranitro diamminecobaltate(III) ion, Co $[(NO_2)_4(NH_3)_2]^{3-}$, 422, (D_4), **55**; the NH_3 groups here are in free rotation. (d) $M(AB_2)_6$, 622, (D_6), **36**. (e) MA_{24}, 432 (O), **26**. [Reproduced by courtesy of Woodhead Publishing, UK.]

3.8.7 Point groups $R2$ and $\bar{1}$

Molecules with point groups of the combination $R2$, considered above, and a centre of symmetry are just four in number, within this scheme, and they are illustrated in Fig. 3.25.

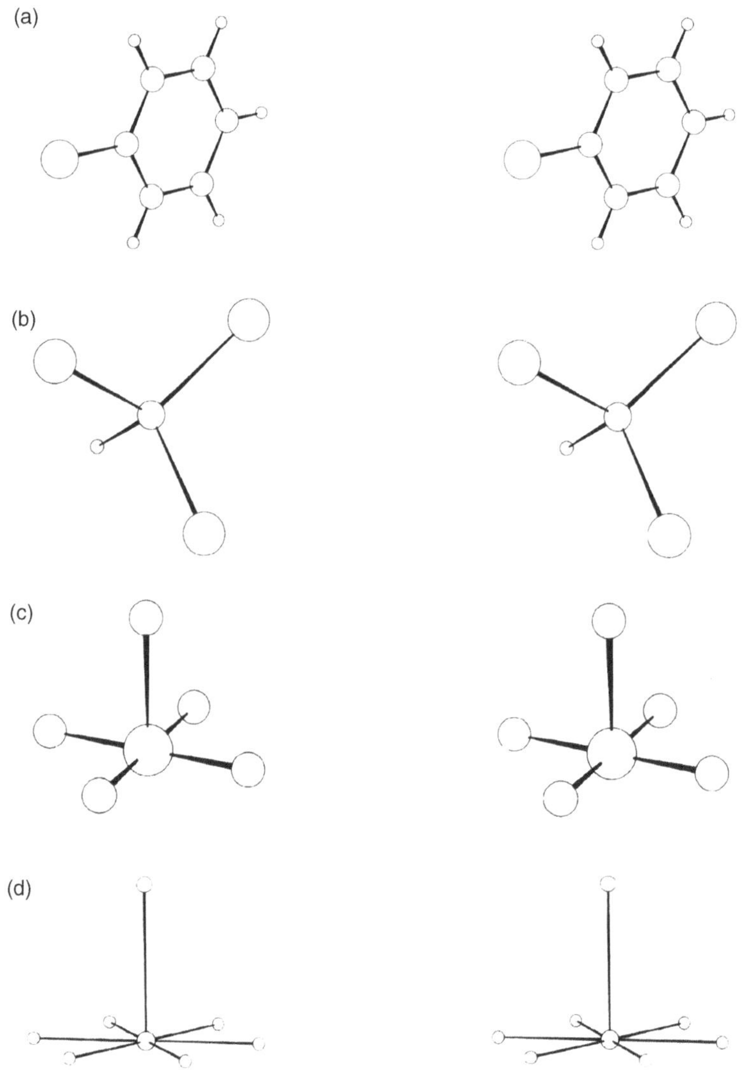

Fig. 3.23
Molecules of point groups *Rm*. (a) Chlorobenzene, $C_6H_5Cl, mm2, (C_{2v})$, **16**. (b) Trichloromethane, $CHCl_3, 3m, (C_{3v})$, **42**. (c) Pentafluoroantimonate(III) ion, $Sb[F_5]^{2-}, 4mm, (C_{4v})$, **55**. (d) $MAB_6, 6mm, (D_6)$, **36**. [Reproduced by courtesy of Woodhead Publishing, UK.]

3.9 Non-crystallographic point groups

There are many molecular species that exhibit point group symmetries other than those of the crystallographic point groups discussed in the previous sections. Some of the commoner molecules will be considered here (Fig. 3.26); other examples can be found in the literature [14]. The elements R and $\overline{R}$ have the actions as before, but R may take on values other that those found in classical crystallography; for, example, linear molecules possess an infinity axis. The program SYMM in the Web Program Suite detects species with an infinity axis but not those with $R = 5$ or $R > 6$, as in Fig. 26c–f.

The stereogram in Fig. 3.27 shows the general form of twenty poles for point group $\overline{10}\,2m$. The molecule of uranium heptafluoride is non-centrosymmetric,

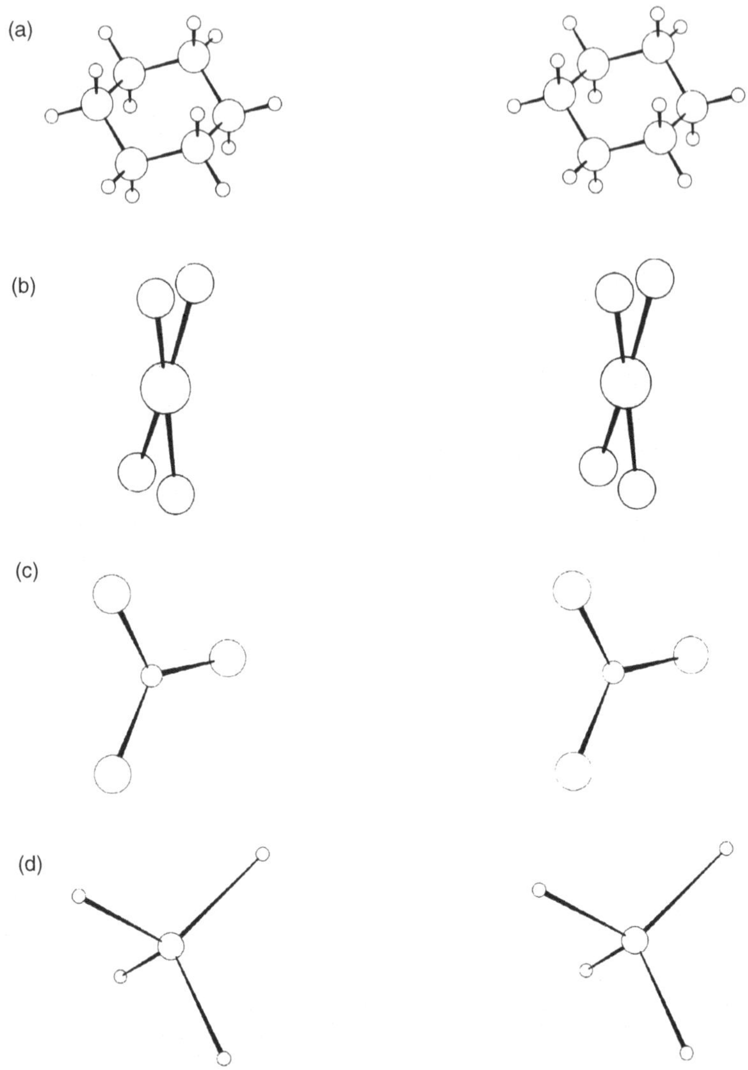

Fig. 3.24
Molecules of point groups $\bar{R}m$. (a) Cyclohexane, C_6H_{12} (chair form), $\bar{3}m, (D_{3d})$, **38**. (b) Thorium tetrabromide, $ThBr_4, \bar{4}2m, (D_{2d})$, **57**. (c) Carbonate ion, $CO_3^-, \bar{6}\ m2, (D_{3h})$, **44**. (d) Methane, $CH_4, \bar{4}3m, (T_d)$, **17**. [Reproduced by courtesy of Woodhead Publishing, UK.]

with the atoms occupying special positions: five on the perimeter and two in the z direction of the stereogram. Bis(η^5-cyclopentadienyl)iron, or ferrocene, is a centrosymmetric molecule, with a staggered conformation and point group $\bar{5}m(D_{5d})$, whereas ruthenocene has the eclipsed form with the point group $\overline{10}m2(D_{5h})$ shown in Fig. 3.27.

3.10 Hermann–Mauguin and Schönflies point group symmetry notations

It is important to be familiar with both symmetry notations as they are in general use. Once symmetry concepts are understood, differences in notation will

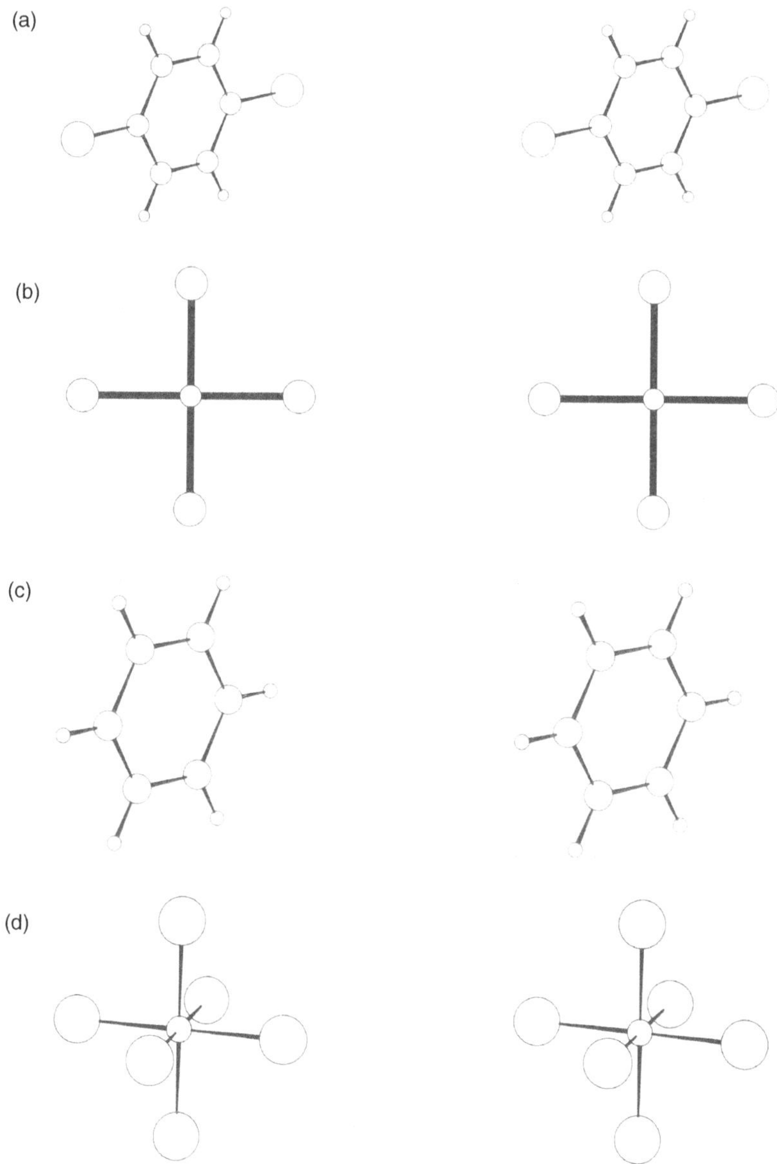

Fig. 3.25
Molecules with point groups $R2$ combined with $\bar{1}$. (d) 1,4-Dichlorobenzene, $mmm, (D_{2h})$, **65**. (b) Tetrabromoaurate (III) ion, $[AuBr_4]^-$, $\frac{4}{m}mm, (D_{4h})$, **49**. (c) Benzene, C_6H_6, $\frac{6}{m}mm, (D_{6h})$, **29**. (d) Hexachloroplatinate(IV) ion, $[PtCl6]^{2-}$, $m\bar{3}\,m, (O_h)$, **3**. [Reproduced by courtesy of Woodhead Publishing, UK.]

present little difficulty, and there is only one difference in the nature of the symmetry elements that they employ: roto-inversion with Hermann–Mauguin and roto-reflection in the Schönflies notation.

The Schönflies notation is used for describing point group symmetry in discussions on theoretical chemistry and spectroscopy but, although both the crystallographic, Hermann–Mauguin (also called international) and Schönflies notations are adequate for point groups, only the Hermann–Mauguin system provides a satisfactory notation for space groups.

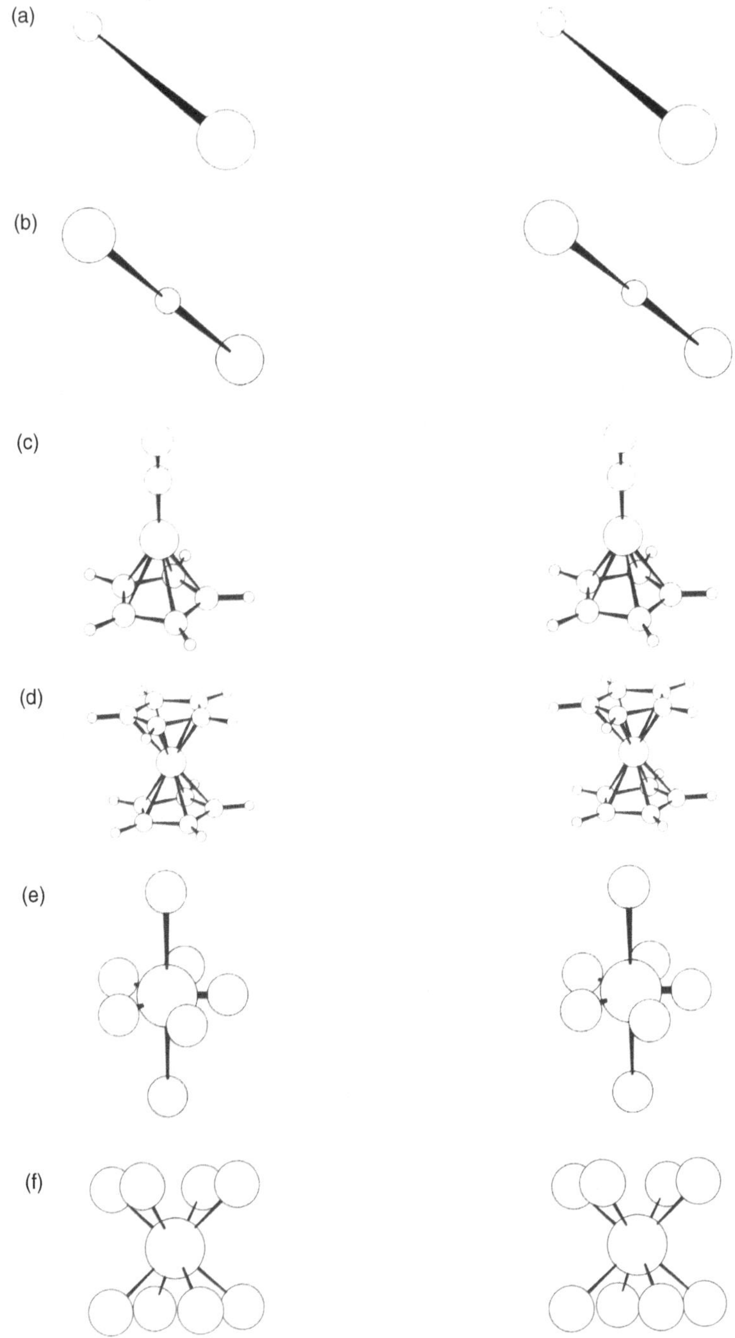

Fig. 3.26
Some non-crystallographic point groups. (a) Iodine monochloride, ICl, $\infty m, (C_{\infty v})$, **101**. (b) Carbon dioxide, $CO_2, \infty/m, (D_{\infty h})$, **102**. (c) ($\eta^5$-cyclopentadienyl)nickel nitrosyl, $\eta^5-C_5H_5$ NiNO, $5m, (C_{5v})$. (d) bis(η^5-cyclopentadienyl)iron, $(C_5H_5)_2$ Fe, $\overline{5}m(D_{5d})$; staggered ring conformation. (e) Uranium heptafluoride, $UF_6, \overline{10}2m$, (D_{5h}). (f) Octacyanotungstate(VI) ion, $\left[W\,(CN)_8\right]^{2-}, \overline{8}2m(D_{4d})$. [Reproduced by courtesy of Woodhead Publishing, UK.]

The Schönflies notation uses the rotation axis and mirror plane symmetry elements that were discussed in Section 3.5.1 and Section 3.5.2, albeit with differing notation, but introduces the roto-reflection axis of symmetry in place of the roto-inversion axis.

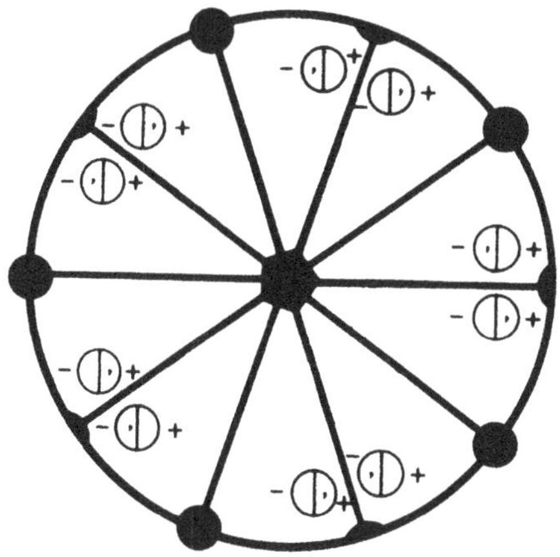

Fig. 3.27
Stereogram of point group $\overline{10}2m$ showing a general form of 20 poles. A special form of five poles, lying on *m* planes as shown, represents the five fluorine atoms in one plane in the molecule of UF_7. The poles for the two remaining fluorine atoms lie at the centre of the stereogram, above and below the plane, and are obscured by the symbol for fivefold rotation; the uranium atom occupies the origin.

3.10.1 Roto-reflection (alternating) axis of symmetry

A crystal is said to have a *roto-reflection*, or an *alternating*, axis of symmetry S_n (Ger. *Spiegel* = mirror) of degree *n*, if it can be brought from one state to another indistinguishable state by the operation of rotation through $(360/n)^\circ$ about the symmetry axis followed by reflection across a plane normal to that axis, overall a *single* symmetry operation. It should be stressed that this plane is *not necessarily* a mirror plane of the point group.

Unlike proper rotations, the S_n operations are not demonstrable with physical models, like mirror reflection and roto-inversion. The stereograms in Fig. 3.12 include the Schönflies symbols in parentheses: it may be noted that among *point groups*, $S_4(\overline{4})$ is unique in that it is not equivalent to any other point group or combination of point groups. The reader should consider what point groups are obtained if the plane of the diagram were a mirror plane in point groups S_n, $n = 2, 4$ and 6. Note, however, that the *operator* $\boldsymbol{S}_4$ is equivalent to the combination of operators $\boldsymbol{C}_4$ and $\boldsymbol{i}$ (see also Section 3.12ff).

Rotation axes in Schönflies' notation are symbolized by C_n (Ger. *Cyclisch* =cyclic); the degree *n* takes the same meaning as *R* in the Hermann–Mauguin system.

The mirror plane symmetry *element* is denoted by σ : σ_h if normal to the C_n axis; σ_v if vertical and containing the C_n axis, when there will be *n* such planes in all; and σ_d if vertical and lying symmetrically between *n* twofold axes that are normal to C_n.

Mirror planes in a point group symbol are indicated by subscripts *h* (horizontal, or normal to C_n), *v* (vertical, or containing C_n) and *d* (also vertical, but lying symmetrically between other symmetry elements. The symbol D_n (Ger. *Diedergruppe* = dihedral group) of degree *n* is introduced for point groups in which there are *n* twofold axes in a plane normal to the principal axis of degree *n*, as in D_{2d}. The cubic point groups are represented through the special symbols *T* (tetrahedral) and *O* (octahedral).

3.10.2 The two symmetry notations compared

The two notations are compared in Table 3.13. The Schönflies symbol equivalent to $\bar{6}$ is C_{3h} ($3/m$). The reason that $3/m$ is not used in the Hermann–Mauguin system is that point groups containing the element $\bar{6}$ describe crystals that belong to the hexagonal system rather than to the trigonal system, and $\bar{6}$ cannot operate on a rhombohedral lattice (see also Section 4.4.7).

3.11 Point group recognition

There are several ways in which one can approach systematically the recognition of the point group of a crystal or molecular model. In the interactive method described here [16], molecules and crystals are divided into four symmetry types (Table 3.14) dependent upon the presence of

- a centre of symmetry and one mirror plane or more, *or*
- a centre of symmetry alone, *or*
- one mirror plane or more but no centre of symmetry, *or*
- neither of these symmetry elements.

Hence, the first step in the scheme is a search for these elements in order to place the model in one of these four classes.

In order to demonstrate the presence of a centre of symmetry, place the given model in any orientation on a flat supporting surface; then, if either the plane through the uppermost atoms, for a chemical species, or the uppermost face, for a crystal, is parallel to the plane through the surface of the

Table 3.13 Schönflies and Hermann–Mauguin point-group symbols.

Schönflies	Hermann–Mauguin	Schönflies	Hermann–Mauguin
C_1	1	D_4	422
C_2	2	D_6	622
C_3	3	D_{2h}	mmm
C_4	4	D_{3h}	$\bar{6}m2$
C_6	6		
C_i, S_2	$\bar{1}$	D_{4h}	$\frac{4}{m}mm$
C_s, S_1	m ($\bar{2}$)		
S_6	$\bar{3}$		
S_4	$\bar{4}$	D_{6h}	$\frac{6}{m}mm$
C_{3h}, S_3	$\bar{6}$		
C_{2h}	$2/m$	D_{2d}	$\bar{4}2m$
C_{4h}	$4/m$	D_{3d}	$\bar{3}m$
C_{6h}	$6/m$	T	23
C_{2v}	$mm2$	T_h	$m\bar{3}$
C_{3v}	$3m$	O	432
C_{4v}	$4mm$	T_d	$\bar{4}3m$
C_{6v}	$6mm$	O_h	$m\bar{3}m$
D_2	222	$C_{\infty v}$	∞m
D_3	32	$D_{\infty h}$	∞/m ($\bar{\infty}$)

Table 3.14 Crystallographic point groups typed by m and/or $\bar{1}$ or neither.

Neither m nor $\bar{1}$	m (no $\bar{1}$)	$\bar{1}$ (no m)	Both m and $\bar{1}$
1, 2, 3, 4, $\bar{4}$, 6	m, $mm2$, $3m$	$\bar{1}$, $\bar{3}$	$2/m$, mmm, $\bar{3}m$
222,32,422 622,23,432	$4mm$, $\bar{4}2m$		$4/m$, $\frac{4}{m}mm$
	$\bar{6}$, $6mm$		$6/m$, $\frac{6}{m}mm$
	$\bar{6}m2$, $\bar{4}3m$		$m\bar{3}$, $m\bar{3}m$

body resting on the support *and* the two planes in question are both equivalent and inverted across the centre of the model, then a centre of symmetry is present.

If a mirror plane is present, it divides the model into enantiomorphic halves. A correct identification of these symmetry elements at this stage places the model into its correct type (Table 3.14). The reader may care to examine a cube or a model of the SF_6 molecule, which shows both a centre of symmetry and mirror planes, and a tetrahedron or a model of the CH_4 molecule, which shows mirror planes but no centre of symmetry. Models of a cube and a tetrahedron may be constructed easily, as described in Appendix A1.2.

Next, the principal rotation axis, the rotation axis of highest degree, must be is identified, together with the number of such axes if there are more than one, the presence and orientations of mirror planes, twofold rotation axes, and so on.

The program SYMM in the Web Program Suite is interactive and the directions on the monitor screen should be followed. If an incorrect response is given during a path through the program, the user will be returned to that question in the program where the error occurred for an alternative response to be made. Two such returns are allowed before the program rejects that particular examination for a further preliminary appraisal. The original program was based on the use of a Krantz set of one hundred wood models, which contained multiple examples of the more commonly encountered point groups. Hence, the numbers in bold type in Figs. 3.19–3.26 may appear as a rather random sequence. Additions to the set of numbers have been made in the program for point groups ∞m and ∞/m. A flow diagram of the program is given in Fig. 3.28, and the reader is encouraged to apply SYMM to the examination of Figs. 3.19–3.26b, which involve all crystallographic point groups and the linear groups.

3.12 Matrix representation of point group symmetry operations

In this section, some of the results developed in Appendix A3 will be required, as they lead to an elegant representation of point groups, and are useful in establishing a relationship between point group and space group symmetry.

A given symmetry operation can act on any aspect of a body, such as the face of a crystal, or an atom or a bond in a molecule. In this discussion, the

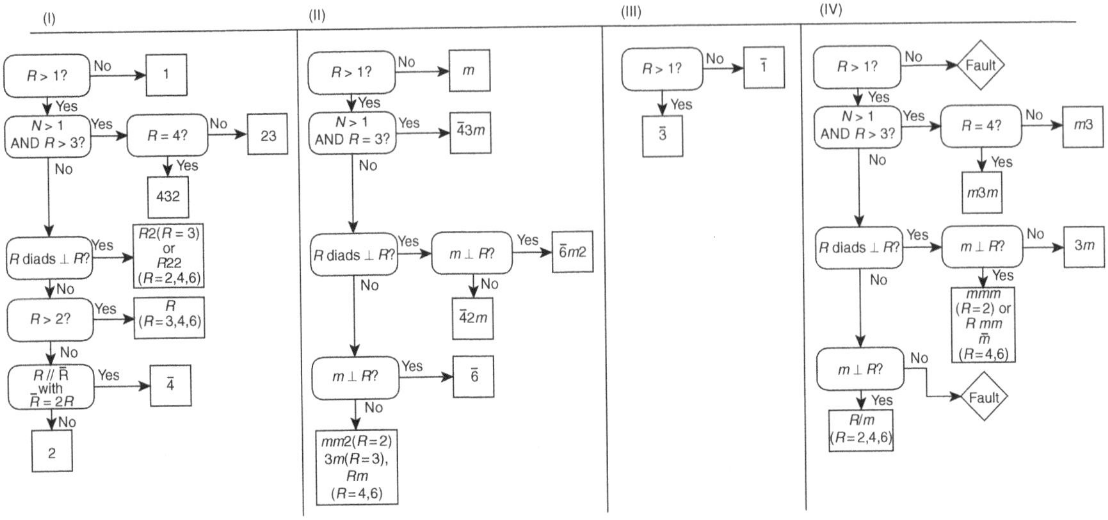

Fig. 3.28
Flow diagram for the Web Program Suite point group recognition program SYMM.

coordinates x, y and z will often be written concisely as the vector **x**, for a point on a three-dimensional object. Note that, for conciseness, $\bar{x}$ is used frequently to represent $-x$, particularly in matrices.

A symmetry operation may be written in the most general manner as

$$\boldsymbol{R}\mathbf{x} + \mathbf{t} = \mathbf{x}' \tag{3.4}$$

where **x** and **x**′ are the vector triplets before and after the action of the symmetry operator $\boldsymbol{R}$, and **t** is a translation vector. By definition of point group, **t** is identically zero, and this condition is achieved as long as all symmetry elements pass through one and the same point, the origin of the reference axes. Were it not so, if two symmetry elements, say, diad axes, were parallel, then the consequence of such an arrangement would be that illustrated in Fig. 3.29.

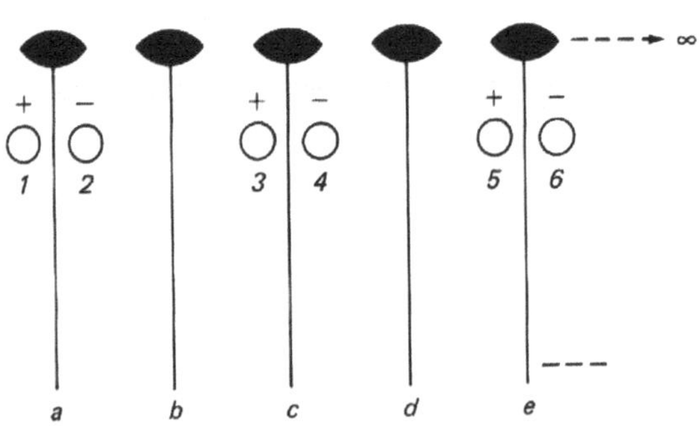

Fig. 3.29
Effect of two parallel diad axes a and b. Point 1 generates point 2 by rotation about axis a. Points 1 and 2 rotated about axis b produce points 4 and 3. But 3 and 4 are now related by another diad, c. The effect of diad c on points 1 and 2 produces points 6 and 5. But these points are related to 3 and 4 by diad d, and to each other by diad e, and so on. Clearly, this progression would lead to an infinite number of parallel, equidistant twofold axes, together with the symmetry related points, an arrangement totally incompatible with a point group.

As an example of Eq. (3.4), let a matrix $\mathbf{R}_1$ represent a twofold rotation in the monoclinic system, in the standard crystallographic orientation, that is about the y axis. Then,

$$\underset{\mathbf{R}_1}{\begin{pmatrix} \bar{1} & 0 & 0 \\ 0 & 1 & 0 \\ 0 & 0 & \bar{1} \end{pmatrix}} \underset{\mathbf{x}}{\begin{pmatrix} x \\ y \\ z \end{pmatrix}} = \underset{\mathbf{x}'}{\begin{pmatrix} \bar{x} \\ y \\ \bar{z} \end{pmatrix}} \tag{3.5}$$

Next, consider the combination of two such operators, $\boldsymbol{R}_1$ followed by $\boldsymbol{R}_2$, where $\boldsymbol{R}_2$ represents reflection across the x, z plane at $y = 0$. Following sequentially from Eq. (3.5),

$$\underset{\mathbf{R}_2}{\begin{pmatrix} 1 & 0 & 0 \\ 0 & \bar{1} & 0 \\ 0 & 0 & 1 \end{pmatrix}} \underset{\mathbf{x}'}{\begin{pmatrix} \bar{x} \\ y \\ \bar{z} \end{pmatrix}} = \underset{\mathbf{x}''}{\begin{pmatrix} \bar{x} \\ \bar{y} \\ \bar{z} \end{pmatrix}} \tag{3.6}$$

where $\mathbf{x}''$ is now related to $\mathbf{x}$ by $\bar{1}$, as expected. Another way of reaching this same result is by first combining matrices $\mathbf{R}_1$ and $\mathbf{R}_2$, and then acting upon $\mathbf{x}$ with $\mathbf{R}_3$:

$$\underset{\mathbf{R}_2}{\begin{pmatrix} 1 & 0 & 0 \\ 0 & \bar{1} & 0 \\ 0 & 0 & 1 \end{pmatrix}} \underset{\mathbf{R}_1}{\begin{pmatrix} \bar{1} & 0 & 0 \\ 0 & 1 & 0 \\ 0 & 0 & \bar{1} \end{pmatrix}} = \underset{\mathbf{R}_3}{\begin{pmatrix} \bar{1} & 0 & 0 \\ 0 & \bar{1} & 0 \\ 0 & 0 & \bar{1} \end{pmatrix}} \tag{3.7}$$

whereupon

$$\underset{\mathbf{R}_3}{\begin{pmatrix} \bar{1} & 0 & 0 \\ 0 & \bar{1} & 0 \\ 0 & 0 & \bar{1} \end{pmatrix}} \underset{\mathbf{x}}{\begin{pmatrix} x \\ y \\ z \end{pmatrix}} = \underset{\mathbf{x}''}{\begin{pmatrix} \bar{x} \\ \bar{y} \\ \bar{z} \end{pmatrix}} \tag{3.8}$$

as before. Compare Eq. (3.5) and Eq. (3.6) with the stereograms for 2 and m (Fig. 3.12a) and their combination, Eq. (3.7), with $\bar{1}$ on the same figure. The symmetry relationship

$$\boldsymbol{m2} \equiv \bar{\boldsymbol{1}} \tag{3.9}$$

is illustrated by their corresponding matrices in Eq. (3.7); this equation is another form of Eqs. (3.2)–(3.3), and Eq. (3.7) is its representation *in extenso* for the three operations involved. It should be noted that Eq. (3.7) implies first applying operation $\boldsymbol{R}_1$ *followed by* operation $\boldsymbol{R}_2$. The order is immaterial with a rotation element $R \le 2$, but it is important with higher symmetries, as the following example shows. Given the orientations 4 along z and m normal to x (and y), what is the nature of the combination of ***4*** followed by $\boldsymbol{m}$?

$$\underset{\mathbf{R}_2}{\overset{m\perp x}{\begin{pmatrix} \bar{1} & 0 & 0 \\ 0 & 1 & 0 \\ 0 & 0 & 1 \end{pmatrix}}} \underset{\mathbf{R}_1}{\overset{4 \text{ along } z}{\begin{pmatrix} 0 & \bar{1} & 0 \\ 1 & 0 & 0 \\ 0 & 0 & 1 \end{pmatrix}}} = \underset{\mathbf{R}_3}{\begin{pmatrix} 0 & 1 & 0 \\ 1 & 0 & 0 \\ 0 & 0 & 1 \end{pmatrix}} \tag{3.10}$$

so that x, y, z would be transformed to y, x, z: $\mathbf{R}_3$ represents a mirror plane operation normal to $[1\bar{1}0]$. Had multiplication been carried in the reverse order,

$$\underset{\mathbf{R}_1}{\overset{4 \text{ along } z}{\begin{pmatrix} 0 & \bar{1} & 0 \\ 1 & 0 & 0 \\ 0 & 0 & 1 \end{pmatrix}}} \underset{\mathbf{R}_2}{\overset{m \perp x}{\begin{pmatrix} \bar{1} & 0 & 0 \\ 0 & 1 & 0 \\ 0 & 0 & 1 \end{pmatrix}}} = \underset{\mathbf{R}_4}{\begin{pmatrix} 0 & \bar{1} & 0 \\ \bar{1} & 0 & 0 \\ 0 & 0 & 1 \end{pmatrix}} \tag{3.11}$$

then $x, y, z \xrightarrow{R_1 R_2} \bar{y}, \bar{x}, z$. The combined operator $\boldsymbol{R}_4$ indicates an m plane symmetry element normal to $[110]$, in the same form as $[1\bar{1}0]$, and related to it by fourfold rotation; therefore, it is one of the eight points related by symmetry $4mm$. Note that $\bar{y}, \bar{x}, z$ is related to y, x, z as x, y, z is to $\bar{x}$, $\bar{y}$, z, namely by a twofold rotation along z; 2 is a subgroup of 4. In general, however, it is always preferable to multiply matrices in the order implied by Eq. (3.10).

3.12.1 Rotation matrices

In Appendix A6, the following matrix **S** is developed for the rotation of a point X, Y, Z by an angle ϕ about an axis z normal to the plane of x and y, where the angle between the x and y axes is γ. The value of Z remains invariant under any rotation about the z axis.

$$S = \begin{pmatrix} (\cos\phi - \cos\gamma\sin\phi/\sin\gamma) & -\left(\sin\gamma\sin\phi + \cos^2\gamma\sin\phi/\sin\gamma\right) & 0 \\ (\sin\phi/\sin\gamma) & (\cos\phi + \cos\gamma\sin\phi/\sin\gamma) & 0 \\ 0 & 0 & 1 \end{pmatrix} \tag{3.12}$$

so that, concisely,

$$\mathbf{X}' = \mathbf{SX} \tag{3.13}$$

Example 3.2

What are the coordinates of a point X, Y, Z after a right-handed 6-fold rotation about the z axis in the hexagonal system, $\gamma = 120°$?

For a 6-fold rotation, $\phi = 60°$. Hence,

$$\mathbf{S} = \begin{pmatrix} 1 & \bar{1} & 0 \\ 1 & 0 & 0 \\ 0 & 0 & 1 \end{pmatrix}$$

Then, from Eq. (3.13)

$$\begin{pmatrix} X' \\ Y' \\ Z' \end{pmatrix} = \begin{pmatrix} 1 & \bar{1} & 0 \\ 1 & 0 & 0 \\ 0 & 0 & 1 \end{pmatrix} \begin{pmatrix} X \\ Y \\ Z \end{pmatrix} = \begin{pmatrix} X - Y \\ X \\ Z \end{pmatrix}$$

so that the column vector $\mathbf{X}$ (X, Y, Z) is rotated to $\mathbf{X}'$ $(X - Y, X, Z)$. Confirmation can be seen in Appendix A3.7.3.

The matrix, Eq. (3.12), will suffice for all rotational operations where the rotation axis is normal to the x, y plane. For a threefold rotation in the cubic system, the stereogram for point group 432, for example (Fig. 3.12c), shows that a fourfold anticlockwise rotation about the x axis followed by a fourfold anticlockwise rotation about the z axis is equivalent to a threefold anticlockwise rotation about the direction [111]. Thus, from Eq. (3.12) for a fourfold rotational operator $\boldsymbol{R}_x$ about the x axis in the cubic system ($\gamma = 90°$), by interchange of axes, the matrix is given by

$$\mathbf{R}_x\mathbf{X} = \begin{pmatrix} 1\,0\,0 \\ 0\,0\,\bar{1} \\ 0\,1\,0 \end{pmatrix} \begin{pmatrix} X \\ Y \\ Z \end{pmatrix} = \begin{pmatrix} X \\ \bar{Z} \\ Y \end{pmatrix} = \mathbf{X}' \qquad (3.14)$$

and for rotation about the z axis,

$$\mathbf{R}_z\mathbf{X}' = \begin{pmatrix} 0\,\bar{1}\,0 \\ 1\,0\,0 \\ 0\,0\,1 \end{pmatrix} \begin{pmatrix} X \\ \bar{Z} \\ Y \end{pmatrix} = \begin{bmatrix} Z \\ X \\ Y \end{bmatrix} = \mathbf{X}'' \qquad (3.15)$$

Hence, $\boldsymbol{R}_{[111]} = \boldsymbol{R}_z\boldsymbol{R}_x$, that is,

$$\mathbf{R}_{3_{[111]}} = \mathbf{R}_z\mathbf{R}_x = \begin{pmatrix} 0\,\bar{1}\,0 \\ 1\,0\,0 \\ 0\,0\,1 \end{pmatrix} \begin{pmatrix} 1\,0\,0 \\ 0\,0\,\bar{1} \\ 0\,1\,0 \end{pmatrix} = \begin{pmatrix} 0\,0\,1 \\ 1\,0\,0 \\ 0\,1\,0 \end{pmatrix} \qquad (3.16)$$

Thus,

$$\underset{\mathbf{R}_{3_{[111]}}}{\begin{pmatrix} 0\,0\,1 \\ 1\,0\,0 \\ 0\,1\,0 \end{pmatrix}} \underset{\mathbf{X}}{\begin{pmatrix} X \\ Y \\ Z \end{pmatrix}} = \underset{\mathbf{X}'}{\begin{pmatrix} Z \\ X \\ Y \end{pmatrix}} \qquad (3.17)$$

3.13 Non-periodic crystals

Each of these topics is concerned with material that is crystalline, in the sense that they produce sharp diffraction maxima by X-irradiation, yet do not possess the periodicity of the crystals that have been discussed.

3.13.1 Quasicrystals

The crystals that have been the subject of the discussion so far are those species which possess an ideally infinite three-dimensional periodic arrangement of atoms, the periodicities being referred to a unit cell with translation vectors **a**, **b** and **c** along the directions of the x, y and z reference axes. In this way, a crystal structure is considered to be composed of such unit cells containing atoms or molecules, and aligned so as to fill completely the space enclosed by the crystal boundaries. Normally, crystal structures can be allocated to one of the 230 space groups, which are assemblies of different types of symmetry elements, as will emerge shortly. X-ray diffraction patterns of these crystals exhibit the symmetry of one of the Laue groups, discussed in Section 3.6.4

From the times that crystal geometry was first discerned, it has been generally accepted that crystals exhibited rotational symmetries of the space filling degrees 1, 2, 3, 4 and 6 only, and were periodic in three dimensions, although examples of incommensurate crystals [17–20] were reported in the early 1970s. These crystalline materials exhibit structures with two or more periodicities that are incommensurate with each other: two periodicities p_1 and p_2 cannot be expressed by an integral ratio m/n $(m, n = 1, 2, 3, \ldots)$; the mineral calaverite $AuTe_2$ is a naturally occurring example of an incommensurate crystal.

Incommensurate structures follow a pattern of long range order, which is a requirement of sharp X-ray diffraction maxima, but the pattern is not of the periodicity of crystals, but rather it follows a pattern reminiscent of a Fibonacci series. This series can be expressed as a set of numbers, 0, 1, 1, 2, 3, 5, 8, 13, . . . , in which the nth term, $\Phi(n)$ is $\Phi(n-1) + \Phi(n-2)$; generally,

$$\Phi(n) = (\varphi^n - \psi^n)/\sqrt{5} \tag{3.18}$$

The parameter φ is the 'golden number', $(1+\sqrt{5})/2$, and ψ is $(1-\sqrt{5})/2$. (see also Appendix A10.2). A Fibonacci sequence may be viewed also on the short diagonals starting from the top (0, not indicated) of a Pascal triangle:

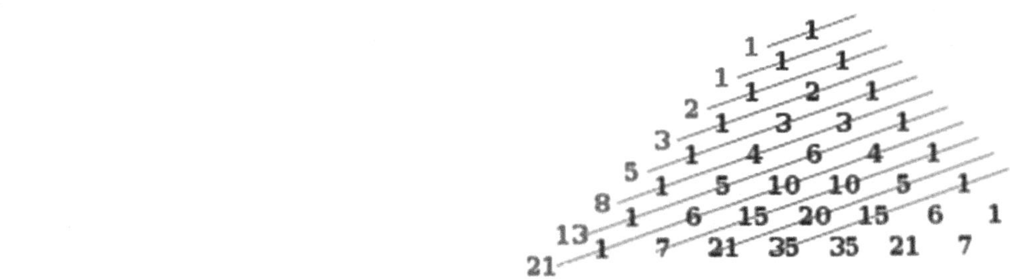

It is interesting that in Eq. (3.18), powers and square roots of irrational numbers always produce in an integer result.

In 1982, Professor Daniel Shechtman, experimenting with an alloy of composition Al_6Mn, noted an unusual appearance on certain areas of the alloy surface (Fig. 3.30); clusters were visible that exhibited outlines or partial outlines of pentagons. Transmission electron microscopy (TEM) photographs of those regions of the surface revealed diffraction spots in patterns of ten, and extending over the area of the photographic record (Fig. 3.31).

Such a result was totally unexpected; attempts were made to explain the results by a process of crystal twinning, but they were unsuccessful. The diagram of Fig. 3.31 is flat film record that suggests the existence of tenfold symmetry, which is actually the combination $5\bar{1}$, because X-ray diffraction always introduces a centre of symmetry into a spectral record; further experimentation established unequivocally the presence of *fivefold* symmetry in the crystalline material [21]. The illustration in Fig. 3.32 is a simulated diffraction pattern along the direction of a fivefold symmetry axis; it is clear that inverting each diffraction spot through the centre would have the appearance of the tenfold symmetry of Fig. 3.31.

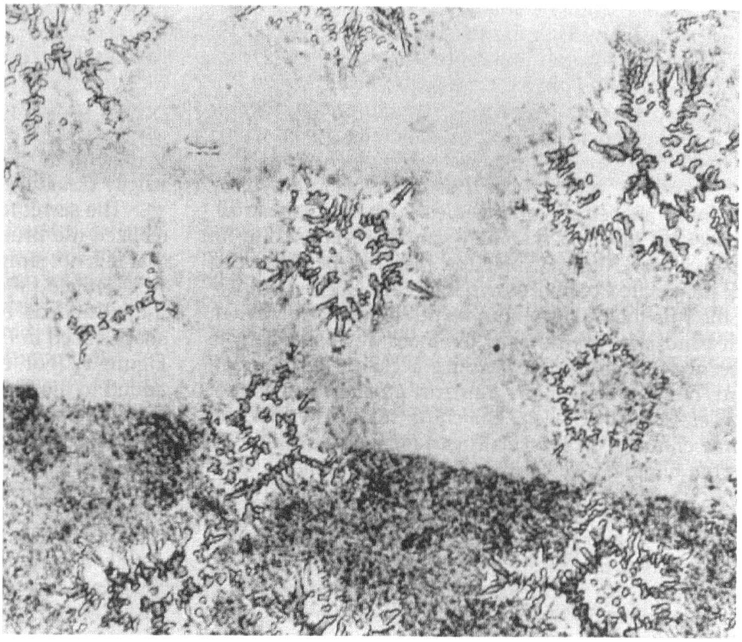

Fig. 3.30
Transmission electron microscopy photograph of an Al_6Mn alloy; the surface of the alloy shows unusual markings. [Reproduced by courtesy of Professor Daniel Shechtman.]

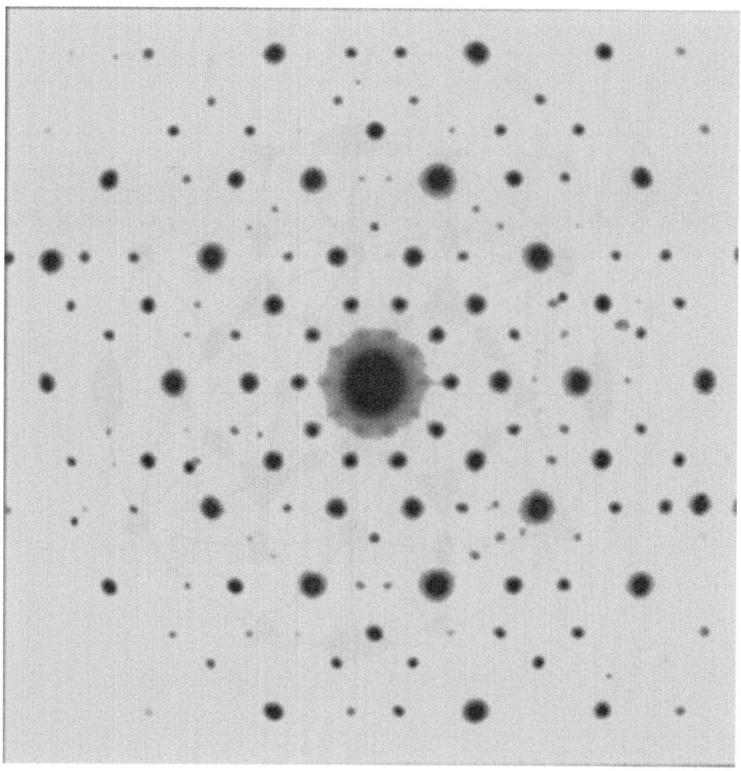

Fig. 3.31
Original photograph of the Al_6Mn surface by transmission electron microscopy, showing a tenfold (*i5*) spot pattern indicative of a crystalline nature, but without the periodicity normally associated with a crystal. [Reproduced by courtesy of Professor Daniel Shechtman.]

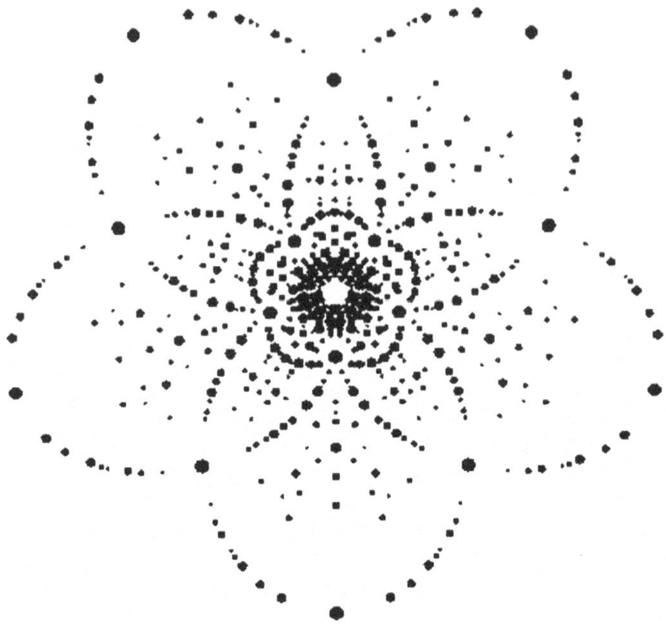

Fig. 3.32
Simulated Laue X-ray photograph of an icosahedral crystal, showing fivefold symmetry. If combined with a centre of symmetry, a tenfold pattern like that in Fig. 3.31 would arise [Reproduced by courtesy of Dr Steffen Weber.]

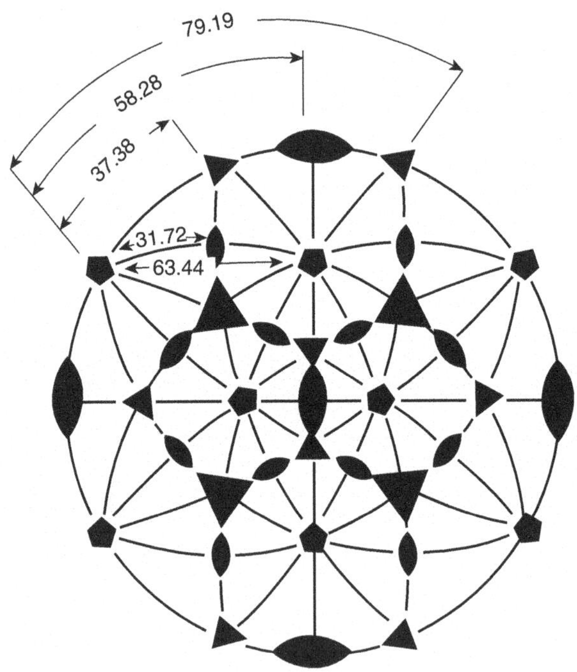

Fig. 3.33
Stereogram showing the rotation symmetry axes of icosahedral 235 (I) symmetry.

The crystal structure of the alloy material exhibits long range order and is space filling in an *aperiodic* manner, that is, without the three-dimensional periodicity that is characteristic of the so-called 'classical' crystals. The aluminium-manganese alloy is termed a *quasicrystal*: it has icosahedral symmetry, point group $m\bar{3}\ \bar{5}$, or I_h in the Schönflies notation; Fig. 3.33 is a stereogram showing the rotational symmetry axes of icosahedral symmetry.

Subsequently, a variety of stable and metastable quasicrystals has been discovered, generally as binary or ternary intermetallic compounds containing *inter alia* aluminium. On the one hand, icosahedral quasicrystals constitute a group of materials that are wholly aperiodic but exhibit sharp diffraction spots, whereas on the other hand, there are polygonal quasicrystals that are aperiodic in two dimensions and show non-crystallographic rotational symmetries: 8-fold (Mn–Fe–Al), 10-fold (Al–Cu–Ni) or 12-fold (Al–Mn–Si). Figure 3.34 is a simulated zero-layer X-ray precession photograph from a polygonal crystal with 12-fold rotational symmetry [22]. Two-dimensional quasicrystals have been reported [23] for rapidly cooled samples corresponding to the compositions V_3Ni_2 and $V_{15}Ni_{10}Si$, and TEM studies showed that they that exhibit 12-fold rotational symmetry but no long-range periodicity.

Two-dimensional quasiperiodical structures occur on decorations in medieval mosques and other similar tilings. Penrose [24] demonstrated the covering of plane space in a non-periodic manner by using just two differently shaped figures or *tiles*, and Fig. 3.35 is an example of a Penrose tiling which shows

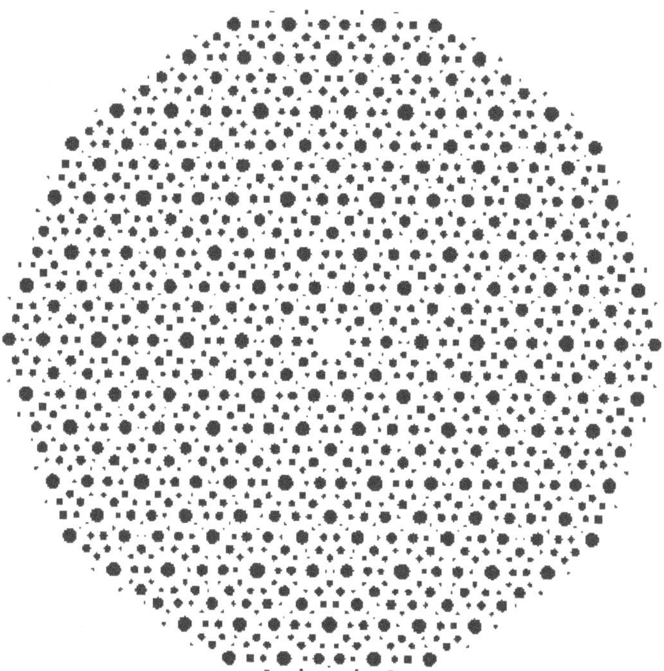

Fig. 3.34
Simulated zero-layer X-ray precession photograph of a polygonal crystal showing tenfold (decagonal) symmetry). [Reproduced by courtesy of Dr Steffen Weber.]

Fig. 3.35
Example of a Penrose tiling that uses just two shapes of tile, namely rhombi. Viewing the diagram edgeways reveals sets of lines of narrow and wide spacings. [Reproduced by courtesy of Jeff Preshing.] **(See Plate 4)**

two layers in a fivefold symmetrical arrangement; the rhombi enclose angles of 72° and 144°. This tiling has been extended to three dimensions[2] and, subsequently, a similarity was discovered between Penrose three-dimensional tiling and icosahedral quasicrystals. Mackay has shown [25] that a diffraction pattern from a Penrose non-periodic plane tiling structure has a two-dimensional Fourier transform consisting of sharp δ-peaks arranged in a fivefold symmetry pattern.

[2] In three dimensions, the angles are 63.43° and 116.57°.

If the Penrose tiling diagram is viewed edgeways on, sets of lines can be seen, lying approximately through the centres of the two types of rhombs and set at 72° to one another, the angle, $2\pi/5$, that is associated with fivefold rotation. The situation with the sets of lines is somewhat similar to a fringe system in an X-ray diffraction pattern (Fig. 3.36), the spacing of which reveals the position of the iodine atoms in the structure from which it was derived [2]. In the tiling pattern, however, the sets of lines are not equally spaced; the separations are either wide (W) or narrow (N), such that the ratio W/N is the golden number discussed above. Thus, the connection with icosahedral symmetry becomes clear: the structure, a sequence of lines, $W\,N\,W\,W\,N\ldots$ (planes in three dimensions), albeit not completely periodic, is the feature that gives rise to the observed sharp diffraction maxima.

As a result of these findings on polygonal crystals and quasicrystals, the International Union of Crystallography revised the definition of 'crystal', as described in Section 1.2, to *a material capable of producing a clear-cut diffraction pattern, with ordering that is either periodic or aperiodic*. It seems to this author, however, that the need for the redefinition of 'crystal' given above

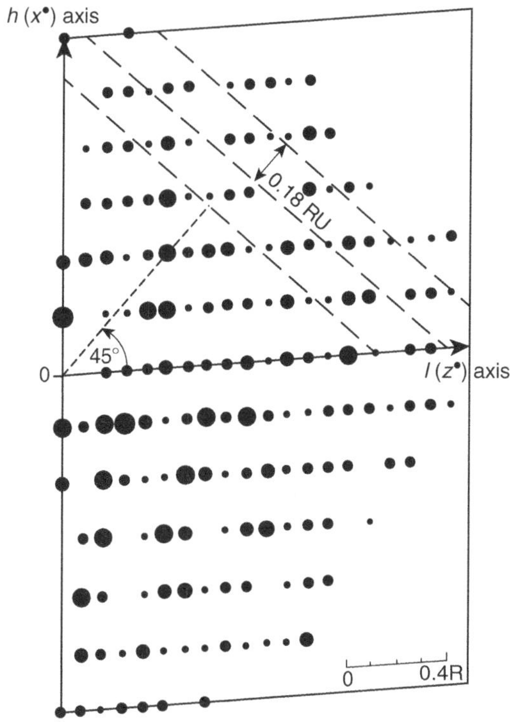

Fig. 3.36
An X-ray photograph of a section of the reciprocal lattice of euphenyl iodoacetate, $C_{32}H_{53}O_2I$, taken with CuK_α radiation ($\lambda = 1.5418$Å); the diffraction spots have been weighted according to their intensities. The regularly spaced fringe system arising from the heavy iodine atoms in the structure can be seen by viewing the diagram edgeways.

is at least debatable: better, perhaps, would be that the term 'crystal' retains its traditional meaning, and that aperiodic or multi-periodic solids such as quasicrystals [26] and incommensurate crystals [17], albeit crystalline, in the sense that they give sharp X-ray diffraction patterns, should be termed 'crystals' but qualified by their appropriate adjectival prefixes, incommensurate- and quasi-, or even quasiperiodic-, which they are.

Naturally occurring quasicrystals have been reported in Russia [27]; in composition they are mostly Cu–Al–Zn minerals with varying amounts of iron, and the crystal fragments were found to be of high crystalline quality [28].

The structures of quasicrystals are derivable by a general mathematical method that treats them as projections of lattices of higher dimensions. Thus, icosahedral quasicrystals were shown to be projections from a six-dimensional hypercubic lattice [29]; five linearly independent vectors, or generalized Miller indices, are needed to index polygonal quasicrystals and six such vectors are required for icosahedral quasicrystals.

The concept of aperiodic crystalline material was introduced as early as 1944, by Schrödinger [30]. He sought to explain how hereditary information is stored: molecules were deemed too small, amorphous solids were plainly chaotic, so it had to be a kind of crystal; and as the periodic structure of a crystal could not encode information, it had to be pattern of another type, a new type of periodicity, namely aperiodicity. Later, DNA was discovered and shown to possess properties similar to those predicted by Schrödinger—a

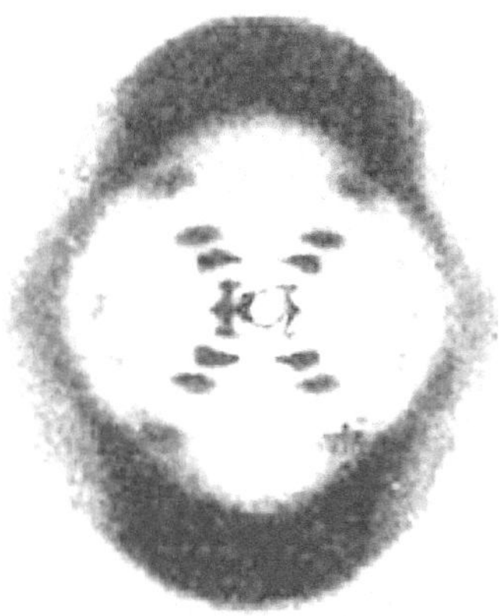

Fig. 3.37
X-ray photograph of DNA, known as 'photograph 51', taken by Rosalind Franklin (1920–1958) and Raymond Gosling, that led to the discovery of the helical structure of DNA. [Reproduced by courtesy of Raymond Gosling—who took the actual photograph.]

regular but aperiodic structure. The X-ray diffraction photograph of crystalline DNA illustrated in Fig. 3.37 led to the realization of a helical structure for DNA, which opened the way to its subsequent complete determination. An on-line account of the DNA story [31] summarizes briefly events over the period 1869 when Miescher extracted DNA from white blood cell material to 1953 when the structure of DNA was finally determined.

3.13.2 Buckyballs

The mass spectrometric examination of the products of a high-energy laser interaction with graphite in a helium atmosphere produced fragments with varying numbers of carbon atoms, dependent on the pressure of helium [32]. A very stable carbon atom cluster was found to correspond to the formula C_{60}, a new allotrope of carbon. The stability arises because a sheet of carbon atoms large enough to form a ball-like structure can satisfy fully the valence requirements for a resonance structure of carbon.

Its structural form is similar to that of a geodesic dome (Fig. 3.38): a geodesic dome is a spherical or a near spherical lattice type surface formed by a network of great circles, or geodesics, on a sphere. The geodesics intersect to form a rigid, stress free triangular structure. The C_{60} molecule is similar in structure, but with interlocking pentagons and hexagons, and completed to a full sphere (Fig. 3.39). This structure was named buckminsterfullerene [33], after R. Buckminster Fuller, who devised the mathematics for the dome and described it in detail.

Molecules that consist entirely of carbon atoms in the form of hollow spheres, ellipsoids and tubes are known as fullerenes. Spherical or near-spherical fullerenes have been given the name *buckyballs*, the simplest stable structure being buckminsterfullerene. This buckyball structure has 32 faces: 20 hexagons and 12 pentagons, related by the icosahedral symmetry discussed in the next section. The bond lengths are 1.458 Å for the bonds fusing five- with six-membered rings and 1.401 Å for the bonds fusing the six-membered rings with one another [32]. Buckyballs and buckytubes are

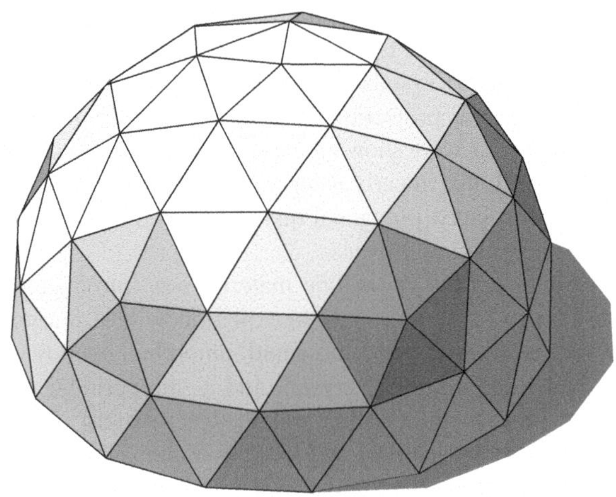

Fig. 3.38
Example of a geodesic dome. [Reproduced by courtesy of Lotus Domes UK.]

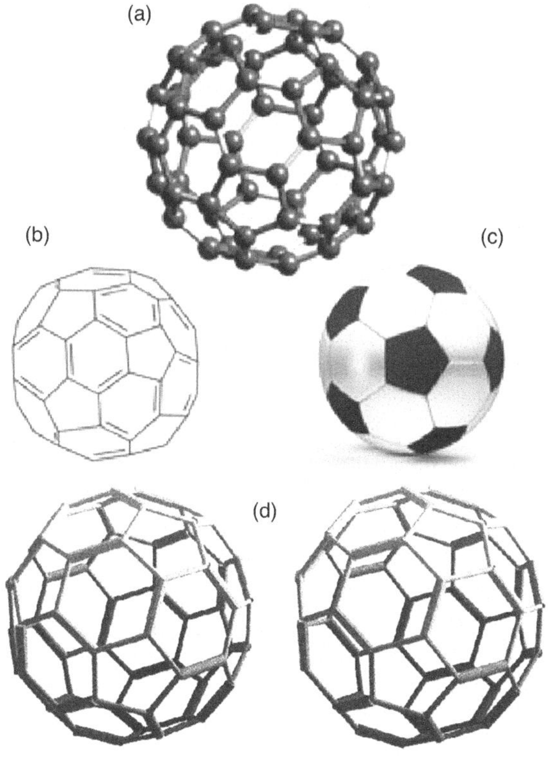

Fig. 3.39
Molecular structure of the C_{60} buckyball in four representations, (a) Ball and spoke model. (b) Resonance structure. (c) Football-style. [Harrison P and McCaw C *Educ. Chem.* 2011; 48: 113; reproduced by permission of The Royal Society of Chemistry.] (d) Stereographic illustration of the C_{60} molecule. [Mooij J. *The vibration spectrum of buckminsterfullerene.* Master's Thesis, Radboud University of Nijmegen, 2003; reproduced by courtesy of Dr Gert Heckman.]

topics of much current research both in pure chemistry, in which fullerenes are manipulated to form compounds, and in technological applications, such as carbon nanotubes.

3.13.3 Icosahedral symmetry

The C_{60} fullerene molecule exhibits icosahedral symmetry. As icosahedral symmetry is not compatible with translational symmetry, there are no associated crystallographic space groups. Nevertheless, icosahedral symmetry can be classified conveniently under point group notation: full icosahedral symmetry has the point group I_h, or $m\bar{3}\,\bar{5}$ $\left(\frac{2}{m}\bar{3}\,\bar{5}\text{ in full}\right)$ in the Hermann–Mauguin notation. This point group comprises the following unique symmetry elements in addition to identity: six C_5 axes through twelve vertices of the icosahedron, ten C_3 axes through twenty triangular faces, fifteen C_2 axes through edges, and a centre of symmetry (i) together with the symmetry elements that it introduces. If you have used the program EULR, it will have revealed three angles in connection with fivefold rotational symmetry, those between the axes in point group 235, as follow:

2 and 5	31.72°
2 and 3	20.91°
3 and 5	37.38°

if all possible angles between the axes of an icosahedron are taken into account, the results set out in Table 3.15 evolve.

Table 3.15 Angles/° between axes in icosahedral symmetry.

Axis	5	3	2
2	31.72	20.91	36
2	58.28	54.74	60
2	90	69.10	72
2	–	90	90
3	37.38	41.81	–
3	79.19	70.53	–
5	63.44	–	–

The icosahedron is one of the five Platonic solids, which are convex regular polyhedra with the same number and type of faces meeting at each vertex. They can be specified by a *vertex type*: thus, the icosahedron is [3.5] (Fig. 3.40a), implying regular *triangular* faces with *five* of them meeting at a vertex. The truncated icosahedron, type {5.6.6], is illustrated by Fig. 3.40b, and shows a striking similarity to the structure of the C_{60} molecule; Fig. 3.40a is a Platonic solid, whereas Fig. 3.40b is an Archimedean solid having polygons of more than one type of face. Both of these solids show the full icosahedral symmetry, I_h. The Platonic solids are illustrated in Fig. 3.41; of these five solids, only the cube is space filling.

The icosahedral group of lower symmetry I, (235) has the symmetry elements of I_h, except for the centre of symmetry and those symmetry elements which are contingent upon its presence; it is of particular interest in biological fields as it can represent a chiral structure. One example is the herpes virus, which has an icosahedral shell. Another classic example is the cowpea chlorotic mottle virus, an RNA virus with a shell composed of 180 protein subunits forming a structure of icosahedral symmetry. Figure 3.42 shows a stereoview of a low temperature TEM reconstruction, showing fivefold and sixfold subunits of structure [34]. A brief but interesting account of virus architecture has been given in the literature [35].

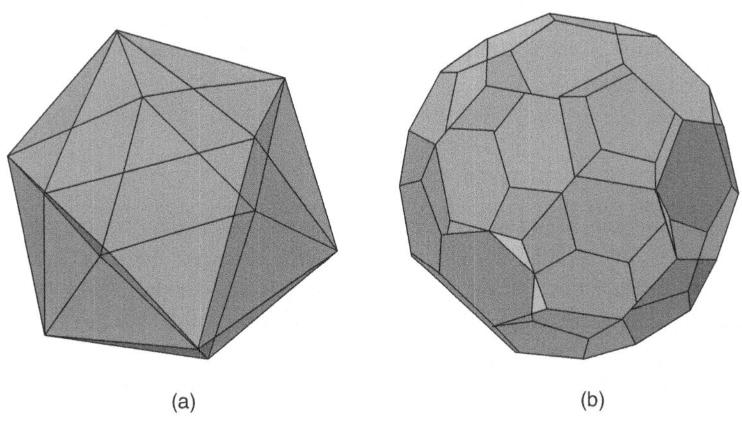

Fig. 3.40
Icosahedral symmetry. (a) Icosahedron of vertex type [3.5]. (b) Truncated icosahedron, vertex type [5.6.6]. Each polyhedron has point group symmetry I_h.

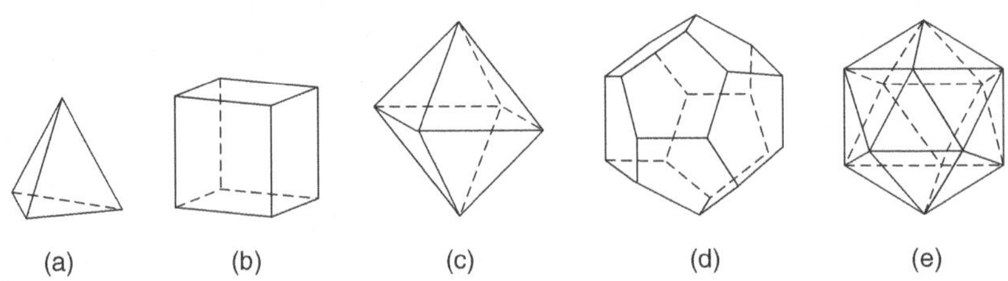

Fig. 3.41
The five Platonic solids. (a) Tetrahedron (Fire). (b) Cube (Earth). (c) Octahedron (Air). (d) Rhombic dodecahedron (Cosmos). (e) Icosahedron (Water). The names in parentheses were given by Plato. [Reproduced by courtesy of Woodhead Publishing, UK.]

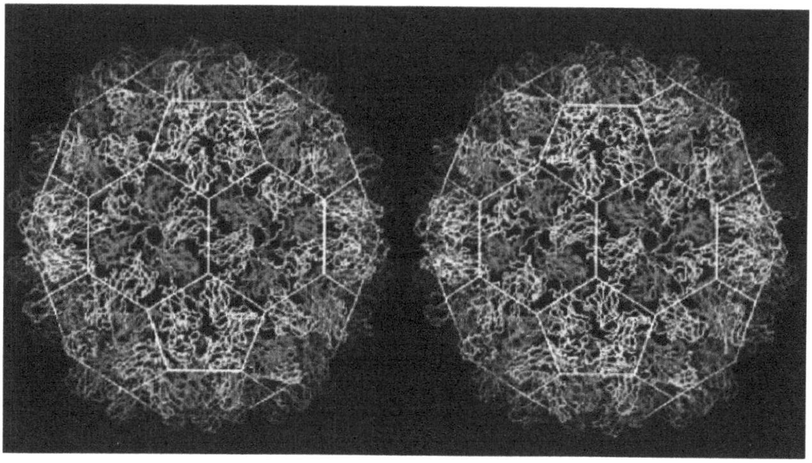

Fig. 3.42
Stereoview of the protein shell of cowpea chlorotic mottle virus. The yellow lines delineate part of the enclosing cage of a truncated icosahedron and reveal the subunits of five-membered and six-membered rings. [Speir JA, et al. *Structure* 1995; 3: 63; reproduced by courtesy of Elsevier.] (**See Plate 5**)

References 3

[1] Plato. *The republic*, Book VII. *ca*. 380 BC.
[2] Ladd M and Palmer R. *Structure determination by X-ray crystallography*. 5th ed. Springer, 2013.
[3] Phillips FC. *Introduction to crystallography*. Longmans, 1986.
[4] Crystallography and minerals arranged by crystal form. <http://webmineral.com/crystall.shtml>.
[5] Smorf Crystal Models. <http://www.smorf.nl/frames480.html>.
[6] Hartshorne NH and Stuart A. *Crystals and the polarizing microscope*. Arnold, 1970.
[7] Lingard RJ and Renshaw AR. *J. Appl. Crystallogr*. 1994; 27: 647.
[8] Hobden MV. *Acta Crystallogr. A* 1968; 24: 676.
[9] Swindells DCM and Gonzalez JL. *Acta Crystallogr. B* 1988; 44: 12
[10] Glazer AM and Stadnicka K. *J. Appl. Crystallogr*. 1986; 19:108.
[11] Nakamoto K. *IR and Raman spectra of inorganic and coordination compounds*. 6th ed. John Wiley & Sons, 2009.
[12] Woodward LA. *Introduction to the theory of molecular vibrations*. Oxford University Press, 1972.
[13] Decius JC and Hexter RM. *Molecular vibrations in crystals*. McGraw-Hill, 1977.
[14] Symmetry @ Otterbein. <http://symmetry.otterbein.edu/gallery/index.html>.
[15] Thewalt U, et al. *Z. Naturforsch*. 1984; 39:1642.
[16] Ladd MFC. *Int. J. Math. Educ. Sci. Tech*. 1976; 7: 395.
[17] van Smaalen S. *Incommensurate crystallography*. Oxford University Press, 2007.
[18] Reithmayer K. *Acta Crystallogr. B* 1993; 49: 6.
[19] Janssen T. *Acta Crystallogr. A* 2012; 68: 667.
[20] Docherty R. *J. Appl. Phys. D* 1981; 24: 89.
[21] Shechtman D, et al. *Phys. Rev. Lett*. 1984; 53: 1951.
[22] Ünal BV, et al. *Phys. Rev. B* 2007; 75: 064205.
[23] Chen H, et al. *Phys. Rev. Lett*. 1984; 60: 1645.
[24] Penrose R. *Sets of tiles for covering a surface*. Patent 413315, USA, 1976.
[25] Mackay AL. *Physica A* 1982; 114: 609.
[26] Quasicrystals. <http://www.jcrystal.com/steffenweber/qc.html#Books%20on%20Quasicrystals>.
[27] Bindi L, et al. *Science* 2009; 324: 1306.
[28] Steinhardt P, et al. *Phil. Mag*. 2011; 91: 2421.
[29] Kramer P, et al. *Acta Crystallogr. A* 1984; 40: 580.
[30] Schrödinger E. *What is life*. Cambridge University Press, 1944.

[31] The DNA story. <http://www.rsc.org/chemistryworld/Issues/2003/April/story.asp>.
[32] Kroto H, et al. *Nature* 1985; 318: 162.
[33] Hedberg K, et al. *Science* 1991; 254: 410.
[34] Zandi R, et al. *Proc. Natl. Acad. Sci.* 2004; 10: 1073.
[35] Principles of virus architecture. <http://web.uct.ac.za/depts/mmi/stannard/virarch.html>.

Problems 3

3.1 (a) Sketch stereograms to show the general form and symmetry symbols for the plane point groups 3, 3*m* and 4*mm*. (b) Why is *m* symmetry not possible in the oblique system, for example, in a general parallelogram?

3.2 Sketch stereograms to show both the symmetry elements and the general form in each of the point groups *mm*2, $\bar{4}2m$ and 23. What are the subgroups for each of these point groups? What point groups are obtained by adding a centre of symmetry to each of the three point groups?

3.3 What is the full, conventional meaning conveyed by the symbols *m* (1*m*1), 422, $\bar{6}m2$ and $\bar{4}3m$?

3.4 Use the equations developed in Section 3.6.3 to derive the angles at which the symmetry axes intersect in point groups 522 and $\bar{8}2m$ ($\bar{8}2\bar{2}$). Hence, give the full meaning to these point group symbols.

3.5 From the symbol 432, write all combinations with rotation and inversion axes for these symbols. What symbols represent crystallographic point groups and what are their standard symbols?

3.6 Consider the cover of a plain matchbox, Fig. P3.1: (a) normal and (b) squashed. What is the point group in each case?

3.7 What are the multiplicities of planes of the forms {010}, {110} and {123} in point groups *mm*2, 4*mm* and 432. Indicate both special and general forms in the results.

3.8 What is the projected point group symmetry for each of the following examples, and what is the Laue class of in each example?

	Point group	Orientation		Point group	Orientation
(a)	$\bar{1}$	{100}	(g)	3	{11$\bar{2}$0}
(b)	*m*	{010}	(h)	3*m*	{10$\bar{1}$0}
(c)	*mmm*	{120}	(i)	$\bar{6}$	{0001}
(d)	$\bar{4}$	{110}	(j)	$\bar{6}m2$	{0001}
(e)	422	{110}	(k)	23	{111}
(f)	$\bar{4}2m$	{001}	(l)	432	{110}

3.9 Determine as much as possible about the point group in each of the following situations:
(a) Laue group $\bar{3}$; optically active.
(b) Optically biaxial; point group 2 in projection on to (010).
(c) Projected symmetry 3*m* on (111) and *m* on (110).
(d) Optically active and uniaxial; *m* symmetry when projected on to (10$\bar{1}$0).

3.10 Write the point group symbol for benzene and each of its thirteen fluoro-derivatives shown in Fig. P3.2.

3.11 What are the point groups of the following molecules?

CH_4, CH_3Cl, CH_2Cl_2, CH_2 (Cl) Br, CHBr (Cl) F

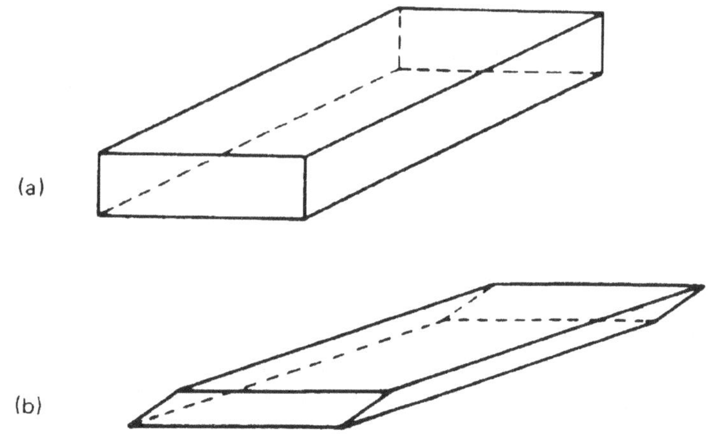

Fig. P3.1
Plain matchbox cover. (a) Normal. (b) Squashed diagonally.

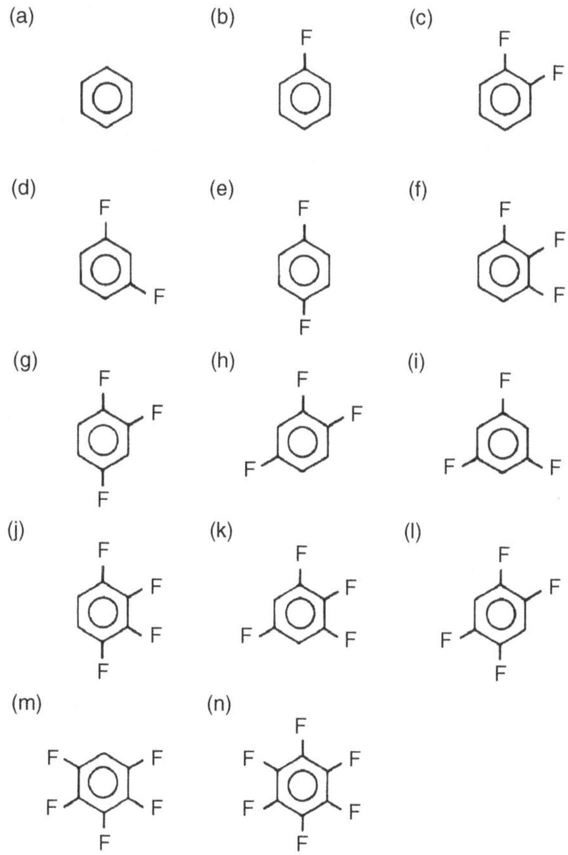

Fig. P3.2
Benzene and its fluoro-derivatives.

3.12 What plane point groups can be obtained by packing together (a) two, (b) four, irregular but identical quadrilaterals?

3.13 Set up matrices for the following symmetry operations: $\bar{4}$ along z; $m \perp x$. Hence, determine the coordinates of a point obtained by operating on x, y, z by $\bar{4}$ and on the resulting coordinates by $\boldsymbol{m}$. What are the nature and orientation of the symmetry operator obtained by the combination $\boldsymbol{m\bar{4}}$?

3.14 Set up matrices for a threefold rotation about the z axis and a mirror plane normal to the y axis of a trigonal crystal referred to hexagonal axes: you may wish to refer to Section 3.12.1. Determine the result of the combination of symmetry operations $\boldsymbol{3}$ followed by $\boldsymbol{m}$, and give the resultant point group. Show how there is no non-trivial meaning to the third position for the point group symbol derived.

3.15 List the total number of subgroups of point groups I_h and I in the Schönflies and Hermann–Mauguin notations.

3.16 Determine the numbers of vertices, faces and edges in (a) the octahedron, (b) the icosahedron and (c) the truncated icosahedron. What is the relationship between vertices, faces and edges in these three polyhedra?

Lattices

4

SYNOPSIS

- Lattices in one, two and three dimensions
- Unit cell and asymmetric unit
- Lattices in the seven crystal systems
- Law of rational intercepts
- Introduction to the reciprocal lattice
- Rotational symmetry of lattices
- Lattice transformations
- Wigner–Seitz cells

4.1 Introduction

The next logical step is the examination of the internal structure of crystals, so their basis, the three-dimensional lattices or Bravais lattices, forms the next topic. The Bravais lattices may be approached by considering first lattices in fewer dimensions. In any dimension, a lattice may be defined as *a regular arrangement of points in space, of infinite extent, such that each point has the same environment as every other point*. Key words here are 'regular', 'infinite', 'environment' and 'point'. The essence of most crystalline solids is regularity, whether in one dimension or more, and a lattice is ideally of infinite extent. That a practical use can be made of an infinite concept follows from the discussion in Section 1.2, and the identical environment of each point ensures the regularity of the pattern built up by the lattice. It is necessary to remember that a lattice is an arrangement of mathematical points: they are joined up, as in children's early drawing books, so as to form a picture, in this case one that allows a ready appreciation of the geometry of the lattice.

4.2 One-dimensional lattice

In order to begin an appreciation of lattices, consider Fig. 4.1a; it is a line of identical, regular bricks placed exactly end-to-end. A fixed point can be identified at the same location on each brick, and so a series of points is

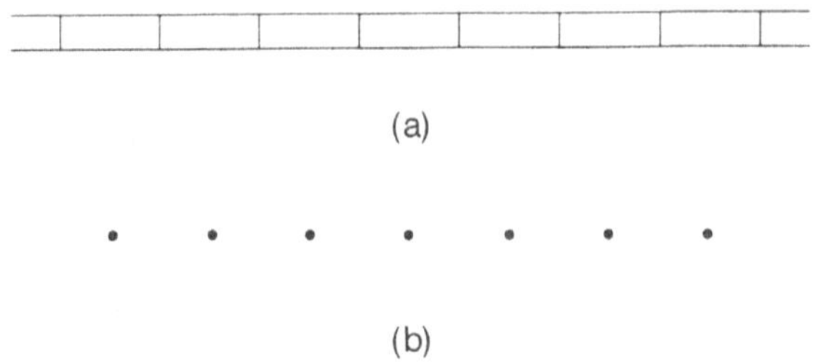

Fig. 4.1
Simulation of a one-dimensional lattice, or *row*. (a) Series of identical bricks placed end-to-end. (b) Line of representative mathematical points forming a one-dimensional lattice; the points relate to the same position on the bricks.

Table 4.1 Symmetry and dimensionality of lattices.

	Dimension		
Symmetry operation	**One (*Row*)**	**Two (*Net*)**	**Three (*Bravais*)**
Reflection	Across a point	Across a line	Across a plane
Rotation	–	About a point	About a line
Inversion	–	–	In a point

defined, representative of the row of bricks: this series of points represents a one-dimensional lattice, or *row*, and the points form a regular, ideally infinite pattern, each point in the same environment as every other point.

In considering the number of lattices, the number of *different arrangements* of points is needed whatever the dimension under consideration: in one dimension, there is only a single lattice, and the only lattice symmetry is that of reflection across any point. The argument in Section 3.5 on the extension of symmetry operations with increased dimensionality may be recalled here; for convenience, the results are summarized in Table 4.1.

4.3 Two-dimensional lattices

If the construction of bricks in Fig. 4.1a is extended to two dimensions, in the traditional manner, a brick wall (Fig. 4.2a) results. As before, each brick can be represented by a point, placed at the same location in each brick, and so build up a two-dimensional lattice, or *net*. Fig. 4.2b is a stack of rows, in this example each row is translated by one half row spacing with respect to successive rows.

A general net is shown in Fig. 4.3. It is convenient to represent the net by a *unit cell*, such as that outlined by vectors **a** and **b** in cell I. The net can be

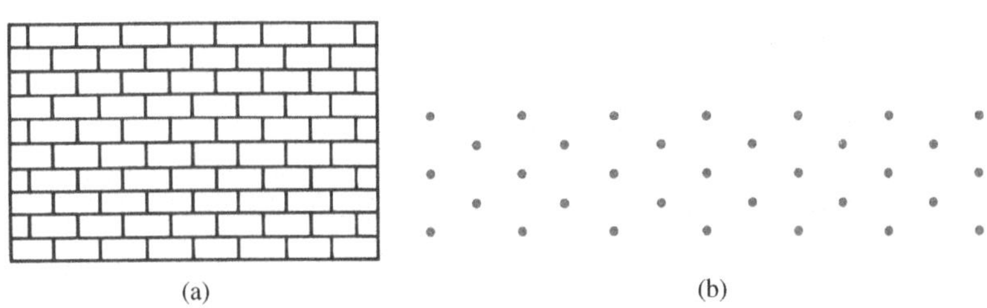

Fig. 4.2
Simulation of a two-dimensional lattice, or *net*. (a) Wall formed by stacking bricks of Figure 4.1a in the traditional manner. (b) Two-dimensional lattice: the points represent the same location on the bricks.

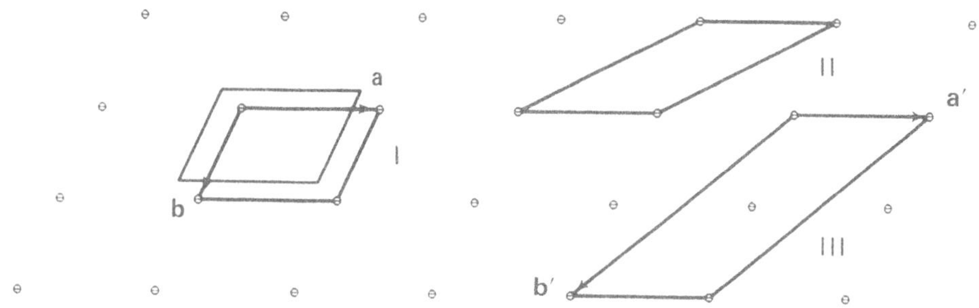

Fig. 4.3
Oblique net: I (conventional unit cell), II and III are three of an infinite number of possible unit cells; I and II are primitive, *p*, and III is centred, *c*. [Reproduced by courtesy of Springer Science+Business Media, New York, © Kluwer Academic/Plenum Publishing.]

then constructed by stacking unit cells side-by-side, in the same orientation and sharing adjacent lattice points. The unit cell I is *primitive*, symbol *p*: each lattice point is associated with the area of one unit cell. This fact may be appreciated either by noting that each point is shared equally by four unit cells, or by translating the unit cell framework by a small distance, as shown by the thin lines in cell I.

Every lattice point is a position of point group symmetry. In one dimension, the symmetry at each point is *m*. In a net, each point has the symmetry of one of the ten two-dimensional point groups (Section 3.4ff). It is desirable not to refer to a unit cell by a point group symbol, because a unit cell is representative of an infinite array of points and, therefore, not truly describable by point group symmetry.

4.3.1 Choice of unit cell

The choice of unit cell is to some extent arbitrary. Three unit cells are shown in Fig. 4.3. The conventional choice is the smallest sized repeat unit, provided that the vectors delineating that unit cell lie on or are parallel to important symmetry directions in the lattice. Thus, a conventional unit cell such as **a**, **b**, in Fig. 4.4b is not always the smallest in its lattice.

4.3.2 Nets in the oblique system

The possible nets are referred conveniently to the two-dimensional systems discussed in Section 3.4.4. In unit cell I (Fig. 4.3), $a \not{c}\, b$ and γ, the angle $\angle\mathbf{ab}$, is $\not{c}\, 90°$ or 120°; angles of these values *in a lattice* would lead to higher symmetry; the symbol ¢ should be read as '*not constrained by symmetry to equal*'. Unit cell II has the same area as cell I, but has a more obtuse, less convenient γ angle. The *centred* unit cell *c* is clearly twice as large as I or II, and would not be chosen. It does not represent a new oblique lattice, as the following transformation shows:

$$\mathbf{a} = \mathbf{a}' \qquad (4.1)$$

$$\mathbf{b}' = \frac{\mathbf{a}'}{2} + \frac{\mathbf{b}'}{2} \tag{4.2}$$

The result is not unexpected, since a lattice is invariant under choice of unit cell.

4.3.3 Nets in the rectangular system

If a, b or γ is specialized in a non-trivial manner, the symmetry at each lattice point will be then greater than 2: a condition such as $a' = 2a$ is trivial, and does not lead to a new arrangement of points.

Consider the net in Fig. 4.4a: the symmetry at each point is $2mm$, and the net is allocated to the rectangular system. Three possible unit cells are outlined, of which cell I is the conventional choice (Section 4.3.1). Unlike in the oblique system, the centred cell **a, b** here forms a new arrangement of points (Fig. 4.4b). The primitive unit cell **a**′, **b**′ is a true unit cell in this net, and is of smaller area. It is not the conventional choice because, in isolation, it does not show clearly the lattice symmetry. The symmetry is still present, as can be inferred from the transformation equations:

$$\left.\begin{aligned} \mathbf{a}' &= \frac{\mathbf{a}}{2} - \frac{\mathbf{b}}{2} \\ \mathbf{b}' &= \frac{\mathbf{a}}{2} + \frac{\mathbf{b}}{2} \end{aligned}\right\} \tag{4.3}$$

whence

$$a' = b' = \left(\frac{a^2}{4} + \frac{b^2}{4}\right)^{1/2} \tag{4.4}$$

Thus, a parallelogram unit cell with equal sides, a rhombus, in a lattice has symmetry $2mm$ at each lattice point: the value of γ', the angle between **a**′ and **b**′, depends on the ratio $\frac{a}{b}$.

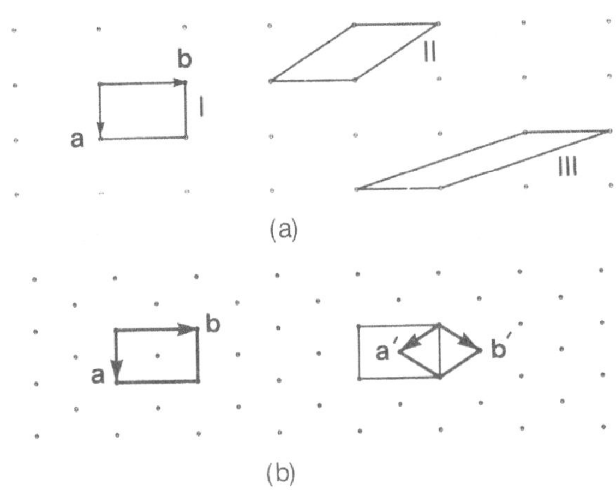

Fig. 4.4
Rectangular nets. (a) In the conventional p unit cell I, **a** and **b** are parallel to the m symmetry lines; p cells II and III both have the same area as I, but neither is simply related to the m lines. (b) A centred c unit cell **a**, **b** is the conventional choice here; the p cell **a**′, **b**′, although of smaller area, is oblique to the m symmetry lines and is not the conventional choice. [Reproduced by courtesy of Springer Science+Business Media, New York, © Kluwer Academic/Plenum Publishing.]

4.3.4 Square and hexagonal nets

Increased specialization in the unit cell parameters leads to nets in the square and hexagonal systems, which will be addressed in Problem 4.1; the five two-dimensional lattices are listed in Table 4.2.

The honeycomb arrangement of points in Fig. 4.5a is not a lattice because the environment of all points is not identical. A true net may be formed by centring the honeycomb: but then each centred honeycomb cell encompasses three primitive (triply primitive) hexagonal unit cells. If one of these primitive cells itself were now centred, the arrangement would no longer represent a hexagonal net: from the above discussion, the symmetry would be degraded to $2mm$ and the net rectangular.

Table 4.2 The five two-dimensional lattices.

System	Unit cell symbol	Point symmetry	Unit cell parameters
Oblique	p	2	a ¢ b; γ ¢ $90^\circ, 120^\circ$
Rectangular	p, c	$2\,mm$	a ¢ b; $\gamma = 90^\circ$
Square	p	$4\,mm$	$a = b; \gamma = 90^\circ$
Hexagonal	p	$6\,mm$	$a = b; \gamma = 120^\circ$

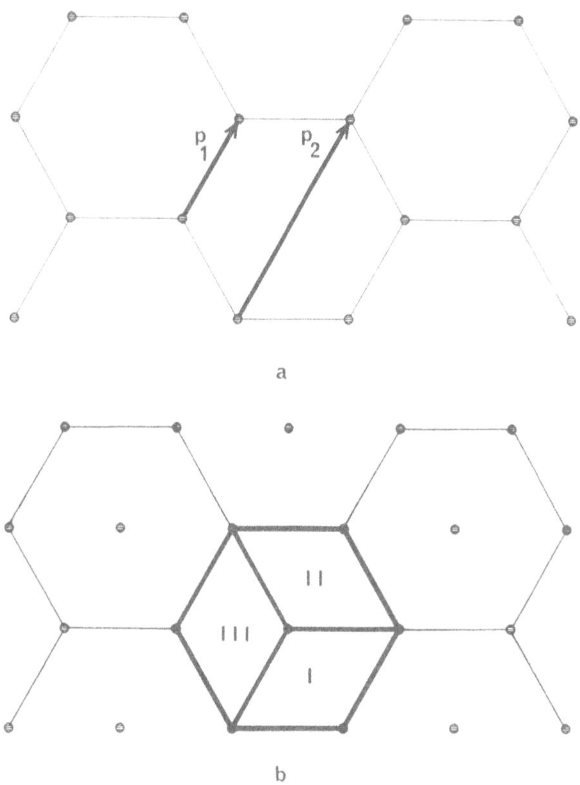

Fig. 4.5
The honeycomb arrangement of points is not a lattice; compare the environment of points $\mathbf{p}_1$ and $\mathbf{p}_2$. (b) True lattice, but now with three p unit cells within the centred hexagon. [Reproduced by courtesy of Springer Science+Business Media, New York, © Kluwer Academic/Plenum Publishing.]

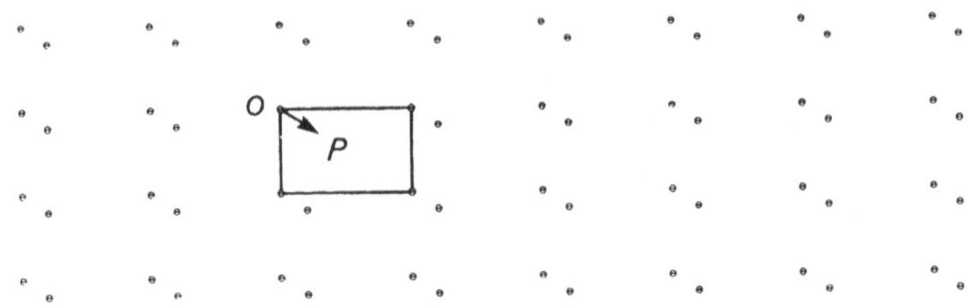

Fig. 4.6
A rectangular net, with two lattice points in the *p* unit cell. Only if all points have the same vector environment can they form a true lattice. Thus, the only position is the centre of the cell, to give a *c* unit cell (see also Section 4.4.6). [Reproduced by courtesy of Woodhead Publishing, UK.]

4.3.5 Unit cell centring

The centred unit cells described here have a lattice point at their geometrical centre. That this is the only site for a centring point in a lattice is clear from Fig. 4.6: if a point were placed at *P*, then an identical vector placed at that point would not terminate on another lattice point. This illustration is a rectangular primitive net with two points per unit cell; alternatively, it could be described as two identical primitive nets superimposed with a vector translation **OP**. There is an exception to this rule, as will emerge with hexagonal and trigonal lattices, in Section 4.4.6 and Section 4.4.7, respectively.

4.4 Three-dimensional lattices

'*If you have to fill a volume with a structure that's repetitive, just keep your wits about you, you don't need to take a sedative! Don't freeze with indecision, there's no need for you to bust a seam! Although the options may seem endless, really there are just fourteen*' [1]. But Frankenheim found 15 [2]! However, Bravais determined that two of Frankenheim's lattices referred to monoclinic *C* [3,4]; thus, the designation *Bravais* lattice is in general use.

Section 4.1 introduced the lattice as the geometrical basis for crystal structure, and gave a definition of it that continues to be applicable in three dimensions. If a number of nets (Section 4.3) is aligned in a regular manner at a spacing *c*, non-collinear with *a* or *b*, a three-dimensional lattice is obtained; Fig. 4.7 shows a stereoview of such a lattice. The fourteen Bravais

Fig. 4.7
Stereoview of a three-dimensional lattice, obtained by stacking nets at a regular spacing, non-collinear with the spacings of the net. [Reproduced by courtesy of Woodhead Publishing, UK.]

lattices are distributed unequally among the seven crystal systems according to their symmetry. A unit cell for each lattice is chosen in the conventional manner (Section 4.3.1) and thus is not always primitive. The lattices in the seven crystal systems will be studied, beginning with the least symmetrical.

4.4.1 Triclinic lattice

The triclinic lattice is illustrated in Fig. 4.8. The unit cell is characterized by the parameters *a* ¢ *b* ¢ *c* and α ¢ β¢ γ¢ 90°, 120°. Similar conditions were given in Table 3.4 for the intercepts of the parametral plane and the interaxial angles, arising from purely morphological considerations.

There is only one triclinic lattice, and its conventional unit cell is primitive, symbol *P* (upper case letters in three dimensions): it contains one lattice point per unit cell and the symmetry at each lattice point is $\bar{1}$; any centred triclinic cell can be reduced to a primitive triclinic cell.

When the unit cell parameters are specialized in a non-trivial manner, higher symmetry always results, and other crystal systems revealed.

4.4.2 Monoclinic lattices

The symmetry at each lattice point in the monoclinic system is $2/m$, and the conventional unit cell takes the standard conditions *a* ¢ *b* ¢ *c*, $\alpha = \gamma =$ 90° ¢ β, with β chosen to be obtuse. The variable angle β corresponds to the characteristic twofold symmetry axis parallel to the *y* axis. In all systems where the unit cell has a typically non-specialized value of a parameter, it is always possible that it may exhibit an uncharacteristic value. Thus, some monoclinic crystals have been shown to have a β angle that is 90° within the limits of experimental measurement: but they remain monoclinic, because the symmetry of the crystal is dictated by the arrangement of the contents of the unit cell.

In three dimensions, there are several types of centring possible for the Bravais lattices, as set out in Table 4.3 with their standard notation. The table lists also the *fractional coordinates* of the centring points.

Fractional coordinates are dimensionless quantities: the fractional coordinate *x* is defined as X/a, where *X* and *a* represent, respectively, the coordinate in

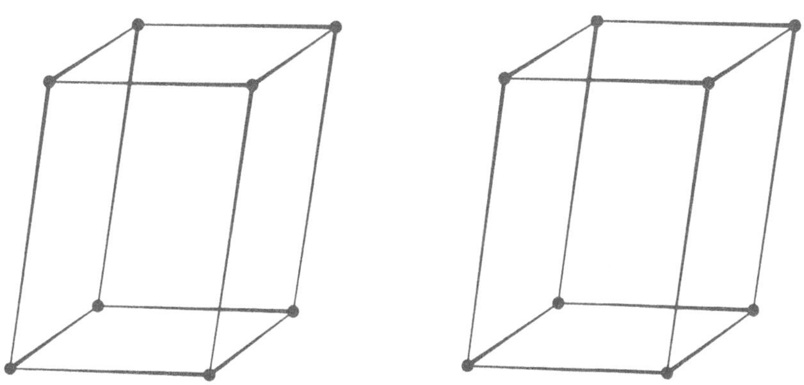

Fig. 4.8
Stereoview of a *P* unit cell in the triclinic lattice. In the illustrations of unit cells herein, the origin is in the bottom left, rear corner, with +*a* towards the reader, +*b* to the right and +*c* upward. The parameters of the conventional unit cells are listed in Table 4.4 (Cp. Table 3.4). [Reproduced by courtesy of Woodhead Publishing, UK.]

Table 4.3 Notation for unit cells of the Bravais lattices.

Centring site/s in the unit cell	Symbol	Miller indices of centred faces	Fraction coordinates of unique lattice points
None	*P*, *R*	—	0, 0, 0
	R_{hex}[a]	—	0, 0, 0; 1/3, 2/3, 2/3; 2/3, 1/3, 1/3
b, *c* faces	*A*	(100)	0, 0, 0; 0, 1/2, 1/2
c, *a* faces	*B*	(010)	0, 0, 0; 1/2, 0, 1/2
a, *b* faces	*C*	(001)	0, 0, 0; 1/2, 1/2, 0
Body	*I*	—	0, 0, 0; 1/2, 1/2, 1/2
All faces	*F*	(100), (010) (001)	0, 0, 0; 0, 1/2, 1/2; 1/2, 0, 1/2; 1/2, 1/2, 0

[a] The points $\pm(1/3, 2/3, 2/3)$ in R_{hex} are as a form of centring.

absolute measure and the corresponding unit cell dimension in the same units; similar definition apply for *y* and *z*. They have the obvious advantage of being independent of the dimensions of the unit cell.

Centring the *B* faces of a monoclinic unit cell is illustrated in Fig. 4.9. In this case another unit cell can be defined by the transformation:

$$\left.\begin{aligned} \mathbf{a}' &= \mathbf{a} \\ \mathbf{b}' &= \mathbf{b} \\ \mathbf{c}' &= -\frac{\mathbf{a}}{2} + \frac{\mathbf{c}}{2} \end{aligned}\right\} \qquad (4.5)$$

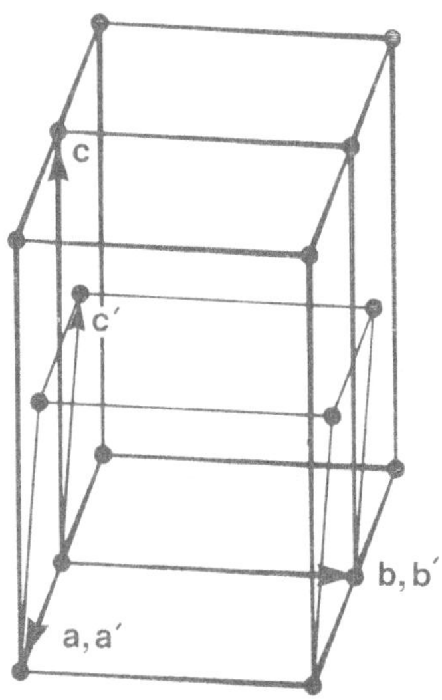

Fig. 4.9
Monoclinic lattice showing two *B* centred unit cells, **a**, **b**; the conventional *P* unit cell **a**′, **b**′ is outlined. [Reproduced by courtesy of Woodhead Publishing, UK.]

The choice of –**a**/2 in **c**′ instead of **a**/2 is often necessary in the monoclinic system, so as to ensure the standard convention that β' is oblique. Now, $a' \not\subset b' \not\subset c'$; and since c' lies in the a', c' plane, $\alpha' = \gamma' = 90°$ but β' is still a general value, consistent with monoclinic symmetry. The transformed unit cell is *P* and has half the volume of the *B* cell, as can be seen by counting the lattice points *per unit cell*, and is the conventional choice.

A *C* unit cell in a monoclinic lattice is shown in stereoview in Fig. 4.10, neglecting the thin lines for the moment. An *A* centred unit cell does not represent a different monoclinic lattice, because the *x* and *z* axes can interchanged, while retaining *b* as the unique twofold symmetry direction, provided that the signs of the directions of the *x* and *z* axes are set so as to preserve a right-handed axial set.

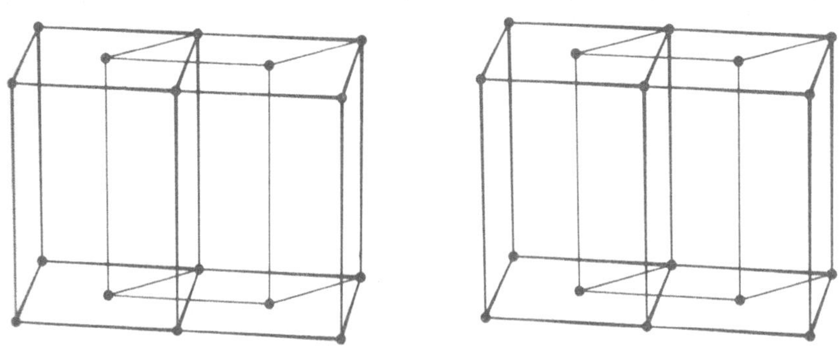

Fig. 4.10
Stereoview of two monoclinic *C* unit cells; the *P* cell outlined is not the conventional choice for this lattice.

So far, the conditions $A \equiv C$ and $B \equiv P$ have been deduced. A different result is obtained if the C unit cell is transformed to P, as shown by the thin lines in Fig. 4.10. A transformation is given by

$$\left.\begin{aligned} \mathbf{a}' &= \frac{\mathbf{a}}{2} - \frac{\mathbf{b}}{2} \\ \mathbf{b}' &= \mathbf{b} \\ \mathbf{c}' &= \mathbf{c} \end{aligned}\right\} \tag{4.6}$$

Now $a' \not\subset b' \not\subset c'$ and $a' = 90°$. However, because the a, b plane is inclined to the b, c plane, γ', as well as β', is different from 90°. These results indicate that monoclinic C is different from monoclinic P in the arrangement of points, so that both descriptors are relevant to this system. Monoclinic I and F unit cells both transform to C, and further discussion of them will be given in the Problems section of this chapter. In carrying out these transformations, it is always necessary to retain both the right-handed system of axes, and, as far as is practicable, an obtuse value for the β angle.

Example 4.1 Practical transformations involve numerical computation. In the transformation $B \rightarrow P$ given by Eq. (4.5), only c' and β' need be calculated.

From the discussion in Appendix A3.2.2, it follows that

$$\left.\begin{aligned} \mathbf{c}' \cdot \mathbf{c}' &= \left(-\frac{\mathbf{a}}{2} + \frac{\mathbf{c}}{2}\right) \cdot \left(-\frac{\mathbf{a}}{2} + \frac{\mathbf{c}}{2}\right) \\ c'^2 &= \frac{a^2}{4} + \frac{c^2}{4} - \frac{ac \cos \angle \mathbf{ac}}{2} \\ c' &= \left(\frac{a^2}{4} + \frac{c^2}{4} - \frac{ac \cos \beta}{2}\right)^{1/2} \end{aligned}\right. \tag{4.7}$$

and

$$\begin{aligned} \mathbf{a}' \cdot \mathbf{c}' &= a'c' \cos \beta' \\ \cos \beta' &= \mathbf{a} \cdot \left(\frac{-\mathbf{a}/2 + \mathbf{c}/2}{ac'}\right) = \frac{1}{ac'}\left(\frac{-a^2}{2} + \frac{ac \cos \beta}{2}\right) \\ \beta' &= \cos^{-1}\left(\frac{-a + c \cos \beta}{2c'}\right) \end{aligned} \tag{4.8}$$

4.4.3 Orthorhombic lattices

The symmetry at a lattice point being that of the appropriate Laue group implies that the point symmetry is *mmm* in any orthorhombic lattice. In the conventional unit cell, $a \not\subset b \not\subset c$ and $\alpha = \beta = \gamma = 90°$. Unit cells P, C, I and F exist for this system. The A-centred and B-centred orthorhombic unit cells can always be transformed to orthorhombic C: in studying space groups, however, A-centring is also required; in such cases, the equivalence of A and B still applies. Fig. 4.11 illustrates a body-centred orthorhombic unit cell. That the P, C, I and F cells represent different orthorhombic arrangements can be demonstrated, following the procedure used with the monoclinic $B \rightarrow P$ transformation:

After centring a P unit cell, the following questions should be considered, in order:

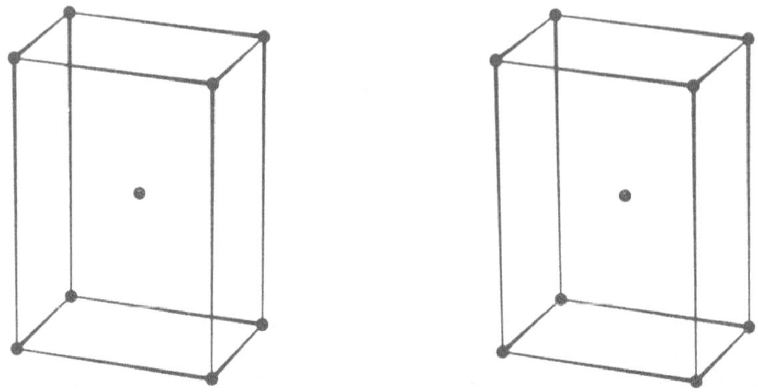

Fig. 4.11
Stereoview of the orthorhombic I unit cell.

- Does the centred cell represent a true lattice?
- If it is a lattice, is the symmetry of the unit cell drawing apparently different from that of the P cell?
- If the symmetry is unchanged, does the centred cell represent a different arrangement of points? It may be judged by comparing the parameters of the two cells in question, which is equivalent to asking if the unit cell has been chosen according to convention.

These questions have been answered implicitly in the monoclinic transformations already studied.

The correct number of lattices can be reached in another way. The orthorhombic system is characterized by three axes, 2 or $\bar{2}$, along a, b and c. From the matrices in Appendix A3.7.1, a twofold rotation of a vector $\mathbf{r}(x, y, z)$ about the axis along a has the effect $x\mathbf{a}$, $y\mathbf{b}$, $z\mathbf{c} \rightarrow x\mathbf{a}$, $\bar{y}\mathbf{b}$, $\bar{z}\mathbf{c}$; similarly, a rotation of the same initial vector about the axis along b has the effect $x\mathbf{a}$, $y\mathbf{b}$, $z\mathbf{c} \rightarrow \bar{x}\mathbf{a}$, $y\mathbf{b}$, $\bar{z}\mathbf{c}$. From the study of point groups, it is evident that the changes in sign show that **a** is perpendicular to **b** and **c**, and that **b** is perpendicular to **c** and **a**. The product of the matrices for these two operations is a matrix that has the effect $x\mathbf{a}$, $y\mathbf{b}$, $z\mathbf{c} \rightarrow \bar{x}\mathbf{a}$, $\bar{y}\mathbf{b}$, $z\mathbf{c}$. Thus, the three twofold axes are mutually perpendicular, as was found earlier, in Section 3.6.3.

Conditions such as this can be confirmed in another way. The scalar product from the first transformation: $x\mathbf{a} \cdot y\mathbf{b} \rightarrow x\mathbf{a} \cdot (-y\mathbf{b})$ or $xyab \cos\gamma \rightarrow -xyab \cos\gamma$. Since the twofold symmetry operation leads to indistinguishability, $xyab\cos\gamma = -xyab\cos\gamma$, or $2\cos\gamma = 0$, which shows that γ is 90°. By considering the product $x\mathbf{a} \cdot z\mathbf{c}$, β is found also to be 90°. If the product $y\mathbf{b} \cdot z\mathbf{c}$ in the first transformation be examined, it evolves as $yzbc\cos\alpha = (-y)(-z)bc\cos\alpha$. This reveals nothing about the angle α, because this transformation concerns only the b, c plane. However, when the argument is applied to the second transformation, α is found to be 90°. Also, since there has been no interchange among the x, y and z coordinates during these transformations, there can be no restriction on the values of a, b and c. Thus, the arrangement of twofold axes determines a system in which $a \neq b \neq c$ and $\alpha = \beta = \gamma = 90°$, and is called orthorhombic. Hence, any reduction of a centred unit cell that infringes conditions so deduced shows that the centred cell is proper to its system. Try it for the monoclinic C cell example (Fig. 4.10). In the crystal systems, the conditions impinging on the conventional unit cells on account of the characteristic symmetry (Table 3.4) can all be deduced in this way.

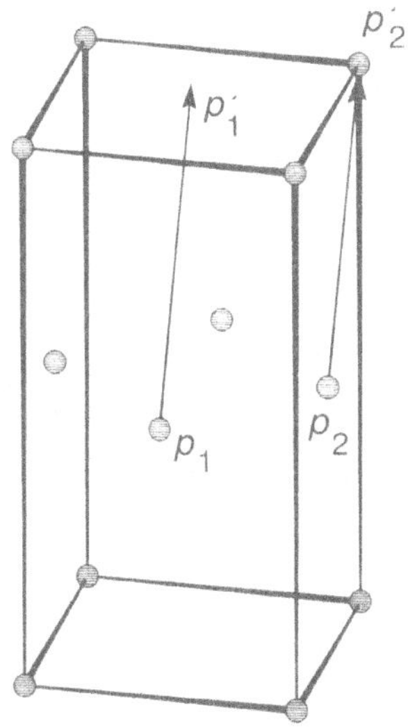

Fig. 4.12
A 'tetragonal' unit cell centred on the A and B faces does not represent a lattice, because of the differing environments of $\mathbf{p}_1$ and $\mathbf{p}_2$.

4.4.4 Tetragonal lattices

There are two tetragonal lattices, symmetry $\frac{4}{m}mm$ at each point, represented by P and I unit cells; their parameters are: $a = b \not\subset c$; $\alpha = \beta = \gamma = 90°$. It is straightforward to show that $C \equiv P$ and $F \equiv I$.

Consider centring the B faces: the unit cell no longer has the characteristic tetragonal symmetry. The symmetry is restored, apparently, by centring the A faces as well. But this arrangement is not a lattice, as Fig. 4.12 shows. Centring now the C faces produces a true tetragonal F unit cell, but this is equivalent to I, and is not a new lattice (Fig. 4.13).

4.4.5 Cubic lattices

There are three cubic lattices, P, I and F, consistent with $m\bar{3}m$ symmetry at each point and unit cell parameters $a = b = c; \alpha = \beta = \gamma = 90°$. If the A faces are centred (Fig. 4.14a), the symmetry of the lattice is reduced $m\bar{3}\,m \rightarrow mmm$. If a cubic I cell is transformed by the equations

$$\left.\begin{array}{l} \mathbf{a}' = \mathbf{a} \\ \mathbf{b}' = \mathbf{b} \\ \mathbf{c}' = \mathbf{a} + \mathbf{c} \end{array}\right\}$$

the new cell is A centred, apparently monoclinic in isolation, and unconventional. However, it still has cubic symmetry as it represents the same lattice (Fig. 4.14b): the symmetry is inherent in the special conditions $a' = b'$, $c' = a'\sqrt{2}, \alpha = \gamma = 90°, \beta' = 45°$.

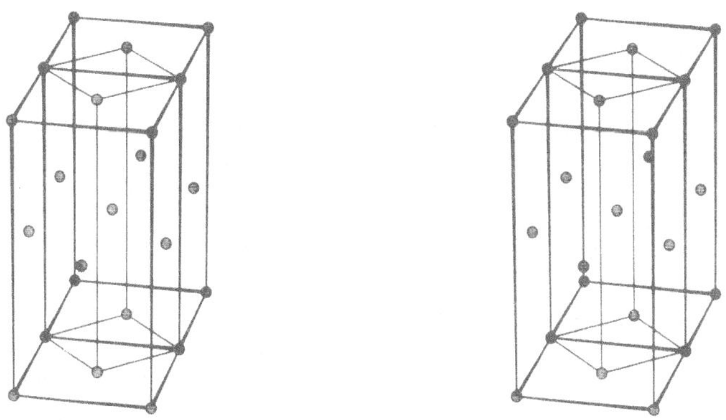

Fig. 4.13
Stereoview of a tetragonal lattice, showing the equivalence of F and I unit cells.

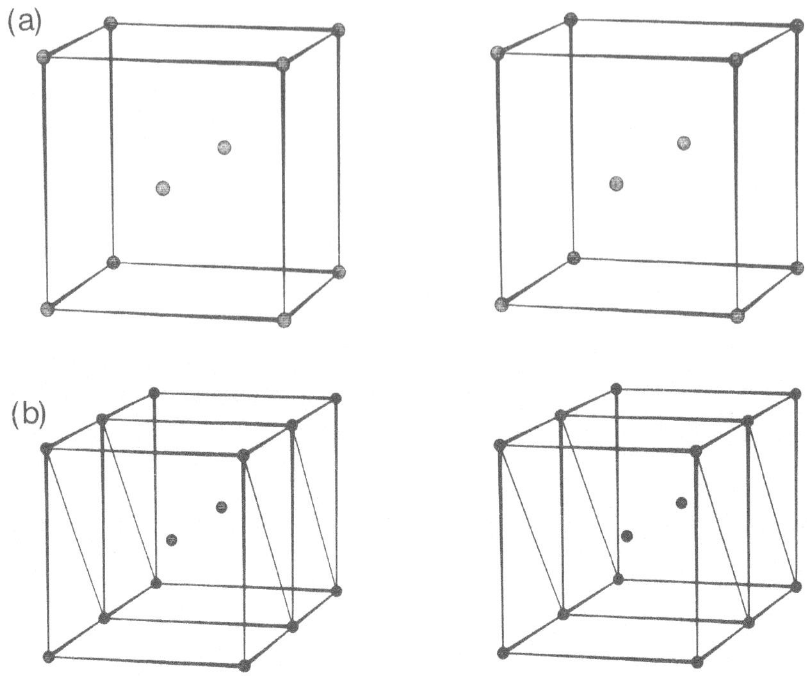

Fig. 4.14
(a) Stereoview of a cubic *P* unit cell degraded to orthorhombic *A* by centring the *A* faces alone. (b) Cubic *I*: the thin lines delineate another unit cell, with $a' = b'$ ¢ $c', \alpha = \gamma = 90^{\circ}, \beta = 45^{\circ}$, apparently monoclinic *C*, but the symmetry at each point remains $m\bar{3}\,m$, so that it is a non-standard representation of the cubic lattice.

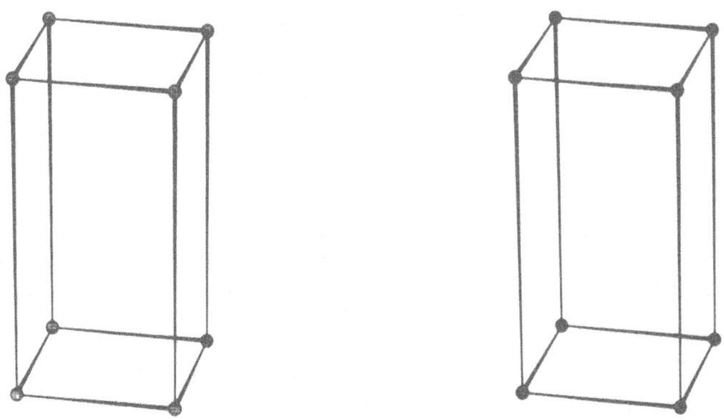

Fig. 4.15
Stereoview of a *P* unit cell in the hexagonal lattice.

4.4.6 Hexagonal lattice

The hexagonal lattice has a *P* unit cell, with $a = b$ ¢ $c; \alpha = \beta = 90^{\circ}, \gamma = 120^{\circ}$; the lattice has point symmetry $\frac{6}{m}mm$, as shown in Fig. 4.15. If the hexagonal unit cell is centred as *C*, *I* or *F*, it is no longer hexagonal. Check this for yourself by drawing, or otherwise. However, if the 'centring' is carried out in the sense of base centring the two trigonal prisms that make up the hexagonal unit cell, that is, at the points $\pm\,(^1\!/_3, ^2\!/_3, 0)$, then the hexagonal symmetry is retained, and a smaller hexagonal unit cell is obtained:

$$|\mathbf{a}'| = |^1\!/_3\mathbf{a} + ^2\!/_3\mathbf{b}|; \quad |\mathbf{b}'| = |-^1\!/_3\mathbf{a} + ^1\!/_3\mathbf{b}|; \ \mathbf{c}' = \mathbf{c},$$

whence $a' = b' = a\sqrt{3}/3$ and $\gamma' = 120^{\circ}$, and a new lattice is not formed. Show that the volume of the 'new' cell is one third that of the 'old' cell.

Another valid lattice is obtained if the hexagonal unit cell is 'centred' at $\pm(^2/_3, ^1/_3, ^1/_3)$, which introduces the rhombohedral lattice.

4.4.7 Lattices in the trigonal system

A hexagonal unit cell is compatible with sixfold and threefold symmetry, and the centring just discussed leads to a lattice in the trigonal system (Fig. 4.16). The lattice no longer has the characteristic 6 or $\bar{6}$ symmetry of the hexagonal system. It is for this reason that the Schönflies point group symbol for $\bar{6}\,m2$ is C_{3h}, and the comment in Section 3.10.2, that the symmetry $\bar{6}$ cannot operate on a rhombohedral lattice should be now clear. Since this trigonal lattice has threefold symmetry axes parallel to z and passing through points with x, y coordinates $^2/_3, ^1/_3$ and $^1/_3, ^2/_3$, the possibility of a triply primitive hexagonal H cell exists. Thus, for some trigonal crystals the unit cell is P, whereas for others it will be R_{hex}, which can be deduced from the systematic absences in an X-ray diffraction record (see Fig. 5.36).

The R_{hex} cell can be transformed to the primitive rhombohedral cell R, for which $a = b = c$ and $\alpha = \beta = \gamma$ ¢ 90° and $<$ 120°. In this R cell, the

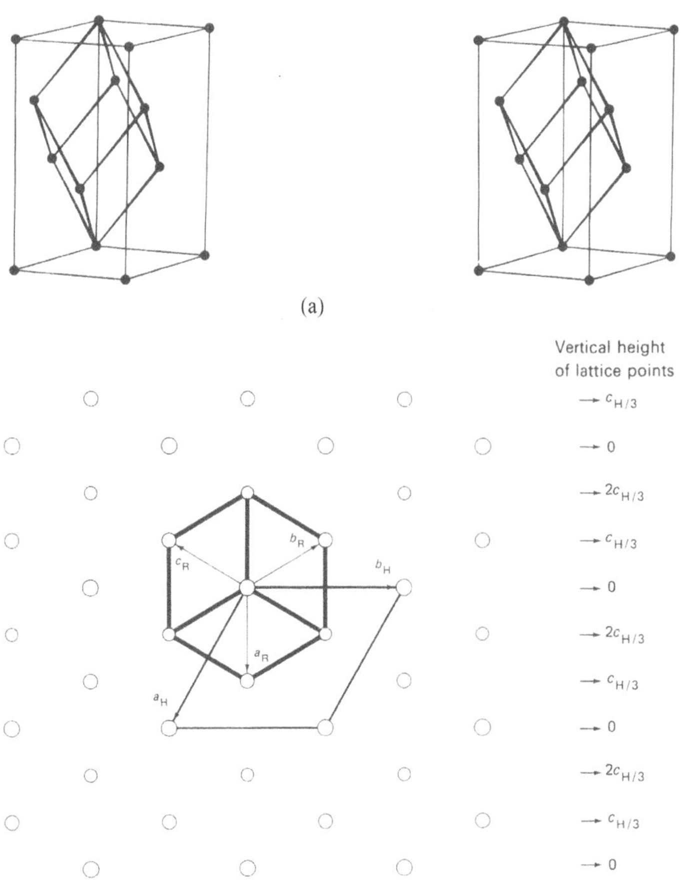

Fig. 4.16
(a) Stereoview of the rhombohedral (trigonal) R unit cell in the *obverse* setting with respect to a hexagonal unit cell. (b) Rhombohedral lattice and R unit cell, $\mathbf{a}_R, \mathbf{b}_R, \mathbf{c}_R$, viewed along [111]. The outlined hexagon delineates the triply primitive hexagonal unit cell, $\mathbf{a}_H, \mathbf{b}_H, \mathbf{c}_H$, and the fractions in the right-hand column refer to heights along z; $\mathbf{c}_H$ is normal to the $\mathbf{a}_H, \mathbf{b}_H$ plane and passes through the origin.

threefold axis is along [111], and its volume takes the simpler expression $V = \sqrt{a^3(1 - 3\cos^2\alpha + 2\cos^3\alpha)}$. The lattice that is represented by a conventional R unit cell is the single, true, trigonal unit cell. A cube extended or compressed along its [111] direction produces a rhombohedral unit cell; in compression it remains trigonal, but the α angle would be greater than 180° and therefore unconventional.

There are two settings of the R cell in relation to the hexagonal cell: the *obverse* setting, shown in Fig. 4.16, and the *reverse* setting, obtained by rotating the R cell clockwise about c_H by 60°; the obverse setting is standard.

It is worth noting that symbols such as P and C do not actually describe lattices, although they are used in this way. A lattice is an infinite array of points, each of the same symmetrical environment. Whether or not it is termed P, C or other designation depends on how the array is viewed, in other words, how the unit cell is chosen. Hence, a designation C, for example, implies that a choice of unit cell type has been made for a particular lattice: other choices are possible in the same lattice, although they are not necessarily conventional [5]. As long as terms such as 'P lattice' and 'C lattice' are used with this proviso, the practice is acceptable. The Bravais lattice data are summarized in Table 4.4.

4.5 Lattice directions

Lattice geometry is based around the three unit cell translation vectors **a**, **b** and **c**. From the definition of lattice, it follows that any point may be reached,

Table 4.4 The Bravais lattice unit cells.

System	Unit cell symbols	Axial relationships[a]	Symmetry[b] at each lattice point
Triclinic	P	$a \not\subset b \not\subset c$ $\alpha \not\subset \beta \not\subset \gamma \not\subset 90°,\ 120°$	$\bar{1}$
Monoclinic (y unique)	P, C	$a \not\subset b \not\subset c$ $\alpha = \gamma = 90°,\ \beta \not\subset 90°,\ 120°$	$2/m$
Orthorhombic	P, C, I, F	$a \not\subset b \not\subset c$ $\alpha = \beta = \gamma = 90°$	mmm
Tetragonal	P, C	$a = b \not\subset c$ $\alpha = \beta = \gamma = 90°$	$\frac{4}{m}mm$
Cubic	P, I, F	$a = b = c$ $\alpha = \beta = \gamma = 90°$	$m\bar{3}m$
Hexagonal	P	$a = b \not\subset c$ $\alpha = \beta = 90°, \gamma = 120°$	$\frac{6}{m}mm$
Trigonal (Hexagonal axes)	P	$a = b \not\subset c$ $\alpha = \beta = 90°, \gamma = 120°$	$\bar{3}m$
Trigonal (Rhombohedral axes)	R	$a = b = c$ $\alpha = \beta = \gamma \not\subset 90°, < 120°$	$\bar{3}m$

[a] Read ¢ as 'not constrained by symmetry to equal'.

[b] A lattice point exhibits the highest symmetry of its system.

starting from any other point, by performing the basic translations, or positive or negative multiples thereof, always in the directions of **a**, **b** and **c**.

Any lattice point may be taken as an origin for the lattice; then, the vector **r** to any other lattice point is given by

$$\mathbf{r} = U\mathbf{a} + V\mathbf{b} + W\mathbf{c} \quad (4.9)$$

where U, V and W are positive or negative integers or zero, and are the *coordinates* of the lattice point. The line joining the origin to the lattice points U, V, W; $2U$, $2V$, $2W$; $\ldots nU, nV, nW$ defines the *direction*, or *directed line*, $[UVW]$. The notation here is similar to that used for zone axes, because a direction, as defined here, is a possible zone axis, since crystal planes are rational. A set of directions related by symmetry defines a form of directions, signified by the notation $<UVW>$.

In a similar way, any general position in the unit cell has fractional coordinates x, y and z. Hence, the vector **d** from the origin of the unit cell to the point x, y, z is given by

$$\mathbf{d} = x\mathbf{a} + y\mathbf{b} + z\mathbf{c} \quad (4.10)$$

The numerical values of **r** and **d** can be evaluated from their scalar products, $\mathbf{d} \cdot \mathbf{d}$ and $\mathbf{r} \cdot \mathbf{r}$, following Appendix A3.2.1.

Example 4.2 Calculate the length of $[\bar{3}1\bar{2}]$ in a trigonal unit cell where $a = 0.473$ nm and $\alpha = 51.22°$.

Using Eq. (4.9) and the scalar product equation, $r^2 = \mathbf{r} \cdot \mathbf{r} = U^2\ a^2 + V^2\ b^2 + W^2\ c^2 + 2VWbc\cos\alpha + 2WUca\cos\beta + 2UVab\cos\gamma$. But $a = b = c$ and $\alpha = \beta = \gamma$. Hence, $r^2 = a^2[(U^2 + V^2 + W^2) + 2\cos\alpha(VW + WU + UV)] =$ Evaluating, $r = 2.094$ nm.

4.6 Law of rational intercepts: reticular density

When the faces of crystals are allocated Miller indices (indexed) in the simplest manner, the indices are small whole numbers; only occasionally does the index of crystal face exceed 5. The law of rational intercepts, also called the law of rational indices, embodies this result, and can be interpreted in terms of lattice theory.

Consider the projection of an orthorhombic lattice, shown in Fig. 4.17. The traces of the families of planes (100), (110) and $(2\bar{3}0)$ are outlined in relation to the **a**, **b** projection of the primitive unit cell. It can be shown that as the Miller indices increase numerically, the reticular densities, D_R, that is, the number of lattice points per unit area, may be expected to decrease. In Fig. 4.17, the (100) planes are the most densely populated, $D_R = \dfrac{1}{ac}$; the (110) planes with $D_R = \dfrac{1}{c\sqrt{a^2 + b^2}}$ are less more densely populated, and the $(2\bar{3}0)$, with $D_R = \dfrac{1}{c\sqrt{9a^2 + 4b^2}}$, less densely still, and so on. The more densely populated planes are those of wider interplanar spacing d in a given material: d is proportional to the reticular density or to 1/reticular area.

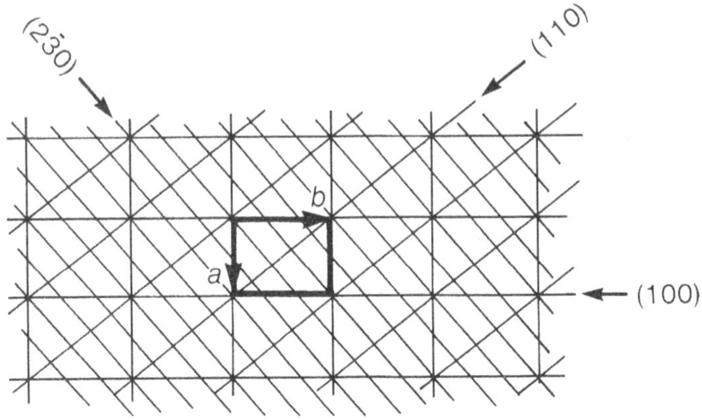

Fig. 4.17
The (100), (110) and (2$\bar{3}$0) families of planes in an orthorhombic lattice. [Reproduced by courtesy of Woodhead Publishing, UK.]

If a lattice is based on an orthorhombic C unit cell, $D_R(100)$ is the same as for the P cell although the interplanar spacing is halved, but $D_R(110)$ is now twice that in the P cell. If, for example, the unit dimensions are $a = 0.4, b = 0.6, c = 1.00\,\text{nm}$, then the D_R ratios $(100) : (110) : (2\bar{3}0)$ are $2.50 : 1.39 : 0.589$. This topic further may be explored a little further, restricting the discussion to the cubic system. In this system

$$d(hkl) = \frac{1}{\sqrt{h^2 + k^2 + l^2}} \propto D_R \tag{4.11}$$

and there are the three lattices to consider, specified by P, I and F. The planes with the greater values of D_R would be expected to correspond to the more stable (lower energy) crystal state. Consider first a P unit cell; the D_R values are listed hereunder:

***P* unit cell**

hkl	100	110	111	210	211	221	310	311	(222)	320	321
D_R	1	0.71	0.58	0.45	0.41	0.33	0.32	0.30	0.29	0.28	0.27

These data indicate a preference for the hexahedral (cubic) {100} form; sodium chlorate, which has a primitive unit cell, develops principally the forms {100}, {110} and {210}. Note that the *hkl* values listed are those found on an X-ray diffraction record. Morphologically, (222) is observed externally as (111) and, in general, $(2p)\,h, (2q)\,k$ and $(2r)\,l$ as $h\,k\,l$, where p, q and r are integers.

In an I unit cell, any index for which $h + k + l$ is not an even number is doubled, as underlined in the table, where the fundamental $h\,k\,l$ is written as a subscript:

***I* unit cell**

hkl	$200_{(100)}$	110	$\underline{222}_{(111)}$	$\underline{420}_{(210)}$	211	$\underline{442}_{(221)}$	310	$\underline{622}_{(311)}$	$\underline{640}_{(320)}$	321	411
D_R	0.5	0.71	0.29	0.22	0.41	0.17	0.32	0.15	0.14	0.27	0.24

In this lattice, there is a preference for dodecahedral {110} forms, as in the garnets and the caesium halides, except caesium fluoride, which show also the {100} and {111} forms. Although caesium chloride, bromide and iodide do not have strictly *I* unit cells because the atom at the centre of the unit cell is different from those at its corners, they simulate *I* in their habits. Rewriting the morphological forms in order of importance:

$$110 > 100 > 211 > 310 > 111 > 321 > 411 > 210 > 221 > 311 > 320$$

For an *F* unit cell, any index not containing mixed odd and even integers is doubled, as in the following record:

	***F* unit cell**										
hkl	$\underline{200}_{(100)}$	$\underline{220}_{(110)}$	111	$\underline{420}_{(210)}$	$\underline{422}_{(211)}$	$\underline{442}_{(221)}$	$\underline{620}_{(310)}$	311	331	511	531
D_R	0.50	0.35	0.58	0.22	0.20	0.17	0.16	0.30	0.23	0.19	0.17

Here, a preference exists for octahedral {111} and {100} forms, as in diamond and calcium fluoride. Re-writing the morphological forms in order of importance:

$$111 > 100 > 110 > 311 > 331 > 210 > 211 > 221/511 > 531 > 310$$

In simple structures, component atoms often occupy positions on lattice points, so that the population of atoms on a given plane may be related in a simple manner to the reticular density. The faces on a crystal represent the terminations of families of planes, and a crystal grows in such a way that the external faces are planes of highest reticular density. This situation produces a more energetically stable system, because of a better balance of interatomic forces than would arise with a surface that contains, on the atomic scale, relatively large holes.

As the planes of higher reticular density are those of lower Miller indices, the law of rational intercepts follows logically; it was developed significantly by Bravais [3]. If the unit cells in a crystal pack in the manner of Fig. 4.18, the external faces will have rational indices. A crystal does not build in fractions of unit cells, so that the apparent steps shown here on {110} are only of atomic dimensions and will not be observed.

A more general discussion on crystal growth and form has been given by Donnay and Harker [6], who took space group symmetry into account. Thus, morphologically, quartz would be expected to have {0001} as the dominant form. In practice, this form is observed rarely. The first three D_R values are $(0001), (10\bar{1}0)$ and $(10\bar{1}1)$; but on account of the 3_1 screw axis in quartz, (0003) takes the place of (0001), so moving this plane down in the D_R order. Subsequent work defined a growth morphology [7, 8] that attempted to explain crystal growth in terms of an attachment energy of a crystal fragment on to an already formed crystal. Equilibrium morphology attempts to determine crystal growth in terms of minimum surface energy [9], the greater stability corresponding to low surface energy and high reticular density. A more recent publication [10] considers the morphology and growth of crystals in detail.

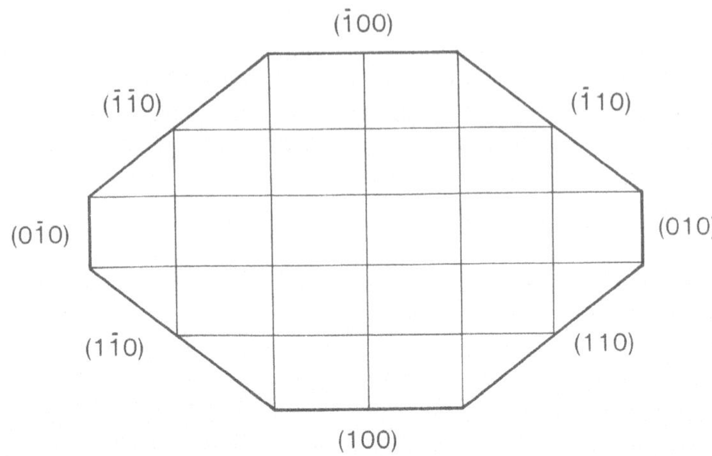

Fig. 4.18
Common shape of the cross-section on an orthorhombic crystal; the zone axis [001] is normal to the diagram. The {110} planes on the macroscopic have rational indices because the apparent steps are microscopic in size. [Reproduced by courtesy of Woodhead Publishing, UK.]

4.7 Reciprocal lattice

It is convenient to introduce here the concept of the reciprocal lattice, which will be needed when considering the diffraction of X-rays in a subsequent chapter. A reciprocal lattice exists for each of the Bravais lattices, and may be derived by the following construction, applied to a monoclinic P lattice.

A projection on to the (010) plane is shown in Fig. 4.19; the primitive unit cell is outlined by vectors **a** and **c**. Lines are constructed from the origin, O, and normal to the families of Bravais lattice planes shown. Recall that, in general, the normal to a lattice plane does not coincide with the direction of the same indices.

Along each line, reciprocal lattice points are defined such that the distances to these points from the origin are inversely proportional to the corresponding interplanar spacings. Thus, in Fig. 4.19, the families of planes (100), (101) and (001) give rise to reciprocal lattice points at distances from the origin that are proportional to $1/d(100)$, $1/d(101)$ and $1/d(001)$, where $d(100) = OP$, $d(101) = OQ$ and $d(001) = OR$. In general,

$$d^* = K/d(hkl) \tag{4.12}$$

where the parameter K is unity (see also Section 6.4ff). The vectors $\mathbf{d}^*(100)$, $\mathbf{d}^*(010)$ and $\mathbf{d}^*(001)$ may be taken to define the translation vectors $\mathbf{a}^*$, $\mathbf{b}^*$ and $\mathbf{c}^*$ of a unit cell in the reciprocal lattice, also termed the reciprocal unit cell. The following equations for the monoclinic reciprocal lattice can now be determined. From Eq. (4.12),

$$d^*(100) = 1/d(100) = a^* \tag{4.13}$$

But $d\,(100)$ is, from Fig. 4.19, $a \sin\beta$. Hence

$$a^* = 1/(a \sin\beta) \tag{4.14}$$

Similarly

$$c^* = 1/(c \sin\beta) \tag{4.15}$$

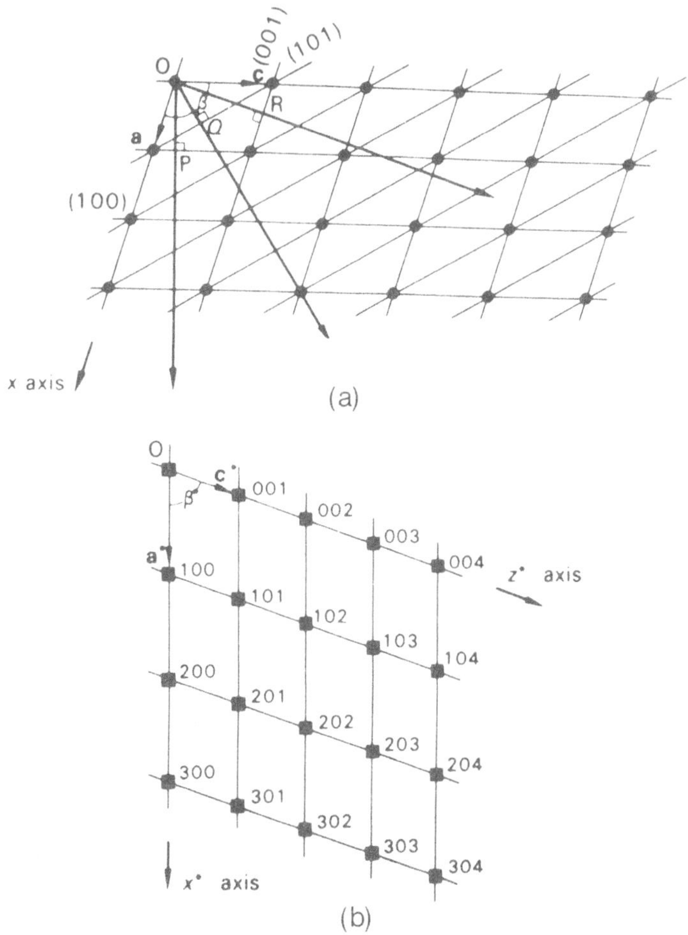

Fig. 4.19
Reciprocal lattice. (a) Monoclinic lattice in projection on to (010), showing P unit cells and the traces of the (100), (101) and (001) families of planes. (b) Monoclinic reciprocal lattice constructed from the lattice in (a), in projection on the a^*, c^* plane.

But

$$b^* = 1/b \tag{4.16}$$

because $d^*(010)$ is normal to the a, c plane. The unique β^* angle is given by

$$\beta^* = (180° - \beta) \tag{4.17}$$

Furthermore,

$$\mathbf{a} \cdot \mathbf{a}^* = aa^* \cos \angle \mathbf{aa}^* = a(1/a)\cos(\beta - 90°) = 1 \tag{4.18}$$

and similarly for $\mathbf{b} \cdot \mathbf{b}^*$ *and* $\mathbf{c} \cdot \mathbf{c}^*$. For the mixed products,

$$\mathbf{a} \cdot \mathbf{b} = ab^* \angle \cos \mathbf{ab}^* = 0 \tag{4.19}$$

and similarly for all other such products. The relationships in Eq. (4.18) and Eq. (4.19) apply to all crystal systems.

In looking next at the reciprocal lattice in a more general manner, the material in Appendix A3 will be useful. In Fig. 4.20, the z^* axis is normal to the plane a, b. Since $\mathbf{c} \cdot \mathbf{c}^* = 1 = cc^* \cos \angle COR$,

$$c^* = |\mathbf{c}^*| = 1/c \cos \angle COR \tag{4.20}$$

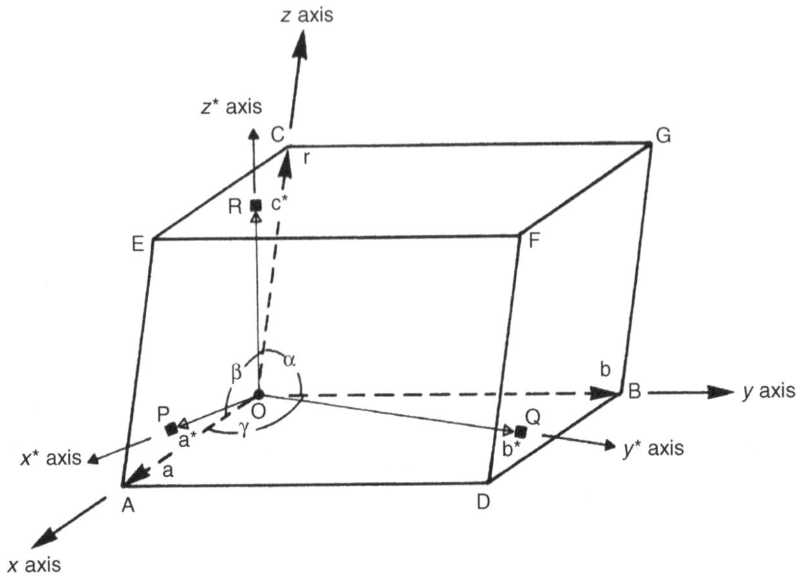

Fig. 4.20
Triclinic unit cell, with vectors **a**, **b** and **c**, and the corresponding reciprocal unit cell vectors $\mathbf{a}^*$, $\mathbf{b}^*$ and $\mathbf{c}^*$. [Reproduced by courtesy of Springer Science+Business Media, New York, © Kluwer Academic/Plenum Publishing.]

But $\mathbf{c}^*$ is normal to both **a** and **b**, so that it lies in the direction of their vector product. Hence,

$$\mathbf{c}^* = \eta(\mathbf{a} \times \mathbf{b}) \tag{4.21}$$

where η is a constant. From Appendix A3, $V = \mathbf{c}^* \cdot (\mathbf{a} \times \mathbf{b})$; forming next the scalar product of Eq. (4.21) and **c**,

$$\mathbf{c} \cdot \mathbf{c}^* = \eta\mathbf{c} \cdot (\mathbf{a} \times \mathbf{b}) = \eta V = 1$$

so that $\eta = \frac{1}{V}$; hence, from Eq. (4.21)

$$\mathbf{c}^* = \frac{(\mathbf{a} \times \mathbf{b})}{\mathbf{c} \cdot (\mathbf{a} \times \mathbf{b})} \tag{4.22}$$

and similarly for $\mathbf{a}^*$ and $\mathbf{b}^*$ by cyclic permutation. In scalar form,

$$c^* = |\ \mathbf{c}^*| = \frac{ab \sin \gamma}{V} \tag{4.23}$$

and similarly for a^* and b^* by cyclic permutation.

The angles of the reciprocal unit cell are obtained through Fig. 4.21. The z^* axis is perpendicular to the plane (001) and is, therefore, coincident with the pole of the great circle containing x and y. Similar arguments apply to x^* and y^*. Thus, the arc $B'C'$ is, at all points on it, 90° from the great circle of which AB is an arc; similarly with $B'C'$ and $C'A'$ *mutatis mutandis*. Hence, the triangle $A'B'C'$ is the polar triangle of triangle ABC.

Referring now to Section 2.6.2, it follows that

$$\gamma^* = 180° - \angle C$$
$$\gamma \ = 180° - \angle C'$$

with similar expressions for A, B', A and B'. Hence, from Eq. (2.33)

$$\cos \gamma^* = \frac{-\cos \gamma + \cos \alpha \cos \beta}{\sin \alpha \sin \beta} \tag{4.24}$$

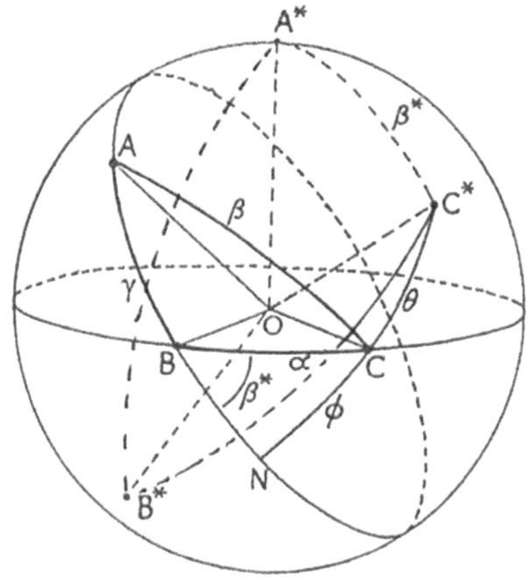

Fig. 4.21
Spherical triangle ABC and its polar triangle $A^*B^*C^*$. [Reproduced by courtesy of Springer Science+Business Media, New York, © Kluwer Academic/Plenum Publishing.]

since A, B and C are here identified with α, β and γ from Euler's construction (Section 3.6.3); the results for $\cos\alpha^*$ and $\cos\beta^*$ may be obtained by cyclic permutation. It remains to show that the reciprocal lattice points, as constructed here, form a true lattice.

From Eq. (A3.39), the vector normal to the plane $(h\,k\,l)$ is $h(\mathbf{b}\times\mathbf{c})+k(\mathbf{c}\times\mathbf{a})+l(\mathbf{a}\times\mathbf{b})$. Dividing by V, and denoting the resulting vector $\mathbf{d}^*(hkl)$,

$$\mathbf{d}^*(hkl)=\left\{h\frac{(\mathbf{b}\times\mathbf{c})}{\mathbf{a}\cdot(\mathbf{b}\times\mathbf{c})}+k\frac{(\mathbf{c}\times\mathbf{a})}{\mathbf{b}\cdot(\mathbf{c}\times\mathbf{a})}+l\frac{(\mathbf{a}\cdot\mathbf{b})}{\mathbf{c}\cdot(\mathbf{a}\cdot\mathbf{b})}\right\}=h\mathbf{a}^*+k\mathbf{b}^*+l\mathbf{c}^* \qquad (4.25)$$

as given in Eq. (A3.5). Since h, k and l are integers, the vectors $\mathbf{d}^*(hkl)$ drawn from the common origin form a lattice, the reciprocal lattice, with translation vectors $\mathbf{a}^*$, $\mathbf{b}^*$ and $\mathbf{c}^*$ and interaxial angles

$$\left.\begin{aligned}\alpha^*&=\angle\mathbf{b}^*\mathbf{c}^*=\angle 010-001\\ \beta^*&=\angle\mathbf{c}^*\mathbf{a}^*=\angle 010-001\\ \gamma^*&=\angle\mathbf{a}^*\mathbf{b}^*=\angle 100-010\end{aligned}\right\} \qquad (4.26)$$

It should be noted that reciprocal lattice points are denoted by the hkl values of the family of planes in the Bravais lattice from which they were derived, but are written without parentheses.

4.8 Rotational symmetry of lattices

Having now studied lattice geometry in some detail, it is relevant to show analytically that rotational symmetry in a periodic lattice is restricted to the degrees 1, 2, 3, 4 and 6, which was demonstrated graphically in Section 3.5.5.

In Fig. 4.22, A and B represent two adjacent points, of repeat distance t, in any row of a lattice, and an R-fold axis acts normal to the plane and through each lattice point. An R-fold anticlockwise rotation ϕ about the axis through A maps B on to B', and a similar but clockwise rotation about the axis through B maps A on to A'; it follows that $A'B'$ is parallel to AB. The lines $A'S$ and $B'T$ are drawn perpendicularly to the line AB. In a lattice, any two points in a row must be separated by an integral multiple of the repeat distance in

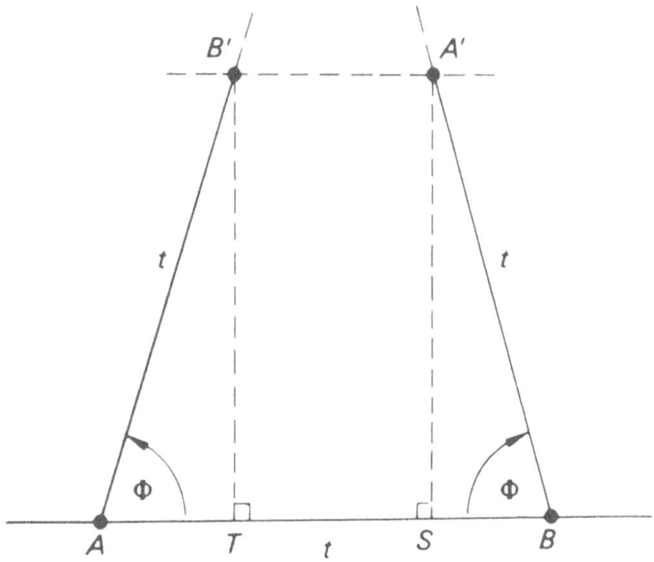

Fig. 4.22
Permitted rotational symmetries in lattices are $1-, 2-, 3-, 4-$ and $6-$ fold, corresponding to rotations of $0°$ (360°), $180°, 120°, 90°$ and $60°$.

the direction of that row. Thus, $A'B' = mt$, where m is an integer. Furthermore, $A'B' = t - (AT + BS) = t - 2t\cos\phi$; hence, $m = 1 - 2\cos\phi$, or

$$\cos\phi = (1 - m)/2 = M/2 \tag{4.27}$$

where M is another integer. Since $|\cos\phi| \leq 1$, the only values of M which are admissible are 2, –2, –1, 0 and 1, which correspond to rotations of $360°, 180°, 120°, 90°$ and $60°$, respectively. This analysis provides a deeper meaning that that given earlier. An alternative proof, which depends upon the fact that the trace of a rotation symmetry matrix lies between –3 and +1, is described in Appendix A3.7.

The 14 Bravais lattices are presented in their entirety by Fig. 4.23. Can you now write the relationships between a, b, c, and α, β, γ for each unit cell shown therein?

4.9 Lattice transformations

In practice, it is often necessary to transform the axes of the unit cell of a crystal that may have been first chosen by experiment. The parameters that may be involved, as well as the unit cell parameters a, b and c, are the Miller indices hkl, the zone symbols or directions UVW, the reciprocal unit cell parameters a^*, b^*, c^* and the coordinates x, y and z in the direct unit cell.

4.9.1 Bravais lattice unit cell vectors

Let $\mathbf{a}$, $\mathbf{b}$ and $\mathbf{c}$ be transformed to $\mathbf{a}'$, $\mathbf{b}'$ and $\mathbf{c}'$, such that

$$\begin{aligned}\mathbf{a}' &= s_{11}\mathbf{a} + s_{12}\mathbf{b} + s_{13}\mathbf{c}\\ \mathbf{b}' &= s_{21}\mathbf{a} + s_{22}\mathbf{b} + s_{23}\mathbf{c}\\ \mathbf{c}' &= s_{31}\mathbf{a} + s_{32}\mathbf{b} + s_{33}\mathbf{c}\end{aligned} \tag{4.28}$$

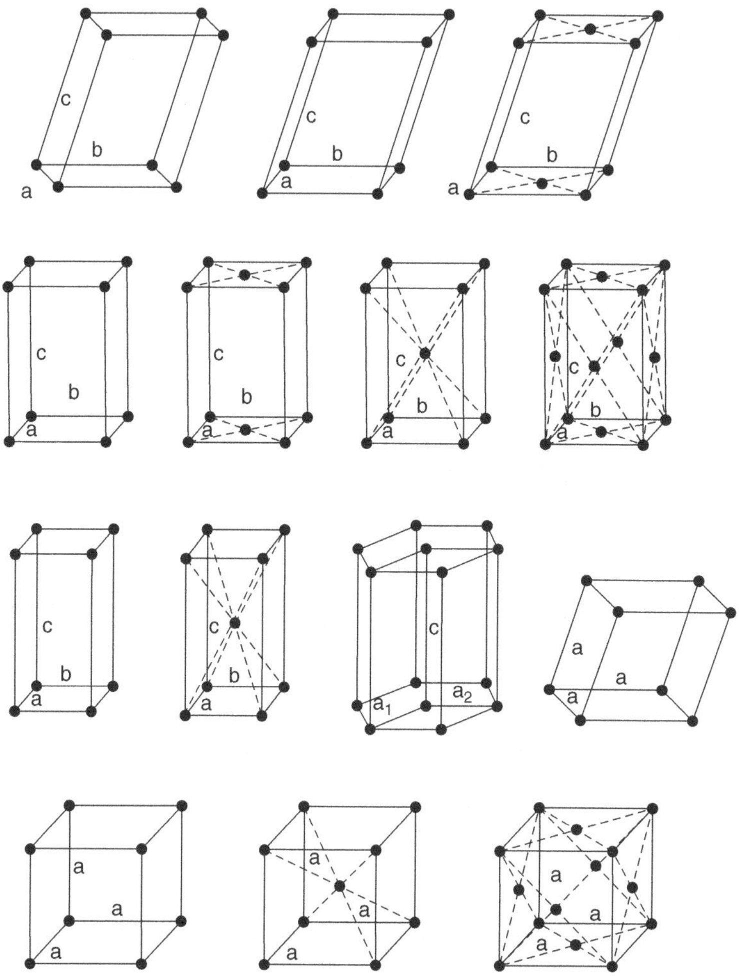

Fig. 4.23
The 14 Bravais lattices. In order, left–right, top–bottom: triclinic *P*, monoclinic *P*, monoclinic *C*, orthorhombic *P*, orthorhombic *C*, orthorhombic *I*, orthorhombic *F*, tetragonal *P*, tetragonal *I*, hexagonal *P* (three unit cells are shown), rhombohedral *R*, cubic *P*, cubic *I*, cubic *F*. [Reproduced by courtesy of Springer Science+Business Media, New York, © Kluwer Academic/Plenum Publishing.]

Following the matrix notation discussed in Appendix A3,

$$\begin{pmatrix} \mathbf{a}' \\ \mathbf{b}' \\ \mathbf{c}' \end{pmatrix} = \begin{pmatrix} s_{11} & s_{12} & s_{13} \\ s_{21} & s_{22} & s_{23} \\ s_{31} & s_{32} & s_{33} \end{pmatrix} \begin{pmatrix} \mathbf{a} \\ \mathbf{b} \\ \mathbf{c} \end{pmatrix} \tag{4.29}$$

or concisely

$$\mathbf{a}' = \mathbf{S}\,\mathbf{a} \tag{4.30}$$

where $\mathbf{a}'$ and $\mathbf{a}$ represent the two column vectors and $\mathbf{S}$ the matrix of elements s_{ij}. The inverse transformation may be written

$$\mathbf{a} = \mathbf{S}^{-1}\,\mathbf{a}' \tag{4.31}$$

where $\mathbf{S}^{-1}$ is the inverse matrix to $\mathbf{S}$. The elements of the inverse matrix, denoted s_{ij}^{-1}, obtained as

$$s_{ij}^{-1} = \frac{1}{\det(\mathbf{S})}(-1)^{i+j}|M_{ji}|$$

where $|M_{ji}|$ is the *minor* determinant of **S**, obtained by striking the row and column containing the ***ji*** element (Appendix A3.4.10). If the matrix is

$$\mathbf{S} = \begin{pmatrix} 2/3 & 1/3 & 1/3 \\ \bar{1}/3 & 1/3 & 1/3 \\ \bar{1}/3 & \bar{2}/3 & 1/3 \end{pmatrix}$$

the value of det(**S**) may be obtained as described in Appendix A3, or more quickly from the cross-multiplication rule:

whence

$$\det(\mathbf{S}) = \frac{2}{27} - \frac{1}{27} + \frac{2}{27} + \frac{1}{27} + \frac{4}{27} + \frac{1}{27} = \frac{1}{3}$$

The element s_{23}, for example, of the inverse matrix can be evaluated from **S** as:

$$s_{23}^{-1} = \frac{1}{1/3}(-1)^{2+3}|M_{32}| \;=\; \frac{(-1)^{2+3}}{1/3}\begin{vmatrix} 2/3 & 1/3 \\ \bar{1}/3 & 1/3 \end{vmatrix} = \frac{1}{1/3}(-1)(1/3) = -1$$

In this way, the inverse matrix can be built up:

$$\mathbf{S}^{-1} = \begin{pmatrix} 1 & \bar{1} & 0 \\ 0 & 1 & \bar{1} \\ 1 & 1 & 1 \end{pmatrix}. \tag{4.32}$$

4.9.2 Zone symbols and lattice directions

It will have been noticed that the normals to faces $(h\,k\,l)$ in the construction of the stereographic projection are also the directions of vectors to points hkl in the reciprocal lattice: they are both normals to lattice planes. The difference lies in the fact that in the stereographic projection the normals lack the linear measurement explicit in the reciprocal lattice.

A direction **r** in a direct unit cell may be written as

$$\mathbf{r} = U\mathbf{a} + V\mathbf{b} + W\mathbf{c} \tag{4.33}$$

and in a transformed cell as

$$\mathbf{r} = U'\mathbf{a}' + V'\mathbf{b}' + W'\mathbf{c}' \tag{4.34}$$

From Eq. (4.33) and Eq. (4.34)

$$(U'V'W')\begin{pmatrix}\mathbf{a}'\\ \mathbf{b}'\\ \mathbf{c}'\end{pmatrix} = (U\;V\;W)\begin{pmatrix}\mathbf{a}\\ \mathbf{b}\\ \mathbf{c}\end{pmatrix} = (U\;V\;W)\,\mathbf{S}^{-1}\begin{pmatrix}\mathbf{a}'\\ \mathbf{b}'\\ \mathbf{c}'\end{pmatrix}$$

or

$$(U'V'W') = (U\,V\,W)\mathbf{S}^{-1}$$

Hence, from Appendix A3.4.12, and concisely,

$$\mathbf{U}' = \mathbf{U}\mathbf{S}^{-1} = (\mathbf{S}^{-1})^{\mathrm{T}}\mathbf{U} \tag{4.35}$$

where $\mathbf{U}$ and $\mathbf{U}'$ are now column vectors.

Since $(\mathbf{S}^{-1})^{\mathrm{T}} = (\mathbf{S}^{\mathrm{T}})^{-1}$, pre-multiplication of both sides of Eq. (4.35) by $\mathbf{S}^{\mathrm{T}}$ leads to,

$$\mathbf{S}^{\mathrm{T}}\mathbf{U}' = \mathbf{S}^{\mathrm{T}}(\mathbf{S}^{\mathrm{T}})^{-1}\mathbf{U} = \mathbf{U}$$

or

$$\mathbf{U} = \mathbf{S}^{\mathrm{T}}\mathbf{U}' \tag{4.36}$$

4.9.3 Coordinates of points in the direct unit cell

For any fractional coordinate x, y, z in the unit cell

$$\mathbf{r} = x\mathbf{a} + y\mathbf{b} + z\mathbf{c}$$

From the transformation in Section 4.9.2, it is evident that these coordinates transform as do zone symbols. Thus,

$$\mathbf{x}' = (\mathbf{S}^{-1})^{\mathrm{T}}\mathbf{x} \tag{4.37}$$

and

$$\mathbf{x} = \mathbf{S}^{\mathrm{T}}\mathbf{x}' \tag{4.38}$$

4.9.4 Miller indices

From Eq. (4.25) and EQ. (4.33),

$$\mathbf{d}^*(hkl)\cdot\mathbf{r} = hU + kV + lW$$

and following Eq. (4.36)

$$\mathbf{d}^*(hkl)\cdot\mathbf{r} = (h\ \ k\ \ l)\begin{pmatrix}U\\V\\W\end{pmatrix} = (h\ \ k\ \ l)\,\mathbf{S}^{\mathrm{T}}\begin{pmatrix}U'\\V'\\W'\end{pmatrix}$$

But in terms of the transformed cell,

$$\mathbf{d}^*(hkl)\cdot\mathbf{r} = \left(h'\ \ k'\ \ l'\right)\begin{pmatrix}U'\\V'\\W'\end{pmatrix}$$

and because $\mathbf{d}^*(hkl)$ and $\mathbf{d}^*(h'k'l')$ are one and the same vector.

$$\left(h'\ \ k'\ \ l'\right) = (h\ \ k\ \ l)\,\mathbf{S}^{\mathrm{T}}$$

Transposing, as before

$$\begin{pmatrix}h'\\k'\\l'\end{pmatrix} = \mathbf{S}\begin{pmatrix}h\\k\\l\end{pmatrix}$$

or

$$\mathbf{h}' = \mathbf{S}\,\mathbf{h} \tag{4.39}$$

By multiplying by $\mathbf{S}^{-1}$,

$$\mathbf{h} = \mathbf{S}^{-1}\mathbf{h}' \tag{4.40}$$

where $\mathbf{h}$ and $\mathbf{h}'$ represent h, k, l and h', k', l'; thus, the Miller indices transform as do unit cell vectors. Miller–Bravais indices transform as Miller indices, using h, k and l from the h, k, i and l values.

4.9.5 Reciprocal unit cell vectors

Equation (4.25) may be written as

$$\mathbf{d}^*(hkl) = \left(\mathbf{a}^* \ \ \mathbf{b}^* \ \ \mathbf{c}^*\right)\mathbf{S}^{-1}\begin{pmatrix} h' \\ k' \\ l' \end{pmatrix}.$$

In the new reciprocal cell

$$\mathbf{d}^*(hkl) = \left(\mathbf{a}^{*\prime} \ \ \mathbf{b}^{*\prime} \ \ \mathbf{c}^{*\prime}\right)\begin{pmatrix} h' \\ k' \\ l' \end{pmatrix}$$

so that

$$\left(\mathbf{a}^{*\prime} \ \ \mathbf{b}^{*\prime} \ \ \mathbf{c}^{*\prime}\right) = \left(\mathbf{a}^* \ \ \mathbf{b}^* \ \ \mathbf{c}^*\right)\mathbf{S}^{-1}$$

By transposition, as earlier

$$\mathbf{a}^{*\prime} = (\mathbf{S}^{-1})^{\mathrm{T}}\mathbf{a}^* \tag{4.41}$$

and

$$\mathbf{a}^* = \mathbf{S}^{\mathrm{T}}\mathbf{a}^{*\prime} \tag{4.42}$$

so that reciprocal unit cell vectors transform as do zone symbols and coordinates in a unit cell of a Bravais lattice. These transformations are summarized in Table 4.5

Table 4.5 Relationships among transformation matrices.

Matrix S	Matrix $\mathbf{S}^{-1}$
Old unit cell translation vectors to new	New unit cell translation vectors to old
Old Miller indices to new	New Miller indices to old
Matrix $(\mathbf{S}^{-1})^{\mathrm{T}}$	**Matrix $\mathbf{S}^{\mathrm{T}}$**
Old reciprocal unit cell translation vectors to new	New reciprocal unit cell translation vectors to old
Old zone symbols to new	New zone symbols to old
Old Bravais unit cell coordinates to new	New Bravais unit cell coordinates to old

4.9.6 Volume relationships

It has been shown earlier that $V = \mathbf{a}\cdot(\mathbf{b}\times\mathbf{c})$. Thus, it follows from Eq. (4.30) that the volume V' of the transformed cell is given by

$$V' = \det(\mathbf{S})V \tag{4.43}$$

Thus, for the matrix

$$\mathbf{S} = \begin{pmatrix} 2/3 & 1/3 & 1/3 \\ \bar{1}/3 & 1/3 & 1/3 \\ \bar{1}/3 & \bar{2}/3 & 1/3 \end{pmatrix}$$

which applies to the transformation $\mathbf{a}\,(\text{rhombohedral}, R_{\text{obv}}) \rightarrow \mathbf{a}\,(\text{hexagonal}, H_{\text{triply primitive}})$, $\det(\mathbf{S}) = 1/3$, so that $V_R = 1/3 V_H$, which is the expected result for the relationship between the triply primitive hexagonal and rhombohedral unit cells.

4.9.7 Reciprocity of *F* and *I* unit cells

In Fig. 4.24, a primitive unit cell has been selected from the face centered unit cell by the transformation

$$\begin{aligned} \mathbf{a}_P &= \frac{1}{2}\mathbf{b}_F + \frac{1}{2}\mathbf{c}_F \\ \mathbf{b}_P &= \frac{1}{2}\mathbf{c}_F + \frac{1}{2}\mathbf{a}_F \\ \mathbf{c}_P &= \frac{1}{2}\mathbf{a}_F + \frac{1}{2}\mathbf{b}_F \end{aligned} \tag{4.44}$$

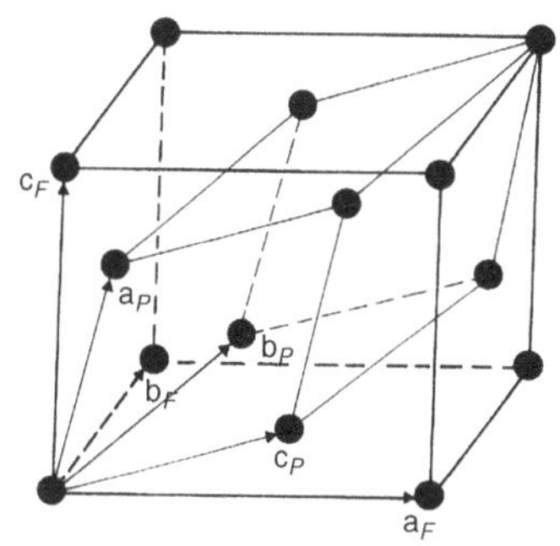

Fig. 4.24
Cubic *F* unit cell with an inscribed rhombohedron.

From Section 4.7, with appropriate cyclic permutation

$$\begin{aligned} \mathbf{a}_P^* = (\mathbf{b}_P \times \mathbf{c}_P)/V_P &= \left[\left(\frac{1}{2}\mathbf{c}_F + \frac{1}{2}\mathbf{a}_F\right) \times \left(\frac{1}{2}\mathbf{a}_F + \frac{1}{2}\mathbf{b}_F\right)\right]/V_P \\ &= \frac{1}{4}[(\mathbf{c}_F \times \mathbf{b}_F) + (\mathbf{c}_F \times \mathbf{a}_F) + (\mathbf{a}_F \times \mathbf{b}_F)]/V_F \\ &= \frac{c_F b_F}{V_F}\sin\alpha + \frac{c_F a_F}{V_F}\sin\beta + \frac{a_F b_F}{V_F}\sin\gamma \end{aligned}$$

since $V_F = 4V_P$. Hence,

$$\mathbf{a}_P^* = -\mathbf{a}_F^* + \mathbf{b}_F^* + \mathbf{c}_F^* \tag{4.45}$$

with similar expressions for $\mathbf{b}_P^*$ and $\mathbf{c}_P^*$. The negative sign in front of $\mathbf{a}_F^*$ in Eq. (4.45) is needed in order to preserve right-handed axes from the product $(\mathbf{c}_F \times \mathbf{b}_F)$.

In the case of the body-centered unit cell, the equations similar to Eq. (4.44) are

$$\mathbf{a}_P = -\frac{1}{2}\mathbf{a}_I + \frac{1}{2}\mathbf{b}_I + \frac{1}{2}\mathbf{c}_I \tag{4.46}$$

with similar expressions for $\mathbf{b}_P$ and $\mathbf{c}_P$. By writing Eq. (4.45) as

$$\mathbf{a}_P^* = -2\mathbf{a}_F^*/2 + 2\mathbf{b}_F^*/2 + 2\mathbf{c}_F^*/2 \tag{4.47}$$

it follows that an *F* unit cell in a Bravais lattice reciprocates into an *I* unit cell in the corresponding reciprocal lattice, where the *I* unit cell is defined by the vectors $2\mathbf{a}_F^*$, $2\mathbf{b}_F^*$ and $2\mathbf{c}_F^*$. If, as is customary in practice, the reciprocal of an I unit cell is defined by vectors $\mathbf{a}_F^*$, $\mathbf{b}_F^*$ and $\mathbf{c}_F^*$, then only those reciprocal lattice points for which each of $h+k, k+l$ (and $l+h$) are even integers belong to the reciprocal of the I unit cell. In other words, Bragg reflections from an F unit cell have indices of the same parity (see also Section 6.6.3).

Example 4.3 A tetragonal C unit cell with $a = 4.774$ and $c = 8.361$ is transformed to a tetragonal P. Calculate the new unit cell dimensions and the position of a point $(0.411, -0.607, -0.193)$ in the new, C unit cell.

The transformation matrix is $\begin{pmatrix} {}^1\!/_2 & -{}^1\!/_2 & 0 \\ {}^1\!/_2 & {}^1\!/_2 & 0 \\ 0 & 0 & 1 \end{pmatrix}$ and the inverse matrix is $\begin{pmatrix} 1 & 1 & 0 \\ \bar{1} & 1 & 0 \\ 0 & 0 & 1 \end{pmatrix}$.

Thus, the P unit cell dimensions are $a' = b' = 3.376, c' = 8.361$ Å, and the coordinates are $1.018\ (\equiv 0.018), -0.196, -0.193$. (The transpose of the inverse is required here.) The results may be checked by a scaled drawing.

4.9.8 Wigner-Seitz cells

Before leaving the topic of lattices and unit cells, it may be noted that any lattice can be represented by a true primitive unit cell, such cells being capable of being stacked by three-dimensional translations so as to generate the lattice from which they were derived. Conventional unit cells are chosen so that the symmetries of their lattices are always self-evident. However, it may be desirable in some solid state studies to work with a primitive unit cell, whatever the corresponding Bravais unit cell might be. Such a cell is the Wigner–Seitz, or Voronoi, cell, the construction of which from a body centred cubic unit cell is shown to be relatively straightforward, as follows.

In a cubic I unit cell, lines are drawn from a lattice point to its nearest neighbour lattice points. Planes are then constructed so as to bisect these lines perpendicularly and the planes then extended as necessary to form the closed polyhedron of smallest volume. In the case of the cubic I cell, the polyhedron is a truncated octahedron (Fig. 4.25a); the directions **a**, **b** and **c** are clearly the normals to {100}, whereas the normals to the hexagonal shaped faces are the directions <111>. For a cubic F cell, the Wigner–Seitz cell is a rhombic dodecahedron. These results are not surprising because the coordination numbers in body- and face-centred structures of identical species are eight and twelve respectively. Fig. 4.25b shows space filled by stacking equal Wigner–Seitz unit cells; each cell displays the full rotation symmetry of its lattice.

A similar type of construction in reciprocal space, or $\boldsymbol{k}$ space, produces the first Brillouin zone, of which there are twenty-four such zones. In some systems more than one Wigner–Seitz cell exist for each Bravais lattice because the shape of the Wigner–Seitz cell can depend also on the axial ratios; thus, in the tetragonal system, for example, two Wigner–Seitz cells are derived from the body centred lattice according as c/a is greater than or less than unity [11].

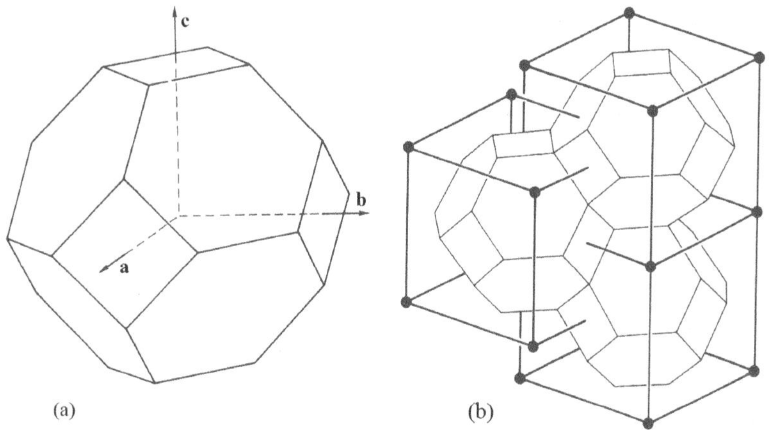

Fig. 4.25
(a) Wigner–Seitz (primitive) unit cell from a cubic lattice represented by an *I* unit cell; in reciprocal space, it corresponds to the first Brillouin zone. (b) Stacking of Wigner–Seitz cells to fill space. [Burns G and Glazer AM. *Space groups for solid state scientists*. 1978; reproduced by courtesy of Elsevier.]

The first Brillouin zone derived from an *I* unit cell is the Wigner–Seitz cell of an *F* unit cell lattice, illustrating again the reciprocity property of *I* and *F* unit cells. Brillouin zones are important in chemistry in the movement from bonds to bands in the quantum mechanical theory of metals [12], and in the physics of lattice dynamics [13]. The working advantage of the Wigner–Seitz cell lies in dealing with the smallest possible number of atoms while maintaining the full symmetry of the structure.

References 4

[1] Smith WF. *The Bravais lattices song*. 2002. [Tune: 'I am the Very Model of a Modern Major General', Gilbert WS and Sullivan A. *Pirates of Penzance*. 1880].
[2] Frankenheim ML. *Nova Acta Acad. Nat. Curr.* 1842; 9: 47.
[3] Bravais A. *Mem. Acad. Roy. Sci. France* 1846; 9: 255.
[4] Bravais A. *J. École Politech.* 1850; 19: 1, 127; 20: 102, 197.
[5] Ladd MFC. *J. Chem. Educ.* 1997; 74: 461.
[6] Donnay JHD and Harker D. *Amer. Mineral.* 1937; 22: 463.
[7] Berkovutch-Yellin Z. *J. Am. Chem. Soc.*, 1985; 107: 8239.
[8] Docherty R, et al. *Phys. D: Appl. Phys.* 1991; 24: 89.
[9] Gibbs JW. *Collected works*. Longmans: 1928.
[10] Sunagawa I. *Crystals: growth, morphology and perfection*. Cambridge University Press: 2005.
[11] Burns G and Glazer AM. *Space groups for solid state scientists*. Elsevier: 2013.
[12] Altmann SL. *Band theory of metals*. Oxford Science Publications: 1970.
[13] Dove MT. *Introduction to lattice dynamics*. Cambridge University Press: 2005.

Problems 4

4.1 Consider two nets: (i) $a = b, \gamma = 90°$, and (ii) $a = b, \gamma = 120°$, (a) What is the plane point group symmetry at each lattice point? (b) To what two-dimensional system does each belong? (c) What is the result of centring the cell in each case? Give transformation matrices as appropriate.

4.2 To what crystal system does a unit cell belong if $a = 0.7, b = 0.6, c = 0.5\ nm, \alpha = 90°, \beta = 120°, \gamma = 90°$? Calculate the length of [123].

4.3 Define Bravais lattice. Which of the following unit cells represent a lattice: (i) Orthorhombic B, (ii) Tetragonal A, (iii) Triclinic I, (d) Cubic $B + C$?

4.4 What are the transformation equations for each of the following?
(a) Monoclinic $I \rightarrow$ Monoclinic C
(b) Rhombohedral $F \rightarrow$ Rhombohedral R
(c) Tetragonal $C \rightarrow$ Tetragonal P

4.5 Do the relationships $a \not\subset b \not\subset c, \alpha \not\subset \beta \not\subset 90°, \gamma = 90°$ represent a diclinic system? If so, how so, and if not, why not.

4.6 Outline an R cell within an F cubic unit cell such that their [111] directions coincide. What is the ratio $\frac{V_F}{V_R}$?

4.7 A rhombohedral I unit cell has the dimensions $\alpha = 7.000$ Å and $\alpha = 50.00°$. Show that the cell obtained by the transformation $\mathbf{a}_R = \mathbf{S}\ \mathbf{a}_I$ is a rhombohedron and calculate its unit cell dimensions; the transformation matrix $\mathbf{S}$ is given by

$$\begin{pmatrix} -1/2 & 1/2 & 1/2 \\ 1/2 & -1/2 & 1/2 \\ 1/2 & 1/2 & -1/2 \end{pmatrix}$$

4.8 An orthorhombic unit cell I is transformed to another unit cell II by the equation $\mathbf{a}_{II} = \mathbf{S}\mathbf{a}_I$, where $\mathbf{S}$ is the matrix

$$\begin{pmatrix} 1 & \bar{1} & 0 \\ 1 & 1 & 0 \\ 0 & 0 & 1 \end{pmatrix}$$

What is the volume of cell II in terms of the volume of cell I? Hence, or otherwise, determine the coordinates of the point (0.123, −0.671, 0.314) when transformed from cell I to cell II.

4.9 What are the relative reticular densities of the planes (100), (220), $(1\bar{3}0)$ and (042) in a cubic I unit cell? Which planes would be expected to form the external faces of a crystal that crystallizes with an I unit cell?

4.10 The monoclinic unit cell of gypsum has been determined in different ways:

	a/Å	b/Å	c/Å	β/°
(I)	10.51	15.15	6.545	151.72
(II)	5.669	15.15	6.545	118.58
(III)	10.51	15.15	6.285	99.30

(a) Outline the three unit cells on a common origin, projected on to the plane (010); a suitable scale is 10 mm = 1 Å.
(b) If cell I is P: (i) What are types II and III? (ii) What are the ratios $\frac{V_{III}}{V_{II}}$ and $\frac{V_{III}}{V_I}$? (iii) Which unit cell would you choose? Give reasons.
(c) Derive the matrices for the transformations from cell I to cell II, cell I to cell III and cell II to cell III.
(d) What are the dimensions of the reciprocal unit cell of the cell that you have chosen?
(e) Carry out the following transformations. (i) $(132)_I \rightarrow (hkl)_{III}$. (ii) $[21\bar{3}]_{II} \rightarrow [UVW]_{III}$. (iii) $(0.600,\ 0.500,\ -0.300)_{III} \rightarrow (x, y, z)_I$

4.11 Most elemental metals crystallize in a close packed cubic, close packed hexagonal or body centred cubic structure. Determine the packing fraction, or space occupied, by equal, spherical atoms in each structures.

4.12 A unit cell of a lattice has a single $\bar{4}$ axis along c. Show that the conditions $a = b \not\subset c$, and $\alpha = \beta = \gamma = 90°$ must obtain.

Space groups

5

SYNOPSIS

- Space groups in one, two and three dimensions
- Limiting conditions by geometry and analytically
- The 17 plane groups
- Space groups in the seven crystal systems
- Translational symmetry
- Derivation of three-dimensional space groups
- Space groups and crystal structures
- Matrix representation of space group symmetry operations
- Black-white and colour symmetry
- The International Tables

5.1 Introduction

The components of a space group pattern are now assembled and the next logical step is to bring them together, so we arrive in this chapter at a study space groups, the ideally infinite, spatial patterns on which crystal structures are based. In Chapter 3, the symmetry of any finite body was described by a point group, and the previous chapter lattices were examined and shown to provide mechanisms for the repetition of a motif by translations parallel to one direction, or two and three non-collinear directions. The periodic structures obtained in this manner generate a repeating pattern which describes the arrangement of atoms and molecules in a crystal; the symmetries of these patterns are described by *space groups*.

In one, two or three dimensions, a space group may be defined as *an infinite set of symmetry elements, the operation of which leaves the pattern to which it refers indistinguishable from its condition before the operation*. Thus, any symmetry operation of a space group maps the space group pattern on to itself.

Space group operations do not have to leave a point unmoved, and it is possible to apply the infinite properties of space groups to crystals of finite extent because the number of repeat units in a crystal of experimental size is very large in relation to the size of the repeat unit itself, as discussed in Section 1.3.

5.2 One-dimensional space groups

A combination of the single one-dimensional lattice and the one-dimensional point groups 1 and m, leads to the two one-dimensional space groups $p1$ and pm. In the crystal structure of magnesium fluoride, MgF_2, the magnesium ions have the fractional coordinates 0, 0, 0 and $^1/_2$, $^1/_2$, $^1/_2$ in the unit cell, and the fluoride ions occupy the sites $\pm(x,\ x,\ 0;\ ^1/_2 - x,\ ^1/_2 + x,\ ^1/_2)$. In the projection of this structure on to the x axis, the cations lie at 0 and $^1/_2$, and the anions at $\pm x$ and $\pm(^1/_2 - x)$. One repeat length along the x axis, from 0 to a, of the electron density projection for magnesium fluoride is illustrated in Fig. 5.1. The space group for this projection may be described as pm, with a projected unit cell length of $a/2$. In general, one-dimensional space groups will not figure largely in this book.

5.3 Two-dimensional space groups

The basic elements of a crystal structure comprise a pattern motif, which itself may or may not exhibit symmetry, and a mechanism for its periodic repetition. The example of a simple wallpaper shown in Fig. 5.2a comprises rectangular unit cells and contents. The unit cell for this pattern is one such rectangle with translation vectors **a** and **b** lying in the directions of the x and y axes, respectively. The wallpaper pattern can be generated from the unit cell in Fig. 5.2b by repeating it with translations $m\mathbf{a}$ and $n\mathbf{b}$ ($m, n = 1, 2, 3, \ldots$). While this procedure suffices to build up an infinite wallpaper pattern, it ignores the symmetry between the two identical flower motifs in the unit cell. If flower 1 is reflected across the dashed line, transiently to position 1′, and then translated parallel to that line by an amount $a/2$, it then occupies the position of flower 2.

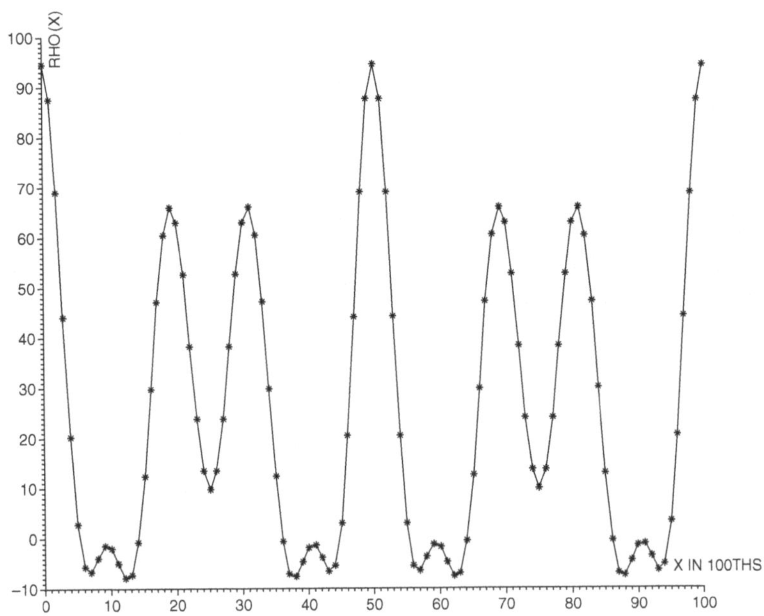

Fig. 5.1
One-dimensional electron density projection $\rho(x)$ for magnesium fluoride as a function of x for the one-dimensional repeat period a, in 100^{ths} of the repeat distance. On account of the m symmetry, the repeat distance in projection is $a/2$.

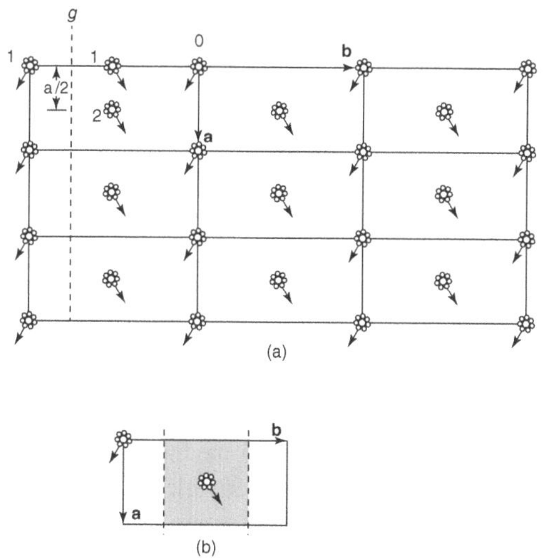

Fig. 5.2
A wallpaper pattern. (a) An extended array; **a** and **b** are translational (repeat) vectors. A glide line g has been inserted into the first few cells. (b) A conventional unit cell for the pattern: the asymmetric unit is either the shaded or other two areas of the cell, and a second interleaving glide line exits in each unit cell. [Reproduced by courtesy of Springer Science + Business Media, New York, © Kluwer Academic/Plenum Publishing.]

Thus, the infinite pattern, of which Fig. 5.2a is a sample, is mapped on to itself by this symmetry operation. In two dimensions, this symmetry operation is a *glide line*, with symbol g and graphic symbol - - - -.

The necessary and sufficient unit of pattern is the area encompassing a single flower, either the shaded or the other two areas of Fig. 5.2b, and may be termed the *asymmetric unit of structure*. Strictly, an asymmetric unit is a *region of space* which when operated on by the space group symmetry generates the lattice of which it is a part. The term used here is a useful extension of that definition to include the true asymmetric unit and its contents:

Asymmetric unit of structure + Space group symmetry → Infinite pattern

Figure 5.3 shows another pattern with the same symmetry as the wallpaper. Can you identify the unit cell, asymmetric unit and symmetry elements?

Fig. 5.3
A pattern in plane group *pg*. [Reproduced by courtesy of the IUCr (E).]

The above discussion leads to a general study of the two-dimensional space groups, also termed *plane groups* or, colloquially, *wallpaper groups*. They can be discussed conveniently in terms of the two-dimensional systems listed in Table 3.1.

5.3.1 Plane groups in the oblique system

The plane groups in the oblique system are $p1$ and $p2$, and the latter will be considered next. Figure 5.4a shows a motif of twofold symmetry and Fig. 5.4b a number of unit cells of a lattice in the oblique system. If the motif is set down in a given orientation at each lattice point, with the symmetry points of the motif and the lattice in coincidence, the pattern of Fig. 5.4c is obtained.

In general, only one unit cell and its immediate environs need be drawn in order to identify the pattern and its symmetry elements, and Fig. 5.5 illustrates such a drawing for $p2$. By convention the origin of the unit cell is the top left-hand corner lattice point on the drawing, with the x (and a) direction running from top to bottom and y (and b) from left to right. As with lattices, the unit cell is a representative portion of an infinite structural array, be it points, atoms or molecules.

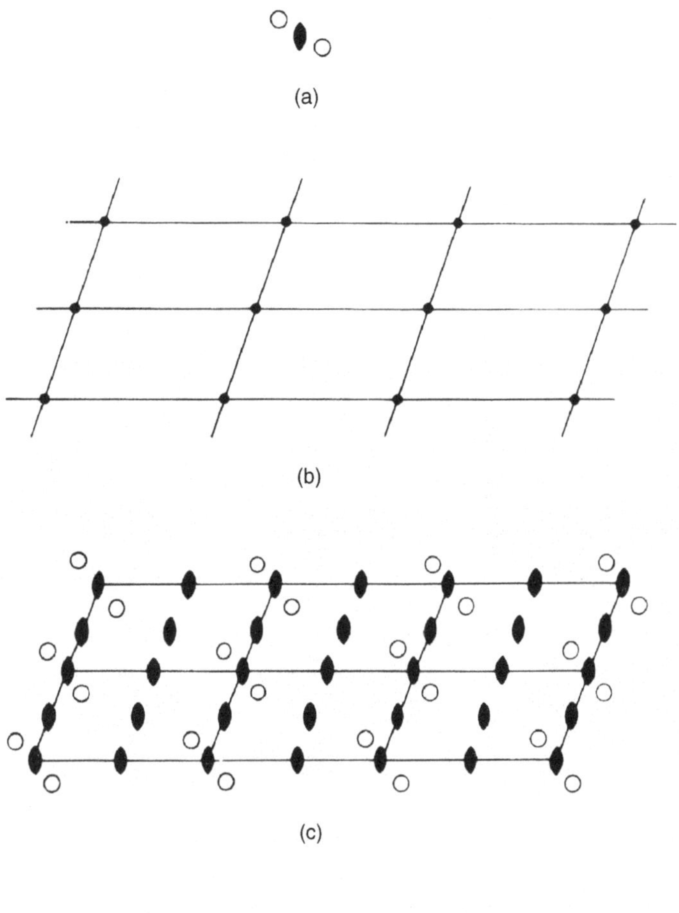

Fig. 5.4
Plane group $p2$. (a) Motif of twofold symmetry. (b) Oblique net, with p unit cells outlined; the symmetry at each point is plane point group 2. (c) Combination of (a) and (b) to form the extended pattern of plane group $p2$. [Reproduced by courtesy of Springer Science + Business Media, New York, © Kluwer Academic/Plenum Publishing.]

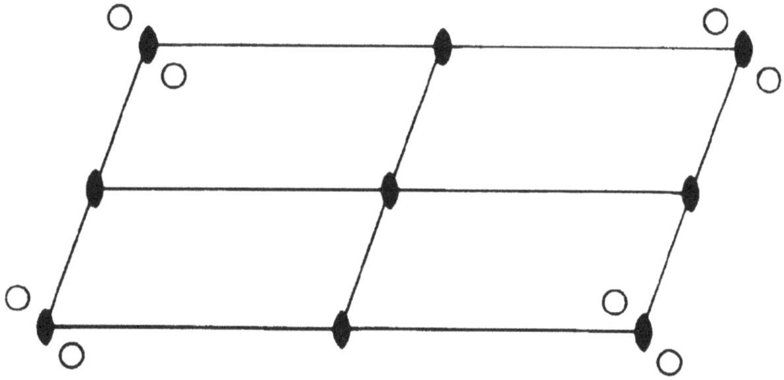

Origin at 2

2	e	1	$x, y; \bar{x}, \bar{y}$.
1	d	2	$\frac{1}{2}, \frac{1}{2}$.
1	c	2	$\frac{1}{2}, 0$.
1	b	2	$0, \frac{1}{2}$.
1	a	2	$0, 0$.

Fig 5.5
The unit cell and environs of plane group $p2$, origin on 2. The general equivalent positions are listed in set (e) and the special equivalent positions are sets (a)–(d). [Reproduced by courtesy of Springer Science + Business Media, New York, © Kluwer Academic/Plenum Publishing.]

The asymmetric unit is a region of space represented conventionally by the symbol ○, one-half of the unit cell in this example, and in a crystal structure will be populated by an atom, a molecule or part of a molecule, which may or may not itself be symmetrical. In a drawing representing a space group, the symbol ○ may be located anywhere within the appropriate fraction of the unit cell, but conveniently, close to the origin, and then repeated by the symmetry $p2$, so as to build up the complete pattern, ensuring that the areas around the four corners of the cell are completed by means of the translations **a**, **b** or $\mathbf{a} \pm \mathbf{b}$. Then, additional twofold rotation points can be identified and added to the drawing. After a little practice, such symmetry elements are identified readily by inspection. However, at first, they may be determined by taking a point x, y and considering how every other point on the diagram may be reached from it by a single symmetry operation of the space group, together with translations of amounts $\pm a$ and/or $\pm b$, as necessary.

The lists of fractional coordinates in Fig. 5.5 are symmetry related sites unique to a single unit cell. The number of such sites that are generated by the plane group symmetry comprises the *general equivalent positions*; they are sites of symmetry 1 in any space group. In $p2$ they are the positions x, y and $\bar{x}$, $\bar{y}$; the negative sign of a single coordinate is usually written above it. The points $1 - x$, $1 - y$ could have used been instead of $\bar{x}$, $\bar{y}$, but it is more usual, and more convenient, to work with a set of coordinates around the origin.

Each line of coordinates lists the number of positions in the set, the Wyckoff classificatory notation for the set in the given plane group, the symmetry at each site in the set, and the fractional coordinates of the sites in the set. If the number of entities in the unit cell is found, by experiment, to be less than the number of general equivalent positions, the entities occupy *special equivalent positions*: they are sites of symmetry greater than1, and the entities must possess the symmetry of that site, or a symmetry of which the site symmetry is a subgroup. The number of sites in a special set is always an integral submultiple of the number in the general set. Exceptions to space group rules occur in disordered structures, in which a given set of equivalent positions, general or special, may be occupied in a random manner by fewer entities that rules indicate: statistically, over a very large number of unit cells a set of equivalent positions can, to the experimental symmetry probe, appear occupied (see also Section 1.3 and Fig. 8.5).

In $p2$ there are four sets of special equivalent positions, (a)–(d), each of point symmetry 2, that are unique to the unit cell; this fact can be confirmed by applying the shifted unit cell action shown on cell I in Fig. 4.3.

5.3.2 Plane groups in the rectangular system

The rectangular system is consistent with plane point groups m and $2mm$, and lattices represented by the descriptors p and c; m symmetry cannot be accommodated in the oblique system.

Consider first the plane groups pm and cm; the m line is chosen normal to the x axis; the full symbol is $p1m1$. The groups (Fig. 5.6) may be constructed in a manner similar to that used for $p2$. The origin is chosen on m, but is not fully defined until the first structural entity is introduced into the unit cell

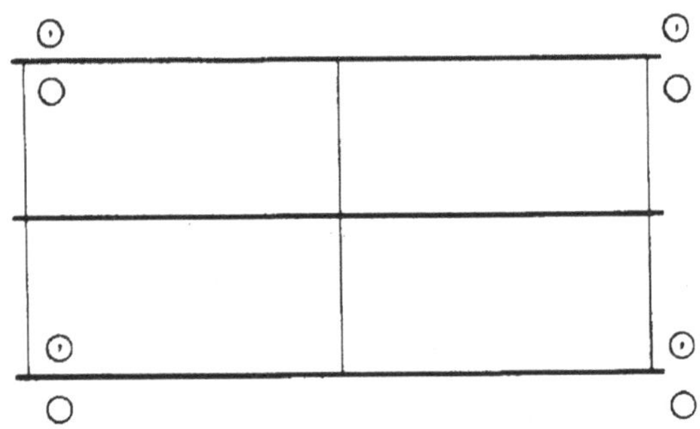

Origin on m

2	c	1	$x, y; \bar{x}, y.$
1	b	m	$\frac{1}{2}, y.$
1	a	m	$0, y.$

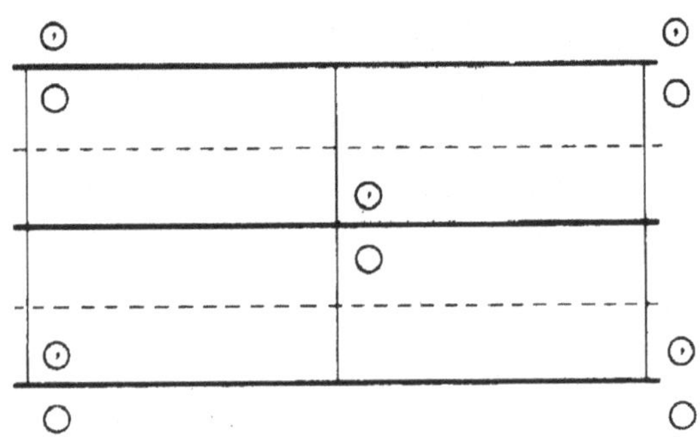

Origin on m

				Limiting conditions
			$(0, 0; \frac{1}{2}, \frac{1}{2})+$	
4	b	1	$x, y; \bar{x}, y.$	$hk: h + k = 2n$
2	a	m	$0, y.$	As above.

Fig. 5.6
Plane groups based on point group m. (a) pm. (b) cm. The lines of coordinates show, in order, the site symmetry, the Wyckoff notation for the group, the symmetry of the sites of the following coordinate set and the limiting conditions. [Reproduced by courtesy of Springer Science + Business Media, New York, © Kluwer Academic/Plenum Publishing.]

and allotted an arbitrary y coordinate. In pm there are two general equivalent positions and two sets of special equivalent sites on the m lines.

Plane group cm introduces certain new features. The coordinate list is headed by the expression $(0,0;\,{}^1\!/\!_2,{}^1\!/\!_2)+$, which means that these translations are added to the coordinates of the positions thereunder. Thus, the full listing of set (c) in cm is $x,\ y;\ \bar{x},\ y;\ {}^1\!/\!_2+x,\ {}^1\!/\!_2+y,\ {}^1\!/\!_2-x,\ {}^1\!/\!_2+y$. Given x, for example, the positions ${}^1\!/\!_2+x$ and ${}^1\!/\!_2-x$ are obtained by moving $a/2$ from the origin in the direction of the $+x$ axis followed by movements of $+x$ and $-x$ respectively, along the same direction.

Centring the unit cell in conjunction with m symmetry introduces glide lines, of the type discussed in Section 5.3, at $x=\pm{}^1\!/\!_4$. The glide operation symbolically is $x,\ y \xrightarrow{g\perp x} {}^1\!/\!_2-x,\ {}^1\!/\!_2+y$. Glide lines arise in two dimensions wherever a centred type of arrangement is coupled with m lines.

Not all plane groups can be obtained from the combination of a point group motif with a lattice. Those that can be, like $p2$, pm and cm have been termed 'point plane groups'. One may now ask what is meant by the symbols pg and cg? The first of these is one of the seventeen plane groups, and is represented by the pattern displayed already in Fig. 5.2a and Fig. 5.3. Plane group cg is an alternative, non-standard name for cm and does not constitute another plane group.

There is only one set of special equivalent positions in cm, in contradistinction to the two sets in pm, because the centring in cm requires that a set of special positions in cm uses both m lines $[0,\ y]$[1] and $[{}^1\!/\!_2, y]$ shown in the unit cell. If two sets are written in cm, by analogy with pm, by specializing the coordinates to the position on the symmetry elements, $0,\ y$, and ${}^1\!/\!_2, y$, there obtains

$$0, y;\ \ {}^1\!/\!_2, {}^1\!/\!_2+y \quad \text{and} \quad {}^1\!/\!_2, y;\ \ 0, {}^1\!/\!_2+y$$

However, these two sets differ only by a change of origin to the point ${}^1\!/\!_2, 0$, and so do not represent different sets of special equivalent positions.

[1] The notation $[p,\ q]$ represents the line $x=p,\ y=q$; $[p,\ q,\ r]$ is used in three dimensions.

5.3.3 Limiting conditions on X-ray reflections

The study of space groups is important in order to gain information about the repeat patterns in crystal structures from the patterns of the hkl indices of an X-ray diffraction record from a crystal, a subject that will be discussed shortly. If a space group contains translational symmetry, certain sets of hkl values will be systematically absent from the diffraction record. This situation exists for plane group cm. While the subject of limiting conditions will be treated in some detail in the next chapter, a short, geometrical explanation is useful at this stage, so that the space group information can be appreciated more fully. Among the plane groups studied so far, *halving* is encountered first in cm; that is, there exist symmetry relationships such as $x \underset{-c}{\overset{+c}{\longleftrightarrow}} {}^1\!/\!_2+x$.

Consider the right-hand diagram in Fig. 4.4b, with **a** and **b** playing the same roles as in the left-hand diagram. The centred cell **a**, **b** is related to the primitive cell **a**′, **b**′ by the transformation

$$\mathbf{a} = \mathbf{a}' + \mathbf{b}'$$
$$\mathbf{a} = -\mathbf{a}' + \mathbf{b}'$$

Since Miller indices transform as do unit cell vectors,

$$h = h' + k'$$
$$k = -h' + k'$$

whence $h + k = 2k'$, which is even for all values of k'. Hence, in a centred group such as *cm*, diffraction data will arise only where the sum of h and k is an even integer. Thus, the principle of halving arises: in two dimensions it may involve general indices *hk*, or special indices *h*0 and *k*0 for halving along lines, such as in Fig. 5.1, where $h0 = 2n$: the diffraction data used in producing this diagram were *h*0, $h = 2, 4, \ldots, 16$. In three dimensions, as well as general halving of *hkl*, special halving can occur where one or two indices are zero. In *cm* both general and special sites are so involved because the halving is general.

The rectangular system includes also plane groups based on point group 2*mm*; nominally, there would appear to be eight plane groups:

$$\begin{array}{llll} p2mm^* & p2mg^* & p2gm & p2gg^* \\ c2mm^* & c2mg & c2gm & c2gg \end{array}$$

However, only those marked * are distinct: *p*2*mg* and *p*2*gm* are equivalent under interchange of the *x* and *y* axes, and all four centred groups have interleaving *m* and *g* planes; plane group *p*2*gg* will be studied.

When considering any space group, it is helpful first to recall the corresponding point group. It is determined readily by removing the unit cell symbol and replacing all translation symmetry elements by the corresponding non-translation elements. Thus, for *p*2*gg*:

$$p2gg \rightarrow \not{p}2gg \rightarrow 2\not{g}\not{g}mm \rightarrow 2mm$$

which is the plane point group for all eight plane group symbols listed above. Then, immediately from Table 3.1 the orientations of the symmetry elements in the point group symbol are known, as this information is carried over from point groups into the realm of space groups.

In 2*mm*, the two *m* lines intersect in the twofold rotation point, from the definition of point group. When glide symmetry is present, it must not be assumed that the glide lines will also intersect the rotation point, and generally they do not.

A correct description of *p*2*gg* may be reached in more than one way. As the two glide lines intersect, one normal to the *x* axis and the other normal to the *y* axis, the unit cell can be constructed first, as shown in Fig. 5.7a. Then, by inserting a point *x*, *y* and repeating it by the symmetry elements, Fig. 5.7b is obtained. From Section 3.6.3, the interaction of two symmetry elements leads to a third intersecting symmetry element, and the twofold rotation points so produced are shown in Fig. 5.7c. Standard practice places the origin on 2, and the revised, standard diagram is illustrated in Fig. 5.7d together with its necessary description.

There are two sets of special positions on symmetry 2, and the coordinates must be paired correctly in order to conform to the symmetry *p*2*gg*. The

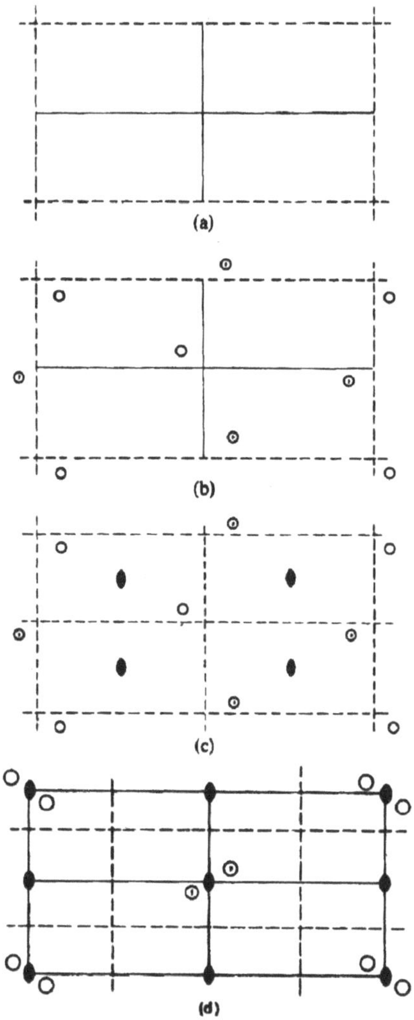

Fig. 5.7
Plane group *p*2*gg* (a) Unit cell framework; origin on *gg*. (b) General equivalent positions added. (c) Additional symmetry elements added. (d) Standard diagram, origin on 2, with the space group description. [Reproduced by courtesy of Springer Science + Business Media, New York, © Kluwer Academic/Plenum Publishing.]

Origin at 2

				Limiting conditions
4	*c*	1	$x, y;\ \bar{x}, \bar{y};\ \frac{1}{2}+x, \frac{1}{2}-y;\ \frac{1}{2}-x, \frac{1}{2}+y.$	*hk*: None *h*0: $h = 2n$ 0*k*: $k = 2n$
2	*b*	2	$\frac{1}{2}, 0;\ 0, \frac{1}{2}.$	As above +
2	*a*	2	$0, 0;\ \frac{1}{2}, \frac{1}{2}.$	*hk*: $h + k = 2n$

selection is fairly self-evident; alternatively, the coordinates of a special position may be inserted into the coordinates of the general positions. Thus, for the twofold rotation point $x = 0$, $y = 0$, the pair 0, 0 and ½, ½ are obtained. What plane group would have been obtained if the rotation points 0, 0 with 0, ½ were paired?

There are no special positions on glide lines, or, in general, on any translational symmetry element. Special positions, as sites for finite entities, must conform to a point group symmetry. For this reason, there are no special positions in plane group *pg* (Fig. 5.2 and Fig. 5.3).

The limiting conditions for general position occupancy in *p*2*gg* only to *h*0 and 0*k*, because halving is one-dimensional in this group, occurring along the *x* and *y* axes. In the special positions, however, halving is general because occupancy of either set (a) or (b) corresponds to a centred arrangement. Note that the general conditions apply additionally.

5.3.4 Plane groups in the square and hexagonal systems

It is desirable to consider a plane group in these systems because some new features arise in the presence of fourfold and sixfold symmetry. Point group 4*mm* has an origin at the intersection of the three symmetry elements; again, Table 3.1 gives their relative orientations. The diagram for plane group *p*4*mm* is constructed with the origin on 4*mm*. A point *x*, *y* is inserted on the diagram and repeated by the symmetry elements as before, taking care always to complete the arrangement of points around the four corners of the unit cell, as shown by Fig. 5.8. If one now asks how any point on the diagram may be reached from the starting point *x*, *y*, by a single symmetry operation, invoking translations *a* and/or *b* as necessary, then by completing the diagram a fourfold rotation point is revealed at ½, ½, twofold rotation points at 0, ½ and ½, 0, and *diagonal g* lines at ½, 0 and 0, ½.

The glide operation here, therefore, comprises reflection across the *g* line coupled with a translation of $a\sqrt{2}/2$, namely, one half of the repeat distance

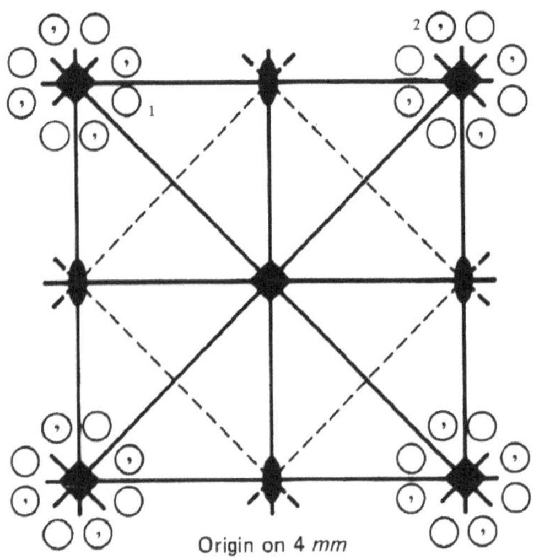

				Limiting conditions
8	*g*	1	$x, y; \bar{x}, \bar{y}; y, \bar{x}; \bar{y}, x; \bar{x}, y; x, \bar{y}; \bar{y}, \bar{x}; y, x$	
4	*f*	*m*	$x, x; \bar{x}, \bar{x}; \bar{x}, x; x, \bar{x}$	
4	*e*	*m*	$x, \frac{1}{2}; \bar{x}, \frac{1}{2}; \frac{1}{2}, x; \frac{1}{2}, \bar{x}$	
4	*d*	*m*	$x, 0; \bar{x}, 0; 0, x; 0, \bar{x}$	
2	*c*	2*mm*	$\frac{1}{2}, 0; 0, \frac{1}{2}$	$hk: h + k = 2n$
1	*b*	4*mm*	$\frac{1}{2}, \frac{1}{2}$	
1	*a*	4*mm*	0, 0	

Fig. 5.8
Plane group *p*4*mm*, origin on 4*mm*, showing the coordinates of the general and special equivalent positions, and the limiting conditions. [Reproduced by courtesy of the IUCr (1).]

along a diagonal of the square, relating the points labelled 1 to 2 on the diagram, for example. The sets of coordinates are listed in the usual manner. The other new feature here, perhaps, is that a right-handed fourfold rotation transforms the point x, y to $\bar{y}$, x. This can be seen readily from the geometry of the plane group diagram; the transformation of $x \rightarrow y$ shows that the relation $a = b$ obtains in this system. One set of special equivalent positions takes a limiting condition, for the reason discussed under *p*2*gg*.

The hexagonal plane groups are approached in the Problems section.

5.3.5 The seventeen plane groups summarized

The 17 plane groups are listed in Table 5.1 and illustrated in Fig. 5.9. After completing the Problems to this chapter, the reader should be able to deal satisfactorily with any of them. Most of the graphic symbols in these diagrams have occurred already in Chapter 3, the only new symbol here being the dashed line for the glide line symmetry. The symbols *p*3*m*1 and *p*31*m* should be noted:

Table 5.1 The 17 plane groups.

System	Point group	Plane groups
Oblique	1	*p*1
	2	*p*2
Rectangular	*m*	*pm*, *pg*, *cm*
	2*mm*	*p*2*mm*, *p*2*mg*, *p*2*gg*, *c*2*mm*
Square	4	*p*4
	4*mm*	*p*4*mm*, *p*4*gm*
Hexagonal	3	*p*3
	3*m*	*p*3*m*1, *p*31*m*
	6	*p*6
	6*mm*	*p*6*mm*

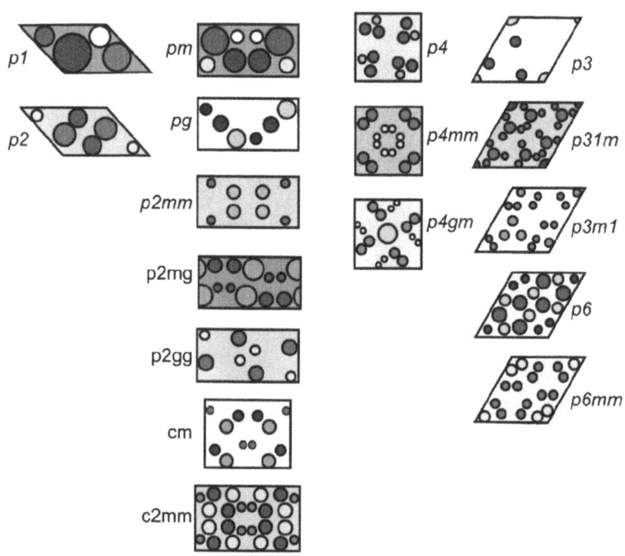

Fig. 5.9
The seventeen plane groups. Can you find the symmetry elements? [Reproduced by courtesy of Dexter Perkins, University of North Carolina.] **(See Plate 6)**

they represent different arrangement in space, and more will be said of this situation in the three-dimensional space groups. The reader may wish to insert the general equivalent positions on a copy of this figure. Similar examples exist in three dimensions with $P\bar{4}m2/P\bar{4}2m$ and $P\bar{6}m2/P\bar{6}2m$.

5.3.6 Comments on notation

Plane group *pm* may be written more fully as $p1m$ (Table 5.1), or even as $p1m1$, so as to emphasize the setting of the *m* line perpendicular to the *x* axis and to give the fullest possible meaning to the plane group symbol. In the rectangular system, *pmm* implies *p2mm*, and it is preferable to use all three positions of any symbol, where they exist. In the two-dimensional hexagonal system, there is only point group of the type 3*m*, but two different settings of the *m* line with respect to the reference axes exist in the corresponding plane groups. Thus, while point groups $3m1$ and $31m$ are equivalent, corresponding only to a rotation of the *x*, *y* and *u* axes by 30° in the *x*, *y* plane, the plane groups $p3m1$ and $p31m$ correspond to different arrangements of points with respect to the symmetry elements, and so are not identical. The reader can investigate this situation further in Problem 5.3.

5.4 Three-dimensional space groups

The three-dimensional space groups, normally referred to just as space groups, present in one respect an extension of the ideas encountered already in the plane groups, but with added intricacies that must be considered carefully.

In 1891, Fyodorov [1] and Schönflies [2], and in 1894 Barlow [3], all independently, described the 230 ordered spatial patterns, or space groups, that represented the ways of arranging infinite arrays of points, or motifs, regularly in space, commensurate with the 14 Bravais lattices. About 90% of the known crystalline substances occur in the triclinic, monoclinic and orthorhombic systems, and space groups in these low symmetry system will be studied first. However, a small number of space groups in the higher symmetry systems will also be discussed in order to bring out features that do not arise with lower symmetries. In this chapter, the original derivations of the space groups will not be a concern. Rather an understanding of space groups will be sought, together with their meaning and use in crystallography. Subsequently, the reader should be able to consult the definitive texts [4, 5] with advantage. *'The important thing in science is not so much to obtain new facts as to discover new ways of thinking about them'* [6].

5.4.1 Triclinic space groups

There are two triclinic point groups, 1 and $\bar{1}$, and one triclinic *P* lattice, so that space groups $P1$ and $P\bar{1}$ could be anticipated correctly for this system; space group $P\bar{1}$ is illustrated in Fig. 5.10.

It is usual, and desirable, to choose the origin of any centrosymmetric space group on $\bar{1}$, or on a point symmetry for which $\bar{1}$ is a subgroup. In this space

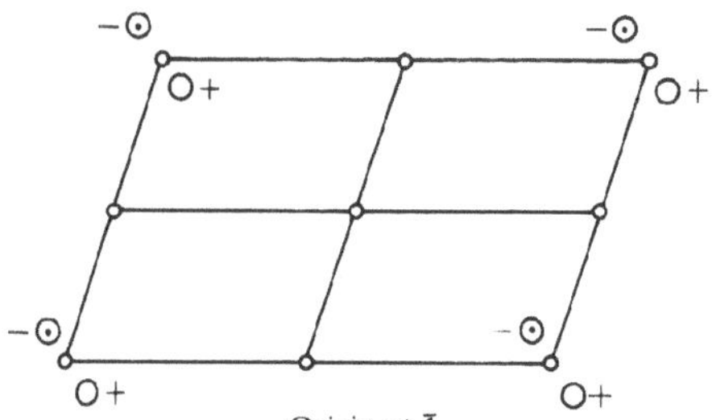

Origin at $\bar{1}$

2	*i*	1	$x, y, z; \bar{x}, \bar{y}, z$
1	*h*	$\bar{1}$	$\frac{1}{2}, \frac{1}{2}, \frac{1}{2}$
1	*g*	$\bar{1}$	$0, \frac{1}{2}, \frac{1}{2}$
1	*f*	$\bar{1}$	$\frac{1}{2}, 0, \frac{1}{2}$
1	*e*	$\bar{1}$	$\frac{1}{2}, \frac{1}{2}, 0$
1	*d*	$\bar{1}$	$\frac{1}{2}, 0, 0$
1	*c*	$\bar{1}$	$0, \frac{1}{2}, 0$
1	*b*	$\bar{1}$	$0, 0, \frac{1}{2}$
1	*a*	$\bar{1}$	$0, 0, 0$

(001) *p*2 (100) *p*2 (010) *p*2

Fig. 5.10
Diagram and description for the triclinic space group $P\bar{1}$. [Reproduced by courtesy of the IUCr (1).]

group, there are two general equivalent positions and eight sets of special equivalent positions of $\bar{1}$ symmetry. In the representations of the three-dimensional space groups, the x and y (or a and b)[2] axes are directed as in the plane groups, and the z (or c)[2] axis is directed upwards from the origin, towards the reader, thus completing the standard, right-handed set of reference axes. In the space group drawings, the + and − signs adjacent to the representative sites refer respectively to z coordinate heights above or below the x, y plane. As well as the four centres of symmetry that are unique to the cell as drawn (a) and (c)–(e),

$$0, 0, 0; \quad {}^1\!/_2, 0, 0; \quad 0, {}^1\!/_2, 0; \quad {}^1\!/_2, {}^1\!/_2, 0$$

there are also four similar centres of symmetry at $z = {}^1\!/_2$, (b) and (f)–(h):

$$0, 0, {}^1\!/_2; \quad {}^1\!/_2, 0, {}^1\!/_2; \quad 0, {}^1\!/_2, {}^1\!/_2; \quad {}^1\!/_2, {}^1\!/_2, {}^1\!/_2$$

A point x, y, z, for example is related to $\bar{x}, \bar{y}, 1 - z$ by $\bar{1}$ at $0, 0, {}^1\!/_2$. This fact may be checked by drawing the space group in either the a, c or the b, c projection. Sometimes a check for special positions in this manner may be needed, but after a while the expectation will become automatic.

The principal projections of $P\bar{1}$ on to (001), (100) and (010) correspond to plane group symmetry $p2$ for each projection since plane point group 2 is the two-dimensional analogue of $\bar{1}$.

[2] The a, b, c notation in this context is used by some authors in lieu of x, y, z.

5.4.2 Monoclinic space groups

In the monoclinic system there are point groups 2, m and $2/m$ and lattices P and C to consider. Here and elsewhere, the commonly used terms such as 'P lattice' have been adopted (see also Section 4.4.7).

5.4.3 Space groups related to point group 2

In the same way that plane group $p2$ was derived from a plane twofold symmetry motif and the oblique p lattice, so the three-dimensional space $P2$ may be realized from a three-dimensional twofold motif

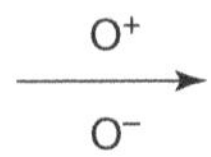

and a monoclinic P lattice, leading to Fig. 5.11 for $P2$; the graphic symbol for a twofold axis lying in plane of projection is $\rightarrow$, but $\rightarrow$ $^1\!/_4$ if lying at $\pm c/4$. The origin of the unit cell is on 2, and is defined only in x and z until the first general position x, y, z in the unit cell is chosen; as a simplification, the word 'equivalent' is often omitted, with understanding, from 'general equivalent positions' and 'special equivalent positions' unless there is a particular call for its use. The general and special positions are derivable readily, and the Wyckoff sets (b) and (d) relate to diad axes at $z = {}^1\!/_2$.

Space group $C2$ is shown in Fig. 5.12: it may be considered as the twofold symmetry motif repeated by the symmetry of a monoclinic C lattice, or as $P2$ with addition of the centring translation $^1\!/_2$, $^1\!/_2$, 0; the parenthetical expression (0, 0, 0; $^1\!/_2$, $^1\!/_2$, 0)+ in Fig. 5.12 corresponds here with the latter statement. There are only two sets of special position in $C2$ compared with four in $P2$, for reasons discussed with pm and cm (Section 5.3.2).

Two new features arise from the study of $C2$. When considering the projection of the space group on to the principal planes containing one of the centring components, that is, on to (100) or (010), there are two translation repeats in the projected unit cell. Thus, in accordance with the rules for choosing a unit cell, the (100) projection would have a length b' equal to $b/2$. When the plane group of the projection cannot be determined by inspection, the projected x, y, z coordinates can be plotted. Thus, for the (100) projection, points y, z; $y, \bar{z}$; $^1\!/_2 + y$, $^1\!/_2 + z$ and $^1\!/_2 + y$, $^1\!/_2 - z$ are plotted on a rectangular unit cell projection (Fig. 5.13a). Then the plane group symmetry for that projection

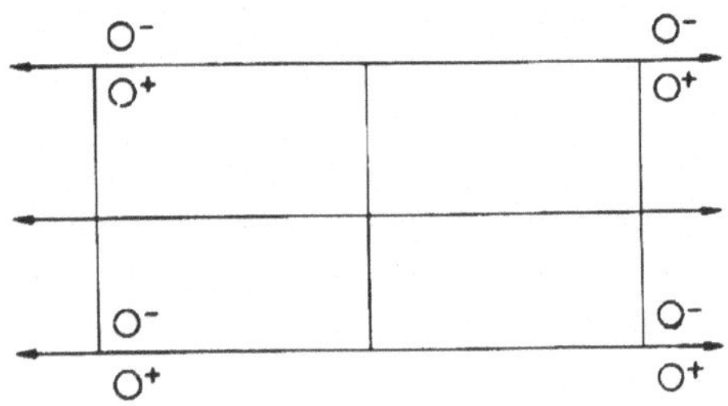

Origin on 2

				Limiting conditions
2	*e*	1	x, y, z; $\bar{x}, y, \bar{z}$.	hkl: $h0l$: $0k0$: } None
1	*d*	2	$\frac{1}{2}, y, \frac{1}{2}$.	
1	*c*	2	$\frac{1}{2}, y, 0$.	
1	*b*	2	$0, y, \frac{1}{2}$.	
1	*a*	2	$0, y, 0$.	

Symmetry of special projections

(001) $pm1(p1m1)$ (100) $p1m(p11m)$ (010) $p2(p211)$

Fig. 5.11
Diagram and description for monoclinic space group $P2$. [Reproduced by courtesy of IUCr (1).]

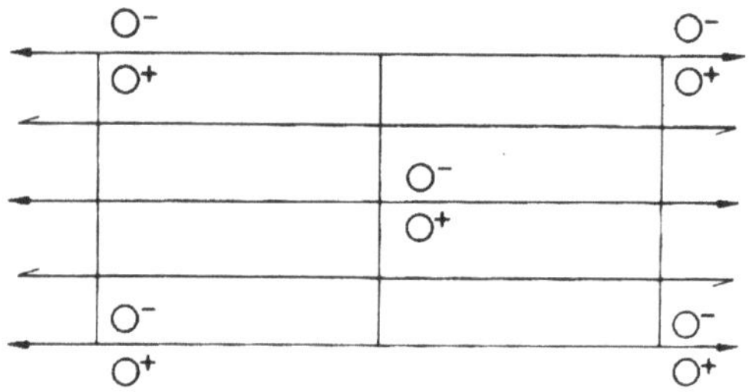

Origin on 2

				Limiting conditions
			$(0, 0, 0;\ \frac{1}{2}, \frac{1}{2}, 0)+$	
4	c	1	$x, y, z;\ \bar{x}, y, \bar{z}.$	$hkl: h+k=2n$
				$h0l: (h=2n)$
				$0k0: (k=2n)$
2	b	2	$0, y, \frac{1}{2}.$	As above
2	a	2	$0, y, 0.$	

Symmetry of special projections

(001) $cm1(c1m1)$ (100) $p1m(p11m)$ $b' = b/2$ (010)$p2(p211)$ $a' = a/2$

Fig. 5.12
Diagram and description for monoclinic space group $C2$: the limiting conditions in parentheses are redundant because of the general conditions $h + k = 2n$.

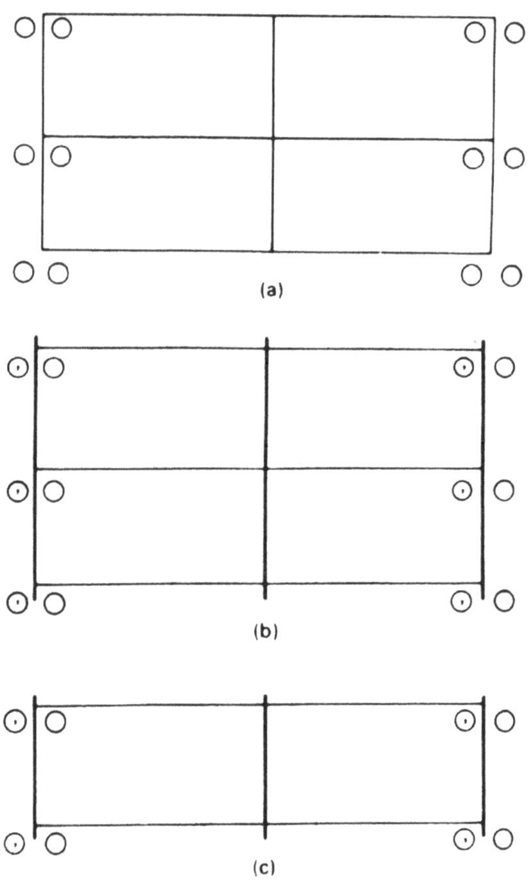

Fig. 5.13
Projection of $C2$ on to (100). (a) The y, z positions from the general equivalent positions set; z axis left to right. (b) Two-dimensional symmetry elements added. (c) One unit cell of plane group $p1m$ ($p11m$), $b' = b/2, c' = c$. Note that in the projection on to (010), $a' = a/2$. What is the symmetry description for the projection on to (001)?

is determined by inspection. With right-handed axes, x' (y) and y' (z) as shown in Fig. 5.13b, the m line is normal to the new y' axis, and the symbol $p11m$ is obtained with b' revised to $b/2$ (Fig. 5.13c). The change-of-hand appearance on projection has been discussed under Section 3.6.5.

The second new feature to emerge from the study of $C2$ is the screw axis. Just as centring in conjunction with reflection symmetry gives rise to glide symmetry (Section 5.3.2), so centring in conjunction with rotation symmetry in three dimensions leads to screw axes.

5.4.4 Screw axes

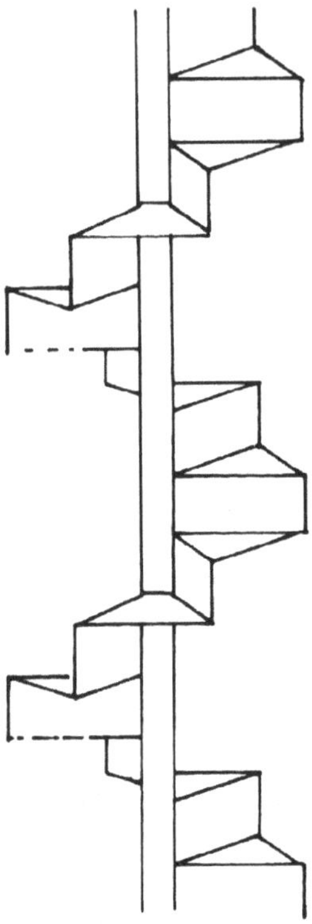

Fig. 5.14
Spiral staircase: an example of 6_1 screw axis symmetry; for a clockwise rotation, the axis would be designated 6_5. [Reproduced by courtesy of Springer Science + Business Media, New York, © Kluwer Academic/Plenum Publishing.]

If, in considering $C2$, the actual symmetry operation that relates the site x, y, z to $\frac{1}{2} - x$, $\frac{1}{2} + y, \bar{z}$ has been sought, then the need for a symmetry operation other than C and 2 will have been realized: it is the screw axis 2_1, graphic symbol ⇀ in the plane of the diagram, or ⇀ ¼ if lying at $c/4$ and $3c/4$.

Screw axes arise in centred space groups, of symmetry higher than triclinic, where rotational symmetry is present in the point group to which it may be said to belong. A screw axis act as a single symmetry operation, but couples both rotational and translational movements. Figure 5.14 illustrates a spiral staircase: imagine the bottom step rotated about the vertical support, or axis, by 60°, coupled with a translation in the direction of the axis of one sixth of the distance between similarly oriented steps, the repeat distance, so that it takes the place of the second step; the second step would move simultaneously into the place of the third step, and so on. Clearly, if the staircase were of infinite length, each operation would map the staircase on to itself, and the operation would be a sixfold screw axis operation, 6_1. A near approximation to a screw axis is the Monument in London; check its screw symmetry at your next visit.

In $C2$, the centring introduces screw diads with a translational component of $b/2$, and interleaving the twofold axes. Following our discussion of pm and cm, the equivalence of $C2_1$ and $C2$ may be seen, but $P2_1$ is a different, new space group (Fig. 5.15). There are no special equivalent positions in $P2_1$, because this space group has only translational symmetry elements. Furthermore, whereas a twofold rotation axis projects parallel to the axis as an m line in two dimensions, a twofold screw axis projected similarly appears as a g line.

A screw axis may be generalized as the operation R_n screw axis, ($n = 2, 3, 4, 6$; $n < R$); it is a rotation of $(360/R)°$ followed by a translation of n/R of the repeat distance in the direction of the screw axis.

5.4.5 Space groups related to point group m: glide planes

If constructions similar to those for pm and cm (Section 5.3.2) are carried out in three dimensions, space groups Pm and Cm are obtained. The centring in cm introduced g lines; in Cm it introduces glide *planes*.

If Fig. 5.2b is considered as a projection of a three-dimensional space group on to the x, y plane, then the dashed line normal to b represents a glide plane. In two dimensions, the translation component of a glide line is unequivocal, but in three dimensions several possibilities arise. In the case of a glide plane

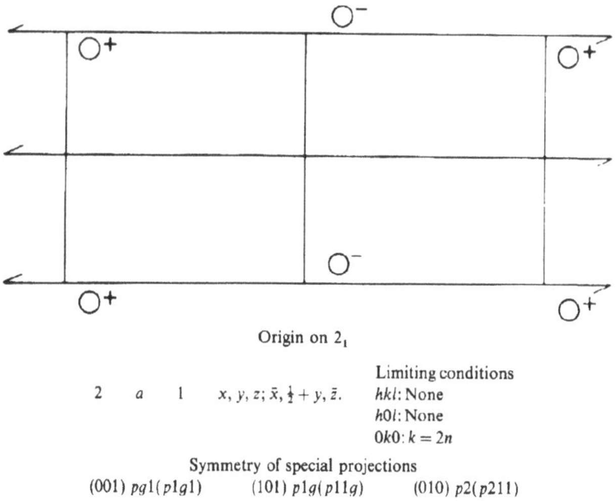

Fig. 5.15
Monoclinic space group $P2_1$; the projection on to (010) is effectively centrosymmetric. [Reproduced by courtesy of Springer Science + Business Media, New York, © Kluwer Academic/ Plenum Publishing.]

normal to c, the translational movement must be in a plane normal to c, and there are five possible situations:

- an a glide plane, with translation $a/2$
- a b glide plane, with translation $b/2$
- a diagonal n glide plane, with translation $(a + b)/2$
- a 'diamond' d glide plane, with translations $(a \pm b)/4$
- a 'double' e glide plane, with translations of $a/2$ and $b/2$

The e glide plane is a recent addition to space group notation [5]. It refers to two perpendicular glide vectors on one and the same plane and related by a unit cell centring operation; it occurs *inter alia* in five centred orthorhombic space groups in which centring exists on only one pair of opposite faces of the unit cell. It does not introduce any new symmetry, but provides a way of looking at related glide planes in a centred group. Thus, the orthorhombic space group *Cmma* would be called *Cmme*, and the following diagram is an example of the e-glide relationship for this space group:

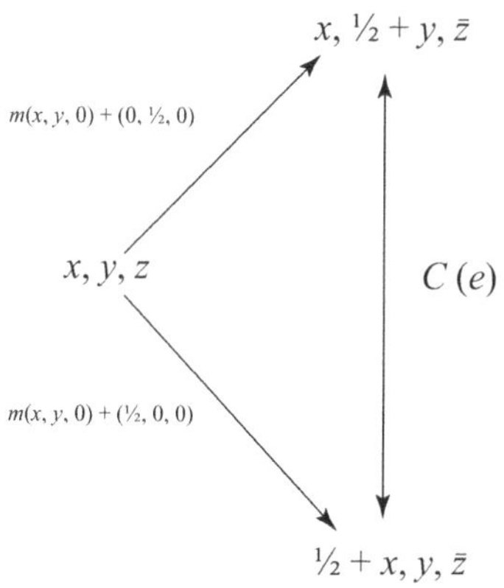

In essence, the *a* glide plane is replace by two *m* planes with translations ½, 0, 0 for one and 0, ½, 0 for the other. The other changes in space group symbols are: $Abm2 \rightarrow Aem2$, $Aba2 \rightarrow Aea2$, $Cmca \rightarrow Cmce$ and $Ccca \rightarrow Ccce$, but the new names do not appear in the literature [5].

All possible different symbols that can be written for a given space group do not necessarily represent differing space groups in any crystal system. Even in the lowest symmetry system, $C1$ is equivalent to $P1$. The space group symbols that might seem to arise from monoclinic point group *m* are:

$$\begin{array}{cccc} Pm^* & Pa & Pc^* & Pn \\ Cm^* & Ca & Cc^* & Cn \end{array}$$

Only those symbols marked * refer to distinct space groups. It may be seen straightaway that *Pa* and *Pc* are equivalent, as are *Cm* and *Ca*, and these space groups will be investigated more fully through the Problems section.

5.4.6 Space groups related to point group $2/m$

By comparison with the monoclinic groups discussed so far, the following scheme may be written for the $2/m$ space groups

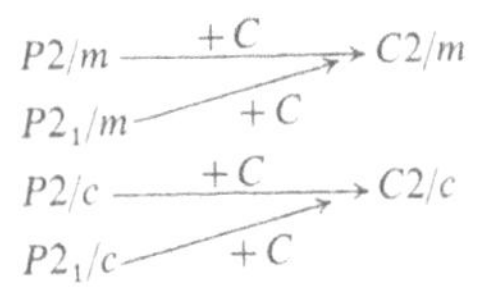

avoiding immediately any equivalent, unnecessary symbols. There is no call to examine such symbols as $B2/m$ or $F2/c$: it has been shown in previous chapter that, in the monoclinic system, the conditions $B \equiv P$ and $F \equiv I \equiv A \equiv C$ exist.

As an example under point group $2/m$, consider space group $P2_1/c$, which represents approximately 36% of known crystal structures. This space group is clearly related to point group $2/m$ and is, therefore, centrosymmetric, but the centre of symmetry does not lie at the intersection of 2_1 and *c*. The space group could be approached after the manner of *p*2*gg*, but a more analytical approach will be adopted which makes use of the full power of the space group symbol. From Table 3.6 and the choice of origin on a centre of symmetry, the following orientations are known:

$\bar{1}$ is at 0, 0, 0
2_1 is the line $[p, y, r]$
c is the plane (x, q, z)

where *p*, *q* and *r* are to be determined. The symmetry operations are carried out according to the following scheme. The essence of this scheme, and all others like it, is to find a path from a starting point *x*, *y*, *z* through the symmetry elements of the group, in positions determined from the full meaning of the space group symbol, and back to the starting point, thereby making use of equivalences such as $\mathbf{2c} \equiv \bar{\mathbf{1}}$.

$$(1)\ x, y, z \xrightarrow{c} (2)\ x, 2q - y, \tfrac{1}{2} + z$$

$$(2) \xrightarrow{2_1} (3)\ 2p - x, \tfrac{1}{2} + 2q - y, 2r - \tfrac{1}{2} - z$$

$$(1) \xrightarrow{\bar{1}} (4)\ \bar{x}\quad \bar{y}\quad \bar{z}$$

The combination of two successive operations (1) → (2) → (3) is equivalent to a third operation performed on the starting point (1) → (4). Thus, by comparing coefficients of points (3) and (4), $p = 0$ and $q = r = {}^1/_4$. Strictly, q evaluates as $-{}^1/_4$ (or ${}^3/_4$), but $+{}^1/_4$ is a crystallographically equivalent position. Alternatively, the y coordinates in the operation (2) → (3) could be written as $2q - y \rightarrow -{}^1/_2 + 2q - y$, making use of the fact that a coordinate can always be moved by ± 1 to a crystallographically equivalent position. These results lead to the desired, standard diagram for space group $P2_1/c$ (Fig. 5.16).

The change in the y coordinate after the c glide operation is illustrated by Fig. 5.17. A similar construction can be applied to any similar situation, and to one coordinate at a time. The details of $P2_1/c$ are completed through the Problem section.

The centred monoclinic space groups in this class need not be discussed in detail here; they present little difficulty once the corresponding primitive space groups have been mastered. The stereoview in Fig. 5.18 shows the ZnI_2 moiety

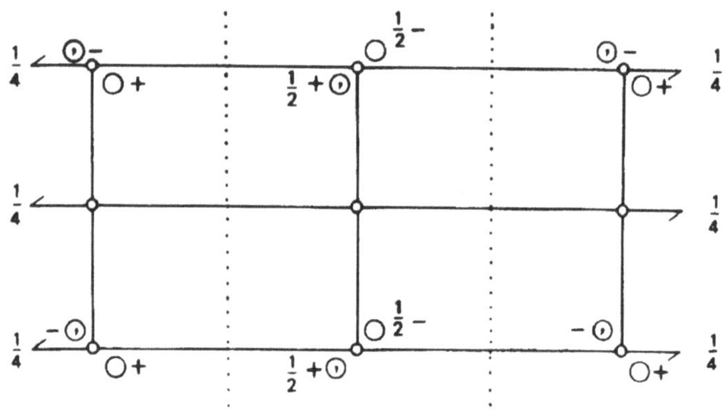

Fig. 5.16
Monoclinic space group $P2_1/c$; origin on $\bar{1}$. The fraction ¼ adjacent to the 2_1 symbols indicates that the screw axes at $z = c/4$ and $3/4$. [Reproduced by courtesy of Springer Science + Business Media, New York, © Kluwer Academic/Plenum Publishing.]

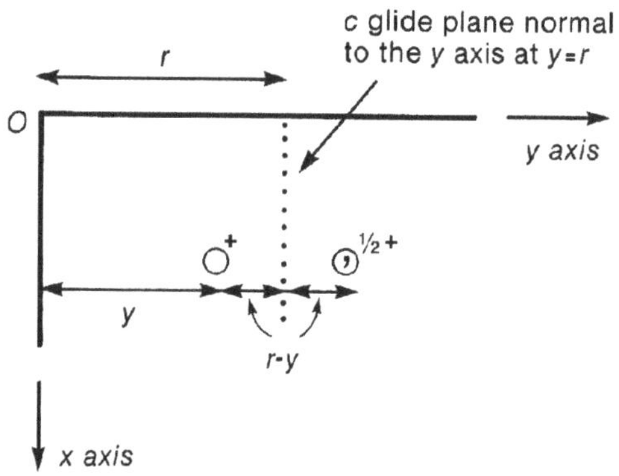

Fig. 5.17
Operation of a c glide plane normal to the y axis and cutting it at a distance r from the origin. Thus, x, y, z is transformed to $x, 2r - y, {}^1/_2 + z$. Thus, if the c glide plane is $(x, {}^1/_4, z)$, x, y, z is transformed to $x, {}^1/_2 - y, {}^1/_2 + z$. This procedure can be applied to any symmetry axis or symmetry plane, and one can always treat one coordinate at a time.

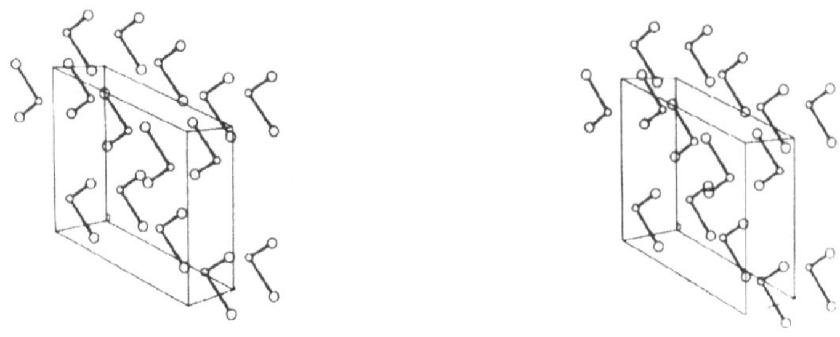

Fig. 5.18
Stereoview of ZnI_2 moieties from the crystal structure of diiodo-(N,N,N',N'-tetra-methylethylenediamine)zinc(II) occupying special positions in space group $C2/c$.

Fig. 5.19
Stereoview of the molecule of diiodo-(N,N,N',N'-tetramethylethylenediamine) zinc(II), $I_2[(CH_3)_2NCH_2CH_2N(CH_3)_2]Zn$.

of the molecule of diiodo-(N,N,N',N'-tetramethylethylenediamine)zinc(II), $I_2[(CH_3)_2NCH_2CH_2N(CH_3)_2]Zn$, lying on the special positions of twofold axes in space group $C2/c$, and Fig. 5.19 illustrates the complete molecular structure [7]. The unit cell dimensions for this crystal are $a = 1.312$, $b = 0.783$, $c = 1.357$ nm, $\beta = 111.4°$ and density 2083 kg m^{-3}. How many molecules occupy one unit cell? Check your result with Eq. (5.1):

$$D_m = \frac{ZM_r m_u}{V} \tag{5.1}$$

where D_m is the crystal density, Z the number of molecules of relative molecular mass M_r in the unit cell, m_u the atomic mass unit (or reciprocal of the Avogadro number) and V the volume of the unit cell.

Example 5.1
A unit cell of dimensions $a = 8.221$, $b = 4.817$, $c = 17.302$ Å and $\beta = 114.71°$ contains an organic chiral species of relative molecular mass 206.52. The conditions limiting X-ray reflections are hkl: none, $h0l$: none, $0k0$: $k = 2n$. Calculate the unit cell volume and crystal density.

The data show that the crystal is monoclinic, space group $P2_1/m$ or $P2_1$; the volume, $abc\sin\beta$, calculates to 622.43 Å^3. Since the species is chiral, the space group is $P2_1$. From Eq. (5.1), $D_m = Z \times 206.52 \times 1.6605/622.43 = 0.5509Z$. For $Z = 2$, $D_m = 1.102$ g cm^{-3}, which is sensible value for an organic substance; $D_m = 2.204$ ($Z = 4$) would be too large, so it may be concluded that $Z = 2$ is correct.

5.4.7 Summary of the monoclinic space groups

The 13 monoclinic space groups are listed in Table 5.2. From time to time, crystal structures are reported with non-standard, albeit valid, space group symbols, such as $P2_1/n$ or $I2$; they can always be transformed to one of the symbols listed in Table 5.2 (see also Section 4.9).

Table 5.2 The 13 monoclinic space groups.

$P2$ $C2$	Pm	Cm	$P2/m$	$P2_1/c$
$P2_1$	Pc	Cc	$P2_1/m$	$C2/c$
			$C2/m$	
		$P2/c$		

5.4.7.1 Symmorphic and non-symmorphic space groups

At this point, it may be noted that space groups that are specified by their symmetry operations acting at a point, those formed *solely* by the combination of a point group and a lattice, are termed *symmorphic* space groups (aka point space groups); otherwise, space groups are *non-symmorphic*. In the latter class of space groups, an origin is specified in relation to a translational symmetry element, as in $P2_1$ (origin on 2_1), $P222_1$ (origin at 212_1) and $P\frac{4}{n}$ (origin at $\bar{4}$ on n, at $-1/4$, $1/4$, 0 from $\bar{1}$ *or* origin at $\bar{1}$ on n, at $1/4$, $-1/4$, 0 from $\bar{4}$), for example. Symmorphic space groups can contain translational symmetry elements, but they are not necessary for the specification of an origin: $C2_1$, for example, is not a non-symmorphic space group; $C2_1 \equiv C2$.

5.4.8 Half-shift rule

In the space groups based on point group $2/m$, there are four possible situations to consider for origins, as shown in Fig. 5.20 where '○' represents the position of $\bar{1}$. Only in (a), corresponding to $P2/m$ and $C2/m$, does the symmetry axis intersect the symmetry plane in the centre of symmetry. In cases (b), (c) and (d), this intersection is translated from the centre of symmetry according to the translational components of the symmetry elements present. From Fig. 5.17, it can be seen that the shift of a symmetry element by an amount r causes a shift in a coordinate of $2r$ from the position with that symmetry element at $r = 0$. Here, it is necessary to work in the opposite manner. In space group $P2_1/m$ let m be the plane (x, 0, z), 2_1 the line [0, y, 0] and $\bar{1}$ the point p, q, r. Then, the following scheme shows that

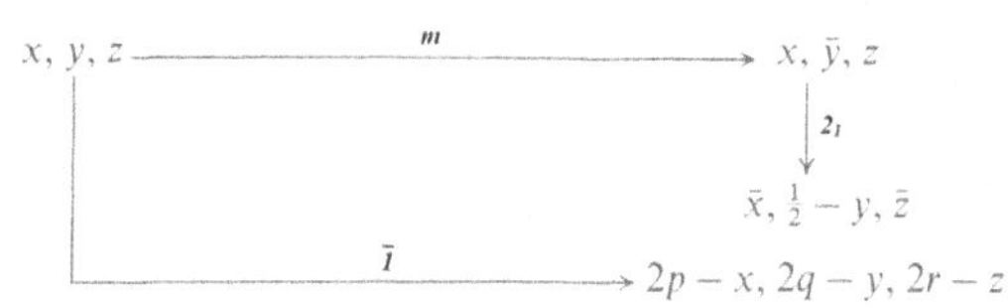

by comparing coefficients $\bar{1}$ lies at the point 0, ¼, 0. In other words, in order to obtain the position of the centre of symmetry with respect to the point of intersection of the symmetry axis and symmetry plane, the total amount of translation is halved in order to determine the amount that the origin has been displaced. The translation here is $b/2$, so the intersection must be moved by $b/4$ in this example in order that $\bar{1}$ occupies the origin, 0, 0, 0 in this group. The reader may wish to apply this *half-shift* rule to the other situations in Fig. 5.20; it will be put to good use in studying space groups.

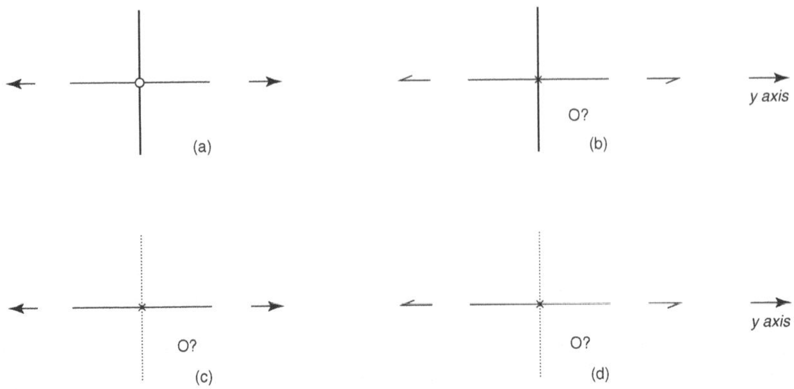

Fig. 5.20
Half-shift rule and position of $\bar{1}$. (a) $P2/m$: $\bar{1}$ at the intersection of 2 and m. (b) $P2_1/m$. (c) $P2/c$. (d) $P2_1/c$. Where should the centre of symmetry ∘ be placed in (b)–(d)?

5.4.9 Orthorhombic space groups

The orthorhombic space groups are a little more complex than those considered so far. Lattices P, C, I and F have to be considered, together with point groups 222, $mm2$ and mmm. Altogether, there are 59 orthorhombic space groups but only a few of them need to be studied, those that occur frequently or which illustrate specific features. In addition, although the lattice descriptors C and A are equivalent, indicating only a change of axes, it will be necessary to consider A centring in combination with the polar point group $mm2$.

5.4.9.1 Space groups related to point group 222

In point group 222, the symmetry axes are mutually perpendicular and along, or parallel to, the x, y and z reference axes; the four orthorhombic lattices lead immediately to four symmorphic space groups:

$P222$	$C222$	$I222$	$F222$
(a)	(b)	(c)	(d)

A 2_1 axis introduced in place of 2 in $P222$ leads to $P222_1$. Since there is no restriction, other than convention, on naming the axes, $P22_12$ and $P2_122$ are equivalent to $P222_1$, the latter being the standard setting. A second 2_1 axis leads to $P2_12_12$, equivalent to $P2_122_1$ and $P22_12_1$.

The introduction of a third 2_1 axis introduces a new and interesting feature. From Euler's theorem (Section 3.6.3), it may be recalled that a third, non-collinear, interacting symmetry element in a permitted orientation gives rise to an operation that is not independent of those given by the other two elements. Thus, if any two perpendicular 2_1 axes intersect, then space group $P2_12_12$, or an equivalent, is produced: since two 2_1 axes introduce half translations of $a/2$ and $b/2$, the twofold axis must pass through the point $(a+b)/2$, as can be confirmed by drawing the space group diagram. In $P2_12_12_1$, however, the three mutually perpendicular axes *do not intersect* (Fig. 5.21).

In type (b), as listed above, it may be recalled that C centring introduces translations of ½ along both a and b, and so leads to 2_1 axes interleaving the twofold axes parallel to the x and y axes. Hence, $C2_122$ and $C22_12$ are equivalent to $C222$, but $C222_1$ is different because the translation of $c/2$ imposed by

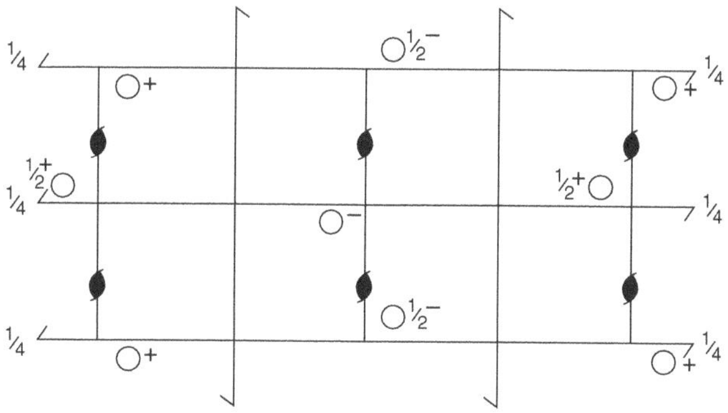

Origin halfway between three pairs of non-intersecting screw axes

			Limiting conditions
a	1	$x, y, z;\ \frac{1}{2}-x, \bar{y}, \frac{1}{2}+z;\ \frac{1}{2}+x, \frac{1}{2}-y, \bar{z};\ \bar{x}, \frac{1}{2}+y, \frac{1}{2}-z.$	hkl:, $0kl$:, $h0l$:, $hk0$: } None
			$h00$: $h=2n$
			$0k0$: $k=2n$
			$00l$: $l=2n$

Symmetry of special projections

(001) $p2gg$ (100) $p2gg$ (010) $p2gg$

Fig. 5.21
Orthorhombic space group $P2_12_12_1$; the origin lies halfway between the three pairs of non-intersecting 2_1 screw axes. [Reproduced by courtesy of Springer Science + Business Media, New York, © Kluwer Academic/Plenum Publishing.]

the 2_1 axis parallel to z is not present in $C222$; $C222_1$ is equivalent to $C2_12_12_1$, and $F222$ is the only unique F-centred space group of type (d).

In $I222$ there are three intersecting diads and three intersecting screw diads. However, if the body centring condition is coupled with $P2_12_12_1$, a space group is obtained with three non-intersecting diads and three non-intersecting screw diads. By convention, the designations $P2_12_12_1$ and $I2_12_12_1$ are applied to the groups with the non-intersecting axes; the 222 space groups are summarized by the scheme in Fig. 5.22.

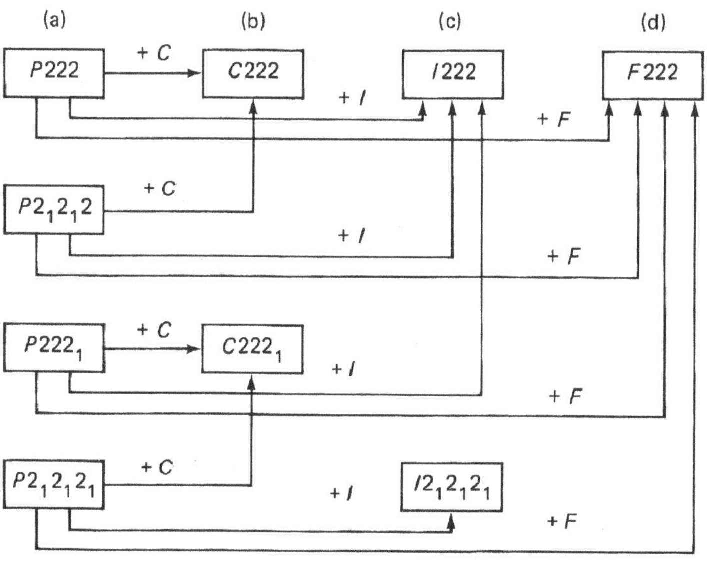

Fig. 5.22
The interrelationships of space groups based on point group 222. In $P2_12_12$, only the 2_1 axes intersect whereas in $P222_1$, a twofold axis intersects the 2_1 axis.

Space group $P2_12_12_1$ occurs to the extent of approximately 10% of known crystal structures. There are more ways of setting this space group with three non-intersecting 2_1 axes than that shown in Fig. 5.21, but that diagram illustrates the standard setting. Although $P2_12_12_1$ is a non-centrosymmetric space group, the three principal projections are centric, exhibiting plane group $p2gg$. In each of the principal projections, (001), (100) and (010), the standard setting of this plane group involves a change of origin from the coordinates of the general positions, so that symmetry 2 is at the origin, as can be deduced from Fig. 5.21.

5.4.9.2 Space groups related to point group *mm*2

In these space groups, two perpendicular symmetry planes intersect in a line which is a symmetry axis or is parallel to such an axis. The space group origin is taken normally on 2 or 2_1, the line $[p, q, z]$, where p and q may or may not equal zero. In space group $Pmn2_1$, however, the standard origin is taken, by convention, along the line of intersection of m and n, as it leads to a simpler expression for the coordinates of the general equivalent positions. In order to illustrate some of the features of the space groups in this class, $Pma2$, $Pma2 + C$ and $Pma2 + A$ will be considered. Table 3.6 shows that the orientations of the symmetry elements are

$$m \perp x,\ a \perp y,\ \text{and 2 along } [0, 0, z]$$

for space groups of the $mm2$ class. Applying the half-shift rule to $Pma2$, the line of intersection of m and a is set off from $[0, 0, z]$ by $a/4$, leading to Fig. 5.23.

The effect of C centring of $Pma2$ is found by applying the translations ½, ½, 0 to the general equivalent positions of $Pma2$. The completion of the diagram reveals further m planes, at $\pm(x, {}^1\!/_4, z)$, b glide planes at $(0, y, z)$ and $({}^1\!/_2, y, z)$, and diads at $\pm\left[{}^1\!/_4, {}^1\!/_4, z\right]$ and $\pm\left[{}^3\!/_4, {}^1\!/_4, z\right]$. By convention the origin is taken on the twofold axis that arises from the C centring at $\left[{}^1\!/_4, {}^1\!/_4, z\right]$, that is, on $mm2$ and the space group is named $Cmm2$ (Fig. 5.24).

Centring the A faces introduces the translations $0, {}^1\!/_2, {}^1\!/_2$ into the coordinate set, which leads to n glide planes at $y = \pm{}^1\!/_4$, with translations of $(a + c)/2$, interleaving the a glide planes; also n glide planes arise at $x = \pm{}^1\!/_4$, *coincident* with the m planes. Screw axes, 2_1, now occur at $\pm\left[0, {}^1\!/_4, z\right]$ and $\pm\left[{}^1\!/_2, {}^1\!/_4, z\right]$ (Fig. 5.25). This space group is different from $Cmm2$ ($Cma2$), showing that A centring is valid and necessary in this class. What would the symbol $Ama2$ become if this space group pattern were referred to a B centred unit cell? The need for A centring here arises because of the polar axis (z).

These few examples serve to show that a centred space group can frequently be treated as the corresponding primitive space group coupled with the appropriate centring condition. The totality of symmetry elements so produced must be sought, along the lines already discussed. The standard space group symbol may be obtained generally through the rule of precedence of naming: where there is more than one symmetry element parallel to a given plane or line, then the precedence is $m > a > b > c > n > d$, and $2 > 2_1$.

*Pma*2 C_{2v}^{4} No. 28 *Pma*2 *mm*2 Orthorhombic

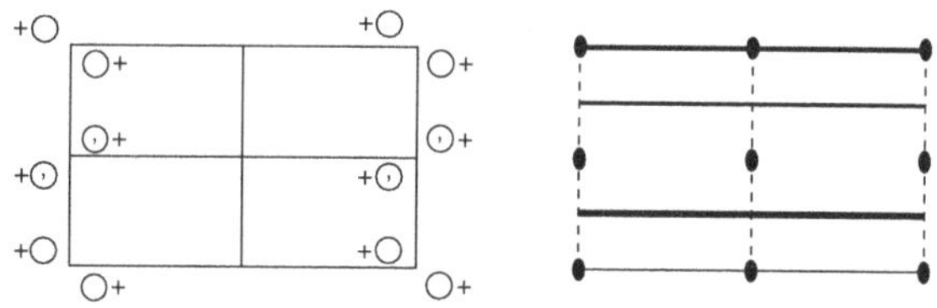

Origin on 2

Number of positions, Wyckoff notation, and point symmetry			Co-ordinates of equivalent positions	Conditions limiting possible reflections
				General:
4	*d*	1	$x,y,z;\ \bar{x},\bar{y},z;\ \tfrac{1}{2}-x,y,z;\ \tfrac{1}{2}+x,\bar{y},z.$	*hkl*: No conditions 0*kl*: No conditions *h*0*l*: $h=2n$ *hk*0: No conditions *h*00: $(h=2n)$ 0*k*0: No conditions 00*l*: No conditions
				Special: as above, plus no extra conditions
2	*c*	*m*	$\tfrac{1}{4},y,z;\ \tfrac{3}{4},\bar{y},z.$	
2	*b*	2	$0,\tfrac{1}{2},z;\ \tfrac{1}{2},\tfrac{1}{2},z.$	} *hkl*: $h=2n$
2	*a*	2	$0,0,z;\ \tfrac{1}{2},0,z.$	

Symmetry of special projections

(001) *pmg*; $a'=a,\ b'=b$ (100) *pml*; $b'=b,\ c'=c$ (010) *plm*; $c'=c,\ a'=a/2$

Fig. 5.23 Space group *Pma*2, origin on 2. [Reproduced by courtesy of the IUCr (1).]

5.4.9.3 Space groups related to point group *mmm*

The centrosymmetric point group *mmm* has the full symbol $\frac{2}{m}\frac{2}{m}\frac{2}{m}$. There are now several more symmetry elements to consider and many combinations are permissible. The intersection of three mutually perpendicular *m* planes defines a point, which is a centre of symmetry. When any translational symmetry element is present in the group, the centre of symmetry will be translated according to the rules already established. In order to obtain the standard setting, the half-shift rule can be applied. However, before tackling a space group in this simple manner, a little analysis may prove instructive.

The scheme in Fig. 5.26 shows how a point *x, y, z* changes in sign according to the seven different types of symmetry operations in point group *mmm*, the basis for the space groups of this class.

Using the full meaning of the space group symbol *Pnma*, for example, the positions of the symmetry planes and axes can be written as follows, taking $\bar{1}$ at the origin. The *n* glide is the plane *x, y* and lies at the fractional distance *p* along *z*; the twofold axis 2_P, rotation or screw according to the value of *P*, is parallel to the *x* axis and lies at $y=B$ and $z=C$, and so on. Thus the full specification becomes:

$$n(p,\ y\ z)\quad m(x,\ q,\ z)\quad a(x,\ y,\ r)\quad 2_P\,[x,\ B,\ C]\quad 2_Q\left[A,y,C'\right]\quad 2_R\left(A',\ B',\ z\right]$$

Orthorhombic $mm2$ $Cmm2$ No. 35 $Cmm2$ C_{2v}^{11}

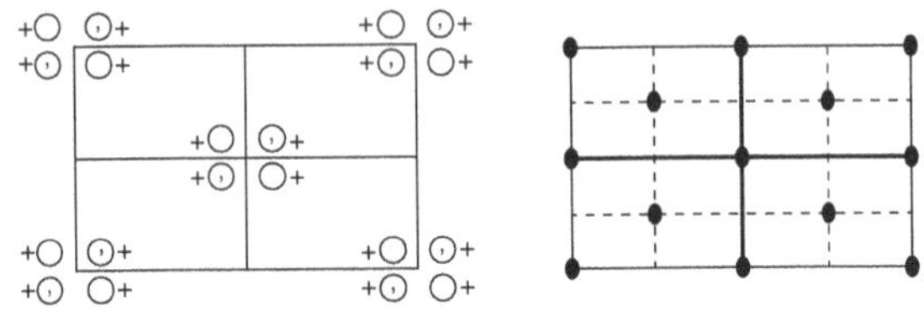

Origin on $mm2$

Number of positions, Wyckoff notation, and point symmetry			Co-ordinates of equivalent positions $(0,0,0;\ \frac{1}{2},\frac{1}{2},0)+$	Conditions limiting possible reflections
				General:
8	f	1	$x,y,z;\ \bar{x},\bar{y},z;\ \bar{x},y,z;\ x,\bar{y},z.$	hkl: $h+k=2n$ $0kl$: $(k=2n)$ $h0l$: $(h=2n)$ $hk0$: $(h+k=2n)$ $h00$: $(h=2n)$ $0k0$: $(k=2n)$ $00l$: No conditions
				Special: as above, plus
4	e	m	$0,y,z;\ 0,\bar{y},z.$	} no extra conditions
4	d	m	$x,0,z;\ \bar{x},0,z.$	
4	c	2	$\frac{1}{4},\frac{1}{4},z;\ \frac{1}{4},\frac{3}{4},z.$	hkl: $h=2n$; $(k=2n)$
2	b	mm	$0,\frac{1}{2},z.$	} no extra conditions
2	a	mm	$0,0,z.$	

Symmetry of special projections

(001) cmm; $a'=a$, $b'=b$ (100) pml; $b'=b/2$, $c'=c$ (010) plm; $c'=c$, $a'=a/2$

Fig. 5.24
Space group $Cmm2$ ($Cma2$), origin on 2. In the $mm2$ class of the orthorhombic system, $Ama2$ is not the same as a transformed $Cmm2$ ($Cma2$), because z (2) is a polar axis in this class. This situation does not arise in classes 222 and mmm. [Reproduced by courtesy of the IUCr (1).]

The eight general equivalent positions may now be written, bearing in mind Fig. 5.17, and the 12 unknown parameters now determined.

(1) x, y, z (2) $2p-x,\ 1/2+y,\ 1/2+z$ (3) $x,\ 2q-y, z$ (4) $1/2+x.y,\ 2r-z$

(5) $\bar{x},\bar{y},\bar{z}$ (6) $2p+x,\ 1/2-y,\ 1/2-z$ (7) $\bar{x},\ 2q+y,\bar{z}$ (8) $1/2-x,\bar{y},\ 2r+z$

The sign changes are in accord with Fig. 5.26: pairs of coordinates such as 1, 2 and 4, 7, or 1, 3 and 2, 8 are related by the same type of symmetry operation, from which it follows that $p=q=r=1/4$. Pairs such as 1, 6 reveal that $P=Q=R=1$; from this same pair, it follows also that $B=C=1/4$. Further such comparisons show that $A=0$, $A'=1/4$, $B'=C'=0$. The nature and orientation of all the unique symmetry elements in $Pnma$ have now been assembled, and the full symbol of this space group is $P\frac{2_1}{n}\frac{2_1}{m}\frac{2_1}{a}$, illustrated by Fig. 5.27.

This analysis makes use of facts illustrated by Fig. 5.26: that reflection produces a *single* sign change, the reflection plane being normal to the coordinate that changes sign; that rotation produces *two* sign changes, which are those coordinates other than that of the direction of the rotation axis; and that a

$Ama2$ C_{2v}^{16}

No. 40 $Ama2$ $mm2$ Orthorhombic

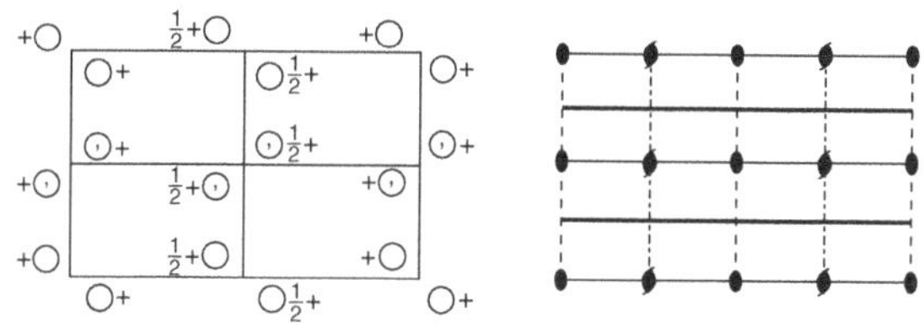

Origin on 2

Number of positions, Wyckoff notation, and point symmetry			Co-ordinates of equivalent positions $(0,0,0;\ 0,\frac{1}{2},\frac{1}{2})+$	Conditions limiting possible reflections
				General:
8	c	1	$x,y,z;\ \bar{x},\bar{y},z;\ \frac{1}{2}-x,y,z;\ \frac{1}{2}+x,\bar{y},z.$	hkl: $k+l=2n$ $0kl$: $(k+l=2n)$ $h0l$: $h+2n$; $(l=2n)$ $hk0$: $(k=2n)$ $h00$: $(h=2n)$ $0k0$: $(k=2n)$ $00l$: $(l=2n)$
				Special: as above, plus
4	b	m	$\frac{1}{4},y,z;\ \frac{3}{4},\bar{y},z.$	no extra conditions
4	a	2	$0,0,z;\ \frac{1}{2},0,z.$	hkl: $h=2n$

Symmetry of special projections

(001) pmg; $a' = a$, $b' = b/2$ (100) $cm1$; $b' = b$, $c' = c$ (010) $p1m$; $c' = c/2$, $a' = a/2$

Fig. 5.25
Space group $Ama2$, origin on 2. [Reproduced by courtesy of the IUCr (1).]

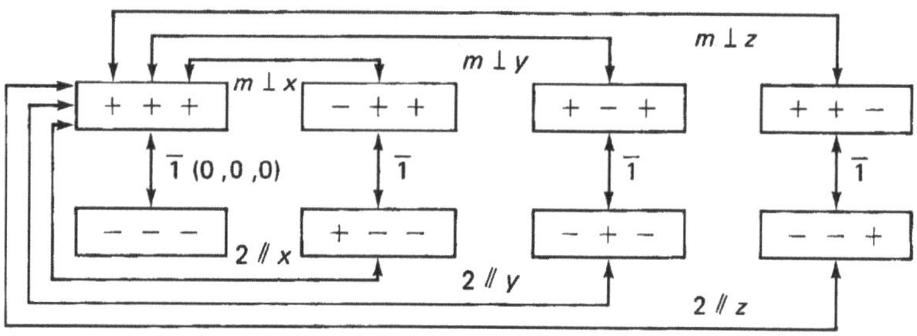

Fig. 5.26
Interrelation of the sign changes for a point x, y, z under the symmetry operations of point group mmm. [Reproduced by courtesy of Springer Science + Business Media, New York, © Kluwer Academic/Plenum Publishing.]

centre of symmetry produces *three* sign changes. The other variations in the coordinates depend upon whether or no translational symmetry elements are present and upon the positions of such symmetry elements with respect to the origin of the reference axes. These rules apply to the triclinic, monoclinic and orthorhombic systems and in other situations where the symmetry elements are either parallel to or contain the reference axes. Thus, if a coordinate x is related by a symmetry element to the position $s - x$, then the symmetry element passes through the point $x = s/2$, which is another expression of Fig. 5.17.

It has been shown that the operations $\boldsymbol{m}_y \boldsymbol{m}_x = \mathbf{2}$, the twofold symmetry being along z. If a centre of symmetry is added, the combination

$$\boldsymbol{m}_z \boldsymbol{m}_y \boldsymbol{m}_x = \bar{\mathbf{1}}$$

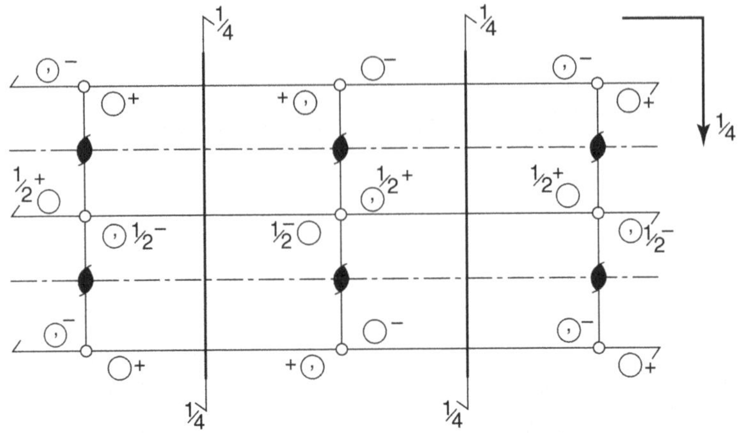

Origin at $\bar{1}$

				Limiting conditions
8	*d*	1	$x, y, z; \frac{1}{2}+x, \frac{1}{2}-y, \frac{1}{2}-z; \bar{x}, \frac{1}{2}+y, \bar{z}; \frac{1}{2}-x, \bar{y}, \frac{1}{2}+z;$ $\bar{x}, \bar{y}, \bar{z}; \frac{1}{2}-x, \frac{1}{2}+y, \frac{1}{2}+z; x, \frac{1}{2}-y, z; \frac{1}{2}+x, y, \frac{1}{2}-z.$	*hkl*: None *0kl*: $k+l=2n$ *h0l*: None *hk0*: $h=2n$ *h00*: $(h=2n)$ *0k0*: $(k=2n)$ *00l*: $(l=2n)$
4	*c*	*m*	$x, \frac{1}{4}, z; \bar{x}, \frac{1}{4}, \bar{z}; \frac{1}{2}-x, \frac{1}{4}, \frac{1}{2}+z; \frac{1}{2}+x, \frac{1}{4}, \frac{1}{2}-z.$	As above
4	*b*	$\bar{1}$	$0, 0, \frac{1}{2}; 0, \frac{1}{2}, \frac{1}{2}; \frac{1}{2}, 0, 0; \frac{1}{2}, \frac{1}{2}, 0.$	As above +
4	*a*	$\bar{1}$	$0, 0, 0; 0, \frac{1}{2}, 0; \frac{1}{2}, 0, \frac{1}{2}; \frac{1}{2}, 0, \frac{1}{2}; \frac{1}{2}, \frac{1}{2}, \frac{1}{2}.$	*hkl*: $h+l=2n; k=2n$)

Symmetry of special projections

(001) *p2gm* (100) *c2mm* (010) *p2gg*

Fig 5.27
Space group *Pnma*, origin on $\bar{1}$.

holds, or in this example

$$\boldsymbol{a\,m\,n} \equiv \bar{\mathbf{1}}$$

with $\bar{1}$ at 0, 0, 0, *provided* that the symmetry planes are in their correct relative positions. In the scheme below, a point *x*, *y*, *z* is moved to $\bar{x}, \bar{y}, \bar{z}$ by three successive reflection operations under space group symmetry *Pnma*, subject to the translational components of each coordinate summing to zero over the cycle:

$$\sum t_x = \sum t_y = \sum t_z = 0$$

$$x, y, z \xrightarrow{n} 2p - x, \tfrac{1}{2}+y, \tfrac{1}{2}+z$$

$$\downarrow m$$

$$\tfrac{1}{2} + 2p - x, 2q - \tfrac{1}{2} - y, 2r - \tfrac{1}{2} - z \xleftarrow{a} 2p - x, 2q - \tfrac{1}{2} - y, \tfrac{1}{2} + z$$

$$\bar{1} \uparrow \text{ (from } \tfrac{1}{2} + 2p - x, 2q - \tfrac{1}{2} - y, 2r - \tfrac{1}{2} - z \text{ to } x, y, z)$$

Finally, an even quicker result obtains from the half-shift rule: the total translation for *Pnma* is $(b/2 + c/2 + a/2)$ and the half shift $(a + b + c)/4$. Thus, the origin ($\bar{1}$) is set off by this amount from the point of intersection of the three symmetry planes, ¼, ¼, ¼, as shown in Fig. 5.27. The rest of the description of this space group follows the lines already discussed.

If one is presented with a centred group, *Cmcm*, for example, it can be treated as *Pmcm* + *C*; *Imma* would be treated as *Pmma* + *I*. In each case, the totality of symmetry elements must be obtained and scrutinized to ensure that the standard setting is chosen.

In the orthorhombic system *d* glide planes are encountered in space groups *Fdd*2 and *Fddd*; the *d* glide translations are $\pm(a+b)/4$ and cyclic permutations thereof. The reader may care to construct a diagram for space group *Fddd*, and then compare it with that of the frontispiece of the book: begin either with the origin on 222, or on $\bar{1}$ and then transform to the origin on 222.

5.4.10 Change of origin

A change of origin may be needed in order to obtain a standard setting of a space group that has been determined by experiment. Consider the projection of $P2_12_12_1$ on to (001); the general equivalent positions in two dimensions are:

$$x,\ y;\ {}^1\!/_2 - x,\ \bar{y};\ {}^1\!/_2 + x,\ {}^1\!/_2 - y;\ \bar{x},\ {}^1\!/_2 + y$$

Let a new origin be set at $^1/_4$, 0; a 2_1 axis projects on to the (001) as a twofold rotation point, and it is desirable that this point be the origin. Next, the coordinates of the new origin are subtracted from the original coordinates:

$$x - {}^1\!/_4,\ y;\ {}^1\!/_4 - x,\ \bar{y};\ {}^1\!/_4 + x,\ {}^1\!/_2 - y;\ \ -x - {}^1\!/_4,\ {}^1\!/_2 + y$$

Then, new variables are assigned such that $x - {}^1/_4 = x_0$ and $y = y_0$, leading to

$$x_0,\ y_0;\ \overline{x_0},\ \overline{y_0};\ {}^1\!/_2 + x_0,\ {}^1\!/_2 - y_0;\ {}^1\!/_2 - x_0,\ {}^1\!/_2 + y_0$$

and, without subscripts, these coordinates now correspond to those derived earlier for plane group *p*2*gg*, with the origin desirably on $\bar{1}$.

5.4.11 Standard and alternative settings of space groups

In certain space groups in the *mmm* class, for example, the origin is chosen on 222 rather than $\bar{1}$ [4,5], the reason being that the expressions for the coordinates of the general equivalent positions are in a simpler form for algebraic manipulation. For our purposes, however, an origin taken on $\bar{1}$ will lead to an equally correct representation of the space groups, and an origin change can be applied as necessary. The orthorhombic space groups are listed in their standard **abc** setting in Table 5.3.

This short discussion on alternative settings will be confined to the monoclinic and orthorhombic space groups. In the monoclinic system, while the unique axis is fixed as *y* (or *b*) by convention, the directions of *a* and *c* are interchangeable. Consider space group $C2/c$. The *a*, *b* plane is centred, the twofold axis is parallel to *b*, and the *c* glide plane is parallel to the *a*, *c* plane, with a translation of $c/2$. Transforming to a new cell, and maintaining right-handed axes,

$$\mathbf{a}' = -\mathbf{c} \quad \mathbf{b}' = \mathbf{b} \quad \mathbf{c}' = \mathbf{a}$$

requires consideration of a new space group symbol. The centred plane *a*, *b* is now b', c' and the new cell is *A* centred. The glide plane was *a*, *c* and is still the same plane, but labelled a', c', but the translational component, previously $c/2$, is now $a'/2$. Thus, the new space group symbol is $A2/a$. Summarizing:

Setting	Symbol
abc	**$\bar{\mathbf{c}}$ba** (**a′b′c′**)
*C*2/*c*	*A*2/*a*

In practice, while recognizing *b* correctly, as it is the symmetry direction, the choice of the *a* and *c* directions is arbitrary. A space group determined in a non-standard setting can easily be transformed to the standard setting. For example, the space group of the ferroelectric phase of bismuth vanadate, $BiVO_4$, has been reported as *I*2/*a*. The transformation of this space group to its standard setting will be examined in the Problems section.

The orthorhombic space groups provide more variations, because the directions *a*, *b* and *c*, although determined by the directions of the 2 or $\bar{2}$ axes, may be chosen in any order subject the right-handed setting of *a*, *b* and *c*.

Consider the standard, **abc**, setting of *Cmcm*. Suppose now that the unit cell had been set as **bca**, that is, through the transformation

$$\mathbf{a}' = \mathbf{b} \qquad \mathbf{b}' = \mathbf{c} \qquad \mathbf{c}' = \mathbf{a}$$

The centred plane *a*, *b* is now *a*′, *c*′, that is, *B* centred, *m* was normal to *a* and is now normal to *c*′; the *c* glide was normal to *b* and is now normal to *a*′ with a translation component of *b*′/2; the second *m* plane was normal to *c* and is now normal to *b*′. Thus, the transformed symbol is *Bbmm*. The transformation is illustrated by Fig. 5.28. It should be now possible to check your deduction of the space group related to *Ama*2 (Section 5.4.9.2). The possible standard settings for space groups are tabulated in the literature [4,5].

It may have been noticed that in the space group diagrams that are reproduced from the International Tables for X-ray Crystallography, the Schönflies space group symbol is also listed. Thus, underneath *Pma*2 is C_{2v}^4 : C_{2v} itself is equivalent to *mm*2, and the superscript 4 implies the fourth space group in the Schönflies list of *mm*2 space groups (Table 5.3)—not a very useful notation for

Table 5.3 The 59 orthorhombic space groups, listed in Schönflies' order.

*P*222	*Pmm*2	*Cmc*2_1	*Pmmm*	*Pbca*
*P*222$_1$	*Pmc*2_1	*Ccc*2	*Pnnn*	*Pnma*
P$2_1 2_1$2	*Pcc*2	*Amm*2	*Pccm*	*Cmcm*
P$2_1 2_1 2_1$	*Pma*2	*Abm*2	*Pban*	*Cmca*
*C*222$_1$	*Pca*2_1	*Ama*2	*Pmma*	*Cmmm*
*C*222	*Pnc*2	*Aba*2	*Pnna*	*Cccm*
*F*222	*Pmn*2_1	*Fmm*2	*Pmna*	*Cmma*
*I*222	*Pba*2	*Fdd*2	*Pcca*	*Ccca*
I$2_1 2_1 2_1$	*Pna*2_1	*Imm*2	*Pbam*	*Fmmm*
	*Pnn*2	*Iba*2	*Pccn*	*Fddd*
	*Cmm*2	*Ima*2	*Pbcm*	*Immm*
			Pnnm	*Ibam*
			Pmmn	*Ibca*
			Pbcn	*Imma*

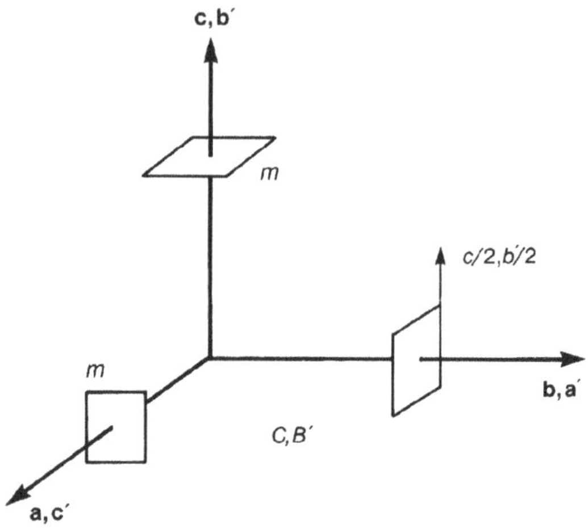

Fig. 5.28
Transformation of the space group setting: the unprimed letters refer to the standard setting **abc** and the symbol *Cmcm*, and the primed symbols to the **bca** setting and symbol *Bbmm*.

crystallographic space groups, although they do show, by their subscripts, the crystal class to which they belong.

5.4.12 Tetragonal space groups

There is a total of 68 tetragonal space groups, and it is necessary to study some of them some in order to illustrate important features of space groups in this crystal system.

5.4.12.1 Space groups within Laue class 4/m

Space groups based on point group 4 are straightforward and their features will be covered through other tetragonal space groups selected in this section.

In space group $I\bar{4}$ an origin may be chosen at the point of inversion on the $\bar{4}$ axis, and then the space group treated as $P\bar{4}+I$. With this information, Fig. 5.29 is produced readily. How are the $\bar{4}$ inversion points (c) at $z = 1/4$ revealed?

In the tetragonal space groups, the projection on to (001) is straightforward: for $I\bar{4}$ it corresponds to plane group $p4$ with $\mathbf{a}' = (\mathbf{a}/2 - \mathbf{b}/2)$ and $\mathbf{b}' = (\mathbf{a}/2 + \mathbf{b}/2)$, or $a' = b' = a\sqrt{2}/2$. On (100), the projected plane group is $c1m1$ with $a' = b$ and $b' = c$, and on (110), the third important plane in the tetragonal system, it is $p1m1$ with $a' = a\sqrt{2}/2$ and $b' = c/2$. The *International Tables for Crystallography* [5] lists the projected plane groups of all space groups.

If the special positions in $I\bar{4}$ are occupied in a crystal structure, they must accommodate a molecule of symmetry 2 or $\bar{4}$, or a symmetry of which 2 or $\bar{4}$ is a subgroup. This situation is realized with crystal structure of pentaerythritol (Fig. 2.44), which crystallizes in $I\bar{4}$ with molecules centred at the sites 0, 0, 0 and ½, ½, ½ in the unit cell. The origin in space groups related to point groups 4 and $\bar{4}$ can be selected on 4 or $\bar{4}$; in $I4_1$, however, an origin may be chosen either on 4_1 or on 2.

The half-shift rule is not always satisfactory for all space groups in the higher symmetry systems, but it is always possible to develop the space group by making use of the full power of the space group symbol. Reference to

$I\bar{4}$
S_4^2

No. 82 $I\bar{4}$ $\bar{4}$ Tetragonal

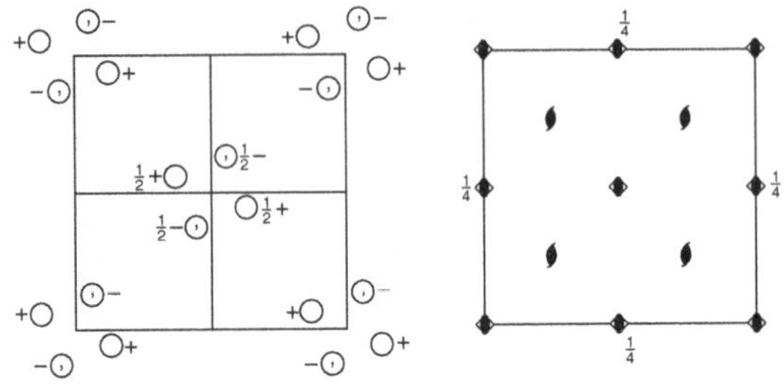

Origin at $\bar{4}$

Number of positions, Wyckoff notation, and point symmetry			Co-ordinates of equivalent positions $(0,0,0;\ \frac{1}{2},\frac{1}{2},\frac{1}{2})+$	Conditions limiting possible reflections
				General:
8	*g*	1	$x,y,z;\ \bar{x},\bar{y},z;\ y,\bar{x},\bar{z};\ \bar{y},x,\bar{z}.$	*hkl*: $h+k+l=2n$
4	*f*	2	$0,\frac{1}{2},z;\ \frac{1}{2},0,\bar{z}.$	Special: as above only
4	*e*	2	$0,0,z;\ 0,0,\bar{z}.$	
2	*d*	$\bar{4}$	$0,\frac{1}{2},\frac{3}{4}.$	
2	*c*	$\bar{4}$	$0,\frac{1}{2},\frac{1}{4}.$	
2	*b*	$\bar{4}$	$0,0,\frac{1}{2}.$	
2	*a*	$\bar{4}$	$0,0,0.$	

Fig. 5.29
Space group $I\bar{4}$, origin on $\bar{4}$. [Reproduced by courtesy of the IUCr (1).]

Fig. 3.12a shows that in point group $4/m$, which is centrosymmetric, two successive operations of ***4*** followed by ***m*** lead to a position that is equivalent to the operation of $\bar{\mathbf{1}}$ on the starting point. Space group $P4_2/n$ has the following intrinsic specification:

$$\bar{1} \text{ at } 0,\ 0,\ 0 \quad 4_2 \text{ along } [p,\ q,\ z] \quad n \text{ parallel to } (x,\ y,\ r)$$

Then, the following scheme can be constructed,

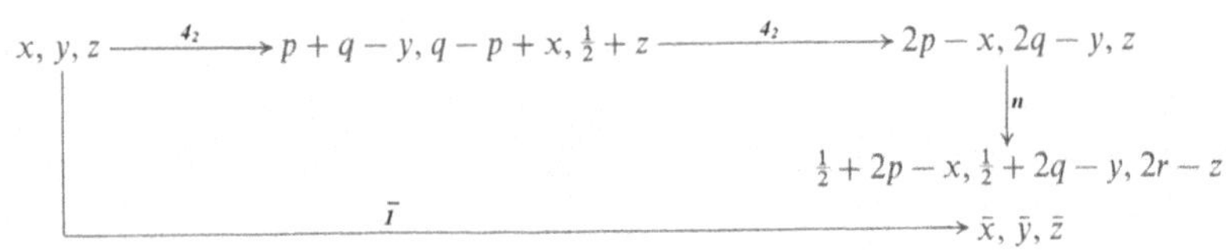

from which $p = q = {}^1/_4$ and $r = 0$, as in Fig. 5.30; the $\bar{4}$ axes are located at $\pm({}^1/_4,\ {}^3/_4,\ z)$ with their inversion points at $\pm({}^1/_4,\ {}^1/_4,\ {}^1/_4)$. The 4_2 axes lie at $\pm({}^1/_4,\ {}^1/_4,\ z)$. The preferred setting with the origin on $\bar{4}$ is shown in Fig. 5.31, obtained by a simple origin shift of Fig. 5.30; it allow a more symmetrical expression of the coordinates of the general equivalent positions. Two successive operations ***4₂*** are equivalent to a twofold rotation about the same axis, so that 2 is a subgroup of 4_2.

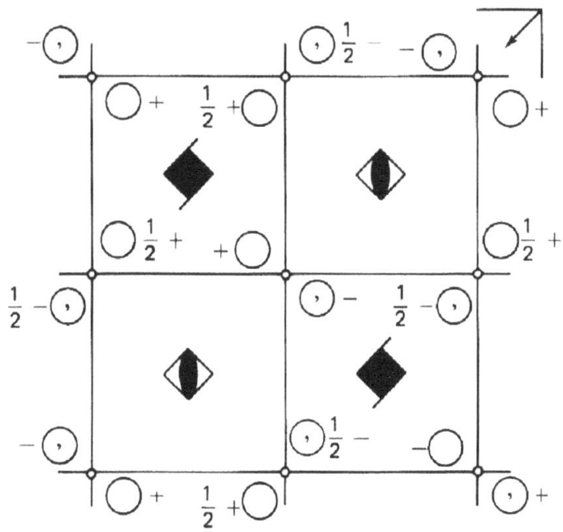

Fig. 5.30
Space group $P4_2/n$, origin on $\bar{1}$.

$P4_2/n$
C_{4h}^4

No. 86 $P\,4_2/n$ $4/m$ Tetragonal

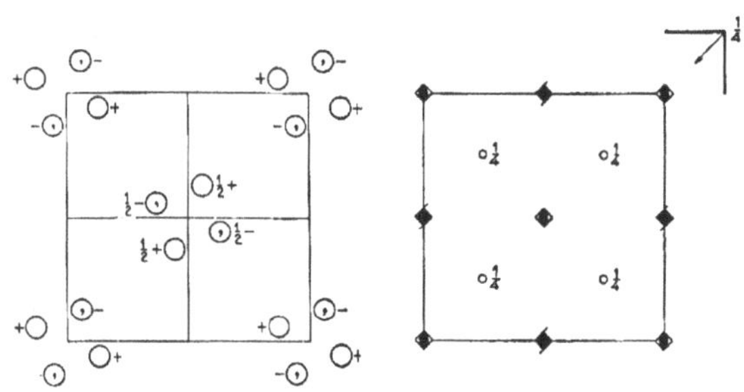

Origin at $\bar{4}$ at $\frac{\bar{1}}{4},\frac{\bar{1}}{4},\frac{\bar{1}}{4}$ from $\bar{1}$ (compare below for alternative origin)

Number of positions, Wyckoff notation, and point symmetry			Co-ordinates of equivalent positions	Conditions limiting possible reflections
				General:
8	*g*	1	x,y,z; $\bar{x},\bar{y},z$; $\frac{1}{2}+x,\frac{1}{2}+y,\frac{1}{2}-z$; $\frac{1}{2}-x,\frac{1}{2}-y,\frac{1}{2}-z$; $\bar{y},x,\bar{z}$; $y,\bar{x},\bar{z}$; $\frac{1}{2}-y,\frac{1}{2}+x,\frac{1}{2}+z$; $\frac{1}{2}+y,\frac{1}{2}-x,\frac{1}{2}+z$.	hkl: No conditions $hk0$: $h+k=2n$ $00l$: $l=2n$
				Special: as above, plus
4	*f*	2	$0,0,z$; $0,0,\bar{z}$; $\frac{1}{2},\frac{1}{2},\frac{1}{2}+z$; $\frac{1}{2},\frac{1}{2},\frac{1}{2}-z$.	hkl: $h+k+l=2n$
4	*e*	2	$0,\frac{1}{2},z$; $0,\frac{1}{2},\frac{1}{2}+z$; $\frac{1}{2},0,\bar{z}$; $\frac{1}{2},0,\frac{1}{2}-z$.	hkl: $l=2n$
4	*d*	$\bar{1}$	$\frac{1}{4},\frac{1}{4},\frac{3}{4}$; $\frac{3}{4},\frac{3}{4},\frac{3}{4}$; $\frac{1}{4},\frac{3}{4},\frac{1}{4}$; $\frac{3}{4},\frac{1}{4},\frac{1}{4}$;	hkl: $h+k=2n$, $k+l=2n$ and $(l+h=2n)$ (c and d)
4	*c*	$\bar{1}$	$\frac{1}{4},\frac{1}{4},\frac{1}{4}$; $\frac{3}{4},\frac{3}{4},\frac{1}{4}$; $\frac{1}{4},\frac{3}{4},\frac{3}{4}$; $\frac{3}{4},\frac{1}{4},\frac{3}{4}$.	
2	*b*	$\bar{4}$	$0,0,\frac{1}{2}$; $\frac{1}{2},\frac{1}{2},0$.	hkl: $h+k+l=2n$ (*a* and *b*)
2	*a*	$\bar{4}$	$0,0,0$; $\frac{1}{2},\frac{1}{2},\frac{1}{2}$.	

Fig. 5.31
Space group $P4_2/n$ origin on $\bar{4}$, $-1/4$, $-1/4$, $-1/4$ from $\bar{1}$; compare Fig. 5.30. [Reproduced by courtesy of the IUCr (1).]

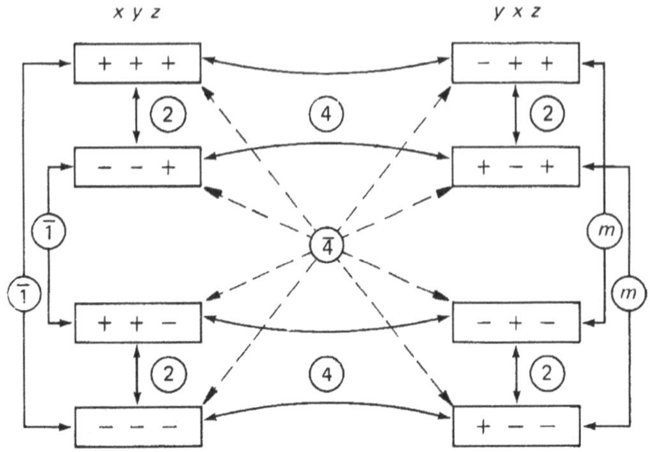

Fig. 5.32
Sign changes for x, y and z in space groups of class 4. The triplets x, y, z and y, x, z apply to their respective columns.

The relationships of the x, y, z coordinates in the space groups of Laue class $4/m$ are shown by the scheme in Fig. 5.32: it can be seen, for example that $x, y, z \xrightarrow{4} \bar{y}, x, z$, and $x, y, z \xrightarrow{\bar{4}} y, \bar{x}, \bar{z}$. The symmetry $\bar{4}$ is present in all space groups based on $4/m$, even though in some cases it is coincident with another symmetry element, as in the symmorphic $P4/m$. It must be remembered that the $\bar{4}$ element is not just a (conceptual) line, but encompasses a *point* through which the inversion is carried out; this point may be the origin or lie at a fraction of c, frequently $\pm{}^1\!/_4$ but sometimes $\pm({}^1\!/_8, {}^3\!/_8)$. Examine the four possible situations for $P4_2/n$; two have been given in Figs. 5.30 and 5.31.

5.4.12.2 Space groups within Laue class $\frac{4}{m}mm$

In point group $4mm$, the fourfold axis lies at the intersection of the two forms of m planes, one normal to the x and y axes, and the other normal to [110] and [$1\bar{1}0$], as listed in Table 3.6. Only one plane in each form need be considered, say, m normal to x, and m normal to [$1\bar{1}0$]. Then a scheme similar to that in Fig. 5.26 may be drawn up, and it can be instructive to do so, and to work out the relationships between the various coordinate triplets. The symbol $P4bm$ implies the standard settings,

4 along [0, 0, z] b parallel to (p, y, z) m parallel to (q, q, z)

Thus,

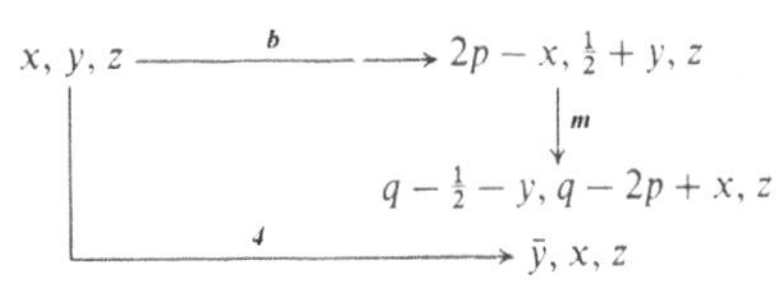

whence $q = {}^1\!/_2$ and $p = q/2 = {}^1\!/_4$, and the space group diagram is shown in Fig. 5.33. The diagonal m planes normal to [$1\bar{1}0$] and [110] make equal intersections on the a and b unit cell edges. The reader may like to confirm geometrically, or otherwise, that the reflection of a point x, y, z across the plane normal to [110] gives rise to the point $q - y$, $q - x$, z, a result that was used above.

$P4bm$
C_{4v}^{2}

No. 100 $P4bm$ $4mm$ Tetragonal

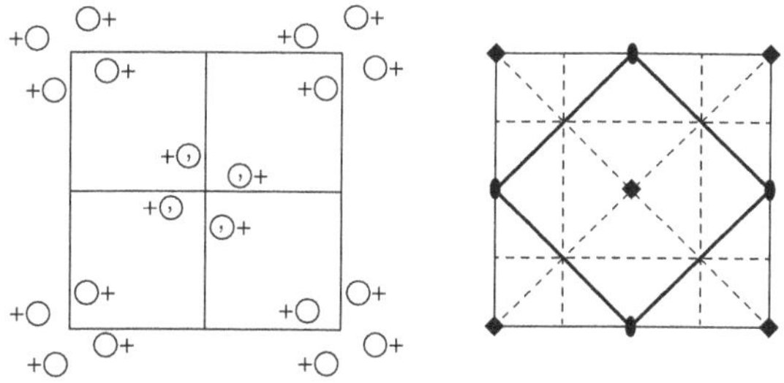

Origin on 4

Number of positions, Wyckoff notation, and point symmetry			Co-ordinates of equivalent positions	Conditions limiting possible reflections
				General:
8	d	1	$x,y,z;\ \bar{x},\bar{y},z;\ \tfrac{1}{2}+x,\tfrac{1}{2}-y,z;\ \tfrac{1}{2}-x,\tfrac{1}{2}+y,z;$ $\bar{y},x,z;\ y,\bar{x},z;\ \tfrac{1}{2}+y,\tfrac{1}{2}+x,z;\ \tfrac{1}{2}-y,\tfrac{1}{2}-x,z.$	hkl: No conditions $0kl$: $k=2n$ hhl: No conditions
				Special: as above, plus
4	c	m	$x,\tfrac{1}{2}+x,z;\ \bar{x},\tfrac{1}{2}-x,z;\ \tfrac{1}{2}+x,\bar{x},z;\ \tfrac{1}{2}-x,x,z.$	no extra conditions
2	b	mm	$\tfrac{1}{2},0,z;\ 0,\tfrac{1}{2},z.$	} hkl: $h+k=2n$
2	a	4	$0,0,z;\ \tfrac{1}{2},\tfrac{1}{2},z.$	}

Fig. 5.33
Space group $P4bm$, origin on 4. [Reproduced by courtesy of the IUCr (1).]

Consider next space group $I4cm$. An attempt to derive $I4cm$ as $P4cm + I$ leads to a result that may appear confusing. First, $P4cm$ does not exist under that name: the c glide planes cause the fourfold symmetry axis to become 4_2, and I centring of $P4_2cm$ then leads to $I4cm$. The space group diagram can be completed fully from this orientation and then, if necessary, the origin changed to the standard setting on 4; $I4cm$ can be obtained also by I centring of $P4bm$. The second point is that if $I4cm$ were named according to the rules of precedence it would be $I4bm$.

Occasional departures from the rules of precedence are made for good reasons: $P2_1/c$ is used instead of $P2_1/a$ because $P2_1/c + C \rightarrow C2/c$, whereas $P2_1/a + C \rightarrow C2/m$. Among orthorhombic space groups, $Ibca$ is used instead of $Ibaa$, because $Pbca + I \rightarrow Ibca$. $I4cm$ implies the existence of 4_2 which might not be so obvious from the equivalent symbol $I4bm$. The logic is sometimes a little difficult to discern. These matters have been discussed elsewhere [4,5].

The reader should, perhaps, devise schemes in order to work through other tetragonal space groups, say, $P4_12_12$ and $P\bar{4}2c$, using the full meaning of the Hermann–Mauguin space group symbols, so as to gain practice with the methods described above. In $P4_12_12$, for example, the 4_1 axis causes the 2_1 axes parallel to x and y to be separated by $c/4$, and similarly for the diagonal twofold axes. Since the latter axes interleave the first set in z, they will be separated from them by $c/8$ and $3c/8$. The origin is not fixed uniquely in z, so that either

2_1 (standard) or 4_1 may be set at 0, 0, z. In $P\bar{4}2c, \bar{4}$ rather than 2, lies at 0, 0, z, and is further located by the point of inversion on the axis, at 0, 0, 0: $x, y, z \xrightarrow{\bar{4}} y, \bar{x}, \bar{z}$.

The space groups, $P\frac{4}{n}bm$ and $I\frac{4_1}{a}md$ in this class will be examined. In these space groups the unique axis can always be set along [0, 0, z], which may result in the centre of symmetry being displaced from the origin. It is always possible to reset a group with $\bar{1}$ at 0, 0, 0. In $P\frac{4}{n}bm$, the fourfold axis is along [0, 0, z], with the perpendicular n glide plane as (x, y, r); then $\bar{1}$ can be set at p, q, 0. It is permissible to choose zero for the z coordinate of either $\bar{1}$ or the n glide plane because the origin is not defined with respect to z by the other symmetry elements. Where the plane normal to the unique axis is an m plane, $p = q = 0$, whereas if it is a glide plane, the values of p and q will depend upon the nature of the glide plane. The space group under discussion has the following settings:

4 along [0, 0, z] n as (x, y, r) $\bar{1}$ at p, q, 0 b as (s, y, z) m as (t, t, z)

where p, q, r, s and t are to be determined. It is worth noting that all this information is derived from the meaning of the symbol, and relates to and shows the importance of Table 3.6. There are two coordinate cycles that can be used here. Applying the first of them (**4**^{2} means two successive fourfold rotational operations):

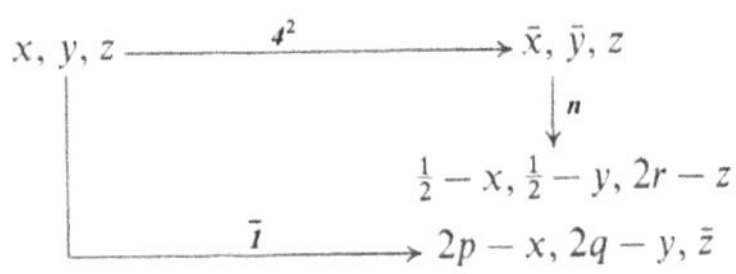

whence $p = q = 1/4$ and $r = 0$. The second cycle is

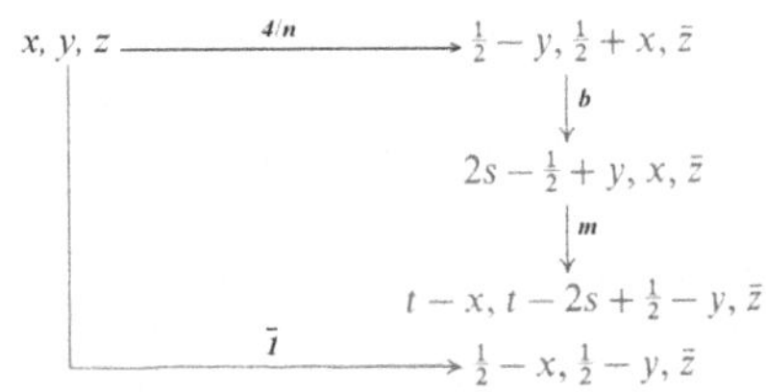

from which $t = 1/2$, and $s = t/2 = 1/4$, and gives the full orientation for $P\frac{4}{n}bm$, leading to the diagram in Fig. 5.34.

Two space groups based on point group $\frac{4}{m}mm$ exhibit d (diamond) glide planes, so called because of their presence in the diamond form of carbon, space group (cubic) $Fd\bar{3}m$; d glide planes occur among orthorhombic, tetragonal and cubic space groups.

In the tetragonal system, two groups with d glides are $I\frac{4_1}{a}md$ and $I\frac{4_1}{a}cd$, related to $I4_1md$ and $I4_1cd$ respectively in class $4mm$, and having translations of types either $(a \pm b)/4$ or $(a \pm b \pm c)/4$. They can be treated by the above methods; preferably, either $\bar{4}$ or $\bar{1}$ may be set at 0, 0, 0. Where $\bar{1}$ is set at the origin, the inversion point along the $\bar{4}$ axis is at $1/8$ and symmetry related points,

Tetragonal $4/mmm$ $P4/n\ 2/b\ 2/m$ No. 125 **$P4/nbm$** D_{4h}^{3}

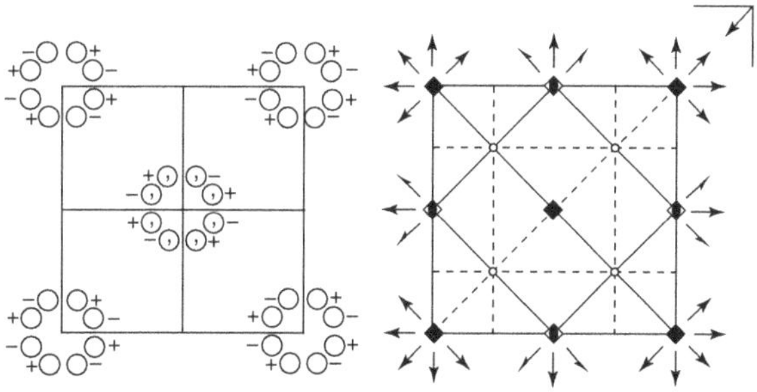

Origin at 422, at $\bar{\tfrac{1}{4}},\bar{\tfrac{1}{4}},0$ from centre ($2/m$) (compare next page for alternative origin)

Number of positions, Wyckoff notation, and point symmetry			Co-ordinates of equivalent positions	Conditions limiting possible reflections
				General:
16	n	1	x,y,z; $\bar{x},\bar{y},z$; $\tfrac12+x,\tfrac12+y,\bar{z}$; $\tfrac12-x,\tfrac12-y,\bar{z}$; $\bar{x},y,\bar{z}$; $x,\bar{y},\bar{z}$; $\tfrac12+x,\tfrac12-y,z$; $\tfrac12-x,\tfrac12+y,z$; $\bar{y},x,z$; $y,\bar{x},z$; $\tfrac12-y,\tfrac12+y,\bar{z}$; $\tfrac12+y,\tfrac12-y,\bar{z}$; $y,x,\bar{z}$; $\bar{y},\bar{x},\bar{z}$; $\tfrac12+y,\tfrac12+x,z$; $\tfrac12-y,\tfrac12-y,z$.	hkl: No conditions $hk0$: $h+k=2n$ $0kl$: $k=2n$ hhl: No conditions
				Special: as above, plus
8	m	m	$x,\tfrac12+x,z$; $\bar{x},\tfrac12-x,z$; $\tfrac12+x,x,\bar{z}$; $\tfrac12-x,\bar{x},\bar{z}$; $x,\tfrac12-x,\bar{z}$; $\bar{x},\tfrac12+x,\bar{z}$; $\tfrac12+x,\bar{x},z$; $\tfrac12-x,x,z$.	no extra conditions
8	l	2	$x,0,\tfrac12$; $\bar{x},0,\tfrac12$; $\tfrac12+x,\tfrac12,\tfrac12$; $\tfrac12-x,\tfrac12,\tfrac12$; $0,x,\tfrac12$; $0,\bar{x},\tfrac12$; $\tfrac12,\tfrac12+x,\tfrac12$; $\tfrac12,\tfrac12-x,\tfrac12$.	
8	k	2	$x,0,0$; $\bar{x},0,0$; $\tfrac12+x,\tfrac12,0$; $\tfrac12-x,\tfrac12,0$; $0,x,0$; $0,\bar{x},0$; $\tfrac12,\tfrac12+x,0$; $\tfrac12,\tfrac12-x,0$.	
8	j	2	$x,x,\tfrac12$; $\bar{x},\bar{x},\tfrac12$; $\tfrac12+x,\tfrac12+x,\tfrac12$; $\tfrac12-x,\tfrac12-x,\tfrac12$; $x,\bar{x},\tfrac12$; $\bar{x},x,\tfrac12$; $\tfrac12+x,\tfrac12-x,\tfrac12$; $\tfrac12-x,\tfrac12+x,\tfrac12$.	hkl: $h+k=2n$ (for l to g)
8	i	2	$x,x,0$; $\bar{x},\bar{x},0$; $\tfrac12+x,\tfrac12+x,0$; $\tfrac12-x,\tfrac12-x,0$; $x,\bar{x},0$; $\bar{x},x,0$; $\tfrac12+x,\tfrac12-x,0$; $\tfrac12-x,\tfrac12+x,0$.	
4	h	mm	$0,\tfrac12,z$; $0,\tfrac12,\bar{z}$; $\tfrac12,0,z$; $\tfrac12,0,\bar{z}$.	
4	g	4	$0,0,z$; $0,0,\bar{z}$; $\tfrac12,\tfrac12,z$; $\tfrac12,\tfrac12,\bar{z}$.	
4	f	$2/m$	$\tfrac14,\tfrac14,\tfrac12$; $\tfrac34,\tfrac34,\tfrac12$; $\tfrac14,\tfrac34,\tfrac12$; $\tfrac34,\tfrac14,\tfrac12$.	hkl: $h,k=2n$ (for f and e)
4	e	$2/m$	$\tfrac14,\tfrac14,0$; $\tfrac34,\tfrac34,0$; $\tfrac14,\tfrac34,0$; $\tfrac34,\tfrac14,0$.	

Fig. 5.34
Space group $P\frac{4}{n}bm$ origin on $\bar{1}$. The fourfold symmetry requires an a glide plane normal to the b glide plane; since the symmetry axis is 4_1, the corresponding operation places the b glide plane at $c/4$ from the a glide plane. [Reproduced by courtesy of the IUCr (1).]

rather than at 0 and ½, and in following the technique used with the space groups above, it is necessary to ensure that the correct direction of the d glide translation is applied to each of x, y and z, so as to bring the final coordinate triplet to the position from where $\bar{1}$ will return it to the original x, y, z position.

Another way of approaching space groups in this class is to begin with a related space group of lower symmetry; in the case of space group $I\frac{4_1}{a}md$, $I\frac{4_1}{a}$ can form a starting point. The scheme hereunder leads to a setting of $I\frac{4_1}{a}$ with the origin on $\bar{1}$; the settings are:

$$\bar{1} \text{ at } 0,\ 0,\ 0; \quad 4_1 \text{ at } [p,q,z]; \quad a(x,y,r)$$

Then, the following scheme may be applied,

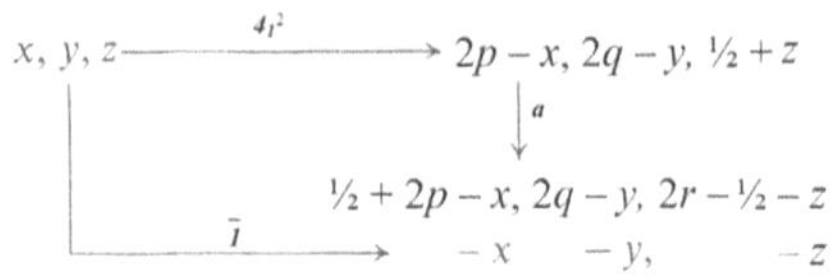

whence $p = {}^1\!/_4$, $q = 0$ and $r = {}^1\!/_4$, leading to the diagram of Fig. 5.35a. The b glide plane normal to z, while symmetry related to the a glide plane, is set off from it by $c/4$ because of the action of the 4_1 axis. Had the scheme used an equivalent b glide in place of the a glide, then the 4_1 axis would have arisen at [0, ¼, z], as though the diagram were rotated by 90°.

From Table 3.6, the m planes in $I\frac{4_1}{a}md$ are known to be perpendicular to x and y. Consider placing an m plane normal to the y axis: of the possible positions 0 (and ½) or ¼ (and ¾), it is clear that an m plane at (x, 0, z), passing through a fourfold screw axis, is not a tenable arrangement, so that these planes must be placed at $y = \pm{}^1\!/_4$. With the total information now to

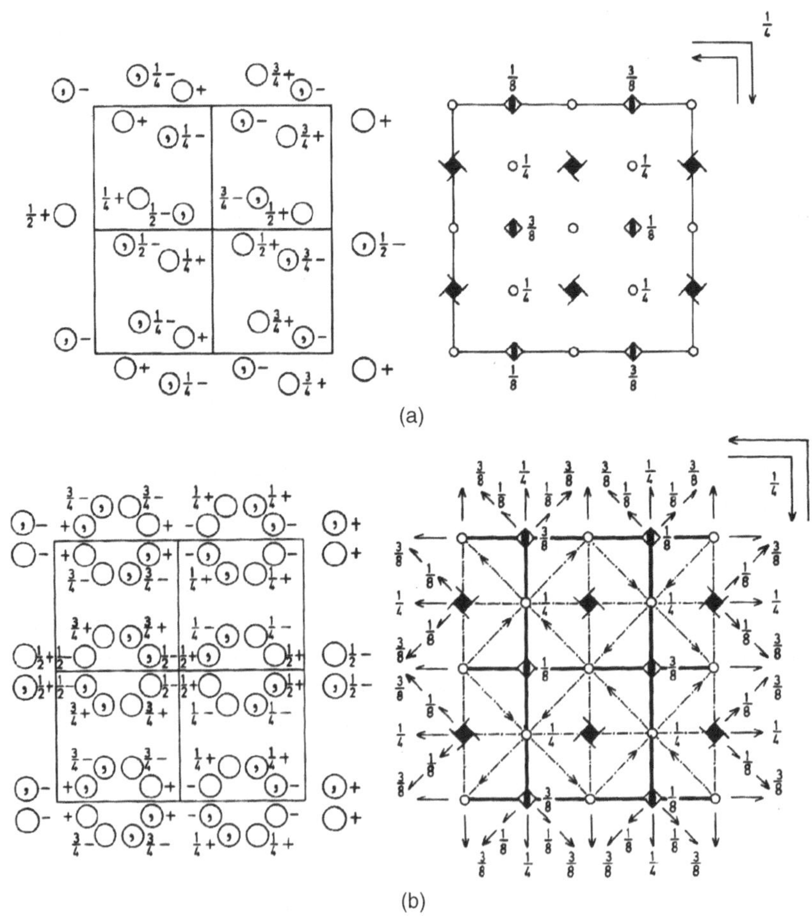

Fig. 5.35
(a) Space group $I\frac{4_1}{a}$, origin on $\bar{1}$, 0, ¼, ⅛ from an alternative origin on $\bar{4}$. a centre of symmetry, $(2/m)$; the tables referenced here give also the alternative diagram. (b) Space group $I\frac{4_1}{a}md$ with the origin on $2/m$, with $\bar{4}2m$ at 0, ¼, ⅜. [Reproduced by courtesy of the IUCr (1).]

hand, it is possible to construct a diagram of the general equivalent positions (Fig. 5.35b), from which the totality of symmetry elements can be identified.

An alternative setting of this space group places $\bar{4}$ at the origin; the corresponding diagram may be obtained by a simple origin shift, but try working this space group, starting with just $\bar{4}$ at the origin and the information from Table 3.6; the resulting diagram can be found in the literature [4,5].

5.4.12.3 *d* Glide planes

The *d* glide planes, of which this space group provides one example, occur in pairs in *I* or *F* centred cells of tetragonal and cubic lattices, and the translations associated with these glide planes are $^1/_4$ along a line parallel to the plane of the projection (the plane of the diagram) combined with $^1/_4$ normal to the plane of the diagram, or three such $^1/_4$-type translations in certain groups; an arrow on the diagram indicates the sense of the translation.

Consider the action of two *d* glide planes *A* and *B* perpendicular to *b* at $y = {}^1/_8$ and $^3/_8$ respectively in an orthorhombic space group. Reflection across plane *A* is followed by a translation of $(c + a)/4$ as is reflection across plane *B*. From a point *x*, *y*, *z* the points $^1/_4 + x, {}^3/_4 - y, {}^1/_4 + z$ and $-{}^1/_4 + x, {}^3/_4 - y, {}^1/_4 + z$ are derived from these *d* glide operations. By continued application of the *d* glide symmetry on these two points, eight points in all are obtained. The reader is invited to generate these points: they will then show the existence of two more *d* glide planes within the unit cell at $y = {}^5/_8$ and $^7/_8$ with translations of $(c + a)/4$ and $(c - a)/4$ respectively. A further discussion on *d* glide planes may be found in the literature [5,8], which latter reference studies the example given as the frontispiece to this book. The symbols for the total of 68 tetragonal space groups are listed in Table 5.4.

5.4.13 Space groups in the trigonal, hexagonal and cubic systems

Of the space groups in these three more complex crystal systems there will be a little less to say, because their occurrence is relatively much less frequent than those of the previous four systems, although they account for the remaining 88 of the 230 space groups. *'For true understanding, comprehension of detail is imperative'* [9].

5.4.13.1 Trigonal space groups

In the trigonal system, it is necessary to consider the threefold symmetry in combination with both the rhombohedral *R* and hexagonal *P* lattices, since both are compatible with a single threefold symmetry axis. The diagrams for space group *R*3*m*, for example, in Fig. 5.36 show the sites of the general equivalent positions and the location of the symmetry elements with respect both rhombohedral axes in the standard, *obverse* setting, and hexagonal axes, the latter leading to the hexagonal triply primitive *H* unit cell. This hexagonal cell has, as its name suggests, three times the volume of the primitive unit cell. It is notable that setting of this space group on hexagonal axes carries a set of limiting conditions that indicate the basic trigonal nature of this space group.

The action of a threefold axis in a rhombohedral unit cell on a point *x*, *y*, *z* can be seen by inspection to transform it cyclically it to *z*, *x*, *y* and then to

Table 5.4 The 68 tetragonal space groups.

$P4$	$P\bar{4}$	$P4/m$	$P422$	$P4mm$	$P\bar{4}2m$	$P\frac{4}{m}mm$
$P4_1$	$I\bar{4}$	$P4_2/m$	$P42_12$	$P4bm$	$P\bar{4}2c$	$P\frac{4}{m}cc$
$P4_2$		$P4/n$	$P4_122$	$P4_2cm$	$P\bar{4}2_1m$	$P\frac{4}{n}bm$
$P4_3$		$P4_2/n$	$P4_12_12$	$P4_2nm$	$P\bar{4}2_1c$	$P\frac{4}{n}nc$
$I4$		$I4/m$	$P4_222$	$P4cc$	$P\bar{4}m2$	$P\frac{4}{m}bm$
$I4_1$		$I4_1/a$	$P4_22_12$	$P4nc$	$P\bar{4}c2$	$P\frac{4}{m}nc$
			$P4_322$	$P4_2mc$	$P\bar{4}b2$	$P\frac{4}{n}mm$
			$P4_32_12$	$P4_2bc$	$P\bar{4}n2$	$P\frac{4}{n}cc$
			$I422$	$I4mm$	$I\bar{4}m2$	$P\frac{4_2}{m}mc$
			$I4_122$	$I4cm$	$I\bar{4}c2$	$P\frac{4_2}{m}cm$
				$I4_1md$	$I\bar{4}2m$	$P\frac{4_2}{n}bc$
				$I4_1cd$	$I\bar{4}2d$	$P\frac{4_2}{n}nm$
						$P\frac{4_2}{m}bc$
						$P\frac{4_2}{m}nm$
						$P\frac{4_2}{n}mc$
						$P\frac{4_2}{n}cm$
						$I\frac{4}{m}mm$
						$I\frac{4}{m}cm$
						$I\frac{4_1}{a}md$
						$I\frac{4_1}{a}cd$

y, z, x. Alternatively, the matrix $\mathbf{3}_{[111]}$ (Appendix A3.7.1.2) may be applied since a rhombohedron may be likened to a cube extended along its [111] direction (compare Appendix A3.7.2). As the points at $\pm$ ($2/3$, $1/3$, $1/3$) imply a form of centring, as discussed in Section 4.4.6, it is not surprising to find the m planes interleaved by glide planes parallel to $\{11\bar{2}0\}$.

The enantiomorphic space groups $P3_1$, origin on 3_1, and $P3_2$, origin on 3_2, are distinct and cannot be interconverted by any change of origin or other translation. The coordinates of the general positions are, for $P3_1$, x, y, z; $\bar{y}, x-y, 1/3+z$; $y-x, \bar{x}, 2/3+z$, and for $P3_2$, x, y, z; $\bar{y}, x-y, 2/3+z$; $y-x, \bar{x}, 1/3+z$, which are clearly different arrangements.

Other trigonal space groups, such as $P321$ and $P312$, are different from each other: the twofold axes are along x, y and u, in the first but perpendicular to the reference axes in the second (Fig. 5.37 and Fig. 5.38). Similar remarks apply to other primitive trigonal space groups, and the groups in this system will be studied further in the Problems section.

5.4.13.2 Hexagonal space groups

In the hexagonal system, the points related to x, y, z by sixfold rotation about z may be obtained directly from the matrix in Appendix A3.7.3, from the

$R3m$
C_{3v}^5

No. 160 $R3m$ $3m$ Trigonal

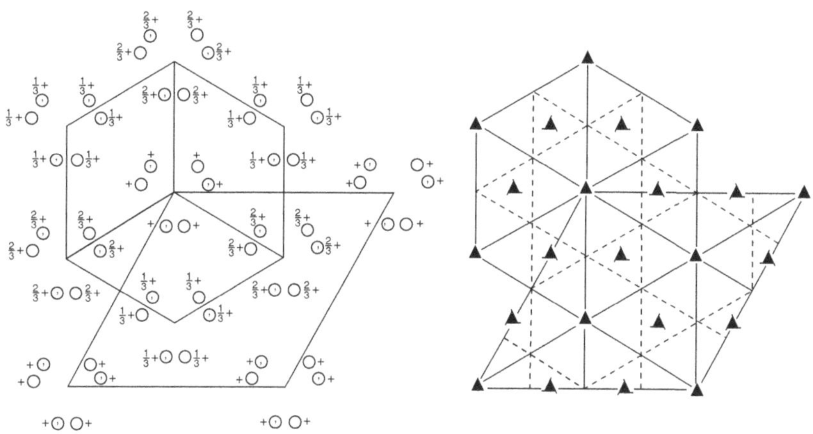

Origin on $3m$

Number of positions, Wyckoff notation, and point symmetry			Co-ordinates of equivalent positions	Conditions limiting possible reflections
			(1) RHOMBOHEDRAL AXES:	General:
6	*c*	1	x,y,z; z,x,y; y,z,x; y,x,z; z,y,x; x,z,y.	No conditions
				Special:
3	*b*	*m*	x,x,z; x,z,x; z,x,x.	No conditions
1	*a*	$3m$	x,x,x.	
			(2) HEXAGONAL AXES: $(0,0,0;\ \frac{1}{3},\frac{2}{3},\frac{2}{3};\ \frac{2}{3},\frac{1}{3},\frac{1}{3})+$	General:
18	*c*	1	$x,y,z;\ \bar{y},x,-y,z;\ y-x,\bar{x},z;$ $\bar{y},\bar{x},z;\ x,x,-y,z;\ y-x,y,z.$	$hkil$: $-h+k+l=3n$ $hh\bar{2h}l$: $(l=3n)$ $h\bar{h}0l$: $(h+l=3n)$
				Special: as above only
9	*b*	*m*	$x,\bar{x},z;\ y,2x,z;\ 2\bar{x},\bar{x},z.$	
3	*a*	$3m$	$0,0,z.$	

Fig. 5.36
Space group $R3m$ on (1) rhombohedral axes and (2) hexagonal axes, showing the sites of the general equivalent positions and the symmetry elements. It can be seen that the 6 positions in the rhombohedral cell transform to 18 when referred to the hexagonal cell. The limiting conditions of the two settings should be noted: a crystal with threefold symmetry indexed as hexagonal may be recognized as trigonal if $-h+k+l=3n$. [Reproduced by courtesy of the IUCr (1).]

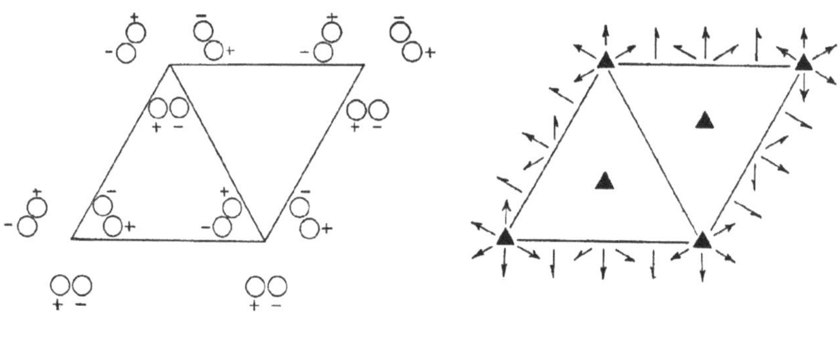

Fig. 5.37
Trigonal space groups with point group 32: $P312$, origin on 312, showing general equivalent positions and symmetry elements.

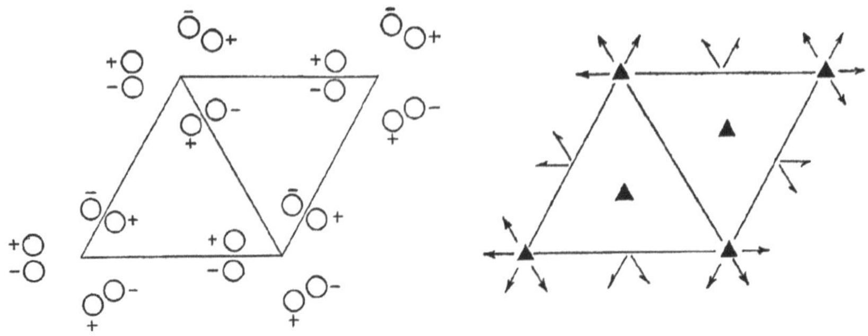

Fig. 5.38
Trigonal space groups with point group 32: *P*321, origin on 321, showing the general equivalent positions and symmetry elements. In this group and that shown in Fig. 5.37, the third position in the symbol, unnecessary in point group 32, is relevant here because of the differing dispositions of the sets of points with respect to the orientations of the twofold axes.

Eq. (A6.6) in Appendix A6, or from the combination of operations $\boldsymbol{23^2} \equiv \boldsymbol{6}$, where the twofold and sixfold axes are coincident, as the following scheme shows:

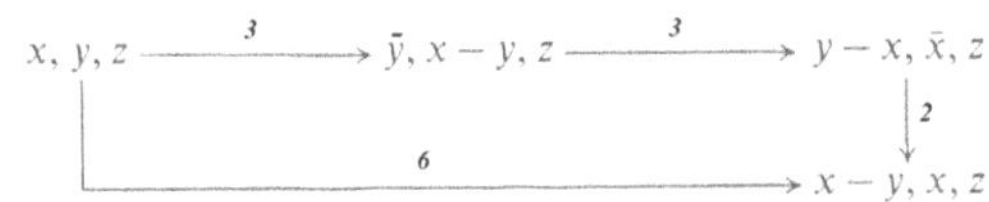

The general orientation of symmetry elements again follows the rules in Table 3.6, the actual positions in any given space group depending on the nature of the symmetry elements of the group; the sixfold symmetry axis, rotation, inversion or screw, is set along the line [0, 0, z].

The space groups in Laue classes $6/m$ are quite straightforward. Other hexagonal space groups can be handled by schemes similar to those used for the lower symmetry systems.

Consider space group $P6_3mc$, a group based on point group $6mm$. From Table 3.6, the second position in the symbol refers to symmetry 2 or $\bar{2}$ along x, y and u which, in $6mm$, means reflection planes perpendicular to x, y and u. A starting point x, y, z is moved around the 6_3 axis to $x-y,\ x,\ {}^1\!/_2+z$, and the point so generated is taken across the m plane at $(x,\ 2x,\ z)$ to $y,\ x,\ {}^1\!/_2+z$. From the earlier study on tetragonal space groups, it is clear that the two points x, y, z and $y, x, {}^1\!/_2+z$, generated are related by a c glide plane at (x, x, z). These relationships are shown in the following scheme:

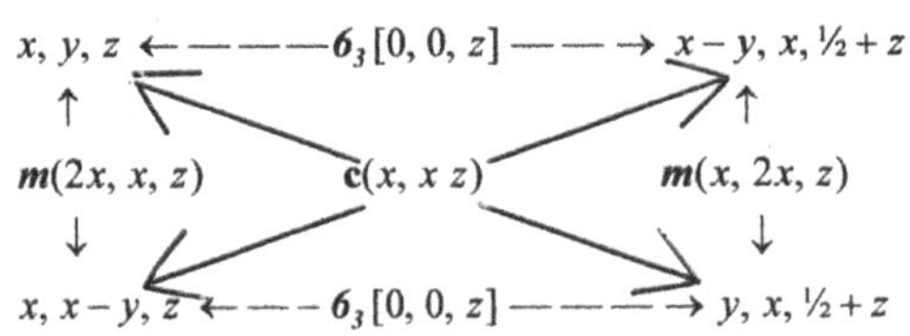

The operation across the plane $(2x,\ x,\ z)$ is symmetry related to that across $(x,\ 2x,\ z)$, and to m at $(x, \bar{x}, z)$, but does not add any new information. These results lead to the coordinates of the general and special equivalent positions (Fig. 5.39a):

				Limiting conditions
				General:
12	d	1	$x,y,z;\ \bar{y},x-y,z;\ y-x,\bar{x},z;$	$hkil$: None
			$\bar{y},\bar{x},z;\ x,x-y,z;\ y-x,y,z;$	$hh2\bar{h}l$: $l=2n$
			$\bar{x},\bar{y},{}^1\!/_2+z;\ y,y-x,{}^1\!/_2+z;\ x-y,x,{}^1\!/_2+z;$	$h\bar{h}0l$: None
			$y,x,{}^1\!/_2+z;\ \bar{x},y-x,{}^1\!/_2+z;\ x-y,\bar{y},{}^1\!/_2+z.$	
				Special:
:6	c	m	$x,\bar{x},z;\ x,2x,z;\ 2\bar{x},\bar{x},z;$	as above
			$\bar{x},x,{}^1\!/_2+z;\ \bar{x},2\bar{x},{}^1\!/_2+z;\ 2x,x,{}^1\!/_2+z$	
2	b	$3m$	${}^1\!/_3,{}^2\!/_3,z;\ {}^2\!/_3,{}^1\!/_3,{}^1\!/_2+z.$	as above +
				If $h-k=3n$, then $l=2n$
2	a	$3m$	$0,0,z;\ 0,0,{}^1\!/_2+z$	as above +
				$hkil$: $l=2n$

The completed diagram allows the limiting conditions to be determined and also reveals the positions of the other symmetry elements of the space group (Fig. 5.39b): threefold rotation axes at ${}^2\!/_3, {}^1\!/_3, z$ and ${}^1\!/_3, {}^2\!/_3, z$, twofold screw axes

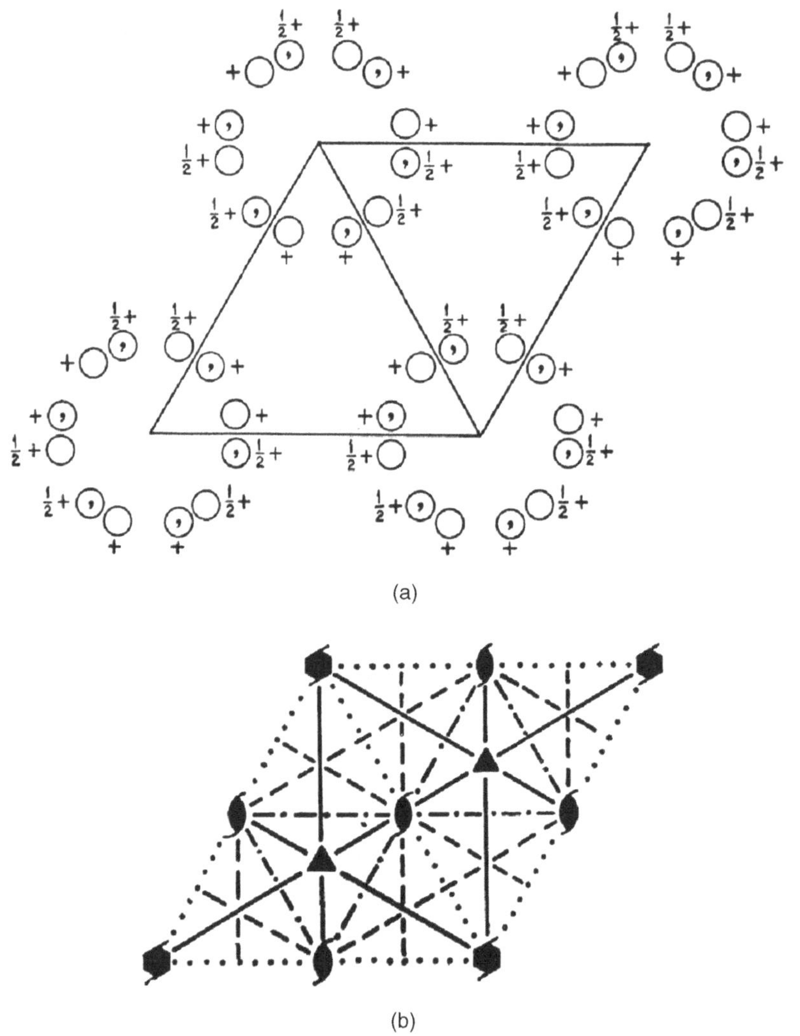

Fig. 5.39
Hexagonal space group $P6_3mc$. (a) General equivalent positions. (b) Symmetry elements.

at $^1/_2$, 0, z and 0, $^1/_2$, z; n glide planes interleaving the c glide planes parallel to $\{10\bar{1}0\}$, and a and b glide planes interleaving the m planes parallel to $\{11\bar{2}0\}$.

In studying the trigonal and hexagonal space groups, it is often helpful to have recourse to an appropriate stereogram, as noted already. Figure 5.40 illustrates the stereograms for point groups (a) $\bar{3}m$ (hexagonal axes) and (b) $\frac{6}{m}mm$.

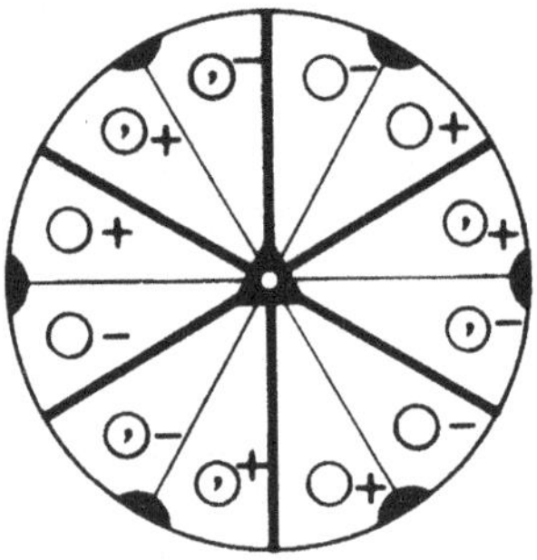

(a)

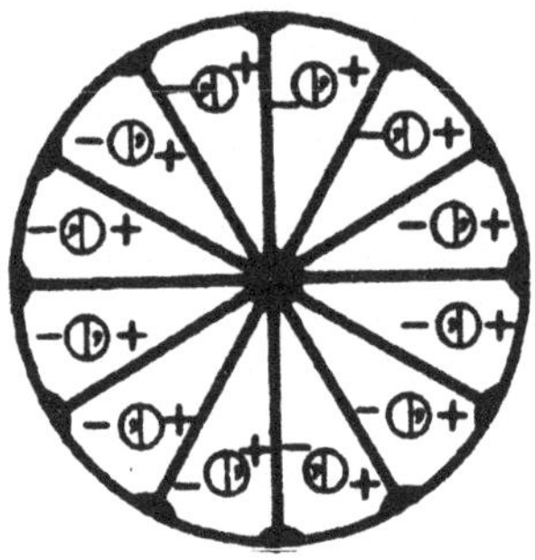

(b)

Fig. 5.40
Trigonal and hexagonal stereograms. (a) Point group $\bar{3}m$ with Miller indices. (b) Point group $\bar{3}m$ with Miller–Bravais indices. (c) Point group $\frac{6}{m}mm$.

Example 5.2
Find the transformation matrix for an obverse rhombohedral unit cell to a triply primitive hexagonal unit cell. Hence, confirm that the Wyckoff (c) positions in $R\bar{3}m$, $\pm(x, x, x)_R$, transform as $\pm(0, 0, z)_H$.

The transformation matrix for $a_{R_{obv}}$ to a_H is $\begin{pmatrix} 1 & \bar{1} & 0 \\ 0 & 1 & \bar{1} \\ 1 & 1 & 1 \end{pmatrix}$ and applying its transformed inverse: $\begin{pmatrix} ^2/_3 & ^{\bar{1}}/_3 & ^{\bar{1}}/_3 \\ ^1/_3 & ^1/_3 & ^{\bar{2}}/_3 \\ ^1/_3 & ^1/_3 & ^1/_3 \end{pmatrix} \begin{pmatrix} \pm x \\ \pm x \\ \pm x \end{pmatrix} = \begin{pmatrix} 0 \\ 0 \\ \pm z \end{pmatrix}$. Thus, the transformation of coordinates is $\pm(x, x, x)_R \rightarrow \pm(0, 0, z)_H$, to which must be added the translations $(^2/_3, ^1/_3, ^1/_3)$ and $(^1/_3, ^2/_3, ^2/_3)$ appropriate to the hexagonal cell.

5.4.14 Cubic space groups

The space groups in the cubic system are more complex still, mainly on account of the increased number of symmetry elements present and the inclination of the threefold symmetry axes to the reference axes. Table 3.6 shows that the first position in the space group symbol refers to the x, y and z directions, and can indicate twofold or fourfold rotational, inverse or screw symmetry. The second position is that of the characteristic threefold symmetry, 3 or $\bar{3}$, along <111>, while the third position refers to the <110> form of directions, with 2 or $\bar{2}$ symmetry. Space groups from both Laue classes will be considered here, and $F\bar{4}3c$ in the next section.

5.4.14.1 Laue class $m\bar{3}$

In space group $P23$, the simplest cubic space group, rotation about [111] has the same cyclic transformation of coordinates as in $R3$, namely

$$x,\ y,\ z \xrightarrow{3_{[111]}} z,\ x,\ y \xrightarrow{3_{[111]}} y,\ z,\ x$$

The remaining nine general positions for this space group can be generated by twofold rotations about the x, y, z axes. One of several such procedures is shown by Table 5.5. Twofold rotations are implicit in a group of four threefold axes set at $\cos^{-1}(1/\sqrt{3})$ to x, y and z:

$$\underset{3_{[\bar{1}11]}}{\begin{pmatrix} 0 & \bar{1} & 0 \\ 0 & 0 & 1 \\ \bar{1} & 0 & 0 \end{pmatrix}} \underset{3_{[111]}}{\begin{pmatrix} 0 & 0 & 1 \\ 1 & 0 & 0 \\ 0 & 1 & 0 \end{pmatrix}} = \underset{2_{[010]}}{\begin{pmatrix} \bar{1} & 0 & 0 \\ 0 & 1 & 0 \\ 0 & 0 & \bar{1} \end{pmatrix}}$$

Table 5.5 General equivalent positions for space group *P*23.

(7) $\bar{x}, y, \bar{z}$	$\overset{3^2_{[1\bar{1}1]}}{\longleftrightarrow}$	(8) $\bar{z}, x, \bar{y}$	$\overset{3^2_{[1\bar{1}1]}}{\longleftrightarrow}$	(9) $\bar{y}, z, \bar{x}$
$\updownarrow 2_y$		$\updownarrow 2_y$		$\updownarrow 2_y$
(1) x, y, z	$\overset{3_{[111]}}{\longleftrightarrow}$	(2) z, x, y	$\overset{3_{[111]}}{\longleftrightarrow}$	(3) y, z, x
$\updownarrow 2_x$		$\updownarrow 2_x$		$\updownarrow 2_x$
(4) $x, \bar{y}, \bar{z}$	$\overset{3^2_{[\bar{1}11]}}{\longleftrightarrow}$	(5) $z, \bar{x}, \bar{y}$	$\overset{3^2_{[\bar{1}11]}}{\longleftrightarrow}$	(6) $y, \bar{z}, \bar{x}$
$\updownarrow 2_z$		$\updownarrow 2_z$		$\updownarrow 2_z$
(10) $\bar{x}, \bar{y}, z$	$\overset{3^2_{[11\bar{1}]}}{\longleftrightarrow}$	(11) $\bar{z}, \bar{x}, y$	$\overset{3^2_{[11\bar{1}]}}{\longleftrightarrow}$	(12) $\bar{y}, \bar{z}, x$

Because of these relationships in the cubic system, the symbol *R* in Table 3.5 refers to one of the symmetry elements 2, 4, $\bar{2}(m)$ or $\bar{4}$; threefold symmetry is always present in the cubic system. The scheme of general equivalent positions in Table 5.5 can form a basis for studying the cubic space groups; the general positions follow from the analysis above, or may be determined readily from an inspection of the stereogram for point group 23 (Fig. 3.12c), since there are no translational symmetry elements in this group (Fig. 5.41).

An interesting feature of space group *P*23 and, indeed, of all cubic space groups, whether or no other screw symmetry is present, is the existence of threefold screw axes inclined to the plane of the diagram, which arise from the interaction of other symmetry elements in the space group with the threefold

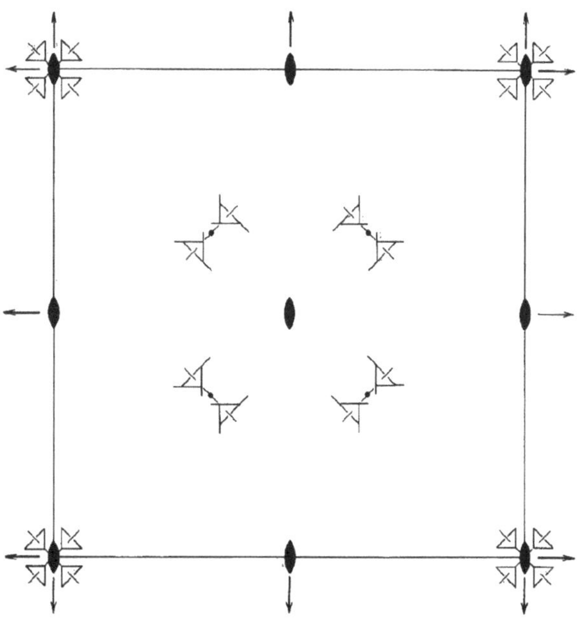

Fig. 5.41
Diagram to show the symmetry elements in *P*23, origin on 2. [Reproduced by courtesy of the IUCr (A).]

axes along <111>. They are not obvious from the lists of equivalent positions, but their presence may be detected, for example, by looking along diagonal directions on a crystal structure model, sodium chloride, for example, space group $Fm\bar{3}m$. That they would exist in the cubic system may be reasoned from the facts that they are present in rhombohedral space groups and that a rhombohedron may be viewed as a cube extended (or compressed, if $\alpha > 180°$ is acceptable) along [111].

In a cubic P unit cell, an anticlockwise rotation of $2\pi/3$, generates 3_1 axes at $(u + {}^1\!/_3)(\mathbf{a} + \mathbf{b} + \mathbf{c})$, where u is any integer, whereas for rotation in a similar sense, but of amount $4\pi/3$, the screw axes are at $(u + 2/3)(\mathbf{a} + \mathbf{b} + \mathbf{c})$. This symmetry can be considered further by reference to Fig. 5.41: 3_1 axes may be seen intersecting the diagram (the dot marks an intersection at $z = 0$) at 1/3, 2/3, 0, and 3_2 axes at 2/3, 1/3, 0, and symmetry related points throughout the lattice; these screw axes neither add new coordinate positions to the general set generated by $P23$, nor create special limiting conditions.

Another example in this class is space group $P2_13$ (Fig. 5.42a), which, unlike $P23$, shows threefold axes other than those along <111>. Consider the threefold axis along [111], marked A in the figure. The twofold screw axis operator at B along $[{}^1\!/_4, 0, z]$, symmetry operation (2) for this space group in ITA, rotates the three-atom group at A around the c direction and translates it by $c/2$ to the position marked C (Fig. 5.42b). Thus, this threefold axis intersects the x, y plane of the unit cell at $[1, {}^1\!/_2, 0]$ which is the point C in the space group diagram (Fig. 5.42a). Thus, a point x, y, z is transformed to $1 - z$, $-{}^1\!/_2 + x$, ${}^1\!/_2 - y$, as can be appreciated in Fig. 5.42a, and which will be discussed further in Section 8.5.4.

The body-centred space groups $I23$ and $I2_13$ are reminiscent of $I222$ and $I2_12_12_1$ in that the first space group of each pair has three mutually perpendicular intersecting twofold and twofold screw axes, whereas in the second group of each pair these sets of axes are non-intersecting. In $I2_13$, as in $I2_12_12_1$, the origin is set at the mid-point of three pairs of non-intersecting twofold axes, and of three non-intersecting twofold screw axes. The reader may wish to compare the diagrams of $I2_12_12_1$ and $I2_13$ [5].

A further not too difficult example is space group $Pn\bar{3}$ (Fig. 5.43). There are two important settings of this space group: the origin may be taken either at point symmetry 23, $-{}^1\!/_4$, $-{}^1\!/_4$, $-{}^1\!/_4$ from $\bar{1}$, or at a centre of symmetry, actually $\bar{3}$, at ${}^1\!/_4$, ${}^1\!/_4$, ${}^1\!/_4$ from 23. Although the first choice gives a simpler expression to the coordinates of the general equivalent positions, the second choice of origin will be adopted here: in the age of sophisticated crystallographic software, manual manipulation of coordinates is seldom needed. The n glide planes are perpendicular to x, y and z, so that only one entry for them arises in the space group symbol. Since each glide plane involves two half translations, the twofold symbol refers to pure rotation, and the full symbol may be written as $P\frac{2}{n}\bar{3}$. The unique centres of symmetry in the unit cell, within $\bar{3}$, are located at $({}^1\!/_2, {}^1\!/_2, {}^1\!/_2) + (0, 0, 0; {}^1\!/_2, 0, 0; 0, {}^1\!/_2, 0; {}^1\!/_2, {}^1\!/_2, 0)$. Thus, the settings chosen are: $\bar{1}$ at 0, 0, 0, 2 along $[p, q, z]$ and n parallel to (x, y, r). A straightforward analysis of the type used previously in this chapter shows that the n glide planes lie at $(0, y, z)$, $(x, 0, z)$ and $(x, y, 0)$, and the twofold axes at $\pm({}^1\!/_4, {}^1\!/_4, z; {}^3\!/_3, {}^1\!/_4, z)$, but try it for yourself and confirm the results for

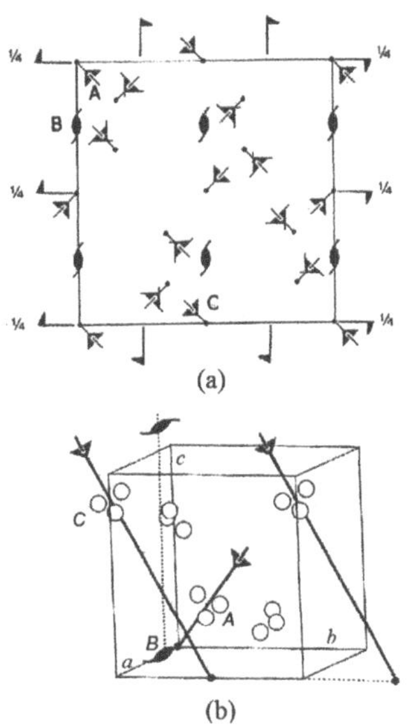

Fig. 5.42
Space group $P2_13$. (a) Diagram of the symmetry elements. (b) Formation of a threefold axis intersecting the plane of projection ($z = 0$) at $x = 1$, $y = {}^1\!/_2$. [Reproduced by courtesy of Professor A. M. Glazer, Clarendon Laboratory.]

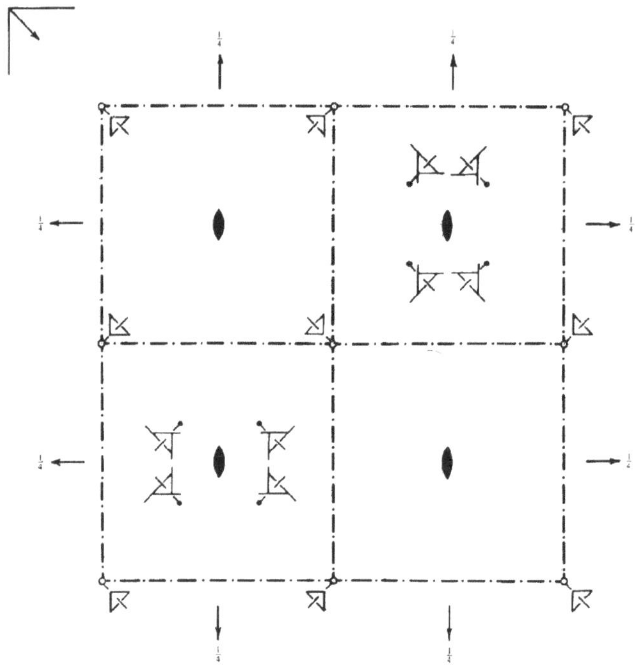

Fig. 5.43
Projection of space group $Pn\bar{3}$, origin on 23, showing the symmetry elements of the group. [Reproduced by courtesy of the IUCr (A).]

the general and special positions. Then, a diagram for the space group can be drawn (Fig. 5.43), and the limiting conditions deduced.

Number of positions Wyckoff notation, and point symmetry			Coordinates of equivalent positions	Limiting conditions
				General:
24	h	1	$\pm(x, y, z;\ x, 1/2-y, 1/2-z;\ 1/2-x, y, 1/2-z;\ 1/2-x, 1/2-y, z;$	hkl : No conditions
			$z, x, y;\ z, 1/2-x, 1/2-y;\ 1/2-z, x, 1/2-y;\ 1/2-z, 1/2-x, y;$	$0kl : k+l=2n$
			$y, z, x;\ y, 1/2-z, 1/2-x;\ 1/2-y, z, 1/2-x;\ 1/2-y, 1/2-z, x)$	and cyclic permutations
				Special: as above +
12	g	2	$\pm(x, 3/4, 1/4;\ 1/4, x, 3/4;\ 3/4, 1/4, x;\ 1/2+x, 1/4, 3/4;\ 3/4, 1/2+x, 1/4;\ 1/4, 3/4, 1/2+x)$	$hkl : h+k+l=2n$
12	f	2	$\pm(x, 1/4, 1/4;\ 1/4, x, 1/4;\ 1/4, 1/4, x;\ 1/2+x, 3/4, 3/4;\ 3/4, 1/2+x, 3/4;\ 3/4, 3/4, 1/2+x)$	$hkl : h+k+l=2n$
8	e	3	$\pm(x, x, x;\ x, 1/2-x; 1/2-x;\ 1/2-x, x, 1/2-x;\ 1/2-x, 1/2-x, x)$	No extra conditions
6	d	222	$\pm(1/4, 3/4, 3/4;\ 3/4, 1/4, 3/4;\ 3/4, 3/4, 1/4)$	$hkl : h+k+l=2n$
4	c	$\bar{3}$	$1/2, 1/2, 1/2;\ 1/2, 0, 0;\ 0, 1/2, 0;\ 0, 0, 1/2$	$hkl : h+k, k+l, (l+h)=2n$
4	b	$\bar{3}$	$0, 0, 0;\ 0, 1/2, 1/2;\ 1/2, 0, 1/2;\ 1/2, 1/2, 0$	$hkl : h+k, k+l, (l+h)=2n$
2	a	23	$1/4, 1/4, 1/4;\ 3/4, 3/4, 3/4$	$hkl : h+k+l=2n$

If it is desired to obtain the coordinates in a more symmetrical form by placing the origin on 23, then a shift of origin by $-1/4, -1/4, -1/4$ will produce the required result (Fig. 5.44). It may be noted that the twofold axis in Fig. 5.43 is not specified in z; thus, the value of $1/4$ for this coordinate is arbitrary, but leads to convenient values for the general positions; note also that the n glide planes are now $(x, y, 1/4)$ and cyclic permutations. Another space group in

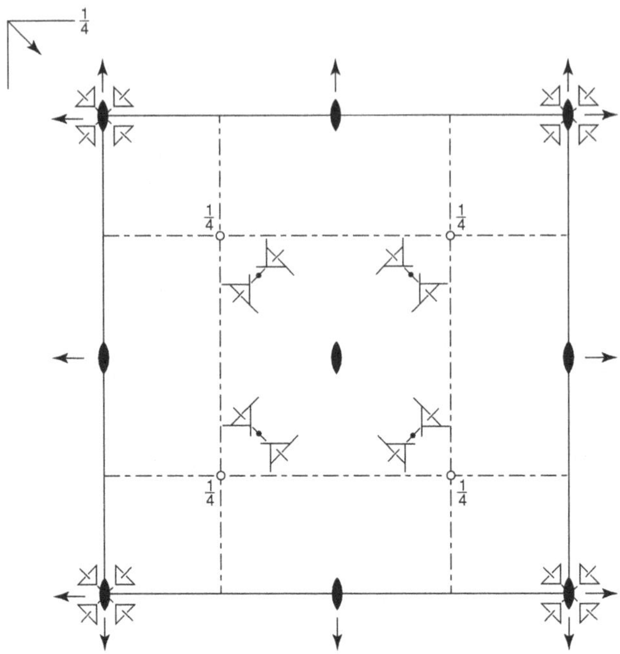

Fig. 5.44
Space group $Pn\bar{3}$, with the origin on $\bar{1}$; the n glide plane are now at $\pm c/4$. [Reproduced by courtesy of the IUCr (A).]

the same point group is $Pa\bar{3}$, which is exhibited by many structures, including pyrite, FeS_2 and the alums (see Section 5.4.15.2). In this space group, there is screw symmetry parallel to each reference axis; hence, the full symbol is $P\frac{2_1}{a}\bar{3}$.

5.4.14.2 Laue class $m\bar{3}m$

The cubic space groups of higher symmetry can be treated by the above methods but are more complicated, and not often encountered in practice. However, a space group based on point group 432 will be described, as certain biological species, such as cowpea chlorotic mottle virus [10] (Fig. 3.42) and a bacteriophage MS2 coat protein dimer [11], both exhibit space group $F432$; crystals of $SrSi_2$ crystallize, perhaps surprisingly, in space group $P4_332$.

This space group has 96 general equivalent positions, but if treated as $P432 + (0, 0, 0; 0, ½, ½; ½, 0, ½, ½, ½, 0)$ there are only 24 unique general positions. However, the translations introduced by the F centring give rise to limiting conditions that are absent in $P432$.

The stereogram for point group 432 is illustrated in Fig. 5.45. The first twenty-four positions can be generated starting from any point on the diagram and using the symmetry operations $\boldsymbol{3}_{[111]}$, $\boldsymbol{4}_{[00z]}$ and $\boldsymbol{2}_{[x00]}$ in succession. The F unit cell translations then takes this number to the complete ninety-six general equivalent positions. The reader may wish to perform this exercise, and then show that, in addition to the general limiting conditions normal for F centring, the other translational symmetry gives rise to conditions for the special equivalent positions: one set of symmetry 2, and sets of symmetries 222 and 23, all having the condition $hkl : h, \ (k, l) = 2n$.

Fig. 5.45
Stereogram of point group 432 in conventional notation.

Table 5.6 Trigonal, hexagonal and cubic space groups.

Trigonal						
$P3$	$P\bar{3}$	$P312$	$P3m1$	$P\bar{3}1m$		
$P3_1$	$R\bar{3}$*	$P321$	$P31m$	$P\bar{3}1c$		
$P3_2$		$P3_112$	$P3c1$	$P\bar{3}m1$		
$R3$*		$P3_121$	$P31c$	$P\bar{3}c1$		
		$P3_212$	$R3m$*	$R\bar{3}m$*		
		$P3_221$	$R3c$*	$R\bar{3}c$*		
		$R32$*				
Hexagonal						
$P6$	$P\bar{6}$	$P6/m$	$P622$	$P6mm$	$P\bar{6}m2$	$P\frac{6}{m}mm$
$P6_1$		$P6_3/m$	$P6_122$	$P6cc$	$P\bar{6}c2$	$P\frac{6}{m}cc$
$P6_5$			$P6_522$	$P6_3cm$	$P\bar{6}2m$	$P\frac{6_3}{m}cm$
$P6_2$			$P6_222$	$P6_3mc$	$P\bar{6}2c$	$P\frac{6_3}{m}mc$
$P6_4$			$P6_422$			
$P6_3$			$P6_322$			
Cubic						
$P23$	$Pm\bar{3}$	$P432$	$P\bar{4}3m$			
$F23$	$Pn\bar{3}$	$P4_232$	$F\bar{4}3m$	$Pn\bar{3}n$		
$I23$	$Fm\bar{3}$	$F432$	$I\bar{4}3m$	$Pm\bar{3}n$		
$P2_13$	$Fd\bar{3}$	$F4_132$	$P\bar{4}3n$	$Pn\bar{3}m$		
$I2_13$	$Im\bar{3}$	$I432$	$F\bar{4}3c$	$Fm\bar{3}m$		
	$Pa\bar{3}$	$P4_332$	$I\bar{4}3d$	$Fm\bar{3}c$		
	$Ia\bar{3}$	$P4_132$		$Fd\bar{3}m$		
		$I4_132$		$Fd\bar{3}c$		
				$Im\bar{3}m$		
				$Ia\bar{3}d$		

* These space groups are listed in the international tables for both rhombohedral and hexagonal axes.

The discussion of the trigonal, hexagonal and cubic space groups is completed by the list thereof in Table 5.6, and a reference list of three-dimensional symmetry symbols is given in Tables 5.6–5.8.

5.4.15 Space groups and crystal structures

At this point, it is useful to consider the importance of knowledge of symmetry in the determination of a crystal structure. All the necessary components of the symmetry of a crystal structure have been discussed in detail, and the following examples indicate how the application of space group symmetry can facilitate a crystal structure analysis.

5.4.15.1 Sodium chloride

A stereo view of a face-centred cubic unit cell is shown in Fig. 5.46a, which represents space group $Fm\bar{3}m$. A motif of pattern, the Na^+Cl^- ion pair is illustrated by Fig. 5.46b; the Na^+–Cl^- distance is 2.814 Å at room temperature. If this ion pair is associated in one and the same given orientation with each point of the lattice, the ideal crystal structure of sodium chloride is obtained (Fig. 5.46c). It is as though the Na^+Cl^- ion pair were 'multiplied' by the

Table 5.7 Notation for symmetry axes in three dimensions.

Symbol	Graphic symbol	Component of translation
1	none	
$\bar{1}$	∘	
2	⬮ ⊥ paper	
	→ // paper	
2_1	⊥ paper	$c/2$
	paper	$a/2$ or b/c
3	▲	
3_1	▲	$c/3$
3_2	▲	$2c/3$
$\bar{3}$	▲	
4	◆	
4_1	◆	$c/4$
4_2	◆	$2c/4 = c/2$
4_3	◆	$3c/4$
$\bar{4}$	◆	
6	⬢	
6_1	⬢	$c/6$
6_2	⬢	$2c/6 = c/3$
6_3	⬢	$3c/6 = c/2$
6_4	⬢	$4c/6 = 2c/3$
6_5	⬢	$5c/6$
$\bar{6}$	⬢	

Table 5.8 Notation for symmetry planes[a] in three dimensions.

Symbol	Graphic symbol	Glide translation
m	⊥ paper // paper	
a	⊥ paper // paper	$a/2$ $a/2$
b	⊥ paper // paper	$b/2$ $b/2$
c	⊥ paper	$c/2$
n	⊥ paper // paper	$(a+b)/2$ or $(b+c)/2$ or $(c+a)/2$; $(a+b+c)/2$ in tetragonal and cubic
d	3/8 1/8 ⊥ paper // paper	$(a \pm b)/4$ or $(b \pm c)/4$ or $(c \pm a)/4$; $(a \pm b \pm c)/4$ in tetragonal and cubic†

[a]The *e* glide plane is not included in this compilation; it is effectively a combination of two unidirectional glides in one and the same plane.

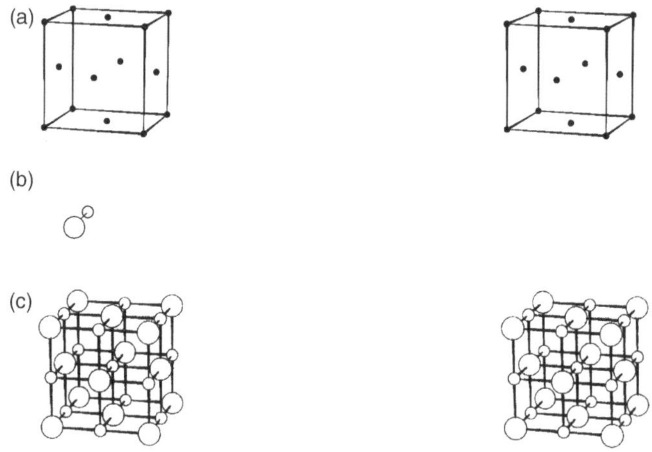

Fig. 5.46
Example of the conceptual construction of a crystal structure: sodium chloride (Na^+Cl^-). (a) Face-centred unit cell in a cubic lattice of symmetry $Fm\bar{3}m$. (b) An Na^+Cl+ ion pair motif of structure. (c) The product (convolution) of the lattice and motif functions—the crystal structure of Na^+Cl^-.

points of the lattice. This particular mathematical process is known as convolution, and has been described in detail elsewhere [12, 13]. Operating on the asymmetric unit of structure by the space group symmetry builds up the entire, ideally infinite crystal: such is the nature of all crystal structures based on a periodic lattice.

5.4.15.2 The alums

The aluminium alums have the general formula MAl $(SO_4)_2.12H_2O$, where M can be NH_4^+, K^+, Rb^+ or Tl^+. They are isomorphous, crystallizing with four formula entities in the unit cell of space group $Pa\bar{3}$, with $a = 12.3$ Å. This structure, excluding the hydrogen atoms, involves the following species in the unit cell: 4M4 Al, 8 S, 32 O, in $[SO_4]^{2-}$ groups, and 48 O as H_2O molecules. In the light of space group $Pa\bar{3}$, the following allocations can be made:

4M	a	0, 0, 0; 0, ½, ½; ½, 0, ½; ½, ½, 0
4Al	b	½, ½, ½; ½, 0, 0; 0, ½, 0; 0, 0, ½
8S	c	$\pm(x, x, x; \tfrac{1}{2}+x, \tfrac{1}{2}-x, \bar{x}; \bar{x}, \tfrac{1}{2}+x, \tfrac{1}{2}-x; \tfrac{1}{2}-x, \bar{x}\ \tfrac{1}{2}+x)$

The asymmetric unit is ½ along x, y and z, or one eighth of the unit cell. The x parameter for sulphur can be determined by an examination along the [111] direction. Since 32 oxygen atoms are associated with the sulphur atoms as a surrounding tetrahedral unit, it seems reasonable to expect that eight of these oxygen atoms would be in Wyckoff (c) positions, with the remaining 24 oxygen atoms completing tetrahedral arrangements but in Wyckoff (d); then the 48 oxygen atoms in the water molecules would most likely occupy two sets of Wyckoff (d) positions around the sulphur atoms. These allocations were confirmed by the full crystal structure analysis. It is clear that a space group symmetry analysis can reduce considerably the task of a structure analysis determination.

5.4.15.3 Copper(I) oxide

A third structure, crystallizing in a space group studied above, is copper(I) oxide. It exhibits space group $Pn\bar{3}$ with two formula entities in the unit cell.

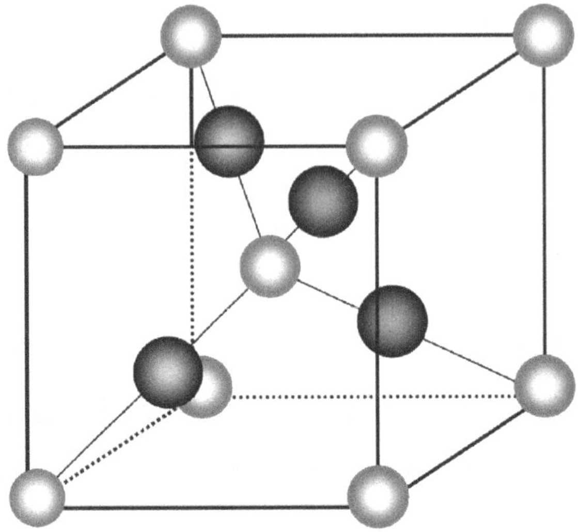

Fig. 5.47
Unit cell of copper(I) oxide, Cu_2O. The oxygen atoms occupy the points of a body centred cubic unit cell, with the copper atoms coordinated about them tetrahedrally.

Considering the space group, with the origin on 23, the following allocation can be made:

$$\begin{array}{lll} 4\,\text{Cu} & a & 0,\ 0,\ 0;\ \tfrac{1}{2},\ \tfrac{1}{2}\ \tfrac{1}{2} \\ 2\,\text{O} & b & \tfrac{1}{4},\ \tfrac{1}{4},\ \tfrac{1}{4};\ \tfrac{3}{4},\ \tfrac{3}{4},\ \tfrac{1}{4};\ \tfrac{3}{4},\ \tfrac{1}{4},\ \tfrac{3}{4};\ \tfrac{1}{4},\ \tfrac{3}{4},\ \tfrac{3}{4} \end{array}$$

Thus, the oxygen atoms adopt a body centred cubic structure, and are surrounded tetrahedrally by four copper atoms (Fig. 5.47); the O–Cu–O linkages are linear, aligned along the [111] directions. The unit cell dimension a is 4.2670 Å, so that the Cu–O bond length is $a\sqrt{3}/4$ or 1.848 Å. The sum of the ionic radii is 1.98 Å and that for the covalent radii 1.78 Å. The result for Cu–O thus indicates a degree of covalent character of the bond, which would be expected in four-coordination.

5.4.15.4 Spinel and inverse spinel structures

Spinel structures have the general formulation AB_2O_4, where A and B are metals in oxidation states II (Mg, Fe, Ni, ...) and III (Al, Fe, Cr, ...) respectively; some of these structures were determined first by Bragg in 1915. Spinel itself is $MgAl_2O_4$: it is cubic with a unit cell dimension of 8.0832 Å and space group $Fd\bar{3}m$; the density D is 3.510 g cm^{-3}. The number of formula entities Z in the unit cell is $Da^3/M_r m_u$: $(3.510 \times 8.0832^3/(139.59 \times 1.6605)$, which evaluates to 7.998, or 8 to the nearest integer. From the space group data for $Fd\bar{3}m$, No. 227, origin at $\bar{4}3m$, at $-\tfrac{1}{8},\ -\tfrac{1}{8},\ -\tfrac{1}{8}$ from a centre ($\bar{3}m$), the following assignments have been made:

$$\begin{array}{llll} & & & (0,\ 0,\ 0;\ 0,\ \tfrac{1}{2},\ \tfrac{1}{2};\ \tfrac{1}{2},\ 0,\ \tfrac{1}{2};\ \tfrac{1}{2},\ \tfrac{1}{2}, 0)+ \\ 8\,\text{Mg} & a & \bar{4}3m & 0,\ 0,\ 0;\ \tfrac{1}{4}\ \tfrac{1}{4},\ \tfrac{1}{4} \\ 16\,\text{Al} & d & \bar{3}m & \tfrac{5}{8},\tfrac{5}{8},\tfrac{5}{8};\ \tfrac{5}{8},\tfrac{3}{8},\tfrac{7}{8};\ \tfrac{7}{8},\tfrac{5}{8},\tfrac{7}{8};\ \tfrac{7}{8},\tfrac{7}{8},\tfrac{5}{8} \\ 32\,\text{O} & e & 3m & \left\{\begin{array}{ll} x, x, x; & \bar{x}, \tfrac{1}{2}-x, \tfrac{1}{2}+x \\ \tfrac{1}{2}-x, \tfrac{1}{2}+x, \bar{x}; & \tfrac{1}{2}+x, \bar{x}, \tfrac{1}{2}-x \\ \tfrac{3}{4}+x, \tfrac{1}{4}+x, \tfrac{3}{4}-x; & \tfrac{1}{4}-x, \tfrac{1}{4}-x, \tfrac{1}{4}-x \\ \tfrac{1}{4}+x, \tfrac{3}{4}-x, \tfrac{3}{4}+x; & \tfrac{3}{4}-x, \tfrac{3}{4}+x, \tfrac{1}{4}+x \end{array}\right. \end{array}$$

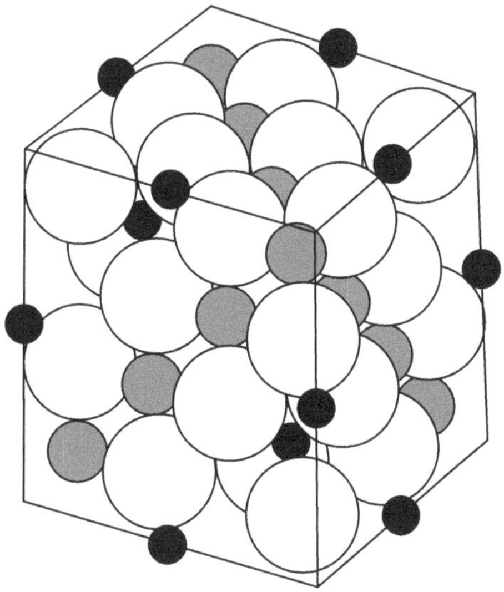

Fig. 5.48
Unit cell of spinel, $MgAl_2O_4$; circles in decreasing order of size represent O, Mg and Al. The structure is very nearly a cubic close pack of oxygen species, with Mg and Al in the tetrahedral and octahedral interstices respectively. The figure shows the unit cell translated by $a/2$ so as to clarify the positions of the octahedral and tetrahedral holes.

Thus, the structure is solved with one unknown parameter; by experiment $x = 0.387$, and is illustrated by Fig. 5.48. It is almost a perfect close packed cubic structure (see also Fig. 2.40a). The magnesium species lie in the tetrahedral holes formed by the close packed oxygen species, with aluminium occupying the octahedral holes. Had the parameter x been 0.375, the structure would have been a perfect close packed cubic.

As well as spinel structures containing varying types of A species, there are also inverse spinel structures, in which the A and B species interchange to a certain degree. Perhaps the best known of them is Fe_3O_4, which is better written as $FeFe_2O_4$. It has been allocated to space group $Fd\bar{3}m$, with $x = 0.379$ in a unit cell of side 8.394 Å. It is well known for its magnetic properties as the mineral lodestone.

5.5 Matrix representation of space group symmetry operations

In Section 3.12, the matrix representation of point group operations was discussed. In that section, a symmetry operation was defined in matrix form, but with the translation vector **t** set at zero, as is proper for a point group. In a space group, however, **t** takes on a value for a symmetry operation that depends upon both the translational symmetry in that operation and the orientation of the symmetry element with respect to the origin of the reference axes.

The general matrix equation is restated here for convenience:

$$\mathbf{x}' = \mathbf{R}\mathbf{x} + \mathbf{t} \tag{5.2}$$

where $\mathbf{x}'$ is the triplet x', y', z' obtained by the action of the matrix $\mathbf{R}$ on the triplet x, y, z, together with the addition of the translation vector $\mathbf{t}$ to $\mathbf{x}'$.

Consider first space group $P2_1/c$. Using the orientations depicted by its symbol, as discussed in Section 5.4.6, the relevant matrices are as follow:

$$\begin{array}{l}\textit{Operation} \rightarrow \quad \boldsymbol{c}_{(x,q,z)} \qquad\qquad\qquad \boldsymbol{2_1}_{[p,y,r]} \qquad\qquad\qquad \bar{\boldsymbol{1}}_{0,0,0}\\
\begin{pmatrix}1&0&0\\0&\bar{1}&0\\0&0&1\end{pmatrix}+\begin{pmatrix}0\\2q\\ 1/2\end{pmatrix}\begin{pmatrix}\bar{1}&0&0\\0&1&0\\0&0&\bar{1}\end{pmatrix}+\begin{pmatrix}2p\\ 1/2\\2r\end{pmatrix}=\begin{pmatrix}\bar{1}&0&0\\0&\bar{1}&0\\0&0&\bar{1}\end{pmatrix}+\begin{pmatrix}0\\0\\0\end{pmatrix}\\
\textit{Matrix} \rightarrow \quad \mathbf{R}_2 \qquad \mathbf{t}_2 \qquad \mathbf{R}_1 \qquad \mathbf{t}_1 \qquad \mathbf{R}_3 \qquad \mathbf{t}_3\end{array} \tag{5.3}$$

that is, $\mathbf{R}_2\mathbf{R}_1 \equiv \mathbf{R}_3$ and $\mathbf{t}_2 + \mathbf{t}_1 \equiv \mathbf{t}_3$. Thus, it follows that $p = 0$ and $q = r = 1/4$, as determined previously; $\mathbf{t}_3$ must be zero because $\bar{1}$ is set at the origin.

A second example is space group $P4bm$, with $\boldsymbol{4}$ acting along $[0, 0, z]$, $\boldsymbol{b}$ parallel to (p, y, z) and $\boldsymbol{m}$ parallel to (q, q, z). Then,

$$\begin{array}{l}\boldsymbol{m}_{(q,q,z)} \qquad\qquad\qquad \boldsymbol{b}_{[p,y,z]} \qquad\qquad\qquad \boldsymbol{4}_{[0,0,z]}\\
\begin{pmatrix}1&0&0\\0&\bar{1}&0\\0&0&1\end{pmatrix}+\begin{pmatrix}0\\2q\\ 1/2\end{pmatrix}\begin{pmatrix}\bar{1}&0&0\\0&1&0\\0&0&\bar{1}\end{pmatrix}+\begin{pmatrix}2p\\ 1/2\\2r\end{pmatrix}=\begin{pmatrix}\bar{1}&0&0\\0&\bar{1}&0\\0&0&\bar{1}\end{pmatrix}+\begin{pmatrix}0\\0\\0\end{pmatrix}\\
\quad \mathbf{R}_2 \qquad \mathbf{t}_2 \qquad \mathbf{R}_1 \qquad \mathbf{t}_1 \qquad \mathbf{R}_3 \qquad \mathbf{t}_3\end{array} \tag{5.4}$$

whence $q = 1/2$ and $p = q/2 = 1/4$, as determined in Section 5.4.12.2.

Finally in this section, consider the cubic space group $F\bar{4}3c$. Previously, centred groups were treated generally as the corresponding primitive group plus the centring condition. However, there is no listed space group $P\bar{4}3c$, its standard name being $P\bar{4}3n$, although it does contain c glide planes in an orientation parallel to those in $F\bar{4}3c$, so one can proceed as follows. Reference to the stereogram for point group $\bar{4}3m$, as shown in Fig. 5.49, shows that the path of operations $(1) \xrightarrow{\bar{4}_{[00z]}} (2) \xrightarrow{3_{[1\bar{1}1]}} (3) \xrightarrow{m_{[01\bar{1}]}} (1)$ completes a suitable cycle. Now, the $\bar{4}$ axis lies along the line $[0, 0, z]$, 3 is the line $[1\bar{1}1]$, and in this space group m becomes the c glide plane (p, p, z), where p is to be determined. Using the matrices developed in Appendix A3 for these three successive stages:

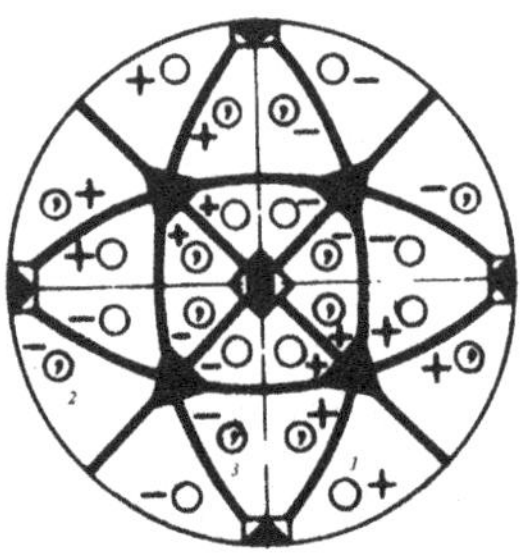

Fig. 5.49
Stereogram of point group $\bar{4}3m$: the path $1 \rightarrow 2 \rightarrow 3 \rightarrow 1$ is described in the texts.

$$(1)\text{—}\bar{\mathbf{4}}_{[0,0,z]} \rightarrow (2) \quad \begin{pmatrix}0&1&0\\\bar{1}&0&0\\0&0&\bar{1}\end{pmatrix}+\begin{pmatrix}0\\0\\0\end{pmatrix}\begin{pmatrix}x\\y\\z\end{pmatrix}=\begin{pmatrix}y\\\bar{x}\\\bar{z}\end{pmatrix}$$

$$(2)\text{—}\mathbf{3}_{[1,\bar{1},1]} \rightarrow (3) \quad \begin{pmatrix}0&\bar{1}&0\\0&0&\bar{1}\\1&0&0\end{pmatrix}+\begin{pmatrix}0\\0\\0\end{pmatrix}\begin{pmatrix}y\\\bar{x}\\\bar{z}\end{pmatrix}=\begin{pmatrix}x\\z\\y\end{pmatrix}$$

$$(3)\text{—}\mathbf{c}_{[p,p,z]} \rightarrow (1) \quad \begin{pmatrix}1&0&0\\0&0&1\\0&1&0\end{pmatrix}+\begin{pmatrix}p\\p\\ 1/2\end{pmatrix}\begin{pmatrix}x\\z\\y\end{pmatrix}=\begin{pmatrix}p+x\\p+y\\z+1/2\end{pmatrix}$$

But the final position is equivalent to x, y, z; hence, $p = q = 0$ and the z coordinate shows that the inversion point on the $\bar{4}$ symmetry element is at $c = 1/4$ (and $3/4$). The findings are confirmed by the space group diagram (Fig. 5.50), and the specification of the space group can be now completed.

Plate 1

A bloom of the orchid *Cattleya walkeriana*, variation *Alba*, showing bilateral reflection symmetry. [Reproduced by courtesy of Greg Allikas.]

Plate 2

Model of Bernini's St. Peter's Colonnade. [Reproduced by courtesy of its author, Enrico Dalbosco; <http://www.enricodalbosco.it>]

Plate 3
Specimen of a naturally occurring quartz, SiO_2, cluster; its pale amethyst colour arises from inclusions of traces of iron and manganese.

Plate 4
Example of a Penrose tiling that uses just two shapes of tile, namely rhombi. Viewing the diagram edgeways reveals sets of lines of narrow and wide spacings. [Reproduced by courtesy of Jeff Preshing.]

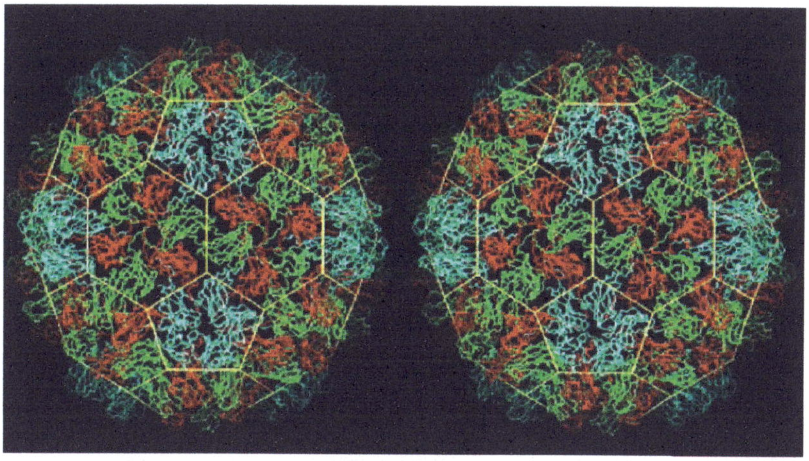

Plate 5

Stereoview of the protein shell of cowpea chlorotic mottle virus. The yellow lines delineate part of the enclosing cage of a truncated icosahedron and reveal the subunits of five-membered and six-membered rings. [Speir JA, et al. *Structure* 1995; 3: 63; reproduced by courtesy of Elsevier.]

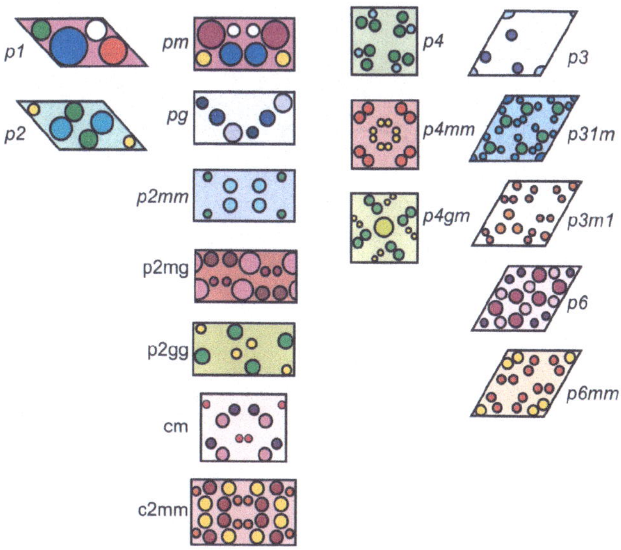

Plate 6

The seventeen plane groups. Can you find the symmetry elements? [Reproduced by courtesy of Dexter Perkins, University of North Carolina.]

Plate 7

Colour symmetry in a Chinese painting.

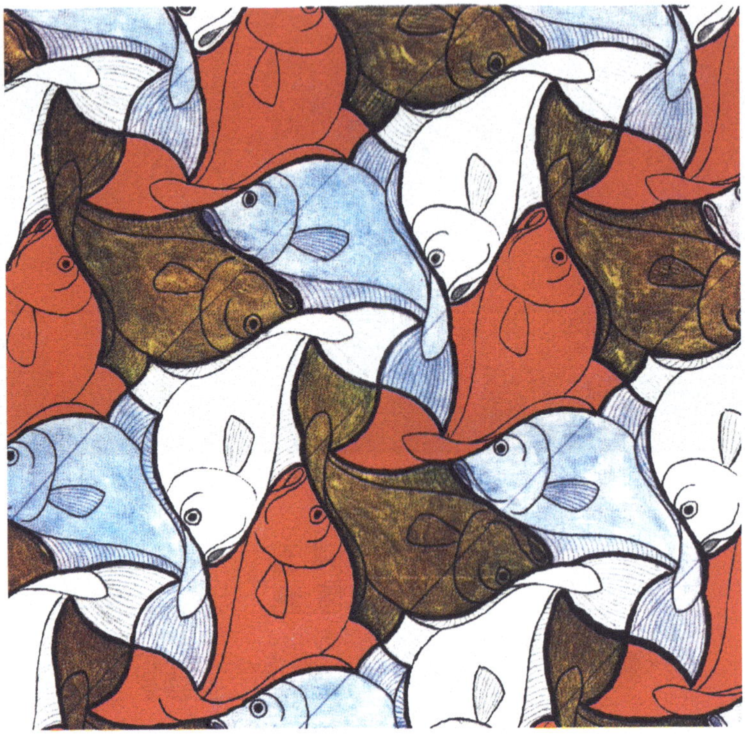

Plate 8

An example of a colour symmetry plane group. [Reproduced by courtesy of the IUCr (E).]

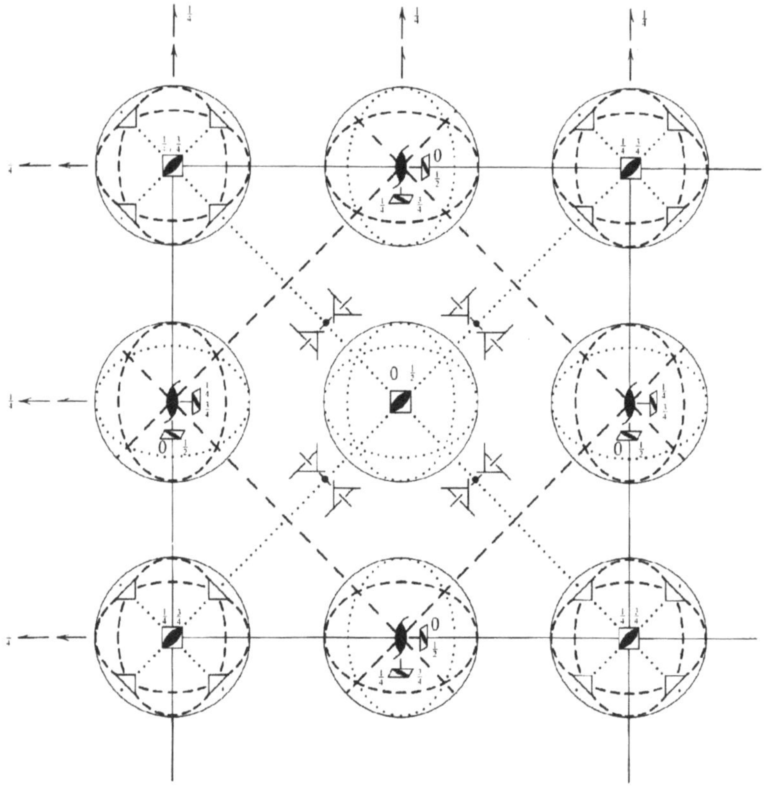

Fig. 5.50
Symmetry elements of space group $F\bar{4}3c$. [Reproduced by courtesy of the IUCr (A).]

In the cubic space groups based on point groups $\bar{4}3m$ and $m\bar{3}m$, there exist inclined symmetry planes, as shown by Fig. 5.50, which may be compared with the great circles on Fig. 5.49. The meaning of the space group symbolism for these and other such planes is described Table 5.9, and further information and the coordinates of general and special equivalent positions have been tabulated in the literature [4,5].

5.6 Black-white and colour symmetry

So far, the symmetry concepts studied have referred to the classical crystallographic groups. A consideration of other objects and patterns, such as wallpapers or the tiled walls and floors of the Alhambra, reveal the existence of colour symmetry, the simplest examples of which are the black-white, or antisymmetric, groups. Antisymmetry is the relationship between a given assemblage having a property such as colour or magnetic moment that could be represented by a positive sign and a similar assemblage related by symmetry to the first and having the same property represented by a negative sign. Concisely, antisymmetric operations involve a + object or pattern and a symmetrically related − object or pattern; hence, as well as the black-white groups, colour groups and magnetic groups arise. There is space here to consider briefly black-white and colour symmetries.

Table 5.9 Symbols and translations for inclined planes in the cubic system, classes $\bar{4}3m$ and $m\bar{3}m$ only.

	Graphical symbol for planes normal to		Glide vector in units of lattice translation vectors for planes normal to		
Symmetry plane	**[011] and [01$\bar{1}$]**	**[101] and [10$\bar{1}$]**	**[011] and [01$\bar{1}$]**	**[101] and [10$\bar{1}$]**	**Printed symbol**
Reflection plane, mirror plane		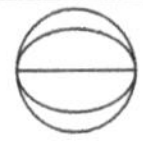	None	None	m
'Axial' glide plane		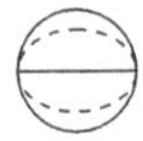	$\frac{1}{2}$ along [100]	$\frac{1}{2}$ along [010]	a or b
'Axial' glide plane	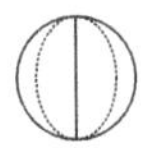	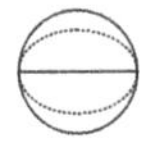	$\frac{1}{2}$ along [01$\bar{1}$] or along [011]	$\frac{1}{2}$ along [10$\bar{1}$] or along [101]	
'Diagonal' glide plane	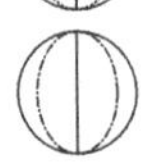	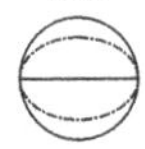	$\frac{1}{2}$ along [11$\bar{1}$] or along [111][a]	$\frac{1}{2}$ along [11$\bar{1}$] or along [111][a]	n
'Diamond' glide plane (pair of planes; in centred cells only)[a]	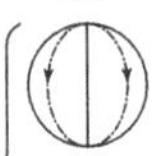	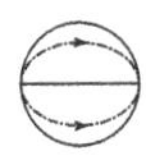	$\frac{1}{2}$ along [1$\bar{1}$1] or along [111][b]	$\frac{1}{2}$ along [$\bar{1}$11] or along [111][b]	d
	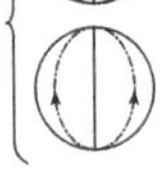	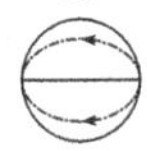	$\frac{1}{2}$ along [$\bar{1}\bar{1}$1] or along [$\bar{1}$11][b]	$\frac{1}{2}$ along [$\bar{1}\bar{1}$1] or along [1$\bar{1}$1][b]	

[a] In $F\bar{4}3m$, $Fm\bar{3}m$ and $Fd\bar{3}m$, the shortest lattice translation vectorss are ½[21 $\bar{1}$] or ½[211] and ½[12 $\bar{1}$] or ½[121], respectively.
[b] The glide vector is one half of the centring vector, that is, one quarter of the conventional unit cell.

Fig. 5.51
Illustration of fivefold rotation and reflection line symmetry, with superimposed tenfold black-white antisymmetry, again, in both rotation and reflection.

5.6.1 Black-white symmetry: potassium chloride

One of the most common examples of black-white symmetry, used by most people every day, is the binary 1 0 code basis used in computing. A little closer to the theme of this book, perhaps, is the example of antisymmetry illustrated by the two-dimensional black-white pattern of Fig. 5.51. Confining attention to the central portion of this figure, about the centre, but not in the figure as a whole, fivefold classical rotation and tenfold black-white rotations exist. In addition, there are ten classical reflection lines through the centres of pairs of both black and white 'kites', and ten black-white reflection lines through the junctions of pairs of black and white kites. Tenfold rotation operates in angular steps of 36° from black to white, but there is also twofold symmetry represented in both fivefold and tenfold operations, black to white.

If the unit cells of the plane groups are centred by points of a different sign, or colour, another five lattices are so represented (Table 5.10); the prime against a unit cell symbol here indicates a colour group, black-white in this section, and the subscripts b and c represent centring of the b edge of the unit cell and its geometrical centre respectively. When combined with the plane point groups, a total of 46 black-white plane groups is obtained; in three dimensions there are 1651 black-white groups.

An example of the classical symmetry that has been studied earlier is shown by Fig. 5.52. At first glance, it might be thought to be a black-white pattern, but this is not the case because the two figures of different colours are not

Table 5.10 Unit cells in the black-white lattices.

System	Unit cell
Oblique	p, p'_b
Rectangular	p, p'_b, p'_c, c, c'
Square	p, p'_c
Hexagonal	p

Fig. 5.52
Classical plane group of symmetry *p4mg* (see also Fig. 5.9). [Reproduced by courtesy of the IUCr (E).]

related by symmetry. At the base of the illustration there are three fourfold rotation points in this two-dimensional pattern. If the centre point of these three is chosen as an origin, then another three points, correctly chosen, form the corners of a plane unit cell set at 45° to the borders of the figure. It may be helpful to make a copy of the figure for this study: twofold rotation points exist at the mid-points of the unit cell edges, as would now be expected, but the fourfold rotation point at the centre of the unit cell is in a different orientation from those at the corners, so it is not a centred group: there are both *m* lines and *g* lines in the pattern, and the plane group is *p4gm*, as in Fig. 5.9. The plane group *p4mg* is equivalent to *p4gm* by interchange of axes, but the standard setting is *p4gm*, as set out here.

A black-white symmetry pattern is shown by Fig. 5.53 which is an assemblage of symmetry related black beetles and white beetles; the same

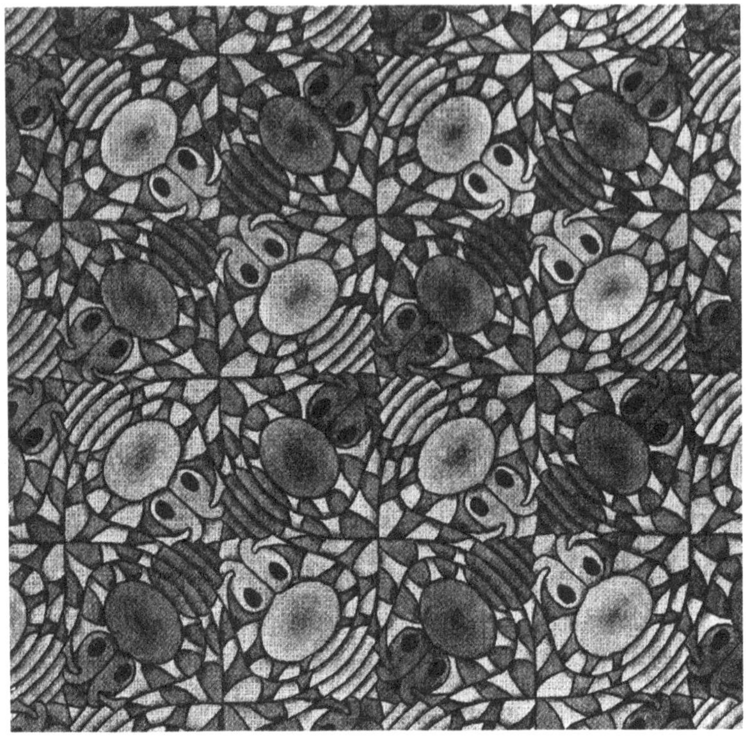

Fig. 5.53
Black-white plane group $p4'gm$. [Reproduced by courtesy of the IUCr (E).]

symmetry elements as in Fig. 5.52 are present in this illustration. A square unit cell may be defined by four fourfold black-white rotation points. The reflection lines are in the same orientations in the unit cell as in Fig. 5.51; however, while the *m* lines in the figure are classical, the *g* lines involve a colour change from white to black and vice versa, as do the fourfold rotation points; this plane group is designated $p4'gm$.

A practical example of black-white symmetry can be found in the early structure analysis of potassium chloride. The crystal structure consists of isoelectronic K^+ and Cl^- ions, and they are arranged on the {100} planes of the cubic unit cell as shown in Fig. 5.54. Because X-rays are scattered by electrons in a crystal structure, each of these species appears almost identical to an X-ray beam. Thus, on first examination, the structure appeared to be based on a cubic *P* unit cell [14], since the resolution of the X-ray diffraction record obtained with the early X-ray spectrometer was not of a high order. An X-ray powder photograph of potassium and sodium chlorides to about the same resolution is illustrated by Fig. 5.55. The lines corresponding to the indices *h*, *k* and *l* all odd, the weak reflections, are just visible on the pattern for sodium chloride but absent on that for potassium chloride.

In the cubic system, the sum

$$h^2 + k^2 + l^2 = N \tag{5.5}$$

takes on a sequence of integer values except where *N* equals one of the forbidden numbers $m^2(8n - 1)$, *m* and *n* being integral. The sequence found for

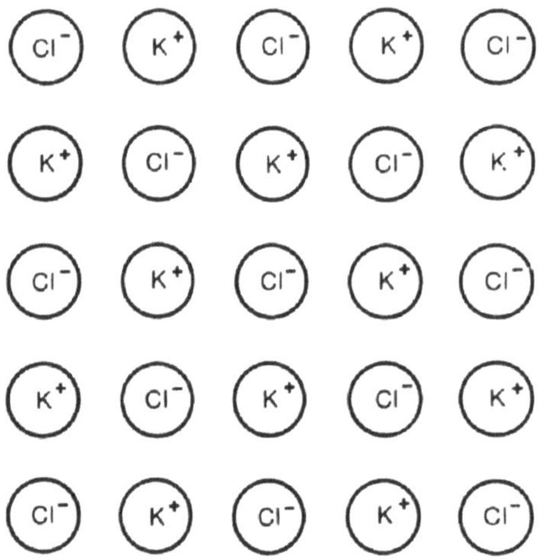

Fig. 5.54
The structure of potassium chloride, space group $Fm\bar{3}m$, as seen in projection on to a cube face. Since the K^+ and Cl^- ions are isoelectronic (18 electrons each), their scattering of X-rays is very closely similar.

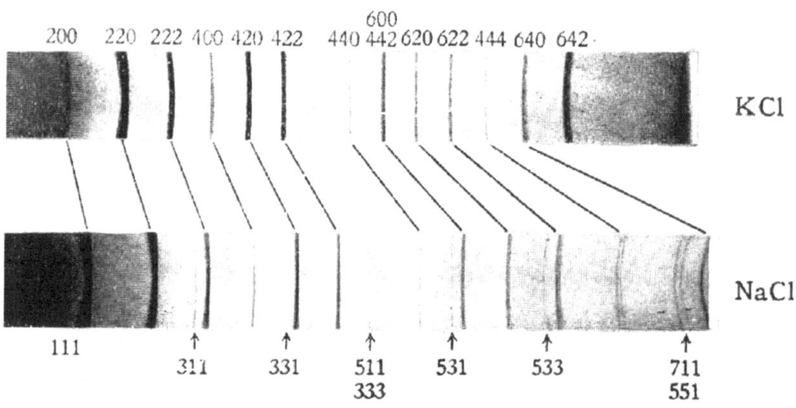

Fig. 5.55
X-ray powder photographs of potassium chloride and sodium chloride. The similarity of the patterns and the absence of the faint lines with h, k and l all odd, visible in the pattern for sodium chloride, in the early photographs for potassium chloride led to the assumption of a P cubic unit cell for this structure: a good example of black-white antisymmetry.

potassium chloride was characteristic of a primitive cubic unit cell with an edge length equal to the apparent repeat distance $Cl^-{-}K^+$ that was actually one half the true value. The X-ray photograph indicates apparent black-white antisymmetry in the pattern, relating *positive* K^+('black') to *negative* Cl^-('white'), the two species possessing 18 electrons each, by fourfold black-white rotation points between any four species, two K^+ and two Cl^- ions.

After other alkali halides, notably sodium chloride, had been examined and their structures found to be cubic F, a more detailed examination of potassium chloride showed that it, too, was cubic F, and its true repeat distance was revealed. The correct repeat period can be found also by neutron scattering, since the scattering powers of the K^+ and Cl^- species differ significantly for neutron radiation.

5.6.2 Colour symmetry

Colour symmetry has been studied extensively as periodic symmetry patterns based on the Bravais lattices, and applications range from the symmetry analysis of Escher's periodic drawings [15] to the study of order-disorder transitions in magnetic crystals [16]. Colour symmetry has been described in numerous other situations: art, architecture, wallpaper, wall and floor paintings and tilings, optical activity in planar arrays of metallic or dielectric gammadions, so named from early decorative figure based on the Greek Γ letter, and in some twin crystals, such as the merohedric twins of quartz.

As a first example of colour symmetry, examine Fig. 5.56 (**Plate 7**), which is a picture of a Chinese painting. If the unit cell is selected correctly, it shows two fourfold rotation points which are mirror images of one another, twofold rotation points and mirror and glide symmetry lines; its plane group symbol is $p4'gm$. Consider next Fig. 5.57 (**Plate 8**): it comprises fish in four different colours and orientations, but all fish of any given colour have identical orientations. The 90° differences in orientation between the pairs white-green, green-red, red-blue and blue-white fish indicate the presence of fourfold colour rotation points. The almost square elements of fins, of sequence white, green, red, blue, at the bottom centre of the figure and three others in similar orientation form the corners of a square unit cell.

Fig. 5.56
Colour symmetry in a Chinese painting. (**See Plate 7**)

The fourfold colour rotation point at the centre of the unit cell, consisting of areas of fish tails, shows the same colour sequence but in a different orientation. The twofold rotation points are again evident at the mid-points of the cell edges. In this pattern, however, the twofold rotations involve a change of colour, as indicated by the motifs at the fourfold rotation points: they are twofold colour rotation points. What is the plane group symbol for this

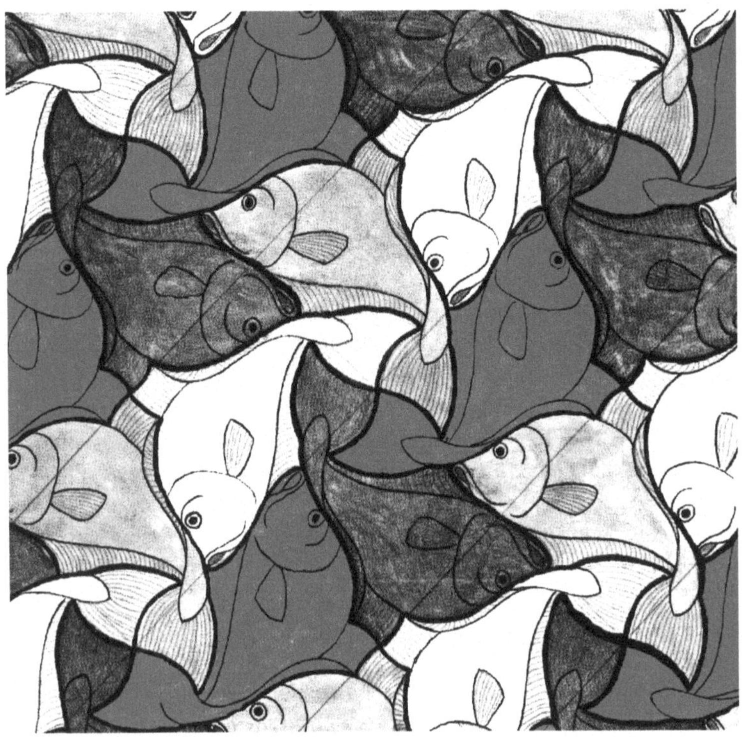

Fig. 5.57
An example of a colour symmetry plane group. [Reproduced by courtesy of the IUCr (E).] (**See Plate 8**)

pattern? Further discussions on black-white and colour symmetry are abundant in the literature [8,17–20].

5.7 The international tables and other crystallographic compilations

The publication of tables that presented the 230 space groups in terms of diagrams and coordinates of equivalent positions appeared first in 1935 as the two volume work *Internationale Tabellen zur Bestimmung von Kristallstrukturen*, edited by Carl Hermann and published by Bornträger, Berlin. In 1944, it was reprinted with corrections by Edwards Brothers Inc., Ann Arbor, and served until the publication of the *International tables for X-ray crystallography*, Volume I, edited by Kathleen Lonsdale and Norman Henry, and published by the Kynoch Press, Birmingham, in 1952; it was reprinted in 1965, 1969 and 1979. This work does not contain diagrams of the cubic space groups, but presents other crystallographic information on these space groups. The current publication on space group symmetry is the *International tables for crystallography*, Volume A, edited by T. Hahn, and Volume A1, edited by H. Wondratschek and U. Müller, and published for the International Union of Crystallography (IUCr) first in 1983 and then in 2004, respectively.

The first of these compilations, the 1935 edition, is now very rare and, at the time of writing, only one copy was found advertised for sale. The second compilation can still be found at a few booksellers, and Volume A (2006), is the current edition. Of the latter two compilations, the 1952 Volume I is perhaps easier to read and comprehend for those new to the subject than is the current Volume A. Volume A1 (2011) treats subgroups in detail as well as presenting the theory of space groups and a discussion of the Wyckoff notation [21]. A study of the material and problems of the present book should prepare the reader to take advantage of the more detailed publications now available on space groups [4,5,8]. For those readers who do not have access to the international tables, a set of space group drawings can be viewed on the internet. This compilation presents *inter alia* space group diagrams and lists of general equivalent positions; readers should be aware that, as the introduction indicates, the drawings are rotated anticlockwise by 90° from the standard orientation [22].

Other compilations of considerable significance to which the attention of the reader is drawn are the Bilbao Crystallographic Server [23] and the *Online dictionary of crystallography* [24], both of which contain *inter alia* important data on point group and space group symmetry.

5.7.1 The international tables for crystallography, Vol. A

The current definitive work on space groups is the *International tables for crystallography*, Vol. A [5], frequently referred to just as ITA, and an example will be chosen for which the space groups diagrams from the earlier tables [4] has already been studied, namely $P\frac{4}{n}bm$, shown in Fig. 5.34, in order to show the range of information that the tables now provide. The importance of the use of the tables in tackling crystal structure analyses, with concomitant saving in personnel and computational time and cost, has been described in Section 5.4.15, and is reviewed further in the literature [12].

Referring now to Fig. 5.58, the two pages show (a) the space group diagrams and symmetry information, and (b) the coordinates of equivalent positions, with additional symmetry information. Some of this data will be discussed in so far as it elaborates the main subject matter of this book, and it is convenient to consider this information line by line.

- The first few lines of Fig. 5.58a need little in addition to the earlier discussions. The *Patterson symmetry* is that of the corresponding symmorphic space group; thus, for space group $P\frac{4}{n}bm$ the Patterson symmetry is $P\frac{4}{m}mm$; for monoclinic space groups with the y axis unique, they are $P1\frac{2}{m}1$ or $C1\frac{2}{m}1$, according as the space group is P or C. All other space groups are treated similarly, and studies on the application of Patterson symmetry in crystal structure analysis can be found in the literature [12]. The space group number, 125 for $P\frac{4}{n}bm$, is important, because some crystallographic software stores space group symmetry data in terms of its number in ITA. The entry 'Origin Choice 1' implies more than one satisfactory origin. In Section 5.4.12.2, origin 1 was used, at '422 at $4/n2\,2/g$, at $-1/4, -1/4, 0$ from centre ($2/m$)' to give it its full ITA specification (here, g indicates diagonal glide planes).
- The diagrams are familiar from earlier reading, although in some low symmetry space groups differing views are provided, and indicate the general equivalent positions and the symmetry elements of the group. They are followed by a statement of the origin position and the selection of the asymmetric unit of the space group.

$P\,4/n\,b\,m$	D^3_{4h}	$4/m\,m\,m$	Tetragonal
No. 125	$P\,4/n\,2/b\,2/m$		Patterson symmetry $P\,4/m\,m\,m$

ORIGIN CHOICE 1

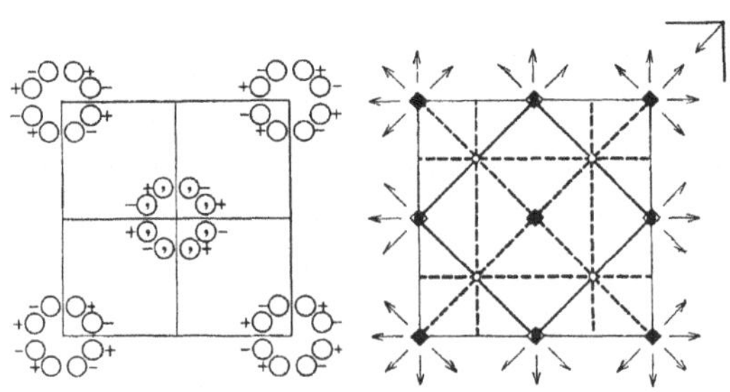

Origin at 422 at $4/n\,2\,2/g$, at $-\frac{1}{4},-\frac{1}{4},0$ from centre ($2/m$)

Asymmetric unit $0\le x\le\frac{1}{2}$; $0\le y\le\frac{1}{2}$; $0\le z\le\frac{1}{2}$; $y\le\frac{1}{2}-x$

Symmetry operations

(1) 1	(2) 2 0,0,z	(3) 4^+ 0,0,z	(4) 4^- 0,0,z
(5) 2 0,y,0	(6) 2 x,0,0	(7) 2 x,x,0	(8) 2 x,$\bar{x}$,0
(9) $\bar{1}$ $\frac{1}{4}$,$\frac{1}{4}$,0	(10) $n(\frac{1}{2},\frac{1}{2},0)$ x,y,0	(11) $\bar{4}^+$ $\frac{1}{2}$,0,z; $\frac{1}{2}$,0,0	(12) $\bar{4}^-$ 0,$\frac{1}{2}$,z; 0,$\frac{1}{2}$,0
(13) a x,$\frac{1}{4}$,z	(14) b $\frac{1}{4}$,y,z	(15) m $x+\frac{1}{2}$,$\bar{x}$,z	(16) $g(\frac{1}{2},\frac{1}{2},0)$ x,x,z

Fig. 5.58
Space group $P\frac{4}{n}bm$, as listed in ITA; compare Fig. 5.34. (a) General equivalent positions, symmetry elements and space group data. (b) Coordinates of equivalent positions, limiting conditions and further space group data. [Reproduced by courtesy of the IUCr (A).]

CONTINUED No. 125 *P*4/*nbm*

Generators selected (1); *t*(1,0,0); *t*(0,1,0); *t*(0,0,1); (2); (3); (5); (9)

Positions

Multiplicity, Wyckoff letter, Site symmetry			Coordinates				Reflection conditions
							General:
16	*n*	1	(1) x,y,z	(2) $\bar{x},\bar{y},z$	(3) $\bar{y},x,z$	(4) $y,\bar{x},z$	$hk0$: $h+k=2n$
			(5) $\bar{x},y,\bar{z}$	(6) $x,\bar{y},\bar{z}$	(7) $y,x,\bar{z}$	(8) $\bar{y},\bar{x},\bar{z}$	$0kl$: $k=2n$
			(9) $\bar{x}+\frac{1}{2},\bar{y}+\frac{1}{2},\bar{z}$	(10) $x+\frac{1}{2},y+\frac{1}{2},\bar{z}$	(11) $y+\frac{1}{2},\bar{x}+\frac{1}{2},\bar{z}$	(12) $\bar{y}+\frac{1}{2},x+\frac{1}{2},\bar{z}$	$h00$: $h=2n$
			(13) $x+\frac{1}{2},\bar{y}+\frac{1}{2},z$	(14) $\bar{x}+\frac{1}{2},y+\frac{1}{2},z$	(15) $\bar{y}+\frac{1}{2},\bar{x}+\frac{1}{2},z$	(16) $y+\frac{1}{2},x+\frac{1}{2},z$	
							Special: as above, plus
8	*m*	*m*	$x,x+\frac{1}{2},z$	$\bar{x},\bar{x}+\frac{1}{2},z$	$\bar{x}+\frac{1}{2},x,z$	$x+\frac{1}{2},\bar{x},z$	no extra conditions
			$\bar{x},x+\frac{1}{2},\bar{z}$	$x,\bar{x}+\frac{1}{2},\bar{z}$	$x+\frac{1}{2},x,\bar{z}$	$\bar{x}+\frac{1}{2},\bar{x},\bar{z}$	
8	*l*	2	$x,0,\frac{1}{2}$	$\bar{x},0,\frac{1}{2}$	$0,x,\frac{1}{2}$	$0,\bar{x},\frac{1}{2}$	hkl : $h+k=2n$
			$\bar{x}+\frac{1}{2},\frac{1}{2},\frac{1}{2}$	$x+\frac{1}{2},\frac{1}{2},\frac{1}{2}$	$\frac{1}{2},\bar{x}+\frac{1}{2},\frac{1}{2}$	$\frac{1}{2},x+\frac{1}{2},\frac{1}{2}$	
8	*k*	2	$x,0,0$	$\bar{x},0,0$	$0,x,0$	$0,\bar{x},0$	hkl : $h+k=2n$
			$\bar{x}+\frac{1}{2},\frac{1}{2},0$	$x+\frac{1}{2},\frac{1}{2},0$	$\frac{1}{2},\bar{x}+\frac{1}{2},0$	$\frac{1}{2},x+\frac{1}{2},0$	
8	*j*	2	$x,x,\frac{1}{2}$	$\bar{x},\bar{x},\frac{1}{2}$	$\bar{x},x,\frac{1}{2}$	$x,\bar{x},\frac{1}{2}$	hkl : $h+k=2n$
			$\bar{x}+\frac{1}{2},\bar{x}+\frac{1}{2},\frac{1}{2}$	$x+\frac{1}{2},x+\frac{1}{2},\frac{1}{2}$	$x+\frac{1}{2},\bar{x}+\frac{1}{2},\frac{1}{2}$	$\bar{x}+\frac{1}{2},x+\frac{1}{2},\frac{1}{2}$	
8	*i*	2	$x,x,0$	$\bar{x},\bar{x},0$	$\bar{x},x,0$	$x,\bar{x},0$	hkl : $h+k=2n$
			$\bar{x}+\frac{1}{2},\bar{x}+\frac{1}{2},0$	$x+\frac{1}{2},x+\frac{1}{2},0$	$x+\frac{1}{2},\bar{x}+\frac{1}{2},0$	$\bar{x}+\frac{1}{2},x+\frac{1}{2},0$	
4	*h*	2.*mm*	$0,\frac{1}{2},z$	$\frac{1}{2},0,z$	$0,\frac{1}{2},\bar{z}$	$\frac{1}{2},0,\bar{z}$	hkl : $h+k=2n$
4	*g*	4..	$0,0,z$	$0,0,\bar{z}$	$\frac{1}{2},\frac{1}{2},\bar{z}$	$\frac{1}{2},\frac{1}{2},z$	hkl : $h+k=2n$
4	*f*	..2/*m*	$\frac{1}{4},\frac{1}{4},\frac{1}{2}$	$\frac{3}{4},\frac{3}{4},\frac{1}{2}$	$\frac{3}{4},\frac{1}{4},\frac{1}{2}$	$\frac{1}{4},\frac{3}{4},\frac{1}{2}$	hkl : $h,k=2n$
4	*e*	..2/*m*	$\frac{1}{4},\frac{1}{4},0$	$\frac{3}{4},\frac{3}{4},0$	$\frac{3}{4},\frac{1}{4},0$	$\frac{1}{4},\frac{3}{4},0$	hkl : $h,k=2n$
2	*d*	$\bar{4}2m$	$0,\frac{1}{2},\frac{1}{2}$	$\frac{1}{2},0,\frac{1}{2}$			hkl : $h+k=2n$
2	*c*	$\bar{4}2m$	$0,\frac{1}{2},0$	$\frac{1}{2},0,0$			hkl : $h+k=2n$
2	*b*	422	$0,0,\frac{1}{2}$	$\frac{1}{2},\frac{1}{2},\frac{1}{2}$			hkl : $h+k=2n$
2	*a*	422	$0,0,0$	$\frac{1}{2},\frac{1}{2},0$			hkl : $h+k=2n$

Symmetry of special projections

Along [001] *p*4*mm*
$\mathbf{a}'=\frac{1}{2}(\mathbf{a}-\mathbf{b})$ $\mathbf{b}'=\frac{1}{2}(\mathbf{a}+\mathbf{b})$
Origin at 0,0,*z*

Along [100] *p*2*mm*
$\mathbf{a}'=\frac{1}{2}\mathbf{b}$ $\mathbf{b}'=\mathbf{c}$
Origin at *x*,0,0

Along [110] *p*2*mm*
$\mathbf{a}'=\frac{1}{2}(-\mathbf{a}+\mathbf{b})$ $\mathbf{b}'=\mathbf{c}$
Origin at *x*,*x*,0

Maximal non-isomorphic subgroups

I	[2]*P*422	1; 2; 3; 4; 5; 6; 7; 8
	[2]*P*4/*n*11 (*P*4/*n*)	1; 2; 3; 4; 9; 10; 11; 12
	[2]*P*4*bm*	1; 2; 3; 4; 13; 14; 15; 16
	[2]*P*$\bar{4}$2*m*	1; 2; 5; 6; 11; 12; 15; 16
	[2]*P*$\bar{4}$*b*2	1; 2; 7; 8; 11; 12; 13; 14
	[2]*P*2/*n*2/*b*1 (*Pban*)	1; 2; 5; 6; 9; 10; 13; 14
	[2]*P*2/*n*12/*m* (*Cmma*)	1; 2; 7; 8; 9; 10; 15; 16

IIa none

IIb [2]*P*4/*nnc* ($\mathbf{c}'=2\mathbf{c}$); [2]*P*4_2/*nbc* ($\mathbf{c}'=2\mathbf{c}$); [2]*P*4_2/*nnm* ($\mathbf{c}'=2\mathbf{c}$)

425

- A new and very useful feature of the tables is the listing of the *symmetry operations* together with their orientations, translations and points of inversion, as appropriate. The coordinates that they generate from any given *x*, *z* follow: thus, in this space group, the symmetry operation entry (11) $\bar{4}^+$ $^1/_2$, 0, *z*; $^1/_2$, 0, 0 implies an anticlockwise (+) operation $\bar{\mathbf{4}}$ along the line [$^1/_2$, 0, *z*] with the point of inversion on that axis at $^1/_2$, 0, 0, that is, at $c/2$. Thus, this operation transforms a point *x*, *y*, *z* to $^1/_2+y$, $^1/_2-x$, $\bar{z}$, point (11) in the list of general equivalent positions in the second page (Fig. 5.58b).

- The *generators* define the symmetry operators and their sequence necessary to produce the coordinates of the general equivalent positions, starting from a point *x*, *y*, *z*, from which all symmetry elements of the diagram to which they lead will become evident. Translations are listed as $t(1, 0, 0)$, for example, and in the presence of centring $t(^1/_2, ^1/_2, 0)$, in the case of *C*, is also given. In space group $P\frac{4}{n}bm$ the generator listing is: (1); $t(1, 0, 0)$; $t(0, 1, 0)$; $t(0, 0, 1)$; (2); (3); (5); (9). Operation (1) implies identity, or selecting the starting point, *x*, *y*, *z*; then, the three translations define the bounds of the unit cell. Operation (2) produces $\bar{x}, \bar{y}, z$, and operation (3) produces $\bar{y}, x, z$, both from the starting point, *x*, *y*, *z*. Operation (5) now rotates all the generated points about the *y* axis. So far, no inversion has been introduced; this is remedied through operation (9), which is a centre of symmetry at ¼, ¼, 0. Proceeding in this way, all 16 Wyckoff (*n*) coordinates are obtained. It is then a small step to the special equivalent positions and limiting conditions, topics that have been examined earlier, together with data on the principal projections. Subgroups and super groups are of importance in physical science, for example, in classifying structures and studying phase transitions. However, they will not be considered beyond what has been treated in this chapter and will be considered further in Chapter 7, but they are discussed in considerable detail in the literature [5,8,21,23,25].

5.7.1.1 Hierarchy in crystal symmetry

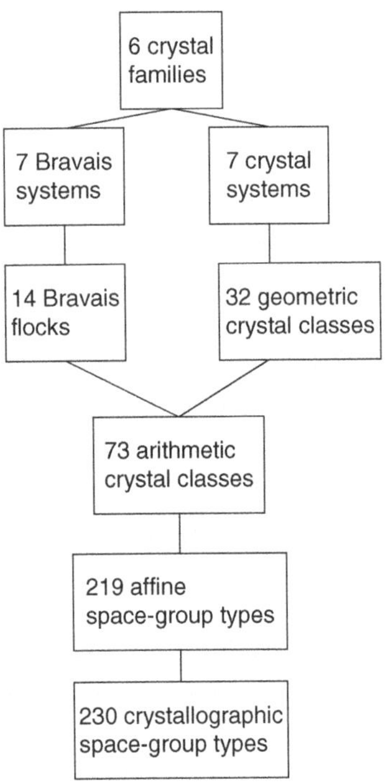

Fig. 5.59
A hierarchical classification of crystal symmetry.

The compilers of the ITA have established a hierarchical classification of crystal symmetry (Fig. 5.59). Some of these subdivisions have been encountered already in different sections of the book.

Of the 230 *crystallographic space groups*, some of which have been studied in this chapter, the 219 *affine space group types* are discussed in Section 7.1.

Arithmetic classes are those for which the Seitz operator is $\{\boldsymbol{R}|\boldsymbol{0}\}$: $P2/m$, $P2_1/m$, $P2/c$ and $P2_1/c$ all belong to the arithmetic crystal class $2/mP$, whereas $C2/m$ and $C2/c$ belong to the arithmetic crystal class $2/mC$. Again, $P31m$ and $P31c$, and $P3m1$ and $P3c1$ form the arithmetic crystal classes $31mP$ and $3m1P$ respectively. Space groups in a given arithmetic crystal class must belong to the same geometric crystal class and the same Bravais flock. They are the 73 symmorphic space groups (Section 5.4.7.1).

The term crystal class corresponds to the *geometric crystal class*: they classify space groups and point groups, and the matrices of their $\boldsymbol{R}$ symmetry operators are identical (Section 3.12 and Section 5.5). Two space groups belong to the same *geometric crystal class* if a matrix **M** exists such that

$$\mathbf{S}' = \mathbf{M}^{-1}\mathbf{S}\mathbf{M} \tag{5.6}$$

where **S** and **S**′ are symmetry operators of the two groups. Consider *Pm* and *Cm*: the mirror and glide matrices for *Pm* and *Cm* are, respectively,

$$\begin{pmatrix} 1 & 0 & 0 \\ 0 & \bar{1} & 0 \\ 0 & 0 & 1 \end{pmatrix} \text{ and } \begin{pmatrix} 0 & 1 & 0 \\ 1 & 0 & 0 \\ 0 & 0 & 1 \end{pmatrix}.$$

If **M** be chosen as $\begin{pmatrix} 1 & \bar{1} & 0 \\ 1 & 1 & 0 \\ 0 & 0 & 1 \end{pmatrix}$, then

$$\begin{pmatrix} 1/2 & 1/2 & 0 \\ \bar{1}/2 & 1/2 & 0 \\ 0 & 0 & 1 \end{pmatrix} \begin{pmatrix} 0 & 1 & 0 \\ 1 & 0 & 0 \\ 0 & 0 & 1 \end{pmatrix} \begin{pmatrix} 1 & \bar{1} & 0 \\ 1 & 1 & 0 \\ 0 & 0 & 1 \end{pmatrix} = \begin{pmatrix} 1/2 & 1/2 & 0 \\ \bar{1}/2 & 1/2 & 0 \\ 0 & 0 & 1 \end{pmatrix} \begin{pmatrix} 1 & 1 & 0 \\ 1 & \bar{1} & 0 \\ 0 & 0 & 1 \end{pmatrix} = \begin{pmatrix} 1 & 0 & 0 \\ 0 & \bar{1} & 0 \\ 0 & 0 & 1 \end{pmatrix}$$

which satisfies Eq. (5.6); hence, *Pm* and *Cm* belong to the same geometric crystal class, namely, *m*.

The *Bravais flock* classification has been devised to take account of accidental relationships among unit cell dimensions that create an apparently higher than true symmetry. For example, a monoclinic crystal structure may have a β angle of 90° within experimental error: its lattice is orthorhombic, but it remains monoclinic because of the underlying structure, as was discussed in Section 4.4.2. In another example, space group $I4_1$ belongs to arithmetic crystal class $4I$, and the possible Bravais classes are $I\frac{4}{m}mm$ and $Im\bar{3}m$. The latter is rejected because of its higher order [5], so that the Bravais class is $I\frac{4}{m}mm$. Even where an accidental equality, such as $a = b = c$ exists, although the lattice symmetry be $Im\bar{3}m$, the Bravais class remains $I\frac{4}{m}mm$.

A *crystal system* contains complete geometric crystal classes of space groups, whereas a Bravais system contains a complete Bravais flock. All Bravais flocks for which the Bravais classes, that describe lattice symmetries, belong to the same (holohedral) geometric crystal class also belong to the same Bravais system. These definitions clarify the correspondence between trigonal and hexagonal crystal classes (Table 5.11).

The *crystal family* is the smallest set of space groups containing, for any of its members, all space groups of both the Bravais flock and the geometric

Table 5.11 Trigonal and hexagonal space groups in crystal systems and Bravais systems.

		Hexagonal Bravais system	Rhombohedral Bravais system
Crystal system	**Crystal class**	**Hexagonal Bravais flock**	**Rhombohedral Bravais flock**
	$\frac{6}{m}mm$	$P\frac{6}{m}mm$, $P\frac{6}{m}cc$, $P\frac{6_3}{m}cm$	
		$P\frac{6_3}{m}mc$	
	$\bar{6}2m$	$P\bar{6}m2$, $P\bar{6}c2$, $P\bar{6}2m$, $P\bar{6}2c$	
Hexagonal	$6mm$	$P6mm$, $P6cc$, $P6_3cm$, $P6_3mc$	
	622	$P622$, $P6_122$, ..., $P6_322$	
	$6/m$	$P6/m$, $P6_3/m$	
	$\bar{6}$	$P\bar{6}$	
	6	$P6$, $P6_1$, ..., $P6_3$	
	$\bar{3}m$	$P\bar{3}1m$, $P\bar{3}1c$, $P\bar{3}m1$, $P\bar{3}c1$	$R\bar{3}m$, $R\bar{3}c$
	$3m$	$P3m1, P31m, P3c1, P31c$	$R3m$, $R3c$
	32	$P312$, $P321$, $P3_112$, $P3_121$	$R32$
		$P3_212$, $P3_221$	
	$\bar{3}$	$P\bar{3}$	$R\bar{3}$
	3	$P3$, $P3_1$, $P3_2$	$R3$

crystal class to which the member belongs. Thus, the types $R3$ and $P6_1$ are of the same crystal family because both $R3$ and $P3$ belong to the same geometric crystal class, 3. Space groups $P3$ and $P6_1$ are members of the same Bravais flock $\frac{6}{m}mm$. Thus, $P3$ acts as a link between $R3$ and $P6_1$. There are, then, four families in two dimensions, and six crystal families in three dimensions: triclinic, monoclinic, orthorhombic, tetragonal, cubic and hexagonal, which contrasts with seven *crystal systems*, in the sense of Section 3.5.7 where seven systems are recognized; in the crystal families, trigonal symmetry is subsumed into hexagonal in this classification.

The material in this section is, perhaps, somewhat esoteric, and not a necessity for pursuance of crystal structure determinations. It is, however, a classification on which future editions of the International Tables will probably be based, and the reader who desires further information on this topic is directed to the definitive literature [5].

References 5

[1] Fyodorov YS. *Zap. Mineral. Obch.* 1891; 28: 1. [In English: *Am. Crystallogr. Assoc. Monograph* 1971; No. 7.]

[2] Schönflies A. *Kristallsysteme und Kristallstruktur*. Leipzig: 1891.

[3] Barlow W. *Z. Kristallogr*. 1894, 23: 1.

[4] Henry NFM and Lonsdale K (Eds.) *International tables for X-ray crystallography*. Kynoch Press, 1965.

[5] Hahn T (Ed.) *International tables for crystallography*, Vol. A. 5th ed. IUCr/Wiley, 2011.

[6] Bragg Sir WH. (1864–1942) *Quotation*.

[7] Htoon S and Ladd MFC.*J. Cryst. Mol. Struct*. 1974; 4: 357.

[8] Burns G and Glazer AM. *Space groups for solid state scientists*. 3rd ed. Elsevier, 2013.

[9] de la Rochefoucald Duc F. *Maxims* 1665; No. 160.

[10] Zandi R, et al. *Proc. Natl. Acad. Sci*. 2004; 10: 1073.

[11] Plevka P, et al. *Protein Sci*. 2008; 17: 1731.

[12] Ladd M and Palmer R *Structure determination by X-ray crystallography*. Springer, 2013.

[13] Woolfson MM. *An introduction to crystallography*. 2nd ed. Cambridge University Press, 1997.

[14] Bragg WL. *The crystalline state. Volume I:A general survey*. G. Bell & Sons, 1949.

[15] Macgillavry CH. *Symmetry aspects of M. C. Escher's periodic drawings*. Oosthoek, 1965.

[16] Schwarzenberger RLE. *Bull. London Math. Soc*.1984; 16: 209.

[17] Shubnikov AV and Belov NV. *Coloured symmetry*. Pergamon Press, 1964.

[18] Senechal M. *Color symmetry*. Elsevier, 2000.

[19] Mackay AL. *ActaCrystallogr*. 1957; 10: 543.

[20] Loeb AL. *Color symmetry*. John Wiley and Sons, 1971.

[21] Wondratschek H and Müller U (Eds.) *International tables for crystallography*, Vol. A1. 2nd ed. IUCr/Wiley, 2010.

[22] *Crystallographic space group diagrams and tables*. <http://img.chem.ucl.ac.uk/sgp/mainmenu.htm>.

[23] Bilbao Crystallographic Server. <http://www.cryst.ehu.es/>.

[24] *Online dictionary of crystallography*. <http://reference.iucr.org/dictionary/A>.

[25] Dresselhaus MS, et al. *Group theory: application to the physics of condensed matter*. New York: Springer, 2008.

Problems 5

5.1 Two nets are described by the unit cells (i) $a = b, \gamma = 90°$, and (ii) $a = b, \gamma = 120°$. In each case: (a) what is the symmetry at each net point? (b) To which two-dimensional system does the net belong? (c) What is the result of centring the unit cell?

5.2 A monoclinic F unit cell has the dimensions $a = 6.000, b = 7.000, c = 8.000$ Å and $\beta = 110.0°$. Show that an equivalent monoclinic C unit cell, with an *obtuse* β angle, can represent the same lattice, and calculate its dimensions. What is the ratio of the volume of the C cell to that of the F cell?

5.3 Carry out the following exercises with drawings of a tetragonal P unit cell. (a) Centre the B faces. Comment on the result. (b) Centre the A and B faces. Comment on the result. (c) Centre all faces. What conclusions can you draw now?

5.4 Calculate the length of $[31\bar{2}]$ for both unit cells in Problem 5.2.

5.5 The relationships *a*¢*b*¢*c*, and *α*¢*β*¢ 90° or 120° and $\gamma = 90°$ may be said to define a diclinic system. Is this an eighth crystal system? Give reasons for your answer.

5.6 (a) Draw a diagram to show the symmetry elements and general equivalent positions in the plane group $c2mm$, origin on $2mm$. Write the coordinates and point symmetry of the general and special positions, in their correct sets, and give the conditions limiting X-ray reflections in a structure of this plane group. (b) Draw a diagram of the symmetry elements in plane group $p2mg$, origin on 2; take care not to put the 2-fold point at the intersection of m and g. (c) Why the caution in (b)? (d) On the diagram, insert each of the motifs P, V and Z in turn, each letter drawn in its most symmetrical manner, using the *minimum* number of motifs consistent with the space-group symmetry. (e) Draw diagrams to show the general equivalent positions and symmetry elements in plane group $6mm$, and list the coordinates of the general equivalent positions.

5.7 Continue the study of space group $P2_1/c$ (Section 5.4.6). (a) Write the coordinates of the general and special positions, in their correct sets, and give the limiting conditions for all sets of positions, and write the plane-group symbols for the three principal projections. (b) Draw a diagram of the space group as seen along the b axis. (c) Biphenyl (Fig. P5.1) crystallizes in space group $P2_1/c$, with two molecules per unit cell. What can be deduced about both the positions of the molecules in the unit cell and the molecular conformation? The benzene rings in the molecule may be assumed to be planar. What difference in conformation might be expected in the free molecule state?

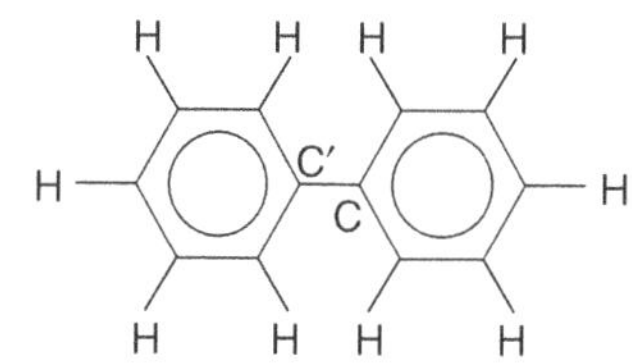

Fig. P5.1
Structural formula for biphenyl, $C_{12}H_{10}$.

5.8 Write the coordinates of the vectors between all pairs of general equivalent positions in $P2_1/c$ with respect to the origin, and note that they are of two types. What is the 'weight', or multiplicity, of each vector set? Remember that $+\frac{1}{2}$ and $-\frac{1}{2}$ in a coordinate are crystallographically equivalent, because 1 can always be added to or subtracted from a fractional coordinate without altering its crystallographic implication.

5.9 The orientation of the symmetry elements in the orthorhombic space group $Pbam$ may be written as follows:

$$\begin{array}{l} \bar{1} \text{ at } 0, 0, 0 \text{ (choice of origin)} \\ \left.\begin{array}{l} b \text{ glide } \parallel (p, y, z) \\ a \text{ glide } \parallel (x, q, z) \\ m \text{ plane } \parallel (x, y, r) \end{array}\right\} \text{ (from the space-group symbol)} \end{array}$$

Determine p, q and r from the following scheme, using the fact that $\boldsymbol{mab} = \bar{\boldsymbol{1}}$;

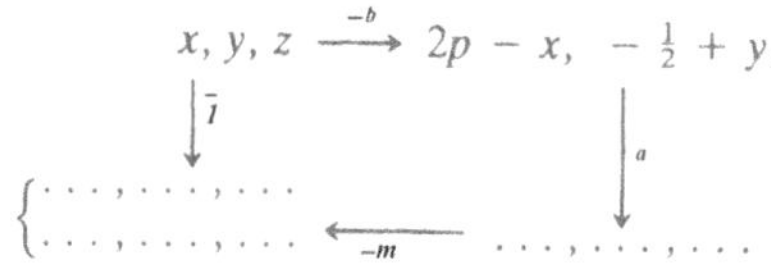

5.10 Construct a spacegroup diagram for *Pbam*, with the origin at the intersection of the three symmetry planes. List the coordinates of both the general equivalent positions and the centres of symmetry. Derive the standard coordinates for the general positions by transforming the origin to a centre of symmetry.

5.11 Show that space groups *Pa*, *Pc*, and *Pn* represent the same pattern, but that *Ca* is different from *Cc*. What is the more usual symbol for space group *Ca*? What would be the space group for *Cc* after an interchange of the x and z axes? Is *Cn* another monoclinic space group?

5.12 For each of the space groups $P2/c$, $Pca2_1$, $Cmcm$, $P\bar{4}2_1c$, $P6_122$, and $Pa\bar{3}$. For each,
(a) Write down the parent point group and crystal system;
(b) List the full meaning conveyed by the symbol;
(c) State the independent conditions limiting x-ray reflections.

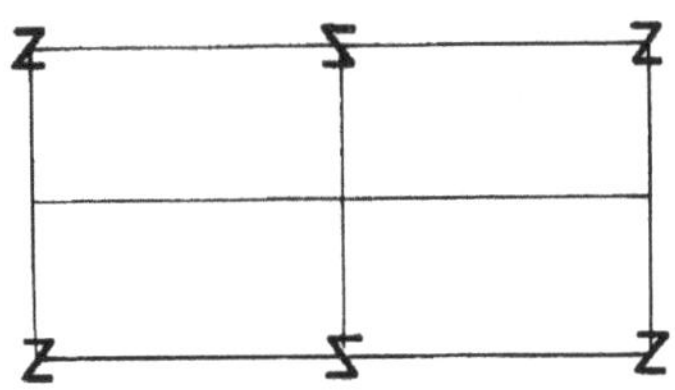

Fig. P5.2
Plane group containing two **Z** species per unit cell.

5.13 Consider the plane group in Fig. P5.2. (a) What is its symbol? (b) What would be the result of constructing this diagram with Z replacing its mirror image?

5.14 (a) Draw a *P* unit cell of a cubic lattice in the standard orientation. (b) Centre the *A* faces. What system and standard unit-cell type now exist? (c) From the position at the end of (b), let the cell side c and all other lines parallel to it be angled backward a few degrees in the a,c plane. What system and standard unit-cell type now exist? (d) From the position at the end of (c), let c and all other lines parallel to it be angled sideways a few degrees in the b, c plane. What system and standard unit-cell type now exist? For (b) to (d), write the transformation equations that take the unit cell as drawn into its standard orientation.

5.15 Set up matrices for the following symmetry operations: $\bar{\boldsymbol{4}}$ along the z axis, $\boldsymbol{m}$ normal to the y axis. Hence, determine the Miller indices of a plane obtained by operating on a plane (*hkl*) first by $\bar{\boldsymbol{4}}$, and on the resulting (*hkl*) planes by $\boldsymbol{m}$. What are the nature and orientation of the symmetry element represented by the given combination of $\bar{\boldsymbol{4}}$ followed by $\boldsymbol{m}$?

5.16 The matrices for an n glide plane normal to a and an a glide plane normal to b in an orthorhombic space group are as follows:

$$\underset{a}{\begin{pmatrix} 1 & 0 & 0 \\ 0 & \bar{1} & 0 \\ 0 & 0 & 1 \end{pmatrix}} + \underset{\mathbf{t}_a}{\begin{pmatrix} \frac{1}{2} \\ 0 \\ 0 \end{pmatrix}} \cdot \underset{n}{\begin{pmatrix} \bar{1} & 0 & 0 \\ 0 & 1 & 0 \\ 0 & 0 & 1 \end{pmatrix}} + \underset{\mathbf{t}_n}{\begin{pmatrix} 0 \\ \frac{1}{2} \\ \frac{1}{2} \end{pmatrix}}$$

What are the nature and orientation of the symmetry element arising from the combination of $\boldsymbol{n}$ followed by $\boldsymbol{a}$? What are the spacegroup symbol and class?

5.17 (a) Determine the matrices for both a $\boldsymbol{6}_3$ rotation operation about [0, 0, z] and a (subsequent) $\boldsymbol{c}$ glide operation normal to the y axis and passing through the origin. Use a hexagonal stereogram to obtain the c glide matrix. (b) What is the matrix for the $\boldsymbol{6}_3$ rotation operation followed by that for the $\boldsymbol{c}$ glide operation? (c) Determine the symmetry represented by the matrix for this combination, and write the spacegroup symbol. (d) Draw diagrams of the general equivalent positions and symmetry elements for the space group. (e) List the number of general equivalent positions, their Wyckoff notation, point symmetry and coordinates. (f) Are there any special equivalent positions? If so, list them as under (c). (g) List the limiting conditions on all sets of equivalent positions.

5.18 A unit cell is determined as $a = b = 3$ Å, $c = 9$ Å, $\alpha = \beta = 90°$, $\gamma = 120°$. Later, it proves to be a triply-primitive hexagonal unit cell. With reference to Fig. 4.16, determine the equations for the unit cell transformation $R_{\text{hex}} \rightarrow R_{\text{obv}}$, and calculate the parameters of the rhombohedral unit cell.

5.19 In relation to Problem 5.18, given the plane (13^*4) and zone symbol $[1\,\bar{2}^*\,3]$ in the hexagonal unit cell, determine these parameters in the obverse rhombohedral unit cell. The symbol * here indicates that the three integers given relate to the x, y and z axes, respectively.

5.20 By means of a diagram, or otherwise, show that a site x, y, z reflected across the plane $(q, \bar{q}, z)$ in the tetragonal system has the coordinates $q + y,\ q + x,\ z$ after reflection.

5.21 Draw the projection of an orthorhombic unit cell on (001), and insert the trace of the (210) plane and the parallel plane through the origin. (a) Consider the transformation $\mathbf{a}' = \mathbf{a}/2,\ \mathbf{b}' = \mathbf{b},\ \mathbf{c}' = \mathbf{c}$. Using the appropriate transformation matrix, write the indices of the (210) plane with respect to the new ($\mathbf{a}'$, $\mathbf{b}'$) unit cell. Draw the new unit cell and insert the planes at the same perpendicular spacing, starting with the plane through the origin. Does the geometry of the diagram confirm the indices obtained from the matrix? (b) Make a new drawing, like the first, but now consider the transformation $\mathbf{a}' = \mathbf{a},\ \mathbf{b}' = \mathbf{b}/2,\ \mathbf{c}' = \mathbf{c}$. What does (210) become under this transformation? Draw the new unit cell and insert the planes as before. Does the geometry confirm the result from the matrix?

5.22 (a) Why are space groups *Cmm*2 and *Amm*2 distinct, yet *Cmmm* and *Ammm* are equivalent? (b) Space group *Abma* relates to the orthorhombic **bca** setting of the reference axes. (i) What is the symbol in the standard, **abc**, setting, and(ii) what would be the symbol using the *e* glide notation?

5.23 A lattice is represented by an experimentally determined monoclinic *C* unit cell, with $a = 8.454,\ b = 10.597,\ c = 11.754$ Å and $b = 111.09°$. Determine by means of the program LEPAGE, or otherwise, the conventional unit cell for the lattice. Give the transformation matrix for $\mathbf{a}'$ in terms of $\mathbf{a}$.

5.24 The ferroelectric phase of bismuth vanadate, $BiVO_4$, has been reported in space group $I2/b$ with $a = 0.620,\ b = 0.509,\ c = 1.114$ nm and $b = 90°$. Transform this space group to its standard form and calculate the dimensions of the transformed unit cell.

5.25 Determine matrices for the symmetry operations in the trigonal space groups $P3_112$ and $P3_121$ referred to hexagonal axes. The matrices may be obtained either by inspection of the stereogram for point group 32 (Fig. 3.12c) or from Appendices A6 or A3. Hence, list the coordinates of the general equivalent positions in order to confirm that the two space groups form two distinct arrangements in space.

5.26 From the discussion on cubic space groups in Section 5.4.14.2 determine the coordinates of the general equivalent positions for space group $P4_232$; select an origin on 23 (think of the implication of 4_2). Hence, list the conditions limiting *hkl* x-ray reflections for this space group.

5.27 What are the plane groups of the projections of (a) *P*4*mm*, (b) *P*4*bm* and (c) $I\frac{4_1}{a}md$ on (i) (001), (ii) (100), (iii) (110), and (d) $P2_13$ on (iv) (001), (v) (111), (vi) (110)?

5.28 Derive the coordinates of the general equivalent positions for space group *P*432. Hence, label the points 1–24 in Fig. P5.3 with their coordinates, taking point (1) as x, y, z.

5.29 Devise a scheme for obtaining both the coordinates of the general equivalent positions and the symmetry elements of space group *Cmce*; set $\bar{1}$ at the origin, and consider just the $\frac{1}{2} + y$ translation for the *e* glide. How does the result differ from that for *Cmca* in Reference [25] may be needed for this part in the absence of ITA.

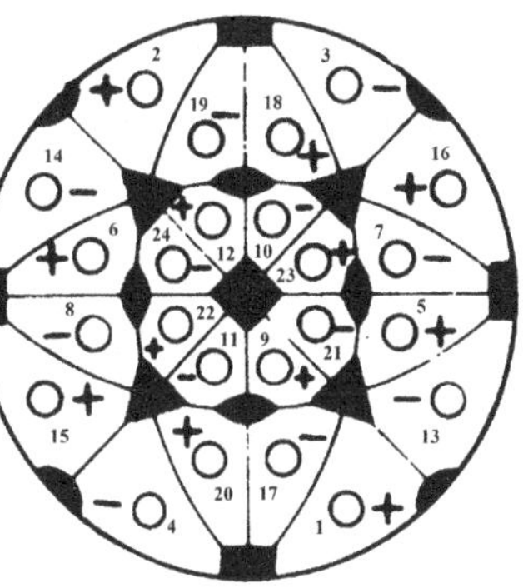

Fig. P5.3
Stereogram for point group 432, with the symmetry-related points labelled from 1 to 24.

6 Symmetry and X-ray diffraction

SYNOPSIS

- Generation and properties of X-rays
- X-ray diffraction
- Collecting diffraction data
- Reciprocal lattice and Ewald's construction
- X-ray photography and diffractometry
- X-ray scattering and structure factor
- Limiting conditions from the structure factor equation
- Space group information from X-ray diffraction data

6.1 Introduction

It is necessary to know how the crystal symmetry and space group data that have been the topics of the earlier chapters are acquired in practice, and for that purpose knowledge of X-rays and X-ray diffraction from crystalline material is required. Only an outline of X-ray crystallography can be provided here, but the modern literature abounds with discussions of this subject.

X-rays are an electromagnetic radiation of short wavelength, produced when electrons that are accelerated by an electric potential become arrested at the surface of a target material, usually metallic copper or molybdenum. If an electron falls through a potential difference of V volt, it gains energy eV, where e is the charge on an electron. In the traditional X-ray tube, a large portion of the energy of the electrons is converted to heat as they strike the target material. This heat must be dissipated by cooling the target, and only about 10% of the electron energy eV leads to useful X-radiation.

The loss of electron energy by collision with the target material involves multiple events. The maximum energy E converted determines the shortest wavelength λ_{min} that can be obtained, according to the equation

$$E = eV = hc/\lambda \qquad (6.1)$$

where h is the Planck constant and c is the speed of light in a vacuum. The higher the accelerating voltage the smaller the minimum wavelength, and a maximum intensity in the continuous, or 'white', radiation occurs at a wavelength of approximately $1.5\lambda_{min}$.

If an X-ray beam of sufficient energy is filtered by passing it through an appropriate metal foil, or better, reflected from the surface of a single crystal such as graphite or quartz, a monochromatic beam of X-rays can be obtained. For a copper target, the mean Cu K_α X-ray wavelength obtained is 0.15418 nm; in the case of copper radiation filtered by a thin nickel foil, a very small percentage of Cu K_β X-radiation, $\lambda = 0.13895$ nm, is also present.

6.2 X-ray diffraction

The distances between atoms in a crystal are commensurate with the wavelengths of X-rays, so that a crystal behaves towards X-radiation like a three-dimensional diffraction grating. All atoms in the unit cells scatter radiation, and strong directional effects arise by the cooperative scattering of a large number of unit cells. The diffraction of X-rays by a crystal may be understood most easily in terms of the Bragg equation, which treats diffraction like reflection from crystal planes.

Consider two adjacent crystal planes in a family of parallel, equidistant planes of interplanar spacing d, and let a parallel X-ray wave train be incident on the planes at a glancing angle θ (Fig. 6.1). The X-rays are in phase along the normal AC to the incident wavefront and also along the normal AD to the reflected wavefront. The path difference the for the two rays is $CB + BD$, which is equal to $2d \sin\theta$, since $\angle DAB = \angle CAB = \theta$. The condition of reinforcement of the reflected rays is a path difference of an integral number of wavelengths; hence,

$$2d \sin\theta = n\lambda \qquad (6.2)$$

where n is an integer. As indicated in Section 2.5.1, each possible family of planes in a lattice has its own individual spacing $d_{nh,nk,nl}$, so that $d_{nh,nk,nl} = d_{hkl}/n$. Thus, the Bragg equation becomes

$$2d_{hkl} \sin\theta_{hkl} = \lambda \qquad (6.3)$$

where θ_{hkl} is the Bragg angle for the specific (hkl) family of planes? If the equation is satisfied for planes 1 and 2 in a family, it will be satisfied also for planes 2 and 3, 3 and 4, and so on for the whole family (hkl) in the crystal specimen.

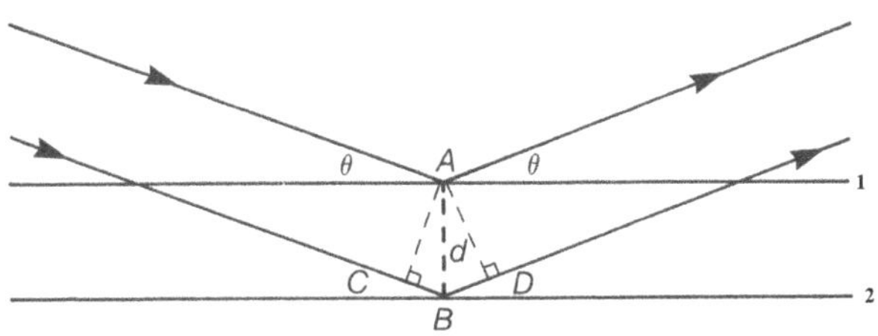

Fig. 6.1
Bragg reflection. Two planes from a family of parallel, equidistant planes of spacing d. Two typical rays are shown; θ is the glancing angle of incidence, also called the Bragg angle, and is equal to the angle of reflection.

6.3 Recording X-ray diffraction spectra

The method used for recording X-ray diffraction data depends upon the particular application in view. The principal methods today use diffractometers with single crystal or powder crystal specimens, but photographic techniques are in use for certain procedures. For our purposes, techniques are needed that reveal the crystal symmetry as clearly as possible, without distortion. In practice, the understanding of diffraction data requires an appreciation of the use of the reciprocal lattice, introduced in Section 4.7. It will be discussed next, and followed by a brief introduction to the Laue and precession methods of the X-ray photography of crystals.

6.4 Reciprocal lattice and Ewald's construction

'*Their moral is this—that a right way of looking at things will see through almost anything*' [1]. The Bragg construction requires us to think about a distribution of planes, which is not always easy. The geometrical interpretation of X-ray diffraction is facilitated by a method introduced by Bernal [2], developed by Ewald [3], and subsequently described as Ewald's construction. In Fig. 6.2, a sphere of radius unity in reciprocal space is constructed on the X-ray beam as

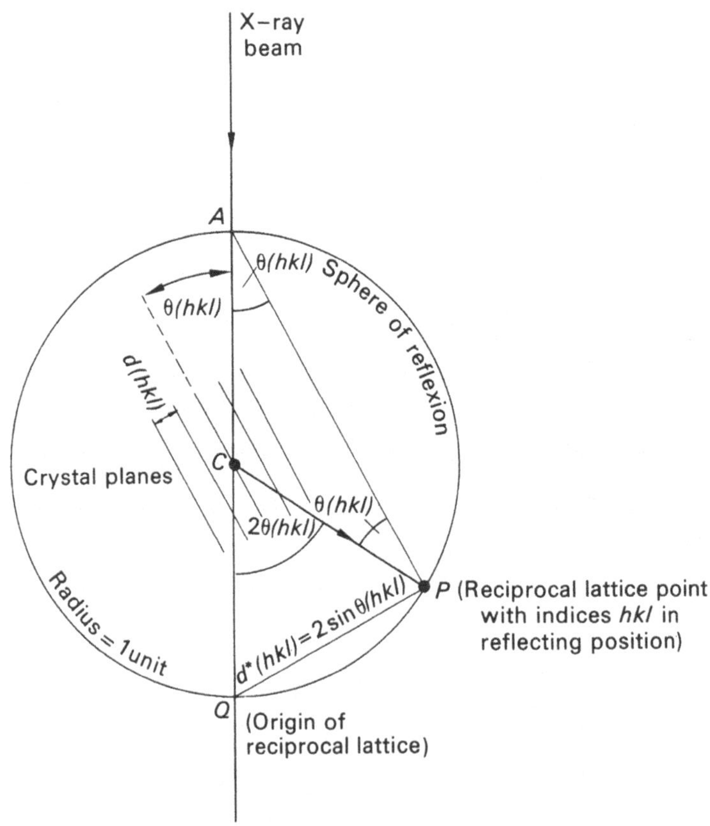

Fig. 6.2
Ewald's construction. An X-ray reflection arises when a reciprocal lattice point P intersects the surface of the Ewald sphere; the direction of the reflected ray is CP. If the x axis in the crystal is normal to the circular section shown and this section is central in the sphere, all reciprocal lattice points such as P will be labelled $0kl$. [Reproduced by courtesy of Springer Science+Business Media, New York, © Kluwer Academic/Plenum Publishing.]

diameter, AQ, where Q is the origin of the reciprocal lattice: the crystal under consideration is at the centre C of the sphere. From the construction, $AQ = 2$ and $\angle APQ = 90°$. Thus, $QP = AQ \sin\theta_{hkl} = 2\sin\theta_{hkl}$ which, from Eq. (6.3) is λ/d_{hkl}, or d^*_{hkl}, from Eq. (4.12) with K now equal to λ, thus determining the size of the reciprocal lattice for that wavelength. Thus, P is the reciprocal lattice point corresponding to the *hkl* family of planes and *CP* is the direction of the reflected X-ray beam. Hence, an X-ray reflection from a family of planes (*hkl*) occurs when the reciprocal lattice point *hkl* intersects the Ewald sphere, or *sphere of reflection*, and the direction of the reflected beam is from the crystal through the point *hkl*. Note that although the term X-ray reflection is used, following the Bragg equation, the process is one of diffraction or, more specifically, of combined diffraction and interference process.

6.5 X-ray intensity data collection

The intensity of an X-ray diffracted beam is recorded either by the blackening of a sensitive film or plate, or by quantum counting. The photographic procedure with a film record has been largely replaced in modern methods by the use of image plates or charge-coupled area detectors. An X-ray diffraction record from a crystal, however obtained, is the Fourier transform of its electron density distribution sampled at the reciprocal lattice points [4]. The variation in the intensity of the diffraction spectra, each spectrum characterized in position by its *hkl* value, is determined by the cooperative scattering of X-rays from the atomic electron density on the (*hkl*) planes in the crystal.

6.5.1 Laue X-ray photography

The earliest X-ray diffraction technique is that due to Laue [5], and Fig. 6.3 shows one of the first Laue X-ray photographs. The basic experimental arrangement for Laue X-ray diffraction photography is simple (Fig. 6.4). A beam of 'white' X-radiation, that is, a continuous spectrum of unfiltered X-rays, impinges on a stationary crystal and the diffraction effects are recorded on a photographic film. In this technique, the parameters d and θ in the

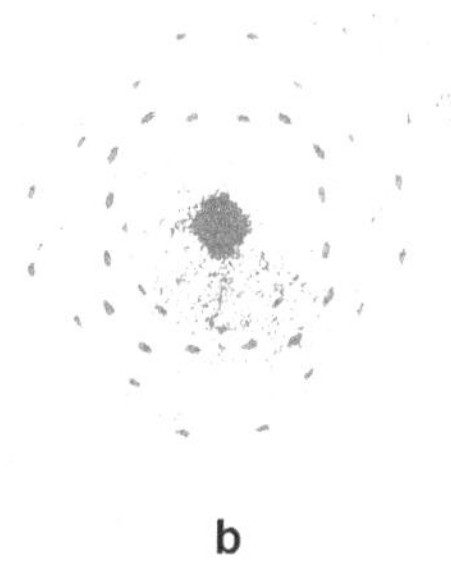

b

Fig. 6.3
One of the first X-ray diffraction photographs: zinc blende, ZnS, showing symmetry 4*mm* on the film.

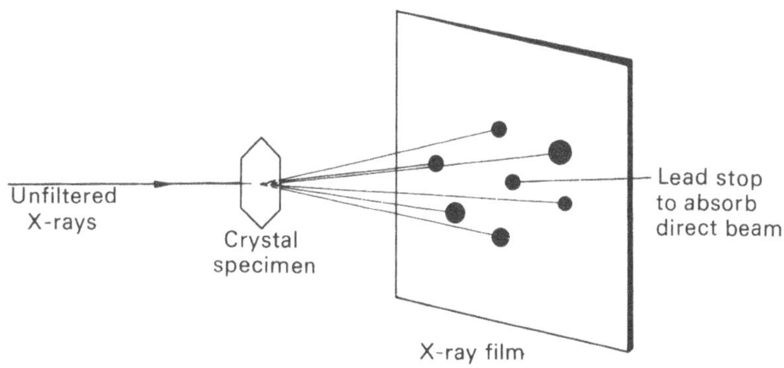

Fig. 6.4
The basic experimental arrangement for Laue X-ray photography.

Bragg equation are fixed, and the equation is satisfied by the crystal effectively selecting from the incident X-ray spectrum the wavelength appropriate for each reflection so as to satisfy the Bragg equation. The distinction between Bragg 'reflection' of X-rays and optical reflection of visible light is clear: in the optical case, as long as incident light meets a reflecting surface a reflection is obtained, but with Bragg reflection, Eq. (6.3) must be satisfied for an X-ray reflection to arise.

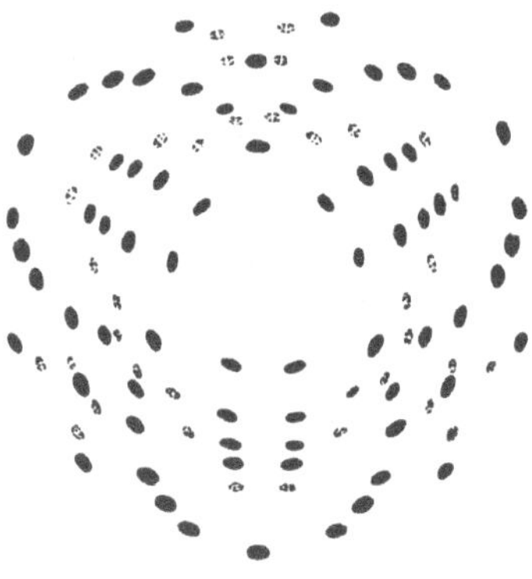

Fig. 6.5
Laue photograph of α-corundum, Al_2O_3, showing symmetry 3*m* on the film.

Another example of a Laue photograph, that of α-corundum (Al_2O_3), is shown by Fig. 6.5. As with Fig. 6.3, the diffraction spots lie on a series of ellipses, with one end of their major axes at the centre of the film. The appearance of the film may be explained by reference to Fig. 6.6. All spots on a given ellipse arise through reflections from planes in one and the same zone. The X-ray beam is along the *X, Y* direction, and a possible zone axis for a given value of θ is *ZZ'*. A reflected ray *R* is the generator of a cone of semi-vertical angle θ, which is imposed on the film as an ellipse, one end of its major axis lying at the centre of the film, *Y*. However, instead of a continuous ellipse of reflection, *vide* optical reflection, the Bragg equation restricts reflections to only those positions on the ellipse satisfying Eq. (6.3).

Figures 6.3 and 6.5 illustrate an important use of the Laue photograph: it reveals the crystal symmetry in the direction of the X-ray beam, 4*mm* and 3*m*, respectively, in the two example photographs. At this point, it is germane to discuss Laue projection symmetry, a topic which was postponed from the discussion in Section 3.6.5.

6.5.2 Laue projection symmetry

Further to the discussions in Section 3.6.4 and Section 3.6.5, *Laue projection symmetry* involves what the X-ray beam 'sees' as it passes through the crystal, and it is this symmetry information that is projected on to the film in the direction of the X-ray beam. The space group of α-corundum is $\bar{3}m$, and so the X-ray beam travelling along the *z* axis sees threefold symmetry and mirror symmetry, which is projected symmetry on (0001) as 3*m* (Fig. 6.5). If an X-ray beam is travelling normal to (110) in a crystal of point group 422 (Fig. 3.12b), the corresponding Laue class, $\frac{4}{m}mm$, must be considered, because X-ray diffraction always introduces a centre of symmetry into the diffraction record—diffraction is identical from both sides of a family of crystal planes.

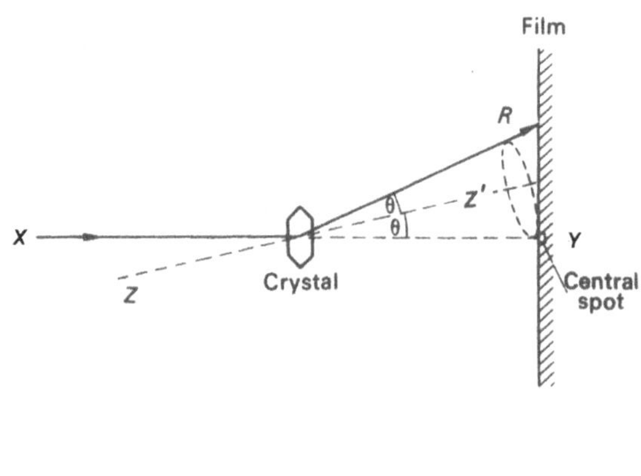

Fig. 6.6
Geometry of the Laue method: *XY*, the X-ray beam direction; *ZZ'*, a zone axis; *R*, the generator of the partial cone of diffracted rays of semi-angle θ, the Bragg angle; *Y*, the central spot on the film and one extremity of the major axes of the partial ellipses of diffraction spots formed by those intersections of the cone with the film when the Bragg equation is obeyed.

Table 6.1 Laue projection symmetry for the 32 crystallographic point groups[a].

Point groups	Laue classes	{ 100 }	{ 010 }	{ 001 }
$1, \bar{1}$	$\bar{1}$	1	1	1
$2, m, 2/m$	$2/m$	m	2	m
$222, mm2, mmm$	mmm	$2mm$	$2mm$	$2mm$
		{ 001 }	**{ 100 }**	**{ 110 }**
$4, \bar{4}, 4/m$	$4/m$	4	m	m
$422, 4mm$				
$\bar{4}2m, \frac{4}{m}mm$	$\frac{4}{m}mm$	$4mm$	$2mm$	$2mm$
		{ 0001 }	**{ $10\bar{1}0$ }**	**{ $11\bar{2}0$ }**
$6, \bar{6}, 6/m$	$6/m$	6	m	m
$622, 6mm$				
$\bar{6}m2, \frac{6}{m}mm$	$\frac{6}{m}mm$	$6mm$	$2mm$	$2mm$
$3, \bar{3}$	$\bar{3}$	3	1	1
$32, 3m, \bar{3}m$	$\bar{3}m$	$3m$	m	2
		{ 100 }	**{ 111 }**	**{ 110 }**
$23, m\bar{3}$	$m\bar{3}$	$2mm$	3	m
$m3$		$2mm$	6	$2mm$
$432, \bar{4}3m, m\bar{3}m$	$m\bar{3}m$	$4mm$	$3m$	$2mm$

[a] The trigonal point groups are here referred to hexagonal axes.

Thus, the X-ray beam encounters twofold symmetry and lies in two *m* planes, and the Laue projection symmetry is $2mm$. Table 6.1 lists the Laue projection symmetry for the principal planes in the 32 crystallographic point groups.

6.5.3 X-ray precession photography

The early methods of recording X-ray data on film presented the reciprocal lattice in a two-dimensional, collapsed, and distorted form, with diffraction spots lying in straight lines, as in the rotation and oscillation methods, or on curves, as with Weissenberg photography. The maximum symmetry of an oscillation or rotation motion is $2mm$; thus, the symmetry revealed by such techniques is $2mm$, or one of its subgroups, m, 2 or 1. For a ready appreciation of crystal symmetry and the unit cell dimensions, an undistorted picture of the reciprocal lattice, at more than one level, is required.

In 1942, a paper was published [6] which presented a method for obtaining an undistorted photograph of the reciprocal lattice, albeit with a doubling on the film of most of the diffraction spots. This deficiency was corrected in the Buerger precession camera: any level of a reciprocal lattice could be recorded on a flat film without distortion provided that the plane of the film oscillated about an axis parallel to the oscillation axis of the crystal, and

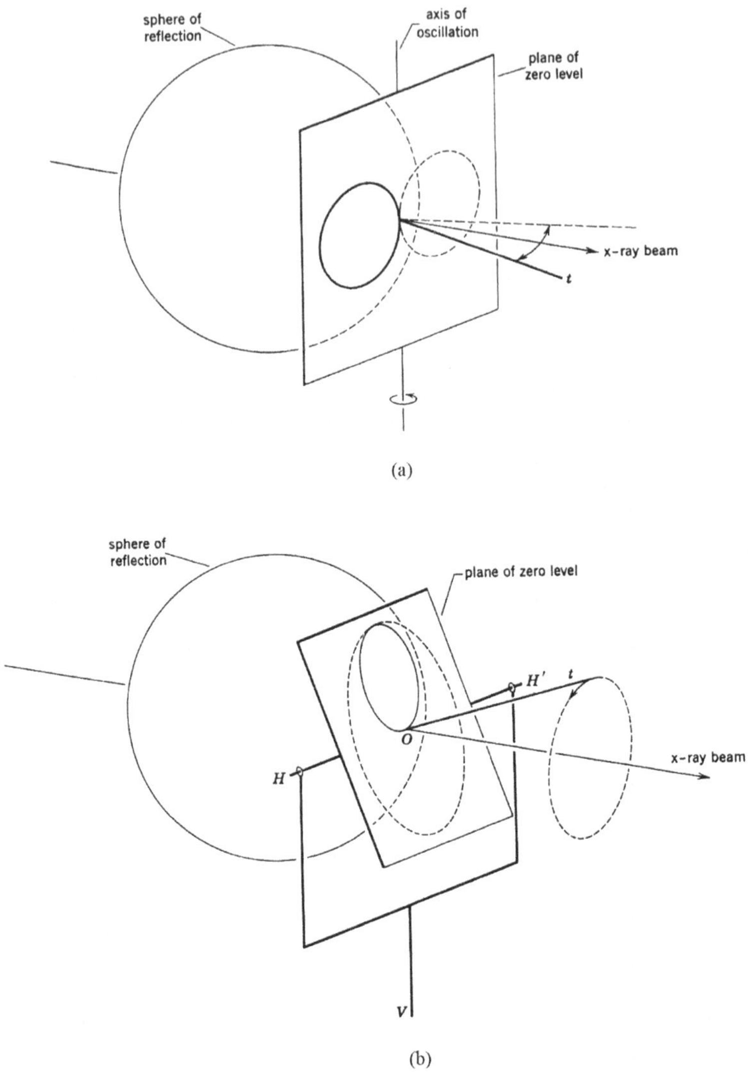

Fig. 6.7
Oscillation and precession geometry compared. (a) Oscillation: the normal t to a zero level reciprocal lattice plane oscillates to equal limits about the direction of the X-ray beam; the maximum symmetry in the direction of t is $2mm$. (b) Precession: the normal t precesses about the X-ray beam; HOH' and OV are horizontal and vertical axes respectively, and the symmetry about t is the true symmetry for that direction. [Reproduced from Buerger MJ. *The precession method in X-ray crystallography*. New York: Wiley, 1964.]

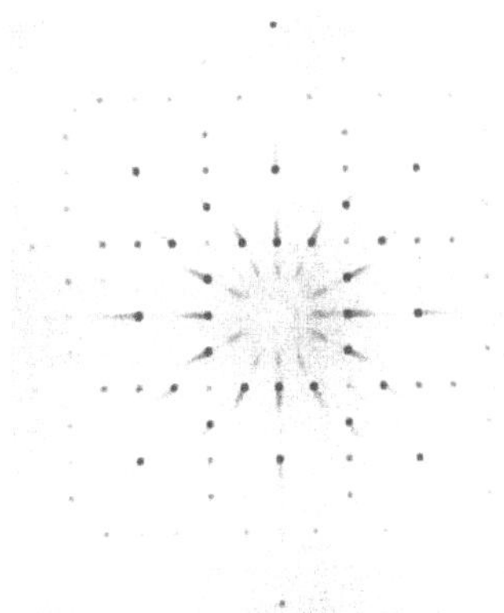

Fig. 6.8
An a axis precession photograph of a garnet taken with Mo K_α radiation, $\lambda = 0.71067$ Å. The two spots at the extreme ends of the central vertical row correspond to $h = \pm 8$. [Reproduced from Buerger MJ. *The precession method in X-ray crystallography*. New York: Wiley, 1964.]

coupled so as to achieve this motion synchronously; such a situation is achieved by ensuring that the film precesses about an axis parallel to the oscillation axis of the crystal. Thus, the film is always tangential to the sphere of reflection at the point where the reflected beam intersects the Ewald sphere. The basic geometry of the precession camera is shown in Fig. 6.7, which compares oscillation and precession geometry, and the subject of X-ray precession photography has been covered comprehensively in the literature [4, 7, 8]. The method is theoretically and mechanically somewhat complicated, but the interpretation of the photographs very straightforward.

The precession photograph in Fig. 6.8 was taken from a garnet crystal, which is cubic, space group $Ia\bar{3}d$; Mo K_α radiation, $\lambda = 0.071067$ Å was used. Figure 6.9 illustrates the calculation of the spacing from this precession

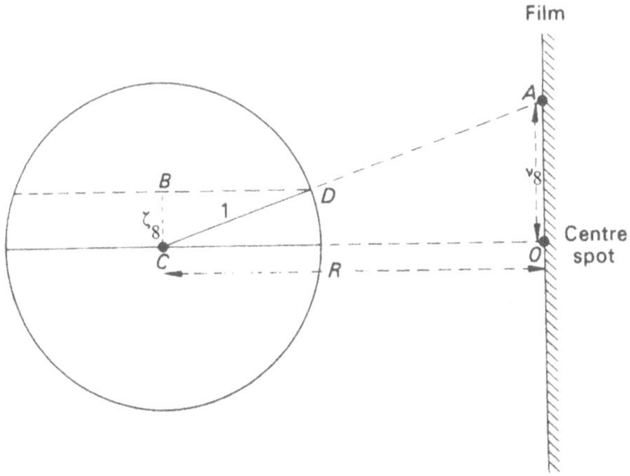

Fig. 6.9
Relationship between the eighth level reciprocal lattice spacing ζ_8 and the corresponding vertical distance v_8 on a flat film; R is the crystal to film distance. The repeat distance along the vertical axis in the crystal is $8\lambda/\zeta_8$.

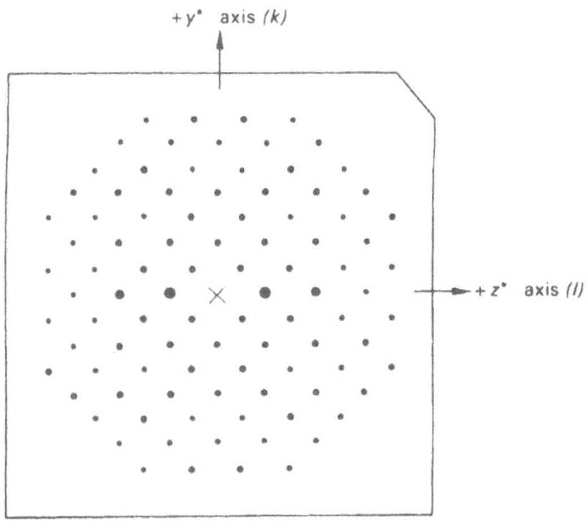

Fig. 6.10
Simulated precession photograph of an orthorhombic crystal precessing about a; Cu K_a radiation, $\lambda = 1.5418$ Å, is assumed. From the undistorted $0kl$ reciprocal net, b^*, c^* and γ^* can be deduced.

photograph, using row 8 from the centre; in practice, the double distance from row $h = \bar{8}$ *to* $h = 8$ is measured, and a is 11.459 Å.

The diagram in Fig. 6.10 is a simulated precession photograph of the $0kl$ level of the reciprocal lattice of an orthorhombic crystal. The limiting conditions are immediately evident:

$$0kl: k + l = 2n; \quad 0k0: (k = 2n); \quad 00l: (l = 2n)$$

Thus, from this photograph, an n glide plane normal to x can be deduced tentatively, but it may be related to a superior condition for a centred unit cell. The conditions for $0k0$ and $00l$ are themselves dependent upon the first condition and do not necessarily indicate the presence of 2_1 axes along y and z. The reciprocal lengths b^* and c^* are measureable on the film, and γ^* is clearly 90°, as expected for an orthorhombic crystal.

An interesting account of the history of the development of the Weissenberg and precession techniques has been given by Karl Weissenberg [9].

Example 6.1

Determine b, c and α from the film and data in Fig. 6.8.

Measurements on the actual film of Fig. 6.10 are $14b^* = 43.7$ mm and $14c^* = 44.1$ mm. The scaling factor measured on the photograph is 0.267, and the crystal to film distance 60.00 mm. From the geometry of the precession photograph, $b^* = (43.7/14)/(0.267 \times 60) = 0.1948$; hence $b = 1.5418/0.1948 = 7.915$Å. Similarly, $c^* = 0.1966$; hence, $c = 7.842$ Å; clearly, $\alpha = 90°$.

Nowadays, the precession camera is not greatly used. Fortunately it is possible to obtain an undistorted reciprocal lattice by diffractometer techniques [10,11]. Data are collected two-dimensionally by diffractometer, one reciprocal lattice layer at a time, and the detected data points transcribed to a plotter so as to obtain an undistorted diagram of the reciprocal lattice.

6.5.4 Diffractometric and image plate recording of X-ray intensities

X-ray diffraction data may be collected by diffractometric procedures; Fig. 6.11 illustrates the essence of a four-circle diffractometer. Computer software orientates the ω, ϕ, χ and 2θ circles so as to bring the crystal specimen into a measuring position for each *hkl* reflection within the recordable portion of reciprocal space, and the intensity of the reflection is determined by quantum counting. Alternatively, data may be recorded by Laue or oscillating crystal techniques and the data collected on an image plate; this method enables a rapid collection of intensity data. The use of high intensity radiation from a synchrotron source is particularly valuable with crystal materials, such as proteins, that are frequently stable over only a short time period. Whatever method be chosen, sophisticated computer software handles the production and collection of the intensity data and its processing to a set of structure factor amplitudes $|F(hkl)|$ which, together with the crystal geometry, forms

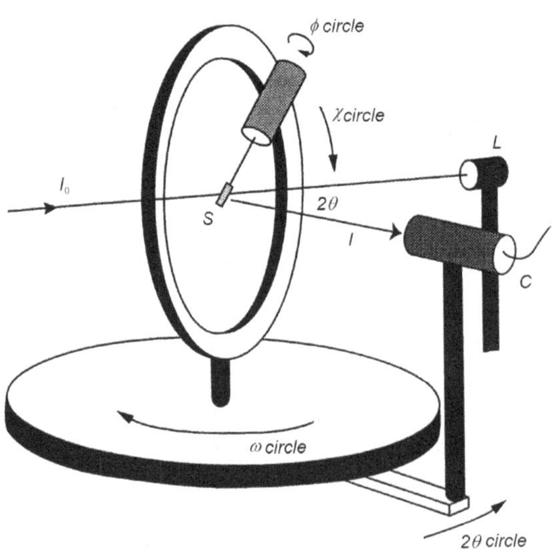

Fig. 6.11
Basic geometry of a four-circle X-ray diffractometer: I_0, incident X-ray beam; S, crystal under examination; I, diffracted X-ray beam; 2θ, angle of scatter (twice the Bragg angle); L, trap for transmitted beam; C, X-ray quantum counting chamber; ω, ϕ, χ and 2θ, the four circles of the diffractometer for orientating the crystal and counter into the measuring position.

the basis of X-ray crystal structure solving process. The procedures involved in structure analysis are well documented, and the reader is directed to a standard text for procedural detail [4].

6.6 X-ray scattering by a crystal: the structure factor

X-rays are scattered by the electrons in an atom. The efficiency of atomic scattering is expressed by the atomic scattering factor, f, which is dependent upon the wavelength of X-radiation and the angle of scatter. Thus, it is tabulated normally as a function of $\sin\theta/\lambda$; Fig. 6.12 illustrates the atomic scattering factor for carbon.

In a crystal, all atoms in the unit cell contribute to scattering, but because of their separations from one another there is a phase difference with respect to a reference point, the origin. The combination of atoms scattering with phase differences has been thoroughly described in the literature [4, 7], and the important results are stated here.

The scattering of the atoms in a unit cell is represented by a structure factor, F(*hkl*), for the *hkl* family of planes in the crystal, and may be written for any reflection as

$$\mathrm{F}(hkl) = \sum_{j=1}^{N} f_{j,\theta} \exp(\mathrm{i}\phi_j) \tag{6.4}$$

where N is the number of atoms in the unit cell, $f_{j,\theta}$ is the scattering factor for the jth atom (frequently written as just f_j), and $\exp(\mathrm{i}\phi_j)$ is the phase[1] of the jth atom with respect to the origin. The equation can be explained by reference to the Argand diagram (Fig. 6.13); the term $\exp(\mathrm{i}\phi_j)$ may be thought of as an operator that rotates the quantity f_j anticlockwise by an angle ϕ with respect to the real axis. The phase angle can be related to the atomic coordinates by the following argument.

[1] $\mathrm{i} = \sqrt{-1}$.

Consider Fig. 6.1, and let plane 1 pass through the origin. Another plane A, parallel to plane 1, lies at a distance d_A from it. From Eq. (2.3) for plane 1, called (*hkl*) here:

$$X\cos\chi + Y\cos\psi + Z\cos\omega = d(hkl)$$

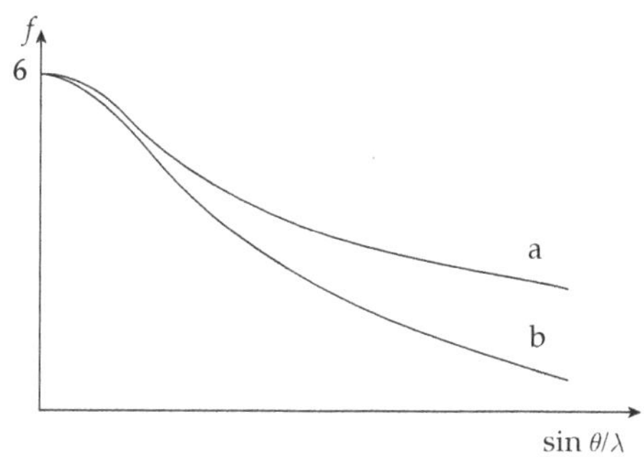

Fig. 6.12
The atomic scattering factor f for carbon; the decrease in f with increasing θ arises from an interference process within the atom itself. For scattering in the forward direction, $\theta = 0$, there is no interference and $f = Z$, the atomic number. (a) Rest values of f. (b) Values of the atomic scattering factor f corrected for thermal vibration.

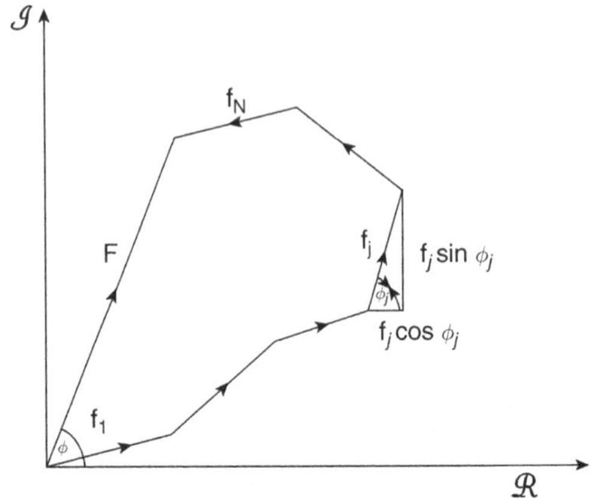

Fig. 6.13
The combined scattering of N atoms ($N = 6$ here) in terms of the real $\mathcal{R}$ and imaginary $\mathcal{I}$ components on an Argand diagram: the resultant F is $\sum_1^6 f_j \exp(\mathrm{i}\phi_j)$.

where X, Y and Z are the coordinates of the point where $d(hkl)$ intersects the plane (hkl). Then, the equation of the plane A, distant d_{ay} from plane 1 is, by analogy

$$X \cos \chi + Y \cos \psi + Z \cos \omega = d_A$$

Since $\cos \chi = \frac{d(hkl)}{a/h}$, and similarly for $\cos \psi$ and $\cos \omega$, the direction cosines can be eliminated:

$$\left(\frac{h}{a}X + \frac{k}{b}Y + \frac{l}{c}Z\right) d(hkl) = d_A$$

If atom A has fractional coordinates $x_A = X_A/a$, $y_A = Y_A/b$, $z_A = Z_A/c$, then

$$d_A = (hx_A + ky_A + lz_A)\, d(hkl)$$

The path difference δ_A for reflection from the planes 1 and A is

$$\delta_A = 2d_A \sin \theta_{hkl} = 2 \sin \theta_{hkl}\, (hx_A + ky_A + lz_A)\, d(hkl)$$

and from the Bragg equation, Eq. (6.3), by elimination of $2d(hkl) \sin \theta_{hkl}$

$$\delta_A = \lambda(hx_A + ky_A + lz_A)$$

The corresponding phase angle ϕ_A is $\frac{2\pi}{\lambda}\delta_A$, which can be expressed as

$$\phi_A = 2\pi\, (hx_A + ky_A + lz_A)$$

Hence, the *structure factor equation* becomes

$$\mathrm{F}(hkl) = \sum_{j=1}^{N} f_{j,\theta} \exp[\mathrm{i}2\pi\, (hx_A + ky_A + lz_A)] \qquad (6.5)$$

The structure factor is related to the intensity I of an X-ray reflection since $I \propto |\mathrm{FF}^*|$, where F^* is the complex conjugate of F, that is, Eq. (6.5) with a negative exponential term. The structure factor may be manipulated by separating it into its real and imaginary parts, following Fig. 6.13. Thus, the real part of F is defined in terms of a cosine function as

$$A'(hkl) = \sum_{j=1}^{N} f_{j,\theta} \cos 2\pi \left(hx_j + ky_j + lz_j\right) \qquad (6.6)$$

and the imaginary part as the sine function

$$B'(hkl) = \sum_{j=1}^{N} f_{j,\theta} \sin 2\pi\left(hx_j + ky_j + lz_j\right) \quad (6.7)$$

and the phase angle in Eq. (6.5) is given by

$$\phi_{hkl} = \tan^{-1}(B'/A') \quad (6.8)$$

If the conjugate of Eq. (6.5) is taken, together with Euler's theorem, $\exp(\pm i\theta) = \cos\theta \pm i\sin\theta$,

$$I \propto |\mathrm{FF}^*| = A'(hkl)^2 + B'(hkl)^2$$

As only the symmetry of the unit cell is of concern here, the summation can be divided into two parts:

$$\sum_{j=1}^{N} = \sum_{r=1}^{n}\sum_{s=1}^{m}$$

where $N = nm$. The sum over r refers to the symmetry independent atoms, and that over s to the symmetry related atoms. Thus, the structure factor equation may be separated:

$$A'(hkl) = \sum_{r=1}^{n} f_r A_r$$
$$B'(hkl) = \sum_{r=1}^{n} f_r B_r$$

and the terms A_r and B_r, being symmetry dependent, will be our concern. For convenience, all subscripts may be dropped in what follows, as only the atomic positions in the unit cell and not the atomic species themselves are of concern here. Thus, the *geometrical structure factor* is defined by its real, $A(hkl)$, and imaginary, $B(hkl)$, components, which may then be treated by normal algebraic processes:

$$A(hkl) = \sum_{s=1}^{N} \cos 2\pi\left(hx_s + ky_s + lz_s\right) \quad (6.9)$$

$$B(hkl) = \sum_{s=1}^{N} \sin 2\pi\left(hx_s + ky_s + lz_s\right) \quad (6.10)$$

6.6.1 Limiting conditions and the structure factor

The development of the structure factor equation is required here in order to appreciate the derivation of the conditions that limit X-ray reflections in any space group, as it is from these experimentally determined conditions that information on the space group of a crystal is obtained. This matter was considered in a simplified way in Section 5.3.3, but now the limiting conditions will be derived more thoroughly, taking into consideration the effects of centrosymmetry, unit cell translational, glide and screw symmetry.

6.6.2 Geometrical structure factor for a centrosymmetric crystal

Friedel's law states that in the absence of anomalous scattering [4], which will be assumed to apply herein,

$$I(hkl) = I(\bar{h}\bar{k}\bar{l}) \tag{6.11}$$

from which

$$|\mathrm{F}(hkl)| = |\mathrm{F}(\bar{h}\bar{k}\bar{l})| \tag{6.12}$$

From Eqs. (6.9)–(6.10), since $\cos(-\theta) = \cos(\theta)$ and $\sin(-\theta) = -\sin\theta$, it follows that in a centrosymmetric crystal *with the origin on* $\bar{1}$,

$$A(hkl) = \sum_{s=1}^{N} \cos 2\pi\,(hx_s + ky_s + lz_s) \tag{6.13}$$

$$B(hkl) = 0 \tag{6.14}$$

These relationships are important when manipulating the structure factor equation for a centrosymmetric crystal, together with the fact that a phase angle, Eq. (6.8), can take only one of the values 0 or π.

6.6.3 Geometrical structure factor for an *I* centred unit cell

In a body-centred unit cell, the atoms are related in pairs as x, y, z and $1/2 + x$, $1/2 + y$, $1/2 + z$. Hence, from Eq. (6.9)–(6.10),

$$\begin{aligned} A(hkl) &= \sum_{s=1}^{N/2} \cos 2\pi\,(hx_s + ky_s + lz_s) + \cos 2\pi\left(hx_s + ky_s + lz_s + \tfrac{h+k+l}{2}\right) \\ &= 2\sum_{s=1}^{N/2} \cos 2\pi\left(hx_s + ky_s + lz_s + \tfrac{h+k+l}{4}\right)\cos 2\pi\left(\tfrac{h+k+l}{4}\right) \end{aligned} \tag{6.15}$$

by the usual trigonometric manipulation (Appendix A7). Similarly,

$$\begin{aligned} B(hkl) &= \sum_{s=1}^{N/2} \sin 2\pi\,(hx_s + ky_s + lz_s) + \sin 2\pi\left(hx_s + ky_s + lz_s + \tfrac{h+k+l}{4}\right) \\ &= 2\sum_{s=1}^{N/2} \sin 2\pi\left(hx_s + ky_s + lz_s + \tfrac{h+k+l}{4}\right)\cos 2\pi\left(\tfrac{h+k+l}{4}\right) \end{aligned} \tag{6.16}$$

where the sums are taken now over the $N/2$ atoms not related by the body centring translations. For the sum $h + k + l = 2n + 1$ ($n = \pm 1, \pm 2, \ldots$), $\cos 2\pi\left(\frac{h+k+l}{4}\right) = 0$, leading to the limiting condition for a body-centred unit cell: $h + k + l = 2n$.

The same situation is often expressed in terms of *systematic absences*, a condition in which reflections are *forbidden* by the space-group symmetry:

$$hkl:\ h + k + l = 2n + 1\ (n = 0,\ \pm 1,\ \pm 2, \ldots)$$

Both terms are in common use, and the reader should distinguish carefully between them.

In this example, the same conclusion could have been reached more speedily from Eq. (6.5): the expression $\{1+\exp[i2\pi(h+k+l)/2]\}$ is equal to $1+\exp(i\pi n)$, where n is the sum of the integers h, k and l, and $[1+\exp(i\pi n)]$ is 2 or 0 for n even or odd, by de Moivre's extension of Euler's theorem.

6.6.4 Geometrical structure factor for space group $P2_1/c$

In this space group (Section 5.4.6), the limiting conditions for both the c glide plane and the 2_1 screw axis may be derived simultaneously. In a centrosymmetric space group with $\bar{1}$ at 0, 0, 0, Eqs. (6.13)–(6.14) may be applied. The general equivalent positions are

$$x,\ y,\ z;\ \bar{x},\ \bar{y},\ \bar{z};\ x,\ {}^1\!/_2-y,\ {}^1\!/_2+z;\ \bar{x},\ {}^1\!/_2+y,\ {}^1\!/_2-z.$$

It may be noted first that taking coordinates in pairs, $\pm(x,\ y,\ z)$,

$$A(hkl)=2\sum_{s=1}^{N/2}\cos 2\pi(hx_s+ky_s+lz_s),$$

where the sum is taken over the atoms not related by the centre of symmetry. Then,

$$A(hkl)=2\left\{\cos 2\pi\left[hx+ky+lz\right]+\cos 2\pi\left[hx-ky+lz+(k+l)/2\right]\right\}$$
$$B(hkl)=0 \qquad (6.17)$$

Combining the two cosine terms

$$A(hkl)=4\cos 2\pi\left[hx+lz+(k+l)/4\right]\cos 2\pi\left[ky-(k+l)/4\right]$$

Separating $A(hkl)$ for $k+l$ even and odd,

$$\begin{aligned} k+l&=2n \qquad & A(hkl)&=4\cos 2\pi(hx+lz)\cos 2\pi(ky)\\ k+l&=2n+1 \qquad & A(hkl)&=-4\sin 2\pi(hx+lz)\sin 2\pi(ky)\end{aligned} \qquad (6.18)$$

Thus, the limiting conditions are

hkl :	None
$h0l$:	$l=2n$ (c-glide normal to y)
$0k0$:	$k=2n$ (2_1 axis $\parallel$ y)

These three classes of reflections are important in monoclinic reciprocal space, because they alone can characterize the systematic absences in the X-ray diffraction patterns that lead to a deduction of the space group of a monoclinic crystal. In this particular example, the hkl diffraction data reveals the true space group unequivocally.

6.6.5 Geometrical structure factor for space group $Pma2$

Another space group discussed earlier (Section 5.4.9.2) is $Pma2$, in the orthorhombic $mm2$ group. From the space group data,

$$\begin{aligned}A(hkl)=&\cos 2\pi(hx+ky+lz)+\cos 2\pi(-hx-ky+lz)\\&+\cos 2\pi(-hx+ky+lz+h/2)+\cos 2\pi(hx-ky+lz+h/2)\end{aligned}$$
$$(6.19)$$

Combining the first and third, and second and fourth terms,

$$A\,(hkl) = 2\cos 2\pi\,(ky + lz + h/4) + \cos 2\pi\,(hx - h/4) \\ +2\cos 2\pi(-ky + lz + h/4)\cos 2\pi\,(hx + h/4) \tag{6.20}$$

Further simplification of this expression requires the separate parts to contain a common factor. It is desirable to return to Eq. (6.19) and make a minor alteration to the term $\cos 2\pi\ (hx - ky + lz + h/2)$. Since h is an integer, this term may be written as the crystallographically equivalent term $\cos 2\pi\,(hx - ky + lz - h/2)$. Another way of looking at this process is that the fourth general equivalent position has been changed to $-{}^1\!/\!_2 + x,\ \bar{y},\ z$, which is equivalent to moving through one repeat a in the negative direction to a crystallographically equivalent position, a perfectly valid and generally applicable tactic. Returning to $Pma2$, Eq. (6.19) now becomes

$$A(hkl) = 2\cos 2\pi\,(ky + lz + h/4)\cos 2\pi\,(hx - h/4) \\ +2\cos 2\pi(-ky + lz - h/4)\cos 2\pi\,(hx - h/4) \tag{6.21}$$

which simplifies to

$$A(hkl) = 2[\cos 2\pi\,(ky + lz + h/4) \\ + \cos 2\pi(-ky + lz - h/4)]$$

Combining again:

$$A(hkl) = 4\ [\cos 2\pi\,(hx - h/4)]\cos 2\pi\,(ky + h/4)\cos 2\pi\,lz \tag{6.22}$$

By a similar argument,

$$B(hkl) = 4[\cos 2\pi\,(hx - h/4)\cos 2\pi\,(ky + h/4)\sin 2\pi\,lz \tag{6.23}$$

The orthorhombic system has seven regions of reciprocal space of particular importance for determining crystal space groups; they are listed on the right-hand side of Fig. 5.27. Separating Eq. (6.22) and Eq. (6.23) for even and odd values of h,

$$h = 2n: \qquad A(hkl) = 4\cos 2\pi\,hx\cos 2\pi\,ky\cos 2\pi\,lz \\ B(hkl) = 2\cos 2\pi\,hx\cos 2\pi\,ky\sin 2\pi\,lz \tag{6.24}$$

$$h = 2n+1: \qquad A(hkl) = -4\sin 2\pi\,hx\sin 2\pi\,ky\cos 2\pi\,lz \\ B(hkl) = -4\sin 2\pi\,hx\sin 2\pi\,ky\sin 2\pi\,lz \tag{6.25}$$

from which it is clear that the only systematic condition arises where $k = 0$ and $h = 2n + 1$, whereupon $A = B = 0$; hence the limiting conditions for $Pma2$ are

$$hkl\text{: None} \qquad h0l\text{: } h = 2n$$

A condition $h00$: $(h = 2n)$ should be considered carefully. One might be excused for thinking at first that it implies the existence of a 2_1 axis parallel to the x axis, but for the knowledge that there are no symmetry axes parallel to the x axis in class $mm2$. This particular limiting condition is dependent upon the $h0l$ condition; $h00$ is in the $h0l$ zone.

It is *emphasized* here that confusion can arise easily if the limiting conditions are interpreted in other than the following hierarchal order of inspection $1 > 2 > 3$:

Table 6.2 Limiting conditions for screw axes.

hkl	Condition	Orientation	Translation	Symbol	Crystal system[a]	
$h00$	$h = 2n$	[100]	$\mathbf{a}/2$	2_1	Orthorhomic, tetragonal	Cubic
				4_2		
	$h = 4n$		$\mathbf{a}/4$	4_1, 4_3		
$0k0$	$k = 2n$	[010]	$\mathbf{b}/2$	2_1	Monoclinic (*b* unique), orthorhonic,tetragonal	Cubic
				4_2		
	$k = 4n$		$\mathbf{b}/4$	4_1, 4_3		
$00l$	$l = 2n$	[001]	$\mathbf{c}/2$	2_1	Orthorhomic	Cubic
				4_2		
	$l = 4n$		$\mathbf{c}/4$	4_1, 4_3	Tetragonal	
$000l$	$l = 2n$	[0001]	$\mathbf{c}/2$	6_3	Hexagonal	
	$l = 3n$		$\mathbf{c}/3$	3_1, 3_2, 6_2, 6_4		
	$l = 6n$		$\mathbf{c}/6$	6_1, 6_5		

[a] The trigonal point groups are here referred to hexagonal axes.

1. *hkl*: for unit cell type
2. $0kl$: for a glide plane $\perp x$
   $h0l$: for a glide plane $\perp y$
   $hk0$: for a glide plane $\perp z$
3. $h00$: for a 2_1 screw axis $\parallel x$
   $0k0$: for a 2_1 screw axis $\parallel y$
   $00l$: for a 2_1 screw axis $\parallel z$

A lower level in this list is considered only after examining the full implications of the conditions at a higher level. Conditions such as that for $h00$ in $Pma2$ are called *redundant* or *dependent*, and are placed in parentheses with the space group information. Reflections involved in such conditions are certainly absent from a diffraction record, but do not contribute to the determination of space-group symmetry. Table 6.2 and Table 6.3 summarize the limiting conditions for screw axis and glide plane, respectively.

6.6.6 Geometrical structure factor for space group $P6_3/m$

Finally here, and in order to show that the manipulations of geometrical structure factors are fundamentally no more difficult with a space group based on non-orthogonal axes, the hexagonal space group $P6_3/m$ is studied; the tetragonal system will be addressed through space group $P4nc$ in the Problems section. The coordinates of the general positions for this space group (can you derive them?) are:

$$\pm(x,\ y,\ z;\ \bar{y},\ x-y,\ z;\ y-x,\ \bar{x},\ z;\ \bar{x},\ \bar{y},\ {}^1\!/_2+z;\ y,\ y-x,\ {}^1\!/_2+z;\ x-y,\ x,\ {}^1\!/_2+z).$$

Since the space group is centrosymmetric, with the origin on $6/m(\bar{1})$, $B(hkl)$ is zero and only the cosine part of the structure factor equation need be

Table 6.3 Limiting conditions for glide planes (excluding the *e* glide).

hkl	Condition	Orientation	Translation	Symbol	Crystal system
$0kl$	$k = 2n$	(100)	**b**/2	*b*	Orthorhombic, tetragonal, cubic
	$l = 2n$		**c**/2	*c*	
	$k + l = 2n$		**b**/2 + **c**/2	*n*	
	$k + l = 4n$ $(k, l = 2n)$[a]		**b**/4 ± **c**/4	*d*	
$h0l$	$l = 2n$	(010)	**c**/2	*c*	Monoclinic (*b* unique); Orthorhombic, tetragonal, cubic
	$h = 2n$		**a**/2	*a*	
	$l + h = 2n$		**c**/2 + **a**/2	*n*	
	$l + h = 4n$ $(l, h = 2n)$[a]		**c**/4 ± **a**/4	*d*	Orthorhombic, tetragonal, cubic
$hk0$	$h = 2n$	(001)	**a**/2	*a*	Orthorhombic, tetragonal, cubic
	$k = 2n$		**b**/2	*b*	
	$h + k = 2n$		**a**/2 + **b**/2	*n*	
	$h + k = 4n$ $(h, k = 2n)$[a]		**a**/4 ± **b**/4	*d*	
$h\bar{h}0l$, $0k\bar{k}l$, $\bar{h}0hl$	$l = 2n$	$(11\bar{2}0)$, $(\bar{2}110)$, $(1\bar{2}10)$ $\{11\bar{2}0\}$	**c**/2	*c*	Hexagonal[e]
$hh.\overline{2h}.l$, $\overline{2h}.hhl$, $h.\overline{2h}.hl$	$l = 2n$	$(1\bar{1}00)$, $(01\bar{1}0)$, $(\bar{1}010)$ $\{1\bar{1}00\}$	**c**/2	*c*	Hexagonal[e]
hhl	$l = 2n$	$(1\bar{1}0)$ $\{1\bar{1}0\}$	**c**/2	*c*, *n*	Rhombohedral[b]
hkk	$h = 2n$	$(01\bar{1})$	**a**/2	*a*, *n*	
hkh	$k = 2n$	$(\bar{1}01)$	**b**/2	*b*, *n*	
hhl, $h\bar{h}l$	$l = 2n$	$(1\bar{1}0)$, (110)	**c**/2	*c*,*n*	Tetragonal[c]; Cubic[d]
	$2h + l = 4n$		**a**/4 ± **b**/4 ± **c**/4	*d*	
hkk, $hk\bar{k}$	$h = 2n$	$(01\bar{1})$, (011)	**a**/2	*a*,*n*	Cubic[d]
	$2k + h = 4n$		±**a**/4 + **b**/4 ± **c**/4	*d*	
hkh, $\bar{h}kh$	$k = 2n$	$(\bar{1}01)$, (101)	**b**/2	*b*,*n*	
	$2h + k = 4n$		±**a**/4 ± **b**/4 + **c**/4	*d*	

[a] Glide planes *d* with orientations (100), (010) and (001) occur only in the orthorhombic and cubic space groups. The combination of the integral reflection condition (*hkl* all odd or all even) with the zonal condition for *d* glides leads to the additional conditions in parentheses.

[b] The three conditions $l = 2n$, $h = 2n$, $k = 2n$ in rhombohedral space groups referred to rhombohedral axes imply interleaving of *c* and *n* glides, *a* and *n* glides, and *b* and *n* glides, respectively. In the Hermann–Mauguin notation, the symbol *c* is always used because *c* glides occur in the corresponding hexagonal descriptions; (see $R3c$ and $R\bar{3}c$).

[c] The two conditions *hhl* and $h\bar{h}l$, $(l = 2n)$ in tetragonal *P* space groups imply interleaving of *c* and *n* glides. In the Hermann–Mauguin notation, *c* is always used irrespective of the nature of the plane passing through the origin (see $P4cc$, $P\bar{4}2c$ and $P\frac{4}{n}nc$).

[d] The three conditions $l = 2n$, $h = 2n$, $k = 2n$ in cubic space groups imply interleaving of *c* and *n* glides, *a* and *n* glides, and *b* and *n* glides respectively. In the Hermann–Mauguin notation, *c* or *n* is used depending upon which of these glides passes through the origin (*see* $P\bar{4}3n$, $Pn\bar{3}n$, $Pm\,\overrightarrow{3}\,n$, $F\bar{4}3c$, $Fm\bar{3}c$ and $Fd\bar{3}c$).

[e] In the hexagonal system, the dots within *hkl* serve to specify the index $2\bar{h}$ and not that it is necessarily a two digit index. Trigonal point groups are here referred to hexagonal axes.

considered. Thus, following Eq. (6.13), with its modification for centrosymmetry already described, the geometrical structure factor is

$$\begin{aligned}A(hkl)/2 = {} & \cos 2\pi(hx + ky + lz) + \cos 2\pi(-hy + k[x - y] + lz) + \\ & \cos 2\pi(h[y - x] - kx + lz) + \cos 2\pi(-hx - ky + lz + l/2) + \\ & \cos 2\pi(hy + k[y - x] + lz + l/2) + \cos 2\pi(h[x - y] + kx + lz + l/2)\end{aligned} \tag{6.26}$$

Combining the cosine terms in pairs, and remembering that $i = -(h + k)$,

$$\begin{aligned}A(hkl) = {} & 4\cos 2\pi(lz - l/4)[\cos 2\pi(hx + ky + l/4) + \cos 2\pi(kx + iy + l/4) \\ & + \cos 2\pi(ix + ky + l/4)]\end{aligned} \tag{6.27}$$

Separating Eq. (6.27) into equations for l even and l odd by expanding the cosine terms (you may need to refer to Appendix A7):

$$l = 2n: \quad A = 4\cos 2\pi lz[\cos 2\pi(hx + ky) + \cos 2\pi(kx + iy) + \cos 2\pi(ix + hy)]$$

$$l = 2n + 1 \quad A = -4\sin 2\pi lz[\sin 2\pi(hx + ky) + \sin 2\pi(kx + iy) + \sin 2\pi(ix + hy)]$$

For l even, there are no reflection conditions; for l odd, $A = 0$ if $h = k = 0$. Thus, the only condition limiting X-ray reflections is $000l = l = 2n$. These calculations for all 230 space groups are available in reference [4] of Chapter 5.

6.7 Using X-ray diffraction information

The derivation of the various limiting conditions for the 230 space groups can now be understood. From these conditions, as determined experimentally from an X-ray diffraction record, information about the space group of the given structure may be obtained. The space group is not always determined unequivocally by these conditions, because Friedel's law shows that X-ray diffraction introduces a centre of symmetry into the diffraction record. Only where the limiting conditions are unique, as in $P2_1/c$, or *Pbca*, for example, can a space group be determined uniquely from the X-ray data; *Pma*2 has the same limiting conditions as $P2_1am$ and *Pmam*, and $P\frac{6_3}{m}$ the same as $P6_3$ and $P6_322$. In practice, given the X-ray diffraction data, with the *hkl* indices assigned to the reflections, the systematic conditions among them can be sought.

Example 6.2

Given: Laue group 2/*m*; *hkl*: No conditions, *h*0*l*: $l = 2n$, 0*k*0: No conditions. What space groups are possible?

The crystal is monoclinic, point group 2, *m* or 2/*m*, *P* unit cell, *c* glide normal to the *y* axis; space group *Pc* or *P*2/*c*.

Example 6.3

Given: Laue class $\frac{4}{m}mm$; *hkl*: No conditions, *hk*0: No conditions, 0*kl*: $k = 2n$, *hhl*: $l = 2n$. What space groups are possible?

The crystal is tetragonal, point group 422, 4*mm*, $\bar{4}2m$, $\frac{4}{m}mm$, *P* unit cell, *b* glide normal to <100>, *c* glide normal to <110>; space groups $P4_2\,bc$, $P\frac{4_2}{m}bc$.

In these two examples the determination of the space group is uncertain because a centre of symmetry cannot be determined from these data. In only 50 of the 230 space groups does the diffraction data define the space group unequivocally. However, statistical tests applied to the reflection data can often determine whether or no their intensity distribution is centric or acentric. A full discussion of these tests has been given elsewhere, to which the reader is referred [4].

Example 6.4
An orthorhombic crystal produced *inter alia* the following X-ray diffraction data:

hkl	*hkl*	*hkl*	*hkl*
111	020	002	200
112	011	102	210
212	022	103	220
312	024	204	401
322	031	303	402
332	040	401	603

Summarizing the data for limiting conditions: hkl, no conditions, $0kl$: $k + l = 2n$, $h0l$: No conditions, $hk0$: $h = 2n$, $h00$: $(h = 2n)$, $0k0$: $(k = 2n)$, $00l$: $(l = 2n)$.

The space group is either $Pnma$ or $Pn2_1a$, the latter being the $\mathbf{a\bar{c}b}$ setting of $Pna2_1$. The symbol 2_1 rather than 2 may be deduced because the translation of $c/2$ arises from the presence of the c glide. The presence of 2_1 from the condition $0k0$: $k = 2n$ cannot be inferred, because this condition is not independent of the n glide, which is why it is set in parentheses.

Only one of the many aspects of X-ray crystallography has been considered here in detail, but sufficient for the purposes in hand; the totality of the subject is well covered in the literature [4,11–14].

References 6

[1] Butler S. *Notebooks*, Vol. V. 1912.
[2] Hodgkin DMC. *Biogr. Mems. Fell. R. Soc.* 1980; 26: 1.
[3] Ewald PP. *Z. Phys.* 1913; 14: 465.
[4] Ladd M and Palmer R. *Structure determination by X-ray crystallography*. 5th ed. New York: Springer Science+Business Media, 2013.
[5] Friedrich W, Knipping P, Laue M. *Münchener Ber.* 1912; 303; *Ann. Phys.* 1913; 41: 917.
[6] de Jong WF and Bouman J.Z. *Kristallogr.* 1938; 98: 456.
[7] Buerger MJ. *X-Ray crystallography*. New York: John Wiley & Sons, 1942.
[8] Buerger MJ. *The precession method in X-ray crystallography*. New York: John Wiley & Sons, 1964.
[9] Karl Weissenberg and the development of X-ray crystallography. <http://weissenberg.bsr.org.uk/2/x-ray%20crystallography.htm>.
[10] Michalski E, et al. *Acta Crystallogr. A* 1995; 51: 548.
[11] Billinge SJL, et al. (eds) *Local structure from diffraction*. Kluwer Academic Publishers, 1998.
[12] Massa W. *Crystal structure determination*. Springer, 2004.
[13] Clegg W, et al. *Crystal structure analysis: principles and practice*. Oxford: IUCr, 2001.
[14] Woolfson MM. *An introduction to X-ray crystallography*. Cambridge University Press, 1997.

Problems 6

6.1 An X-ray generator is operated at 30 kV. What is the minimum wavelength of X-rays obtainable with these experimental conditions?

6.2 X-rays are absorbed by all materials according to the law $I = I_0 \exp(-\mu t)$, where I_0 and I are, respectively, the intensities of the incident and transmitted X-rays, and μ is the linear absorption coefficient. If Cu K_α X-rays are passed through a 0.02 mm thick nickel foil of μ equal to 44 mm^{-1}, what fraction of the incident beam is transmitted?

6.3 Potassium chloride is cubic, $a = 0.6278$ nm. Calculate the magnitude of $d(123)$. At what value of the Bragg angle would reflection occur from the (123) family of planes, using copper radiation of wavelength 0.15418 nm?

6.4 The atomic scattering factor data for Na^+, K^+ and Cl^- are listed below. The space group of NaCl and Cl is $Fm\bar{3}m$, with the Cl^- species at 0, 0, 0; 0, ½, ½; ½, 0, ½; ½, ½, 0, and Na^+ or K^+ at ½, 0, 0; 0, ½, 0; 0, ½, 0; ½. ½, ½. The unit cell dimension a is 5.627 Å for NaCl and 6.278 Å for KC1. (a) Calculate the ideal intensities, $|F|^2$, for the reflections 111, 200, 311, 400, 422, 420, 511 (333). (b) Show how KC1 was thought initially to be P cubic. (c) Why are the reflections 511 and 333 superimposed on the powder photograph in Fig. 5.55? Values for f can be calculated from the equation

$$f(\sin\theta/\lambda) = \sum_{j=1}^{4} a_j \exp[-b_j(\sin\theta/\lambda)^2] + c_j,$$

using the following data:

	a_1	b_1	a_2	b_2
Na^+	3.2565	2.6671	3.9362	6.1153
K^+	7.9578	12.6331	7.4917	0.76740
Cl^-	18.2915	0.00660	7.2084	1.1717

	a_3	b_3	a_4	b_4	c
Na^+	1.3998	0.20010	1.0032	14.039	0.40400
K^+	6.3590	−0.00200	1.1915	31.9128	−4.9978
Cl^-	6.5337	19.5424	2.3386	60.4486	−16.378

The calculated structure factors should be multiplied by a temperature factor, as f applies to the atom at rest. The simplest factor is $\exp[-B(\sin\theta/\lambda)^2]$, where B is the Debye–Waller overall isotropic temperature factor for the substance; values of B are 1.5Å^2 for NaC1 and 2.2Å^2 for KCl.

6.5 Figure P6.1 shows a Laue photograph on a flat film place normal to a beam of 'white' X-rays. (a) What is the Laue symmetry shown by the film? (b) If the crystal is cubic, what direction in the crystal is parallel to the X-ray beam? (c) What are the possible point groups for the crystal?

6.6 Use the structure factor equation to determine the conditions limiting reflections in an A centred unit cell.

6.7 In a P crystal unit cell, the atoms present are related by a screw axis parallel to c at [¼, 0, z]. Deduce what reflections would be systematically absent on account of this symmetry?

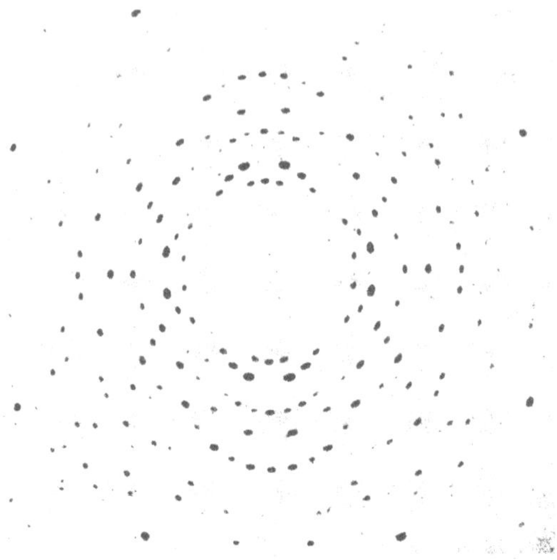

Fig. P6.1
Flat film Laue photograph, taken with the beam of 'white' X-rays normal to the film.

6.8 Determine the limiting conditions for an n glide plane normal to the z axis of a crystal.

6.9 What space groups are indicated by the following data?
Monoclinic
(a) hkl: None; $h0l$: None; $0k0$: $k = 2n$.
(b) hkl: $h + k = 2n$; $h0l$: $(h = 2n)$; $0k0$: $(k = 2n)$.
Orthorhombic
(c) hkl: None; $0kl$: $k = 2n$; $h0l$: None; $hk0$: None; $h00$: None; $0k0$: $(k = 2n)$; $00l$: None.
(d) hkl: $h + k = l = 2n$; $0kl$: $k = 2n, l = 2n$; $h0l$: $l = 2n, h = 2n$; $hk0$: $h = 2n, k = 2n$; $0k0$: $k = 2n$; $00l$: $l = 2n$ The dependent conditions here have not been parenthesized.
Tetragonal
(e) hkl: $h + k = l = 2n$; $hk0$: h (or k) $= 2n$; $00l$: $l = 4n$.
(f) hkl: None; $0kl$: $l = 2n$; hhl: $l = 2n$; Laue class $\frac{4}{m}mm$.
(g) How would the deduction in (a) be affected if the crystal contained two molecules of a steroid molecule in the unit cell?
(h) Rewrite the conditions in (d), putting the redundant conditions in parentheses.

6.10 By evaluating the geometrical structure factors, establish the limiting conditions for space group Cc, origin on c.

6.11 What is the effect on the structure factor equation for a centrosymmetric crystal by changing the origin $\bar{1}$ from 0, 0, 0 to ½, 0, ½? What is the implication for the phase angle?ϕ_{hkl}?

Elements of group theory

7

SYNOPSIS

- Group theory rules and definitions
- Group multiplication tables
- Subgroups, cosets and invariance
- Similarity transformations and symmetry classes
- Representations and character tables
- Transformations of atomic orbitals
- Irreducible representations
- The great orthogonality theorem
- Constructing a character table
- Direct products

7.1 Introduction

In previous chapters, implicit use has been made of some concepts from group theory. Now it is desirable to treat that subject with more rigour, and to increase the scope of its application to structure and symmetry in chemical systems.

A group is a set of objects which, for the purposes in hand, are mathematical objects, or *members* (aka *elements*), that are interrelated according to a number of clearly defined rules. A member of the group need not be a symmetry operation, although in this discussion it will most often be a proper or an improper rotation.

The combination of two members is called their *product*, although it may not always refer to the normal process of algebraic multiplication. The product of two symmetry operations is an operation of the first kind, a proper rotation, or of the second kind, an improper rotation, according as the two operations are respectively of the same kind or of different kinds. Group theory is concerned with the ramifications of these combinations and their developments.

An *affine* group is a general group that preserves collinearity in various geometric transformations, including *inter alia* the operations of translation, rotation and reflection or combinations thereof; points lying on a line initially remain on that line after the transformation. Under this definition, there are 219 affine space groups [1], the eleven enantiomorphous, or chiral, pairs

Table 7.1 Crystallographic point groups in the Schönflies notation.

Type	Triclinic	Monoclinic	Trigonal	Tetragonal	Hexagonal	Cubic
C_R	C_1	C_2	C_3	C_4	C_6	T
S_R	C_i^{a}	C_s^{b}	S_6	S_4		
C_{Rh}		C_{2h}	C_{3h}	C_{4h}	C_{6h}	T_h
		Orthorhombic				
D_R		D_2	D_3	D_4	D_6	O
C_{Rv}		C_{2v}	C_{3v}	C_{4v}	C_{6v}	
D_{Rd}			D_{3d}	D_{2d}		T_d
D_{Rh}		D_{2h}	D_{3h}	D_{4h}	D_{6h}	O_h

[a] May also be written as S_2 or i.
[b] May also be written as S_1.

such as $P3_1$and $P3_2$ are not distinguished. The 230 crystallographic space groups and point groups are subsets, or subgroups, of real affine groups (Appendix 13): they include the chiral pairs, because the determination of the absolute configuration of a chemical species requires them to be distinguished (see also Section 5.7.1.1 and Appendix 13). Thus, the structure of tris(trimethylsilyl)chromium pentacarbonyl was determined and refined only in space group $P3_2$ [2].

Rotation axes of symmetry, whether of the first or second kind, including symmetry planes and the centre of symmetry, when given a specific orientation relative to reference axes are termed *symmetry operators*, and symbolized in bold italic (see also Section 3.2). It is conventional to discuss group theory when applied to crystals and molecules in terms of the Schönflies notation rather than the Hermann-Mauguin symmetry notation, although there is no reason why the latter could not be so used; the reader may wish to review Section 3.10 at this stage. Table 7.1 sets out the crystallographic point groups in the Schönflies notation in the same manner as that in Table 3.5, with which it may be compared. The Schönflies notation emphasizes the proper rotation axis of highest degree: thus $C_{3h}(\overline{6})$ and $D_{3h}(\overline{6}m2)$ appear under the trigonal heading although, crystallographically, they are hexagonal.

7.2 Group requirements

'*The growth of observation consists in a continual analysis of facts of rough and general observation into groups of facts more precise and minute*' [3]. A group may be symbolized as $G\{\boldsymbol{E}, \boldsymbol{A}, \boldsymbol{B}, \boldsymbol{C}, \boldsymbol{D}, \ldots\}$, where the members $\boldsymbol{E}, \boldsymbol{A}, \boldsymbol{B}, \boldsymbol{C}, \boldsymbol{D}, \ldots$ of the group are governed by the following rules:

1. *The product of two members of a set is also a member of the set*. Thus, in $\boldsymbol{G}\{\boldsymbol{A}, \boldsymbol{B}, \ldots\}$, if $\boldsymbol{AB} = \boldsymbol{C}$, then $\boldsymbol{C}$ is also a member of the set G; this property of a group is termed *closure*. If the law of combination is multiplication, then the group is said to be *closed* with respect to multiplication. A set of negative integers cannot form a closed group under multiplication because the product of two such integers is not negative. A set of cardinal numbers is closed under addition and multiplication but not under subtraction and division. A set of three-dimensional vectors is closed under

vector multiplication (cross product) but not under scalar multiplication (dot product). If $\boldsymbol{A}$ and $\boldsymbol{B}$ are two members of a group, their combination is either $\boldsymbol{AB}$ or $\boldsymbol{BA}$. In general, such combinations do not *commute*, that is, $\boldsymbol{AB} \neq \boldsymbol{BA}$, unlike ordinary algebra where $X \times Y = Y \times X$.

2. *The associative law of combination must hold*, that is, $\boldsymbol{A}(\boldsymbol{BC}) = (\boldsymbol{AB})\boldsymbol{C}$. This law holds for any continued number of members of the group.
3. *The group contains an identity member* $\boldsymbol{E}$(Ger. *Einheit* = unity), such that $\boldsymbol{XE} = \boldsymbol{EX} = \boldsymbol{X}$ for any member $\boldsymbol{X}$ of the group.
4. *Each member* $\boldsymbol{A}$ *of a set has an inverse* $\boldsymbol{A}^{-1}$ that is also a member of the set such that $\boldsymbol{AA}^{-1} = \boldsymbol{E}$; $\boldsymbol{E}$ is its own inverse. As a corollary, the inverse of a product of two or more members of a group is equal to the product of their inverses in reverse order: Let $\boldsymbol{ABC} = \boldsymbol{D}$. Post-multiply each side of this equation by $\boldsymbol{C}^{-1}\,\boldsymbol{B}^{-1}\,\boldsymbol{A}^{-1}$. Then,

$$\boldsymbol{AB\,CC}^{-1}\boldsymbol{B}^{-1}\boldsymbol{A}^{-1} = \boldsymbol{DC}^{-1}\boldsymbol{B}^{-1}\boldsymbol{A}^{-1}$$

which becomes

$$\boldsymbol{AB\,EB}^{-1}\boldsymbol{A}^{-1} = \boldsymbol{DC}^{-1}\boldsymbol{B}^{-1}\boldsymbol{A}^{-1} = \boldsymbol{AE\,A}^{-1} = \boldsymbol{E}$$

since $\boldsymbol{BEB}^{-1} = \boldsymbol{E}$ and $\boldsymbol{AA}^{-1} = \boldsymbol{E}$; thus, $\boldsymbol{E} = \boldsymbol{DC}^{-1}\boldsymbol{B}^{-1}\boldsymbol{A}^{-1}$, and it follows that $\boldsymbol{C}^{-1}\boldsymbol{B}^{-1}\boldsymbol{A}^{-1}$ is the inverse of $\boldsymbol{D}$; thus,

$$\boldsymbol{D}^{-1} = (\boldsymbol{ABC})^{-1} = \boldsymbol{C}^{-1}\boldsymbol{B}^{-1}\boldsymbol{A}^{-1} \tag{7.1}$$

7.3 Group definitions

(a) A group may be *finite*, the number of members being the *order* of the group, or it may be *infinite.*

(b) An *abstract* group is concerned only with the relationships among its members, there being no particular interpretation attached to any member of the group.

(c) A group in which all members commute is termed *Abelian*, and has already been encountered in earlier chapters where the degree of rotational symmetry was not greater that two. Thus, $\boldsymbol{2}\,\overline{\boldsymbol{1}} = \overline{\boldsymbol{1}}\,\boldsymbol{2}$, or in Schönflies notation $\boldsymbol{C}_2\boldsymbol{i} = \boldsymbol{i}\boldsymbol{C}_2$, so that the group C_{2h} is an Abelian group.

(d) A group consisting of a single member $\boldsymbol{A}$ and its powers $\boldsymbol{A}^2, \boldsymbol{A}^3, \ldots \boldsymbol{A}^p$ is a *cyclic* group of order p. Thus, C_4 is a cyclic group of order 4 containing the members $\boldsymbol{C}_4$, $\boldsymbol{C}_4^2\,(=\boldsymbol{C}_2)$, $\boldsymbol{C}_4^3$, $(=\boldsymbol{C}_4^{-1})$, $\boldsymbol{C}_4^4$, $(=\boldsymbol{E})$. It was noted in Section 3.2 that a given symmetry element may be associated with more than one operation, and that property is evident here. Note that C_4 symbolizes the group (or a symmetry element) whereas $\boldsymbol{C}_4$ symbolizes a symmetry operator in that group.

(e) A subset of members of a group that can itself form a group is a *subgroup* of the original group. Thus, C_2 is a subgroup of C_4(see also Section 3.4.5).

(f) Every solid body may be characterized by a set of symmetry operations that constitute a point group. A point group is a finite group, of *order* equal to the number of members of the group. The order is also the number of general equivalent points generated by the point group. It was indicated in Chapter 5 that a search for a relationship between one point on a symmetry diagram and all other points on that diagram can ensure that all symmetry elements have been detected. A stereogram for point group D_{2h} (Fig. 3.12a) of order eight can be produced given only three mutually perpendicular, intersecting mirror planes. By seeking each point on the diagram from a single point, the other symmetry elements are revealed. There are as many symmetry operations in a group as general equivalent points on its diagram (don't forget the identity operation). What are the symmetry operators in D_{2h}—in the Schönflies notation?

(g) The properties of the members of a group may be presented conveniently as a *group multiplication table*. The members of the group are listed in the top row, identity first, and also in the first column. The products are not, in general, commutative, so that a rule for forming a product is formulated, namely, *column member* × *row member*, and the product member lies in the table at the point of intersection of the row and column for the two members forming the product, as in matrix multiplication.

(h) Two groups G and G' are isomorphous if there is a one to one correspondence between the members of the group. Thus, the product $\boldsymbol{C} = \boldsymbol{AB}$ in the group $G\,\{\boldsymbol{E\,A\,B\,C}\}$ implies $\boldsymbol{C}' = \boldsymbol{A}'\boldsymbol{B}'$ in the group $G'\,\{\boldsymbol{E}'\boldsymbol{A}'\boldsymbol{B}'\boldsymbol{C}'\}$.

(i) Two groups $G\{\boldsymbol{A}, \boldsymbol{B}, \boldsymbol{C}, \ldots\}$ and $G'\{\boldsymbol{\alpha}, \boldsymbol{\beta}, \boldsymbol{\chi}, \boldsymbol{\delta}, \boldsymbol{\varepsilon}, \boldsymbol{\phi}, \boldsymbol{\varphi}, \boldsymbol{\gamma}, \boldsymbol{\eta}, \ldots\}$ are homomorphic if to one member of G two or more members of a group G' are associated:

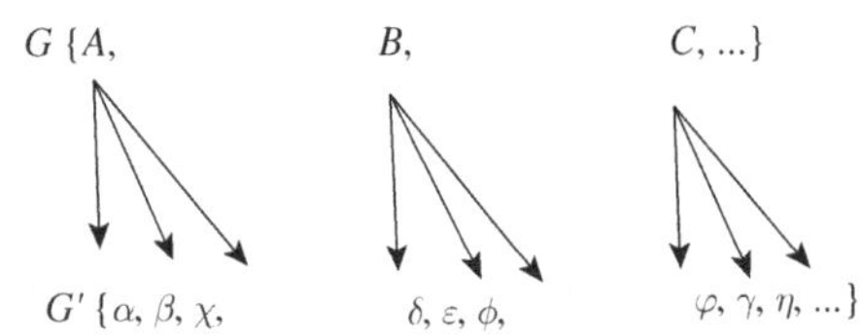

In terms of group multiplication table, homomorphism can be illustrated thus:

ONES	1	1	$\bar{1}$	$\bar{1}$
1	1	1	$\bar{1}$	$\bar{1}$
1	1	1	$\bar{1}$	$\bar{1}$
$\bar{1}$	$\bar{1}$	$\bar{1}$	1	1
$\bar{1}$	$\bar{1}$	$\bar{1}$	1	1

+1 links with $\boldsymbol{E}$ and $\boldsymbol{C}_2$

C_{2v}	$\boldsymbol{E}$	$\boldsymbol{C}_2$	$\boldsymbol{\sigma}_v$	$\boldsymbol{\sigma}'_v$
$\boldsymbol{E}$	$\boldsymbol{E}$	$\boldsymbol{C}_2$	$\boldsymbol{\sigma}_v$	$\boldsymbol{\sigma}'_v$
$\boldsymbol{C}_2$	$\boldsymbol{C}_2$	$\boldsymbol{E}$	$\boldsymbol{\sigma}'_v$	$\boldsymbol{\sigma}_v$
$\boldsymbol{\sigma}_v$	$\boldsymbol{\sigma}_v$	$\boldsymbol{\sigma}'_v$	$\boldsymbol{E}$	$\boldsymbol{C}_2$
$\boldsymbol{\sigma}'_v$	$\boldsymbol{\sigma}'_v$	$\boldsymbol{\sigma}_v$	$\boldsymbol{C}_2$	$\boldsymbol{E}$

−1 links with $\boldsymbol{\sigma}_v$ and $\boldsymbol{\sigma}'_v$

The group multiplication tables give identical results for members linked by homomorphism: $\boldsymbol{\sigma}'_v\boldsymbol{C}_2 = \boldsymbol{\sigma}_v$, so that $(-1)1 = -1$.

7.4 Examples of groups

Symmetry groups are finite groups except for those describing the symmetry of linear molecules, such as hydrogen fluoride, HF, or carbon dioxide, CO_2. An example of an infinite group is the series

$$-\infty, \ldots, -n, \ldots, -3, -2, 1, 0, 1, 2, 3, \ldots, n, \ldots, \infty$$

With addition as the law of combination, the group is Abelian group, $1+2=2+1$; the associative law holds, $1+(2+3)=(1+2)+3$; the identity member is $0, 0+n=n$; and it contains the inverse $-n$ of each element n; $-n+n=0$; see also Fig. 7.1 and text. $G\{\mathit{1},\ 2,\ \boldsymbol{m},\ \bar{\mathit{1}}\}$ is the symmetry group $2/m$ in the Hermann-Mauguin notation; check it for closure, association, identity and inverse. In the Schönflies notation, which will be used in this work, it is written $C_{2h}\,\{\boldsymbol{E},\ \boldsymbol{C}_2,\ \boldsymbol{\sigma}_h, \boldsymbol{i}\}$.

7.4.1 Group multiplication tables

A group of h members is completely and uniquely defined when all possible h^2 products are known. An important property of a group multiplication table, also known as a Cayley table, is that *each row and each column lists each group member once only*, which may be proved as follows.

Let $\boldsymbol{E},\ \boldsymbol{A}_1,\ \boldsymbol{A}_2, \ldots, \boldsymbol{A}_h$ be members of a group of order h. If $\boldsymbol{A}_k$ is any group member, then it can be shown that the set of members

$$\boldsymbol{A}_k\boldsymbol{E},\ \boldsymbol{A}_k\boldsymbol{A}_1,\ \boldsymbol{A}_k\boldsymbol{A}_2, \ldots,\ \boldsymbol{A}_k\boldsymbol{A}_h \tag{7.2}$$

contains each member of the group once only. For if $\boldsymbol{Z}$ is any member of the group, there will be a member $\boldsymbol{A}_j=\boldsymbol{A}_k^{-1}\boldsymbol{Z}$. Then, $\boldsymbol{A}_k\boldsymbol{A}_j=\boldsymbol{A}_k\boldsymbol{A}_k^{-1}\boldsymbol{Z}=\boldsymbol{Z}$, so that $\boldsymbol{Z}$ will always be present after the appropriate multiplications, and so all such members will be contained by the group. Furthermore, $\boldsymbol{Z}$ occurs once only. For if $\boldsymbol{Z}$ actually appears twice in Eq. (7.2), say as $\boldsymbol{A}_k\boldsymbol{A}_r$ and $\boldsymbol{A}_k\boldsymbol{A}_s$, then by premultiplication with $\boldsymbol{A}_k^{-1}$, it follows that $\boldsymbol{A}_r=\boldsymbol{A}_s$, which implies two identical elements in the group, contrary to the original proposition. This argument is known as the *rearrangement theorem*, which may be stated: *each row and each column lists each member once only.*

(a) Consider first an abstract group G_3 of members $\boldsymbol{E}, \boldsymbol{A}, \boldsymbol{B}$ for which $\boldsymbol{A}^2=\boldsymbol{B}$, and $\boldsymbol{AB}=\boldsymbol{E}$: the group multiplication table will be

G_3	E	A	B
E	E	A	B
A	A	B	E
B	B	E	A

From the table, it can be deduced that $\boldsymbol{B}^2=\boldsymbol{A}$; $\boldsymbol{A}^2\boldsymbol{B}=\boldsymbol{B}^2=\boldsymbol{A}(\boldsymbol{AB})=\boldsymbol{AE}=\boldsymbol{E}$.

(b) A group G_3' consisting of threefold rotations comprises the members $\boldsymbol{E}$, $\boldsymbol{C}_3$ and $\boldsymbol{C}_3^2$. Instead of $\boldsymbol{C}_3^2$, $\boldsymbol{C}_3^{-1}$ the inverse of $\boldsymbol{C}_3$, may be written, since $\boldsymbol{C}_3^{-1}\,\boldsymbol{C}_3=\boldsymbol{C}_3^2\boldsymbol{C}_3=\boldsymbol{E}$. If $\boldsymbol{C}_3$ represents a right-handed, or anticlockwise, threefold rotation, then $\boldsymbol{C}_3^{-1}$ is a clockwise rotation of the same degree. The multiplication table is

1st Operation

2nd Operation

G'_3	E	C_3	C_3^2
E	E	C_3	C_3^2
C_3	C_3	C_3^2	E
C_3^2	C_3^2	E	C_3

and the order of operations is important.

(c) Finally, consider the group G''_3 of integers modulo 3, under addition as the law of combination; the group multiplication table is

G''_3	0	1	2
0	0	1	2
1	1	2	0
2	2	0	1

These three groups are isomorphous. There is a one to one correspondence between the members of the groups:

$$\boldsymbol{A} \rightarrow \boldsymbol{C}_3 \rightarrow 1$$
$$\boldsymbol{B} \rightarrow \boldsymbol{C}_3^2 \rightarrow 2$$
$$\boldsymbol{E} \rightarrow \boldsymbol{C}_3^3 \rightarrow 0$$

The groups are also Abelian; this nature can be detected by the symmetry of the entries across the main diagonal of the table (not including the axial row and column).

There is only one crystallographic group of order unity, namely, C_1. The three groups of order two are C_i, C_2 and C_s, and the only crystallographic group of order three is the cyclic group C_3.

The point groups C_{3v} and D_3 are isomorphous and can be given a common group multiplication table. In this table, the symbols a, b and c represent three $\boldsymbol{\sigma}_v$ operators in C_{3v} and three $\boldsymbol{C}_2$ operators in D_3. These groups are of interest as the lowest order non-Abelian crystallographic groups; it should be noted that $\boldsymbol{ab} = \boldsymbol{C}_3$, whereas $\boldsymbol{ba} = \boldsymbol{C}_3^{-1}$.

C_{3v}/D_3	E	C_3	C_3^{-1}	a	b	c
E	E	C_3	C_3^{-1}	a	b	c
C_3	C_3	C_3^{-1}	E	b	c	a
C_3^{-1}	C_3^{-1}	E	C_3	c	a	b
a	a	c	b	E	C_3^{-1}	C_3
b	b	a	c	C_3	E	C_3^{-1}
c	c	b	a	C_3^{-1}	C_3	E

Fig. 7.1
Persian archers under King Darius, *ca.* 500 BC.

An important group in crystallography is the translation group. The unit cell translation vectors, $n_1\mathbf{a} + n_2\mathbf{b} + n_3\mathbf{c}$, $(n_1, n_2, n_3 = 0, \pm1, \pm2, \ldots, \pm\infty)$ are integral, and form a group that is infinite under the law of addition; zero stands for the identity vector. Figure 7.1 shows a portion of a translation group, but not extending to infinity.

Space groups are infinite groups. Every space group based on a given point group is homomorphic with that point group. Consider point group D_2: the general equivalent positions have the coordinates x, y, z; $\bar{x}, \bar{y}, z$; $\bar{x}, y, \bar{z}$; $x, \bar{y}, \bar{z}$. The equivalent positions for all space groups listed in Fig. 5.22 follow this pattern but are modified by translations arising from the presence of screw axes and by displacements of the symmetry axes from the space group origin.

The power of a group multiplication table is that it allows the table to be written without necessarily knowing first all relationships between its members. Consider a C_2 axis and two mirror planes, σ_v and σ_v', intersecting in that axis. A partial table can be written immediately:

C_{2v}	E	C_2	σ_v	σ_v'
E	E	C_2	σ_v	σ_v'
C_2	C_2			
σ_v	σ_v			
σ_v'	σ_v'			

The table can be enhanced because each of the operators $\boldsymbol{C_2}$ and $\boldsymbol{\sigma}$ is its own inverse. Hence,

C_{2v}	E	C_2	σ_v	σ'_v
E	E	C_2	σ_v	σ'_v
C_2	C_2	E		
σ_v	σ_v		E	
σ'_v	σ'_v			E

Is $\boldsymbol{C}_2\boldsymbol{\sigma}_v = \boldsymbol{\sigma}_v$ or $\boldsymbol{\sigma}'_v$? If the former, then $\boldsymbol{\sigma}_v$ would appear twice in row 3 (do not count the axial column or row), so the table may be completed as

C_{2v}	E	C_2	σ_v	σ'_v
E	E	C_2	σ_v	σ'_v
C_2	C_2	E	σ'_v	σ_v
σ_v	σ_v	σ'_v	E	C_2
σ'_v	σ'_v	σ_v	C_2	E

This table is Abelian and occurs frequently; examples include H_2O, SF_4 and C_6H_5Cl. There is a second group of order 4: it is cyclic and, therefore, Abelian. Can you derive its group multiplication table?

7.4.2 Reference axes in group theory

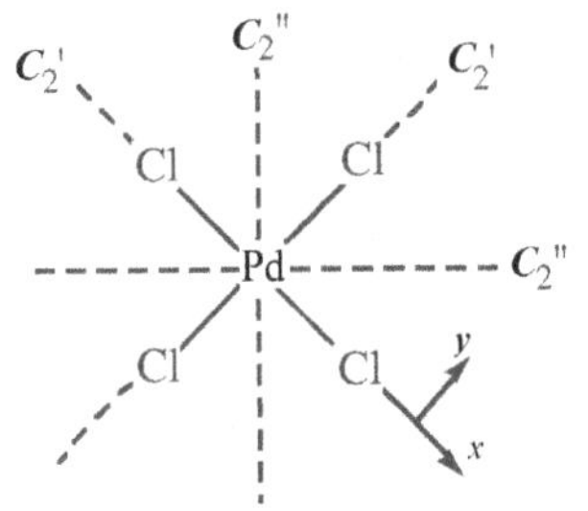

Fig. 7.2
The square planar anion $[PdCl_4]^{2-}$; point group D_{4h}.

The setting of reference axes in group theory is somewhat different from that in crystallography. Here, the principal *proper* rotation axis is assigned to the z direction, and the x and y axes aligned perpendicular to z and to each other. In setting the reference axes in a molecule, the z axis is chosen as above: if there are several equivalent symmetry axes, z is set to pass through the greatest number of atoms. If the molecule is planar and z lies in that plane, the x axis is normal to that plane. If the molecule is non-planar, the largest planar fragment is used to set the x axis within the above rules. In other cases, the selection of the x axis is arbitrary, subject to perpendicularity with z. In all cases, the y axis is normal to both the x and z axes, and the three axes form a right-handed set; Fig. 7.2 shows an example of axes for a chemical species; z is normal to the plane and passes through the Pd atom.

7.4.3 Subgroups and cosets

Subgroups have been mentioned briefly in Section 7.3, but a little more needs to be said about them. The group table above for C_{2v}, indicates three subgroups: the identity group $E\{\boldsymbol{E}\}$ or $C_1\{\boldsymbol{E}\}$, the group $C_2\{\boldsymbol{E}, \boldsymbol{C}_2\}$ and the group $C_s\{\boldsymbol{E}, \boldsymbol{\sigma}\}$. If all reference to the operator $\boldsymbol{C}_2$ is removed from the C_{2v} table, there remains

C_s	E	σ
E	E	σ
σ	σ	E

which is the subgroup C_s. There is only one way in which C_s maybe extracted from C_{2v}; it is, therefore, an *invariant* subgroup of its *supergroup* C_{2v}. If within a group there is a subset of members that can form a group that satisfies the four group requirements (Section 7.2), that set is a subgroup of the original group. Not all subgroups are invariant: thus, C_{2v} is an invariant subgroup of D_{2d}, C_{4h} and D_{2h} but not of C_{6v} or D_{3h}. The order g of a subgroup is an integral submultiple of the order h of its supergroup, that is, $h/g = n$, where n is an integer greater than unity.

The point group, D_{2h} may be written

$$D_{2h}\{\boldsymbol{E}, \boldsymbol{C}_2(z), \boldsymbol{C}_2(y),\ \boldsymbol{C}_2(x),\ \boldsymbol{i},\ \boldsymbol{\sigma}(xy), \boldsymbol{\sigma}(zx), \boldsymbol{\sigma}(yz)\}$$

One subgroup is $D_2\{\boldsymbol{E}, \boldsymbol{C}_2(z), \boldsymbol{C}_2(y), \boldsymbol{C}_2(x)\}$, and the product of each member of D_2 with a member of D_{2h} not included in D_2, say $\boldsymbol{\sigma}(xy)$ may be formed. By post-multiplication, remembering the order of operations, a *right coset of* D_2 is formed:

$$\begin{aligned} \boldsymbol{D}_2\boldsymbol{\sigma}(xy) &= \boldsymbol{E}\boldsymbol{\sigma}(xy), \boldsymbol{C}_2(z)\boldsymbol{\sigma}(xy),\ \boldsymbol{C}_2(y)\boldsymbol{\sigma}(xy), \boldsymbol{C}_2(x)\boldsymbol{\sigma}(xy) \\ &= \boldsymbol{\sigma}(xy),\ \boldsymbol{i}, \boldsymbol{\sigma}(yz),\ \boldsymbol{\sigma}(zx) \end{aligned}$$

Thus, the group D_{2h} can be represented by the combination of D_2 and the right coset of its subgroup D_2 under addition:

$$\{\boldsymbol{E}, \boldsymbol{C}_2(z),\ \boldsymbol{C}_2(y),\ \boldsymbol{C}_2(x)\} + \{\boldsymbol{\sigma}(xy), \boldsymbol{i},\ \boldsymbol{\sigma}(yz),\ \boldsymbol{\sigma}(zx)\}$$

that is,

$$D_{2h} = D_2 + D_2\boldsymbol{\sigma}(xy) \qquad (7.3)$$

Left cosets can be constructed in an analogous manner.

7.4.4 Similarity transformations, conjugates and symmetry classes

As well as in subgroups, members of a group may be segregated into subsets called *symmetry classes*. The term class in this context pertains to group theory, and should not be confused with the *crystal class* discussed in earlier chapters, or with the *arithmetic crystal class* in Section 5.7.1.1. There are, unfortunately, duplicate uses of some terms and symbols, but their use is conventional and will be adopted herein.

If $\boldsymbol{A}, \boldsymbol{B}, \boldsymbol{C}, \ldots, \boldsymbol{X}, \ldots$ be members of a group, then the members $\boldsymbol{A}$ and $\boldsymbol{B}$, for example, are *conjugate* if, for some member $\boldsymbol{X}$ in the group,

$$\boldsymbol{A} = \boldsymbol{X}^{-1}\boldsymbol{B}\boldsymbol{X} \qquad (7.4)$$

This equation expresses the *similarity transformation* of $\boldsymbol{B}$ by $\boldsymbol{X}$; the member $\boldsymbol{A}$ is the similarity transform of $\boldsymbol{B}$ by $\boldsymbol{X}$, which is a feature of conjugancy.

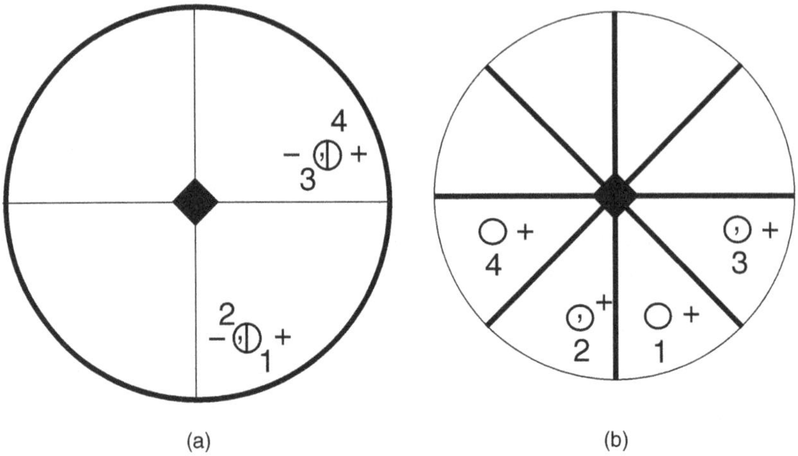

Fig. 7.3
Partial stereograms to illustrate similarity transformations. (a) C_{4h}: $\boldsymbol{\sigma}_h \boldsymbol{C}_4 \boldsymbol{\sigma}_h$, (b) C_{4v}: $\boldsymbol{\sigma}_v \boldsymbol{C}_3^4 \boldsymbol{\sigma}_v$, showing that C_4 and C_4^3 belong to the same symmetry class in C_{4h}, but to different symmetry classes in C_{4v}.

The operation $\boldsymbol{\sigma}$ is its own inverse. Thus, a similarity transformation may be performed on $\boldsymbol{C}_4$ by finding the operation that is equivalent to $\boldsymbol{\sigma C_4 \sigma}$; the result may be sought graphically or by matrix manipulations. Let a C_4 axis lie along the z axis in Fig. 7.3a, which shows a part of a stereogram for point group C_{4h}; σ_h lies normal to z and the operator $\boldsymbol{\sigma}_h^{-1}$ is equivalent to $\boldsymbol{\sigma}_h$. The sequence (1) → (2) → (3) → (4) represents the similarity transformation $\boldsymbol{\sigma}_h^{-1}\boldsymbol{C}_4\boldsymbol{\sigma}_h$, and is equivalent to the operation $\boldsymbol{C}_4$ performed directly on point (1). Convince yourself that the transformation on $\boldsymbol{C}_4^3$ leads only to $\boldsymbol{C}_4^3$. Thus, $\boldsymbol{C}_4$ and $\boldsymbol{C}_4^3$, *in this point group*, do not belong to the same symmetry class.

An alternative procedure is by matrix multiplication:

$$\underset{\boldsymbol{C}_4}{\begin{pmatrix} 0 & \bar{1} & 0 \\ 1 & 0 & 0 \\ 0 & 0 & 1 \end{pmatrix}} \underset{\boldsymbol{\sigma}_h}{\begin{pmatrix} 1 & 0 & 0 \\ 0 & 1 & 0 \\ 0 & 0 & \bar{1} \end{pmatrix}} = \underset{\boldsymbol{C}_4\boldsymbol{\sigma}}{\begin{pmatrix} 0 & \bar{1} & 0 \\ 1 & 0 & 0 \\ 0 & 0 & \bar{1} \end{pmatrix}}$$

$$\underset{\boldsymbol{\sigma}}{\begin{pmatrix} 1 & 0 & 0 \\ 0 & 1 & 0 \\ 0 & 0 & \bar{1} \end{pmatrix}} \underset{\boldsymbol{C}_4\boldsymbol{\sigma}}{\begin{pmatrix} 0 & \bar{1} & 0 \\ 1 & 0 & 0 \\ 0 & 0 & \bar{1} \end{pmatrix}} = \underset{\boldsymbol{\sigma C_4 \sigma} = \boldsymbol{C}_4}{\begin{pmatrix} 0 & \bar{1} & 0 \\ 1 & 0 & 0 \\ 0 & 0 & 1 \end{pmatrix}}$$

The same procedure applied to $\boldsymbol{C}_4^3$ confirms that $\boldsymbol{C}_4$ is not in the same class as $\boldsymbol{C}_4^3$ in point group C_{4h}.

If the same procedure be applied to point group C_{4v}, then it follows from the stereogram for the partial point group in Fig. 7.3b, or otherwise, that $\boldsymbol{\sigma}_v^{-1}\boldsymbol{C}_4\boldsymbol{\sigma}_v = \boldsymbol{C}_4^3$, since $\boldsymbol{\sigma}_v^{-1} = \boldsymbol{\sigma}_v$, and that $\boldsymbol{\sigma}_v^{-1}\boldsymbol{C}_4^3\boldsymbol{\sigma}_v = \boldsymbol{C}_4$, showing that $\boldsymbol{C}_4$ and $\boldsymbol{C}_4^3(\equiv \boldsymbol{C}_4^{-1})$ belong to one and the same symmetry class in this point group. What of $\boldsymbol{C}_4^2$ in the same point group? The similarity transform $\boldsymbol{C}_4^{-1}\boldsymbol{C}_4^2\boldsymbol{C}_4 = \boldsymbol{C}_4^2 \equiv \boldsymbol{C}_2$ shows that $\boldsymbol{C}_4^2$ it is not of the same class as $\boldsymbol{C}_4$, and is usually referred to as $\boldsymbol{C}_2$; like $\boldsymbol{\sigma}$, $\boldsymbol{C}_2$ is its own inverse.

In C_{4v}, there is a second form of mirror planes, usually designated σ_d and σ_d', of a different class from σ_v and σ_v'. It is straightforward to show that

σ_d and σ'_d are in one and the same class, and different from σ_v; the symmetry classes for this point group are written:

$$E, C_4, C_4^3(=C_4^{-1}), \sigma_v, \sigma'_v, \sigma_d, \sigma'_d$$

or concisely

$$E, 2C_4, C_2, 2\sigma_v, 2\sigma_d$$

It is not always necessary to work through a similarity transformation in full in order to assign symmetry operations to classes. Operations belong to the same class if they are equivalent by symmetry. Thus, the carbonate ion, $[CO_3]^{2-}$, symmetry D_{3h}, has the operations

$$E, C_3, C_3^2, C_2, C'_2, C''_2, S_3, S_3^2, \sigma_h, \sigma_v, \sigma'_v, \sigma''_v$$

which may be arranged in their classes as

$$E, 2C_3, 3C_2, 2S_3, \sigma_h, 3\sigma_v$$

Consider next point group D_{3h}, of which boron trifluoride, BF_3, is an example. In Fig. 7.4, the three fluorine atoms are numbered 1, 2 and 3, at the apices of the equilateral triangle. The sequence of operations $C_3^{-1}\sigma_v C_3$ (recall that the sequence is C_3 first) is the route (a) → (b) → (c) → (d). The same position (d) can be reached from position (e), which is equivalent to the starting position (a), by the operation of σ'_v, an operation related to σ_v by a C_3 operation. What is the effect of $(C_3^2)^{-1}\sigma_v C_3^2$? The three σ_v operations, related by another symmetry operator of the group, are *conjugate* one to the other; such sets of operations *define* symmetry class.

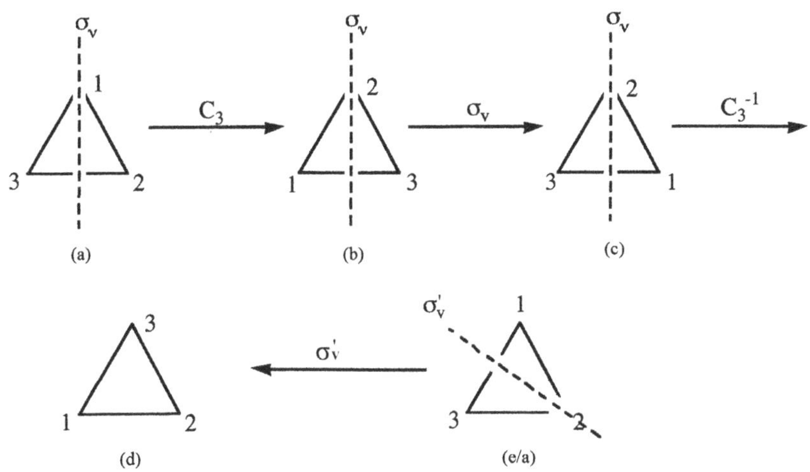

Fig. 7.4
Symmetry equivalent operations in BF_3, point group D_{3h}. All operations associated with equivalent σ_v symmetry elements belong to the same class, which may be verified by similarity transformations such as $C_3^{-1}\sigma_v C_3$. Likewise, all C'_2 operations are in one and the same class.

In the square planar species of point group D_{4h} shown by the example in Fig. 7.2, two pairs of twofold symmetry axes are present in the plane of the ion: one pair, C'_2, passes through opposite chlorine atoms, and the other pair, C''_2, lies at 45° to the C'_2 axes. The C'_2 and C''_2 axes are not interconvertible by any symmetry operator of the group: they are not conjugate members of the D_{4h} group, and belong to different symmetry classes. The following properties of conjugates may be noted:

- For every member $\boldsymbol{A}$ of a group, there exists another member $\boldsymbol{X}$ such that $\boldsymbol{X}^{-1}\boldsymbol{AX}=\boldsymbol{A}$, that is, every member is conjugate with itself. Pre-multiplying by $\boldsymbol{A}^{-1}$,

$$\boldsymbol{A}^{-1}\boldsymbol{A}=\boldsymbol{E}=\boldsymbol{A}^{-1}\boldsymbol{X}^{-1}\boldsymbol{AX}=(\boldsymbol{XA})^{-1}\boldsymbol{AX}$$

  Thus, $\boldsymbol{A}$ and $\boldsymbol{X}$ must commute, and $\boldsymbol{X}$ is either $\boldsymbol{E}$ or another member of the group that commutes with $\boldsymbol{A}$.
- If $\boldsymbol{A}$ and $\boldsymbol{B}$ are members of a group and $\boldsymbol{A}$ is conjugate with $\boldsymbol{B}$, then $\boldsymbol{B}$ is conjugate with $\boldsymbol{A}$. Thus, if

$$\boldsymbol{A}=\boldsymbol{X}^{-1}\boldsymbol{BX} \tag{7.5}$$

  there must be a another group member $\boldsymbol{Y}$ that satisfies

$$\boldsymbol{B}=\boldsymbol{Y}^{-1}\boldsymbol{AY} \tag{7.6}$$

  For $\boldsymbol{XAX}^{-1}=\boldsymbol{XX}^{-1}\boldsymbol{BXX}^{-1}=\boldsymbol{B}$; then if $\boldsymbol{Y}$ is the inverse of $\boldsymbol{X}$, that is, $\boldsymbol{Y}=\boldsymbol{X}^{-1}$, then $\boldsymbol{YX}=\boldsymbol{X}^{-1}\boldsymbol{X}=\boldsymbol{E}$. Thus, if $\boldsymbol{Y}^{-1}=\boldsymbol{X}$, it follows that $\boldsymbol{B}$ must be given by $\boldsymbol{B}=\boldsymbol{Y}^{-1}\boldsymbol{AY}$.
- If $\boldsymbol{A}$ is conjugate with $\boldsymbol{B}$ and $\boldsymbol{C}$, then $\boldsymbol{B}$ and $\boldsymbol{C}$ are conjugate with each other. This result will be addressed in the Problems section of the chapter.

Another example is G_3', for which the multiplication table has been given in Section 7.4.1. Let G_3' be the point group C_3:

C_3	E	C_3	C_3^2
E	E	C_3	C_3^2
C_3	C_3	C_3^2	E
C_3^2	C_3^2	E	C_3

A similarity transformation may be performed on each member of the group, with the following results:

$$\boldsymbol{E}:\ \boldsymbol{E}^{-1}\boldsymbol{EE}=\boldsymbol{EE}=\boldsymbol{E};\qquad \boldsymbol{C}_3^{-1}\boldsymbol{EC}_3=\boldsymbol{C}_3^{-1}\boldsymbol{C}_3=\boldsymbol{E};\ (\boldsymbol{C}_3^2)^{-1}\boldsymbol{EC}_3^2=(\boldsymbol{C}_3^2)^{-1}\boldsymbol{C}_3^2=\boldsymbol{E}$$

$$\boldsymbol{C}_3:\ \boldsymbol{E}^{-1}\boldsymbol{C}_3\boldsymbol{E}=\boldsymbol{C}_3\boldsymbol{E}=\boldsymbol{C}_3;\quad \boldsymbol{C}_3^{-1}\boldsymbol{C}_3\boldsymbol{C}_3=\boldsymbol{EC}_3=\boldsymbol{C}_3;\quad (\boldsymbol{C}_3^2)^{-1}\boldsymbol{C}_3\boldsymbol{C}_3^2=(\boldsymbol{C}_3^2)^{-1}\boldsymbol{E}=\boldsymbol{C}_3$$

$$\boldsymbol{C}_3^2:\ \boldsymbol{E}^{-1}\boldsymbol{C}_3^2\boldsymbol{E}=\boldsymbol{C}_3^2\boldsymbol{E}=\boldsymbol{C}_3^2;\quad \boldsymbol{C}_3^{-1}\boldsymbol{C}_3^2\boldsymbol{C}_3=\boldsymbol{C}_3^2\boldsymbol{E}=\boldsymbol{C}_3^2;\ (\boldsymbol{C}_3^2)^{-1}\boldsymbol{C}_3^2\boldsymbol{C}_3^2=\boldsymbol{EC}_3^2=\boldsymbol{C}_3^2$$

Thus, this cyclic group contains the three members, $\boldsymbol{E}$, $\boldsymbol{C}_3$ and $\boldsymbol{C}_3^2$. It is left as an exercise to the reader to show that point group C_{3v}, to which ammonia belongs, contains the symmetry operators $\boldsymbol{E}$, C_3, $\boldsymbol{C}_3^2$, $\boldsymbol{\sigma}_v$, $\boldsymbol{\sigma}_v'$ and $\boldsymbol{\sigma}_v''$ and to assign them to their classes; the group multiplication table is presented hereunder:

C_{3v}	E	C_3	C_3^2	σ_v	σ'_v	σ''_v
E	E	C_3	C_3^2	σ_v	σ'_v	σ''_v
C_3	C_3	C_3^2	E	σ'_v	σ''_v	σ_v
C_3^2	C_3^2	E	C_3	σ''_v	σ_v	σ'_v
σ_v	σ_v	σ''_v	σ'_v	E	C_3^2	C_3
σ'_v	σ'_v	σ_v	σ''_v	C_3	E	C_3^2
σ''_v	σ''_v	σ'_v	σ_v	C_3^2	C_3	E

As a final example in this section, point group C_{3v} will be studied further. The top left hand corner of the group multiplication table forms a subset $H\{\boldsymbol{E},\ \boldsymbol{C}_3,\ \boldsymbol{C}_3^2\}$ that is finite and closed under multiplication, and so forms a subgroup, C_3, of C_{3v}. Following the procedure leading to Eq. (7.3), the members $\boldsymbol{\sigma}_v$, $\boldsymbol{\sigma}'_v$ and $\boldsymbol{\sigma}''_v$ of $\boldsymbol{C}_{3v}$ can form a coset H'. It is straightforward to show that right and left cosets $\boldsymbol{\sigma}_v H$ and $H\boldsymbol{\sigma}_v$ are equal, and this equality forms a *definition of invariance*, in this case of the subgroup H of C_3. Any member of H' multiplied by the inverse of another member of H' gives a member of the subgroup. Thus, $(\boldsymbol{\sigma}'_v)^{-1}\boldsymbol{\sigma}_v = \boldsymbol{\sigma}'_v\boldsymbol{\sigma}_v = \boldsymbol{C}_3$, since $\boldsymbol{\sigma}$ is its own inverse. It follows that if $\boldsymbol{\sigma}_v H\,\sigma_v = H$, then subgroup H is invariant. For if H is $\{\boldsymbol{E},\ \boldsymbol{C}_3,\ \boldsymbol{C}_3^2\}$, then $\boldsymbol{\sigma}_v\{\boldsymbol{E},\ \boldsymbol{C}_3,\ \boldsymbol{C}_3^2\}\boldsymbol{\sigma}_v = \boldsymbol{\sigma}_v\{\boldsymbol{\sigma}_v,\ \boldsymbol{\sigma}'_v, \boldsymbol{\sigma}''_v\} = \{\boldsymbol{E},\ \boldsymbol{C}_3^2,\ \boldsymbol{C}_3\}$. The subset $H'' = \{\boldsymbol{E},\ \boldsymbol{\sigma}_v\}$ is finite, and also closed under the group operation; therefore, it is another subgroup, C_s, of C_{3v}, and may be formed in more than one way, so that it is not an invariant subgroup: $\{\boldsymbol{E},\ \boldsymbol{\sigma}_v\}\boldsymbol{C}_3 = \{\boldsymbol{\sigma}''_v,\ \boldsymbol{C}_3\}$, whereas $\boldsymbol{C}_3\{\boldsymbol{E},\ \boldsymbol{\sigma}_v\} = \{\boldsymbol{C}_3,\ \boldsymbol{\sigma}'_v\}$. A further brief discussion of cosets arises in the Problems section of this chapter.

7.5 Representations and character tables

The symmetry of a group may be described by a set of transformation matrices, one for each symmetry operation in the group. The set of such matrices is called a *representation* of the group. The matrix representations act on a basis property or function, and the actual matrices depend on that basis set. Symmetry operators, such as $\boldsymbol{C}_3$ have an algebraic significance that leads to precise conclusions in numerical terms. Then, it is possible to link molecular symmetry with practical conclusions, such as infrared and Raman spectral activity. There are several ways in which a representation may be constructed.

7.5.1 Representations on position vectors

One procedure for obtaining a representation is through the effect of symmetry operations on a position vector within a body. If $\mathbf{p}$ is the position vector from the origin of the axes to a point x, y, z in the body, then

$$\mathbf{p} = \mathbf{i}x + \mathbf{j}y + \mathbf{k}z \tag{7.7}$$

In matrix notation, Eq. (7.7) is written as

$$\mathbf{p} = (\mathbf{i}\,\mathbf{j}\,\mathbf{k}) \begin{pmatrix} x \\ y \\ z \end{pmatrix} \tag{7.8}$$

where the unit vectors form a 1 × 3 row and the coordinates a 3 × 1 column. An anticlockwise rotation by an angle θ of the vector **p** about the z axis is given, following Eq. (A6.6), by

$$\begin{pmatrix} x' \\ y' \\ z' \end{pmatrix} = \begin{pmatrix} \cos\theta & -\sin\theta & 0 \\ \sin\theta & \cos\theta & 0 \\ 0 & 0 & 1 \end{pmatrix} \begin{pmatrix} x \\ y \\ z \end{pmatrix} \tag{7.9}$$

or more concisely

$$\mathbf{p}' = \mathbf{D}(\boldsymbol{R})\mathbf{p} \tag{7.10}$$

where $\mathbf{D}(\boldsymbol{R})$ is the matrix representation (Ger. *Darstellung* = representation) of the operator $\boldsymbol{R}$.

Example 7.1
Find the coordinates of a position vector, initially at $x = 0.1000$, $y = 0.2000$, $z = 0.3000$ with respect to orthogonal axes, after a right handed rotation of 60° about the z axis.
From the foregoing, or Eq. (A6.6),

$$\mathbf{D}(\boldsymbol{C}_6) = \begin{pmatrix} 1/2 & -\sqrt{3}/2 & 0 \\ \sqrt{3}/2 & 1/2 & 0 \\ 0 & 0 & 1 \end{pmatrix}$$

whereupon $x = -0.1232$, $y = 0.1866$ and $z = 0.3000$.

The water molecule, point group C_{2v}, provides an example of forming a representation based on a position vector. Figure 7.5 illustrates the setting of the reference axes for this molecule; the y and z axes lie in the molecular plane, with the x axis towards the observer and normal to the y and z axes. The transformation matrices, or **D**-matrices, are set out in Eq. (7.11):

$$\mathbf{D}(\boldsymbol{E}) = \begin{pmatrix} 1 & 0 & 0 \\ 0 & 1 & 0 \\ 0 & 0 & 1 \end{pmatrix} \quad \mathbf{D}(\boldsymbol{C}_2) = \begin{pmatrix} \bar{1} & 0 & 0 \\ 0 & \bar{1} & 0 \\ 0 & 0 & 1 \end{pmatrix}$$
$$\mathbf{D}(\boldsymbol{\sigma}_v) = \begin{pmatrix} 1 & 0 & 0 \\ 0 & \bar{1} & 0 \\ 0 & 0 & 1 \end{pmatrix} \quad \mathbf{D}(\boldsymbol{\sigma}'_v) = \begin{pmatrix} \bar{1} & 0 & 0 \\ 0 & 1 & 0 \\ 0 & 0 & 1 \end{pmatrix} \tag{7.11}$$

Fig. 7.5
Water molecule, H_2O: the C_2 axis is along z (in the plane of the diagram), σ_v is the vertical (x, z) plane and σ'_v is the molecular (y, z) plane; hydrogen atom H_a has been given the coordinates x, y, z in the discussion (Section 7.6). [Reproduced by courtesy of Woodhead Publishing, UK.]

The expression of the combination of symmetry operations under multiplication by equations such as $\boldsymbol{\sigma}_v\boldsymbol{\sigma}'_v = \boldsymbol{C}_2$ takes on a numerical significance through the **D**-matrices:

$$\mathbf{D}(\boldsymbol{\sigma}_v)\mathbf{D}(\boldsymbol{\sigma}'_v) = \mathbf{D}(\boldsymbol{C}_2) \tag{7.12}$$

The set of matrices, Eq. (7.11), is one representation of the symmetry operations of C_{2v}, and forms a group because it satisfies the group requirements in Section 7.2. Equalities such as Eq. (7.12) *in extenso* mirror a group multiplication table, and a table for the **D** matrices may be formulated that is homomorphic with the table for the symmetry operations:

C_{2v}	E	C_2	σ_v	σ_v'
E	E	C_2	σ_v	σ_v'
C_2	C_2	E	σ_v'	σ_v
σ_v	σ_v	σ_v'	E	C_2
σ_v'	σ_v'	σ_v	C_2	E

$D(C_{2v})$	$\mathbf{D}(E)$	$\mathbf{D}(C_2)$	$\mathbf{D}(\sigma_v)$	$\mathbf{D}(\sigma_v')$
$\mathbf{D}(E)$	$\mathbf{D}(E)$	$\mathbf{D}(C_2)$	$\mathbf{D}(\sigma_v)$	$\mathbf{D}(\sigma_v')$
$\mathbf{D}(C_2)$	$\mathbf{D}(C_2)$	$\mathbf{D}(E)$	$\mathbf{D}(\sigma_v')$	$\mathbf{D}(\sigma_v)$
$\mathbf{D}(\sigma_v)$	$\mathbf{D}(\sigma_v)$	$\mathbf{D}(\sigma_v')$	$\mathbf{D}(E)$	$\mathbf{D}(C_2)$
$\mathbf{D}(\sigma_v')$	$\mathbf{D}(\sigma_v')$	$\mathbf{D}(\sigma_v)$	$\mathbf{D}(C_2)$	$\mathbf{D}(E)$

The **D**-matrices are orthogonal (Appendix A3.4.11), so that the inverse of each matrix is equal to its transpose: in this example the matrices are also symmetric: $\mathbf{A} = \mathbf{A}^{\mathrm{T}}$. Thus, for C_{2v}, $\mathbf{D}(\boldsymbol{R}) = \mathbf{D}^{\mathrm{T}}(\boldsymbol{R}) = \mathbf{D}^{-1}(\boldsymbol{R})$.

7.5.2 Representations on basis vectors

A set of orthogonal basis vectors can be used to form a representation. Ammonia, NH_3, point group C_{3v} may be used as an example. In Fig. 7.6, the three hydrogen atoms shown may be considered to form the base of the molecule. The basis vectors are **i, j** and **k**, where +**k** is directed upwards. The C_3 operator rotates **i** and **j** to **i**′ and **j**′ respectively: operating next on **i**′ and **j**′ with σ_v leads to **i**″ and **j**″. Operating on **i** and **j** directly with σ_v'' leads to **i**″ and **j**″; thus,

$$\boldsymbol{\sigma}_v C_3 = \boldsymbol{\sigma}_v'' \text{ and } C_3\boldsymbol{\sigma}_v = \boldsymbol{\sigma}_v' \tag{7.13}$$

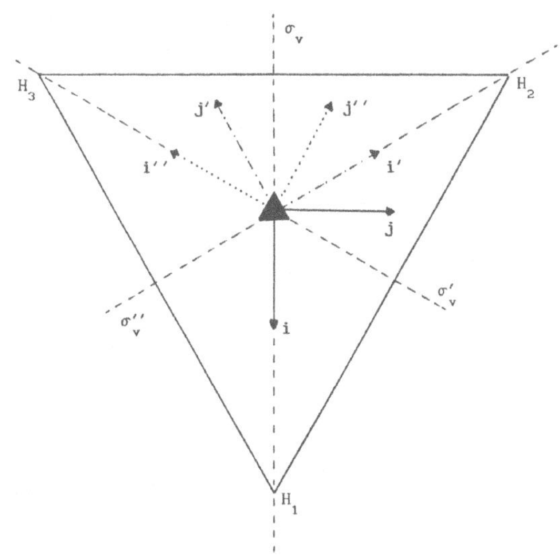

Fig. 7.6
Trigonal-pyramidal base of the three hydrogen atoms of the ammonia molecule, NH_3, point group C_{3v}: the C_3 axis is normal to the diagram; **i**, **j** and **k** (along z) are orthogonal unit vectors, **k** being invariant under all operations of C_3. [Reproduced by courtesy of Woodhead Publishing, UK.]

as expected. By simple geometry, or with the matrix $\mathbf{D}(\boldsymbol{R})$ in Eqs. (7.9)–(7.10), and remembering that $\mathbf{i}$, $\mathbf{j}$ and $\mathbf{k}$ form a 1×3 row vector,

$$\left.\begin{aligned} \mathbf{i}'(= C_3\mathbf{i}) &= -(^1\!/\!_2)\mathbf{i} + (\sqrt{3}/2)\mathbf{j} \\ \mathbf{j}'(= C_3\mathbf{j}) &= -(\sqrt{3}/2)\mathbf{i} - \mathbf{j}/2 \\ \mathbf{k}'(= C_3\mathbf{k}) &= \mathbf{k} \end{aligned}\right\} \tag{7.14}$$

From arguments similar to that just pursued for $\boldsymbol{C}_{2v}$, all six $\mathbf{D}$-matrices for C_{3v} can be written as follow: it should be noted that the vector coefficients on the right hand sides of Eq. (7.14) now appear as columns in Eq. (7.15); this representation may be called Γ_1.

$$\left.\begin{aligned} &\mathbf{D}(\boldsymbol{E}) = \begin{pmatrix} 1 & 0 & 0 \\ 0 & 1 & 0 \\ 0 & 0 & 1 \end{pmatrix} \quad \mathbf{D}(C_3) = \begin{pmatrix} -^1\!/\!_2 & -\sqrt{3}/2 & 0 \\ \sqrt{3}/2 & -^1\!/\!_2 & 0 \\ 0 & 0 & 1 \end{pmatrix} \quad \mathbf{D}(\boldsymbol{C}_3^2) = \begin{pmatrix} -^1\!/\!_2 & \sqrt{3}/2 & 0 \\ -\sqrt{3}/2 & -^1\!/\!_2 & 0 \\ 0 & 0 & 1 \end{pmatrix} \\ &\mathbf{D}(\boldsymbol{\sigma}_v) = \begin{pmatrix} 1 & 0 & 0 \\ 0 & \bar{1} & 0 \\ 0 & 0 & 1 \end{pmatrix} \quad \mathbf{D}(\boldsymbol{\sigma}_v') = \begin{pmatrix} -^1\!/\!_2 & \sqrt{3}/2 & 0 \\ \sqrt{3}/2 & ^1\!/\!_2 & 0 \\ 0 & 0 & 1 \end{pmatrix} \quad \mathbf{D}(\boldsymbol{\sigma}_v'') = \begin{pmatrix} -^1\!/\!_2 & -\sqrt{3}/2 & 0 \\ -\sqrt{3}/2 & ^1\!/\!_2 & 0 \\ 0 & 0 & 1 \end{pmatrix} \end{aligned}\right\} \tag{7.15}$$

An alternative representation of C_{3v}, again using the ammonia molecule example, may be formed in terms of unit vectors along the three N–H bonds : $\mathbf{r}_1$ (N–H_1), $\mathbf{r}_2$ (N–H_2) and $\mathbf{r}_3$ (N–H_3), as shown in Fig. 7.7. The C_3 rotation axis is set along $\mathbf{k}$, and the σ_v planes oriented as in the figure, with σ_v as the $\mathbf{i}$, $\mathbf{k}$, $\mathbf{r}_1$ plane. Under $\boldsymbol{C}_3, \mathbf{r}_1 \to \mathbf{r}_2$, and similarly for $\mathbf{r}_2$ and $\mathbf{r}_3$. Hence, $\mathbf{r}'$, the new $\mathbf{r}_1$, is $\boldsymbol{C}_3\,\mathbf{r}_3$, and similarly for $\mathbf{r}_2$ and $\mathbf{r}_3$.

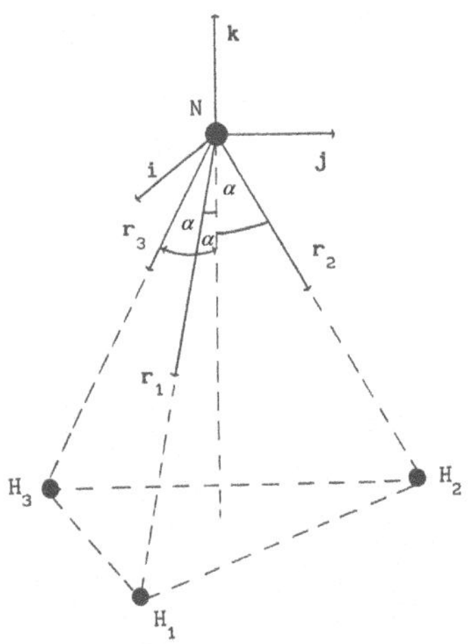

Fig. 7.7
Representation on the N–H bond vectors of the ammonia molecule: $\mathbf{i}$, $\mathbf{j}$ and $\mathbf{k}$ are unit vectors; $\mathbf{i}$, $\mathbf{k}$ and $\mathbf{r}_1$ are coplanar, thus defining the position of σ_v. [Reproduced by courtesy of Woodhead Publishing, UK.]

$$\left.\begin{aligned} \boldsymbol{C}_3\mathbf{r}_1 &= \mathbf{r}_2 \\ \boldsymbol{C}_3\mathbf{r}_2 &= \mathbf{r}_3 \\ \boldsymbol{C}_3\mathbf{r}_3 &= \mathbf{r}_1 \end{aligned}\right\} \qquad \left.\begin{aligned} \boldsymbol{\sigma}_v\mathbf{r}_1 &= \mathbf{r}_1 \\ \boldsymbol{\sigma}_v\mathbf{r}_2 &= \mathbf{r}_3 \\ \boldsymbol{\sigma}_v\mathbf{r}_3 &= \mathbf{r}_2 \end{aligned}\right\} \tag{7.16}$$

These equations lead to two $\mathbf{D}$-matrices, remembering that the right hand sides of Eq. (7.16) become columns in the matrices. Following through this procedure, a total of six $\mathbf{D}$-matrices may be formed, which may be called Γ_2:

$$\left.\begin{aligned} &\mathbf{D}(\boldsymbol{E}) = \begin{pmatrix} 1 & 0 & 0 \\ 0 & 1 & 0 \\ 0 & 0 & 1 \end{pmatrix} \quad \mathbf{D}(C_3) = \begin{pmatrix} 0 & 1 & 0 \\ 0 & 0 & 1 \\ 1 & 0 & 0 \end{pmatrix} \quad \mathbf{D}(\boldsymbol{C}_3^2) = \begin{pmatrix} 0 & 0 & 1 \\ 1 & 0 & 0 \\ 0 & 1 & 0 \end{pmatrix} \\ &\mathbf{D}(\boldsymbol{\sigma}_v) = \begin{pmatrix} 1 & 0 & 0 \\ 0 & 0 & 1 \\ 0 & 1 & 0 \end{pmatrix} \quad \mathbf{D}(\boldsymbol{\sigma}_v') = \begin{pmatrix} 0 & 1 & 0 \\ 1 & 0 & 0 \\ 0 & 0 & 1 \end{pmatrix} \quad \mathbf{D}(\boldsymbol{\sigma}_v'') = \begin{pmatrix} 0 & 0 & 1 \\ 0 & 1 & 0 \\ 1 & 0 & 0 \end{pmatrix} \end{aligned}\right\} \tag{7.17}$$

It can be shown that these two representations, Eq. (7.15) and Eq. (7.17), are equivalent. Referring again to Fig. 7.7, the angles between the $\mathbf{k}$ and each of $\mathbf{r}_1, \mathbf{r}_2$ and $\mathbf{r}_3$ are equal, say α. Then, the component of each bond vector resolved along $\mathbf{k}$ is $-\mathbf{k}\cos\alpha$, and using also Fig. 7.6, the following equations may be obtained:

$$\left.\begin{aligned} \mathbf{r}_1 &= \mathbf{i}\sin\alpha && - \mathbf{k}\cos\alpha \\ \mathbf{r}_2 &= -\tfrac{1}{2}\mathbf{i}\sin\alpha - (\sqrt{3}/2)\mathbf{j}\sin\alpha && - \mathbf{k}\cos\alpha \\ \mathbf{r}_3 &= -\tfrac{1}{2}\mathbf{i}\sin\alpha + (\sqrt{3}/2)\mathbf{j}\sin\alpha && - \mathbf{k}\cos\alpha \end{aligned}\right\} \qquad (7.18)$$

A convenient substitution is $p = \sin\alpha$ and $q = \cos\alpha$, and Eq. (7.18) solved for $\mathbf{i}$, $\mathbf{j}$ and $\mathbf{k}$:

$$\left.\begin{aligned} \mathbf{i} &= (2\mathbf{r}_1 - \mathbf{r}_2 - \mathbf{r}_3)/(3p) \\ \mathbf{j} &= (\mathbf{r}_3 - \mathbf{r}_2)/(p\sqrt{3}) \\ \mathbf{k} &= (-\mathbf{r}_1 - \mathbf{r}_2 - \mathbf{r}_3)/3q \end{aligned}\right\} \qquad (7.19)$$

En passant, the vectors $\mathbf{r}_1, \mathbf{r}_2$ and $\mathbf{r}_3$ are not orthogonal because their scalar products are non-zero:

$$\mathbf{r}_1 \cdot \mathbf{r}_2 = \mathbf{r}_2 \cdot \mathbf{r}_3 = \mathbf{r}_3 \cdot \mathbf{r}_1 = -\tfrac{1}{2}p^2 + q^2 \qquad (7.20)$$

Orthogonality (Gk. *orthos* = straight; *gonia* = angle) of two vectors $\mathbf{u}$ and $\mathbf{v}$ is achieved if $\mathbf{u} \cdot \mathbf{v} = 0$; in this case, when $p/q = \tan\alpha = \sqrt{2}$; that is, when $\alpha = \cos^{-1}(1/\sqrt{3})$, or one half of the tetrahedral angle.

If the two representations are equivalent, it should be possible to retrieve a given matrix, say $\mathbf{D}(C_3)$ in Γ_2, from $\mathbf{D}(C_3)$ in Γ_1 with the similarity transformation

$$\mathbf{Q}^{-1}\mathbf{D}_{\Gamma_2}(C_3)\mathbf{Q} = \mathbf{D}_{\Gamma_1}(C_3) \qquad (7.21)$$

where the $\mathbf{D}$-matrices are taken from Eq. (7.17) and Eq. (7.15); $\mathbf{Q}$ is formed from Eq. (7.19), and $\mathbf{Q}^{-1}$ is its inverse [4]:

$$\mathbf{Q} = \begin{pmatrix} 2/3p & 0 & -1/3q \\ -1/3p & 1/p\sqrt{3} & -1/3q \\ -1/3p & -1/p\sqrt{3} & -1/3q \end{pmatrix} \quad \mathbf{Q}^{-1} = \begin{pmatrix} p & -p/2 & -p/2 \\ 0 & p\sqrt{3}/2 & -p\sqrt{3}/2 \\ -q & -q & -q \end{pmatrix} \qquad (7.22)$$

from which the results in Eq. (7.15) follow. While it is a slightly difficult manipulation to obtain $\mathbf{Q}^{-1}$ from $\mathbf{Q}$, it is easy to show that $\mathbf{Q}\mathbf{Q}^{-1} = \mathbf{E}$. In general, two representations of a point group by matrices $\mathbf{D}_\alpha(\boldsymbol{R})$ and $\mathbf{D}_\beta(\boldsymbol{R})$ are equivalent if a similarity transformation, Eq. (7.21), is obeyed for all symmetry operators $\boldsymbol{R}$ of the group.

7.5.3 Representations on atom vectors

Refer again to Fig. 7.5 for the water molecule. Imagine three new sets of orthogonal components, designated x_o, y_o, and z_o for the oxygen atom, and x_a, y_a, z_a and x_b, y_b, z_b for hydrogen atoms H_a and H_b respectively. These axial sets are conveniently, though not necessarily, parallel, and by applying the symmetry operations of C_{2v} to them they lead to a set of four 9×9 matrices. For example, $\boldsymbol{C}_2$ applied to the oxygen atom has the effects:

$$
\begin{aligned}
x_o &\rightarrow -x_o \\
y_o &\rightarrow -y_o \\
z_o &\rightarrow z_o
\end{aligned}
$$

whereas $\boldsymbol{\sigma}_v$ operating in the x,z plane, on the H_a atoms leads to

$$
\begin{aligned}
x_a &\rightarrow x_a \\
y_a &\rightarrow -y_a \\
z_a &\rightarrow z_a
\end{aligned}
$$

Proceeding in this way, four matrices, Eqs. (7.23)–(7.26), are developed. Application of the rules for matrix multiplication shows that these matrices reproduce the transformations of coordinates listed above, the equivalent transformations for hydrogen H_b, and the other symmetry operations of the group; the χ values are the traces, or *characters*, of the matrices.

$$
\begin{pmatrix}
1 & 0 & 0 & 0 & 0 & 0 & 0 & 0 & 0 \\
0 & 1 & 0 & 0 & 0 & 0 & 0 & 0 & 0 \\
0 & 0 & 1 & 0 & 0 & 0 & 0 & 0 & 0 \\
0 & 0 & 0 & 1 & 0 & 0 & 0 & 0 & 0 \\
0 & 0 & 0 & 0 & 1 & 0 & 0 & 0 & 0 \\
0 & 0 & 0 & 0 & 0 & 1 & 0 & 0 & 0 \\
0 & 0 & 0 & 0 & 0 & 0 & 1 & 0 & 0 \\
0 & 0 & 0 & 0 & 0 & 0 & 0 & 1 & 0 \\
0 & 0 & 0 & 0 & 0 & 0 & 0 & 0 & 1
\end{pmatrix}
\begin{pmatrix} x_o \\ y_o \\ z_o \\ x_a \\ y_a \\ z_a \\ x_b \\ y_b \\ z_b \end{pmatrix}
=
\begin{pmatrix} x'_o \\ y'_o \\ z'_o \\ x'_a \\ y'_a \\ z'_a \\ x'_b \\ y'_b \\ z'_b \end{pmatrix}
$$

$\mathbf{D}(\boldsymbol{E})$: $\chi = 9$ (7.23)

$$
\begin{pmatrix}
-1 & 0 & 0 & 0 & 0 & 0 & 0 & 0 & 0 \\
0 & -1 & 0 & 0 & 0 & 0 & 0 & 0 & 0 \\
0 & 0 & 1 & 0 & 0 & 0 & 0 & 0 & 0 \\
0 & 0 & 0 & 0 & 0 & 0 & -1 & 0 & 0 \\
0 & 0 & 0 & 0 & 0 & 0 & 0 & -1 & 0 \\
0 & 0 & 0 & 0 & 0 & 0 & 0 & 0 & 1 \\
0 & 0 & 0 & -1 & 0 & 0 & 0 & 0 & 0 \\
0 & 0 & 0 & 0 & -1 & 0 & 0 & 0 & 0 \\
0 & 0 & 0 & 0 & 0 & 1 & 0 & 0 & 0
\end{pmatrix}
\begin{pmatrix} x_o \\ y_o \\ z_o \\ x_a \\ y_a \\ z_a \\ x_b \\ y_b \\ z_b \end{pmatrix}
=
\begin{pmatrix} x'_o \\ y'_o \\ z'_o \\ x'_a \\ y'_a \\ z'_a \\ x'_b \\ y'_b \\ z'_b \end{pmatrix}
$$

$\mathbf{D}(\boldsymbol{C}_2)$: $\chi = -1$ (7.24)

Each of the 9×9 matrices is *block-factored*: a matrix is termed block factored if all the non-zero elements lie in blocks along the diagonal direction of the matrix. Two similarly blocked matrices when multiplied produce a matrix with similar blocking. Elements in the product matrix are determined solely by the corresponding blocks in the original matrix. Thus, each block may be treated independently. Corresponding blocks can be multiplied together: taking the first block, for example, from each of $\mathbf{D}(\sigma_v)$ and $\mathbf{D}(C_2)$, Eqs. (7.23)–(7.24), it is evident that

$$\mathbf{D}(C_2)\mathbf{D}(\sigma_v) = \mathbf{D}(\sigma_v')$$

and similar results arise from all other pairs of corresponding blocks. The sets of blocks themselves are also representations of C_{2v}. The 9×9 matrices are said to be *reducible representations*, and the representations formed by the blocks are termed *irreducible representations* of the group.

$$\begin{pmatrix} 1 & 0 & 0 & 0 & 0 & 0 & 0 & 0 & 0 \\ 0 & -1 & 0 & 0 & 0 & 0 & 0 & 0 & 0 \\ 0 & 0 & 1 & 0 & 0 & 0 & 0 & 0 & 0 \\ 0 & 0 & 0 & 0 & 0 & 0 & 1 & 0 & 0 \\ 0 & 0 & 0 & 0 & 0 & 0 & 0 & -1 & 0 \\ 0 & 0 & 0 & 0 & 0 & 0 & 0 & 0 & 1 \\ 0 & 0 & 0 & 1 & 0 & 0 & 0 & 0 & 0 \\ 0 & 0 & 0 & 0 & -1 & 0 & 0 & 0 & 0 \\ 0 & 0 & 0 & 0 & 0 & 1 & 0 & 0 & 0 \end{pmatrix} \begin{pmatrix} x_o \\ y_o \\ z_o \\ x_a \\ y_a \\ z_a \\ x_b \\ y_b \\ z_b \end{pmatrix} = \begin{pmatrix} x'_o \\ y'_o \\ z'_o \\ x'_a \\ y'_a \\ z'_a \\ x'_b \\ y'_b \\ z'_b \end{pmatrix}$$

$\mathbf{D}(\sigma_v)$: $\chi = 1$ (7.25)

$$\begin{pmatrix} -1 & 0 & 0 & 0 & 0 & 0 & 0 & 0 & 0 \\ 0 & 1 & 0 & 0 & 0 & 0 & 0 & 0 & 0 \\ 0 & 0 & 1 & 0 & 0 & 0 & 0 & 0 & 0 \\ 0 & 0 & 0 & -1 & 0 & 0 & 0 & 0 & 0 \\ 0 & 0 & 0 & 0 & 1 & 0 & 0 & 0 & 0 \\ 0 & 0 & 0 & 0 & 0 & 1 & 0 & 0 & 0 \\ 0 & 0 & 0 & 0 & 0 & 0 & -1 & 0 & 0 \\ 0 & 0 & 0 & 0 & 0 & 0 & 0 & 1 & 0 \\ 0 & 0 & 0 & 0 & 0 & 0 & 0 & 0 & 1 \end{pmatrix} \begin{pmatrix} x_o \\ y_o \\ z_o \\ x_a \\ y_a \\ z_a \\ x_b \\ y_b \\ z_b \end{pmatrix} = \begin{pmatrix} x'_o \\ y'_o \\ z'_o \\ x'_a \\ y'_a \\ z'_a \\ x'_b \\ y'_b \\ z'_b \end{pmatrix}$$

$\mathbf{D}(\sigma_v')$: $\chi = 3$ (7.26)

The representation obtained from Eqs. (7.23)–(7.26) may be written as

C_{2v}	E	C_2	σ_v	σ_v'
Γ_{3n}	9	−1	1	3

Table 7.2 Vector shift contributions to χ_R in the water molecule.

	O			H$_a$			H$_b$		
	x	y	z	x	y	z	x	y	z
E	1	1	1	1	1	1	1	1	1
C_2	−1	−1	1	0	0	0	0	0	0
σ_v	1	−1	1	0	0	0	0	0	0
σ'_v	−1	1	1	−1	1	1	−1	1	1

where the numbers are the traces of the 9×9 matrices, and n refers to the number of atoms in the molecule, so that $3n$ is the number of basis vectors.

7.5.3.1 Unshifted atom contributions to a representation

The Γ_{3n} type of representation deduced above is important in studying molecular vibrations: some general relationships that govern the characters of the **D**-matrices can be deduced in terms of the vector shifts under the symmetry operations of a point group.

If a symmetry operator $\boldsymbol{R}$ shifts an atom, or any of its attached orthogonal vectors, then the diagonal terms of the **D**-matrix for the operator $\boldsymbol{R}$ indicate its contribution to $\chi_{\boldsymbol{R}}$. The contribution N_u of an unshifted vector component was indicated by $+1$ or -1 according as the sign of the component either remains the same or is negated. Table 7.2 lists the result of this analysis for the water molecule.

The process can be generalized by the equations:

$$\begin{aligned} \text{For } C_n : N_\text{u} &= 2\cos(360/n) + 1 \\ \text{For } S_n : N_\text{u} &= 2\cos(360/n) - 1 \end{aligned} \qquad (7.27)$$

From these relationships, Table 7.3 was constructed, so as to encompass all crystallographic symmetry operations of concern in this work.

Table 7.3 Contributions to χ_R for an unshifted atom in Γ_{3n}.

Operation R	χ_R	Operation R	χ_R
$E(\equiv C_1)$	3	$\sigma(\equiv S_1)$	1
C_2	−1	$i(\equiv S_2)$	−3
C_3, C_3^2	0	S_3, S_3^2	−2
C_4, C_4^3	1	S_4, S_4^3	−1
C_6, C_6^5	2	S_6, S_6^5	0

Example 7.2

A Γ_{3n} representation is to be be deduced for the ammonia molecule, point group C_{3v}.

From Fig. 7.6 or Fig. 7.7, the unshifted atoms are determined under each operation of C_{3v} and then Table 7.3 used to obtain the contributions of $\chi_{\boldsymbol{R}}$. Only one symmetry operation in each class need be considered because all operations in one and the same class behave alike, Thus, since $\Gamma_{3n} = N_u \chi_{\boldsymbol{R}}$

	E	$2C_3$	$3\sigma_v$
Unshifted atoms	4	1	2
Γ_{3n}	12	0	2

(The reduction of representations will be considered shortly.)

7.5.4 Representations on functions

Mathematical functions may be used as a basis for generating representations. The angular components of the $2p$ functions (Section 2.9.5) may be written in the form:

$$\begin{aligned} p_x &= N \sin\theta \cos\phi \\ p_y &= N \sin\theta \sin\phi \\ p_z &= N \cos\theta \end{aligned} \tag{7.28}$$

where N is the normalizing constant of $\sqrt{3/4\pi}$. The **D**-matrices for the transformations under $\boldsymbol{C}_3$ operations may be determined as follows. From Fig. A8.1 it is evident that θ is invariant under all operations of $\boldsymbol{C}_3$, but that ϕ is changed by $2\pi/3$ for each $\boldsymbol{C}_3$ operation. Hence,

$$\sin\theta' = \sin\theta \tag{7.29}$$

and using the general rotation matrix

$$\begin{aligned} \begin{pmatrix} \cos\phi' \\ \sin\phi' \end{pmatrix} &= \begin{pmatrix} \cos 2\pi/3 & -\sin 2\pi/3 \\ \sin 2\pi/3 & \cos 2\pi/3 \end{pmatrix} \begin{pmatrix} \cos\phi \\ \sin\varphi \end{pmatrix} \\ \cos\phi' &= -(1/2)\cos\phi - (\sqrt{3}/2)\sin\phi \\ \sin\phi' &= (\sqrt{3}/2)\cos\phi - (1/2)\sin\phi \end{aligned} \tag{7.30}$$

Then, setting N to unity as only the angular relationships are of interest,

$$\boldsymbol{C}_3 p_x = -(1/2)\sin\theta\cos\phi - (\sqrt{3}/2)\sin\theta\sin\phi = -(1/2)p_x - (\sqrt{3}/2)p_y \tag{7.31}$$

and

$$\boldsymbol{C}_3 p_y = (\sqrt{3}/2)\sin\theta\cos\phi - (1/2)\sin\theta\sin\phi = (\sqrt{3}/2)p_x - (1/2)p_y \tag{7.32}$$

Similarly, it can be shown that

$$\left.\begin{aligned} \boldsymbol{\sigma}_v p_x &= \sin\theta\cos\phi = p_x \\ \boldsymbol{\sigma}_v p_y &= -\sin\theta\sin\phi = -p_y \end{aligned}\right\} \tag{7.33}$$

Clearly $\boldsymbol{E}p_x = p_x$ and $\boldsymbol{E}p_y = p_y$, and the following matrices are obtained:

$$\boldsymbol{E} : \begin{pmatrix} 1 & 0 \\ 0 & 1 \end{pmatrix} \chi_E = 2 \quad \boldsymbol{C}_3 : \begin{pmatrix} -1/2 & -\sqrt{3}/2 \\ \sqrt{3}/2 & -1/2 \end{pmatrix} \chi_{C_3} = -1 \quad \boldsymbol{\sigma}_v \begin{pmatrix} 1 & 0 \\ 0 & -1 \end{pmatrix} \chi = 0 \tag{7.34}$$

Since the characters for $\boldsymbol{C}_3^2$ and $\boldsymbol{C}_3$ are the same, as also are those for $\boldsymbol{\sigma}_v$, $\boldsymbol{\sigma}'_v$ and $\boldsymbol{\sigma}''_v$, the x, y representation may be written as

C_{3v}	E	$2C_3$	$3\sigma_v$
$\Gamma_{x,y}$	2	−1	0

From Appendix A3, the matrix for each operation on p_z is (1), so that $\chi_{p_z} = 1, 1, 1$. Reference to Eq. (7.15) shows that Eqs. (7.34) are portions of block-factored matrices; for example,

$$\mathbf{D}(E) = \begin{pmatrix} 1 & 0 & 0 \\ 0 & 1 & 0 \\ 0 & 0 & 1 \end{pmatrix} \tag{7.35}$$

if the representation for p_z is added, $\Gamma_{x,y,z}$ becomes recognizable as the totality of Eq. (7.15):

C_{3v}	E	$2C_3$	$3\sigma_v$
Γ	3	0	1

7.6 A first look at character tables

Consider again Fig. 7.5, confining attention to the hydrogen atom H_a of the water molecule. The symmetry operations act on x, y and z in the following way:

$$\mathbf{E}: \text{No change in } x,\ y \text{ or } z \}$$

$$C_2: \left.\begin{array}{l} x \rightarrow -x \\ y \rightarrow -y \\ \text{No change in } z \end{array}\right\}$$

$$\sigma_v: \left.\begin{array}{l} y \rightarrow -y \\ \text{No change in } x \text{ or } z \end{array}\right\}$$

$$\sigma'_v: \left.\begin{array}{l} x \rightarrow -x \\ \text{No change in } y \text{ or } z \end{array}\right\}$$

These results may be written in the following manner:

C_{2v}	E	C_2	σ_v	σ'_v
x	1	−1	1	−1
y	1	−1	−1	1
z	1	1	1	1
xy	1	1	−1	−1

The effect of the symmetry operations on the product functions x^2, y^2, z^2, yz and zx can be obtained in like manner. Thus, x^2, y^2 and z^2 transform like z under all operations of the group, yz transforms like y and xz like x: xy transforms differently; therefore, it appears in the above table. There are four ways in which functions may transform under C_{2v}; they are tabulated with the usual names given to the representations in the form of a *character table*:

C_{2v}	E	C_2	σ_v	σ'_v		
A_1	1	1	1	1	z	x^2, y^2, z^2
A_2	1	1	−1	−1	R_z	xy
B_1	1	−1	1	−1	x, R_y	xz
B_2	1	−1	−1	1	y, R_x	yz

There are no other ways in which new functions can transform; for example, x^3 transforms like x and x^4 like x^2. The four transformation names A_1, A_2, B_1 and B_2 are labels for the *irreducible representations* of C_{2v}, and the numbers in the body of the table are the *characters of the irreducible representations*, or just *characters* for the point group. In the sixth column, the symbols x, y and z (sometimes listed as T_x, T_y and T_z) refer to coordinate axes, or translational movements with respect to these axes. If p orbital functions are transformed, then because they are named for their axes, bearing in mind the comment in Section 2.9.5, it follows that p_x, p_y and p_z transform under C_{2v} as B_1, B_2 and A_1, respectively. The final column of the table lists the binary products of the coordinates. More will be said of character tables in a later chapter.

The parameter x is said to belong to, or *span*, the B_1 representation of C_{2v}; again, x is *symmetric* (+1) with respect to $\boldsymbol{E}$ and $\boldsymbol{\sigma}_v$, and *antisymmetric* (−1) with respect to $\boldsymbol{C}_2$ and $\boldsymbol{\sigma}'_v$; z is *totally symmetric* under all operations of point group C_{2v}. If another function, such as $x^3 + y^2 + z$ is formed, it is clear that it does not belong to any one reducible representation in C_{2v}. However, since x^3 transforms like x and y^2 like z, the given function spans $2A_1 + B_1$.

There is a certain dependence of the spanning upon the choice of axes. On the one hand, if for C_{2v} σ_v is chosen as the x,z plane (which is usual) and σ'_v as y, z plane, then the above character table is obtained. On the other hand, if the x and y axes are interchanged, x would span B_2, y would span B_1, and so on; the character of B_1 under $\boldsymbol{\sigma}_v$ remains as +1 (see Section 7.6.3).

7.6.1 Transformation of atomic orbitals

The orbital functions for a species transform in accordance with its point group symmetry. It was shown in Section 2.9 that a wave function may be written in terms of polar coordinates and is separable into radial and angular parts. The radial component of the function is unaffected by point group symmetry operators. An s orbital has no angular component; it is spherically symmetric and transforms under the totally symmetric A_1 irreducible representation. The transformation of p, d and other orbital functions depends on how their angular components change with each point group symmetry operation; the s, p and d angular functions are listed in Table 2.8.

The notation of the d orbital functions can be deduced in the same manner as the p functions: taking the d angular function $\sin\theta\cos\theta\cos\phi$ from Table 2.8, for example:

$$\sin\theta\cos\theta\cos\phi = (x/r)\cos\theta = xz/r^2 \propto xz \qquad (7.36)$$

using Appendix A8; hence, the notation d_{xz} is applied to this angular function.

7.6.2 Orthonormality and orthogonality

Consider any irreducible representation, such as B_1 in C_{2v}. If the sum of the products of the characters, each with itself, under the symmetry operators is formed and the result normalized by dividing by h, the order of the group, the result is

$$(1/4)\{(1\times 1) + [(-1)\times(-1)] + (1\times 1) + [(-1)\times(-1)]\} = 1$$

whereas if the same process be carried out taking the product of each character with the corresponding character in another irreducible representation, say B_2, then

$$(1/4)\{(1\times 1) + [(-1)\times(-1)] + [(1\times(-1)] + [(-1)\times 1)]\} = 0$$

Similar results would be obtained with other pairs of irreducible representations, and these equations are expressions of *orthonormality* relationships. In general, each individual produced would be multiplied by the symmetry class multiplicity, but they are all unity in C_{2v}.

The term 'orthonormal' combines the terms 'orthogonal' and 'normal'. Two coordinate axes or two functions are orthogonal when there is no component of either one on the other. Thus, mutually perpendicular x, y, and z axes are orthogonal. Normalized functions have equal weight, frequently unity; in the above examples $1/h$ acts as a normalizing constant.

Example 7.3

Determine whether or no the following two functions $f_1 = \cos\nu$ and $f_2 = \sin\nu$ are orthonormal over the range $\pm\pi$.

The normalization condition is set up with N as the normalizing constant:

$$N_1^2\int_{=\pi}^{\pi}\cos^2\nu\,\mathrm{d}\nu = N_1^2\int_{=\pi}^{\pi}{}^1\!/\!_2(1+\cos 2\nu)\,\mathrm{d}\nu = N_1^2\pi = 1$$

Hence, $N_1 = 1/\sqrt{\pi}$. A similar result obtains for f_2. Furthermore, for the product of f_1 and f_2,

$$N_1N_2\int_{=\pi}^{\pi}\cos\nu\ \sin\nu\,\mathrm{d}\nu = N_1N_2\int_{=\pi}^{\pi}\sin 2\nu\,\mathrm{d}\nu = 0\,.$$

Thus, f_1 and f_2 are *orthogonal*, and become *orthonormal* if each is multiplied by $1/\sqrt{\pi}$.

7.6.3 Notation for irreducible representations

The symbols in the first column of a character table, below the point group symbol, are given in the Mulliken notation [5] that is employed in this context. For a two-dimensional (doubly degenerate) representation, the symbol E

is used; take care not to confuse this symbol E with $\boldsymbol{E}$, the identity operator. Multidimensional irreducible representations arise when two or more directions are related by symmetry: thus, in C_{4v}, for example, the x and y directions are related by the fourfold axis, and in the character table for this point group they appear coupled, against the E label.

The symbol A is used for an irreducible representation that is totally symmetric, that is, all characters are +1, with respect to the $\boldsymbol{C}_n$ or $\boldsymbol{S}_n$ operation of highest degree in the given point group. If the irreducible representation is antisymmetric, that is, the characters $= -1$, under the same rule, it is labelled B. In the groups C_1, C_s and C_i, where $\boldsymbol{C}_n$ operations with n greater than unity do not appear, all irreducible representations are of type A.

A centrosymmetric group G_i may be regarded as the direct product of groups G and C_i. For a given type of irreducible representation, A, B, ..., the characters of those symmetry operations that are *not* in both G_i and G are, in order, of opposite sign in the irreducible representation of subscript u, and of the same sign where the subscript is g; the signs of the characters in G that are common to both G and G_i, are determined by the signs in G. This statement is illustrated below for the A_2 type irreducible representations in D_3 and D_{3d}:

G/D_3	E	$2C_3$	$3C_2'$	i	$2S_6$	$3\sigma_d$
A_2	1	1	−1	—	—	—

G_i/D_{3d}	E	$2C_3$	$3C_2'$	i	$2S_6$	$3\sigma_d$
A_{2g}	1	1	−1	1	1	−1
A_{2u}	1	1	−1	−1	−1	1

If a point group contains the operator $\boldsymbol{\sigma}_h$ but no operator $\boldsymbol{i}$, the irreducible representation labels are singly primed if the character is +1 under $\boldsymbol{\sigma}_h$ and doubly primed otherwise, as in C_s and D_{3h}, for example. With the A and B representations, the subscripts 1 and 2 show, respectively, a symmetric (+1) or an antisymmetric (−1) character with respect to a C_2 axis *perpendicular* to the principal C_n axis or, in the absence of this symmetry, to a σ_v plane, as in D_4 and C_{3v}, for example. For multidimensional representations, the subscripts 1, 2,..., are added in order to distinguish between non-equivalent irreducible representations that are not separated by the above rules, as in D_{6h}, for example.

7.6.4 Complex characters

In certain point groups complex characters arise: with C_3, for example, which is also Abelian, the characters ε and its complex conjugate ε^* exist, where ε is $\exp(i2\pi/3)$; the term $\exp(i2\pi/3)$ may be regarded as an operator that rotates a vector anticlockwise by the angle $2\pi/3$, or 120°, in the complex plane.

For this particular value of ε, $\varepsilon + \varepsilon^* = -1$; in any point group, $\varepsilon\varepsilon^* = 1$. Two irreducible representations involving complex characters are usually bracketed together under the doubly degenerate label E, as in C_3, for example:

C_3	$\boldsymbol{E}$	C_3	C_3^2		$\varepsilon = \exp(i2\pi/3)$
A	1	1	1	z, R_z	$x^2 + y^2, z^2$
E	$\left\{\begin{matrix}1\\1\end{matrix}\right.$	$\begin{matrix}\varepsilon\\\varepsilon^*\end{matrix}$	$\left.\begin{matrix}\varepsilon^*\\\varepsilon\end{matrix}\right\}$	$(x, y), (R_x, R_y)$	$(x^2 - y^2, xy), (yz, xz)$

7.6.5 Linear groups

Linear molecules belong to point group $C_{\infty v}$ (HF, COS) or to $D_{\infty h}(H_2, CO_2)$; the character table for $C_{\infty v}$ is represented hereunder. The operator $\boldsymbol{C}_\infty^\phi$ indicates a rotation by an angle ϕ of any value, including an infinitesimal rotation. An infinite number of such rotations is possible, together with an infinite number of vertical mirror symmetry planes. What is the significance of the number 2 in $2\boldsymbol{C}_\infty^\phi$?

$C_{\infty v}$	$\boldsymbol{E}$	$2\boldsymbol{C}_\infty^\phi$	...	$\infty\boldsymbol{\sigma}_v$		
$A_1(\Sigma^+)$	1	1	...	1	z	$x^2 + y^2, z^2$
$A_2(\Sigma^-)$	1	1	...	$\bar{1}$	R_z	
$E_1(\Pi)$	2	$2\cos\phi$	...	0	$(x, y), (R_x, R_y)$	(xz, yz)
$E_2(\Delta)$	2	$2\cos 2\phi$	...	0		$(x^2 - y^2, xy)$
$E_3(\Phi)$	2	$2\cos 3\phi$	...	0		
...	...	...	...	...		

The linear group $D_{\infty h}$ (Appendix A10.2) can be formed by the direct product of $C_{\infty v}$ and C_i. However, it should be noted that, in the linear groups, the signs of the characters for the pairs of irreducible representations $\Pi_g, \Pi_u, \ldots,$ and $\Delta_g, \Delta_u, \ldots$ do not follow the rules given for other groups [6]. In addition, Greek symbols are often used in place of the Mulliken notation, and the primes on the symbols for irreducible representations are changed to + and – signs, as determined by the sign of the character under the $\boldsymbol{\sigma}_v$ operation.

Example 7.4

Find the irreducible representations for a molecule of the type H–Cl.

The point group for this molecule is $C_{\infty v}$. Following Section 7.5.3, orthogonal x, y, z vectors are set up on each atom. A 6×6 matrix applies to each symmetry operation. Under $\boldsymbol{E}$ the character is clearly 6. For $\boldsymbol{C}_\infty^\phi$, z is invariant, and both x and y are rotated by an angle ϕ, thus contributing $2(1 + 2\cos\phi) = (2 + 4\cos\phi)$ to the representation, and for $\boldsymbol{\sigma}_v$ the character is 1 for each atom. Thus, the reducible representation Γ_r is

$C_{\infty v}$	E	$2C_{\infty}^{\phi}$	$\infty\sigma_v$
Γ_r	6	$2+4\cos\phi$	2

A term of the form $n\cos\phi$ in Γ must arise from the Π irreducible representation, 2Π in this example, with characters 4, $4\cos\phi$, 0. If they are subtracted, there remain the characters 2, 2, 2, which must refer to Σ^+, since translational movements are the concern here. Thus, Γ_r is reducible to $2\Sigma^+ + 2\Pi$.

7.6.6 Some properties of character tables

There are five important rules for irreducible representations and their characters, which are described here with reference to the character table for point group C_{4v}:

C_{4v}	E	$2C_4$	C_2	$2\sigma_v$	$2\sigma_d$		$h=8$
A_1	1	1	1	1	1	z	z^2; x^2+y^2
A_2	1	1	1	$\bar{1}$	$\bar{1}$	R_z	
B_1	1	$\bar{1}$	1	1	$\bar{1}$		x^2-y^2
B_2	1	$\bar{1}$	1	$\bar{1}$	1		xy
E	2	0	$\bar{2}$	0	0	(x,y) ; (R_x,R_y)	

1. The order of an irreducible representation matrix is called the dimension of the irreducible representation; the sum of the squares of all the dimensions is equal to the order h of the group. Thus, under E : $1^2+1^2+1^2+1^2+2^2=8$; under $\boldsymbol{\sigma}_d$: $\left[1^2+(-1)^2+(-1)^2+1^2\right]\times 2=8$. Note that the factor of two is needed because the symmetry number k is 2 for $\boldsymbol{\sigma}_d$.
2. The sum of the squares of an irreducible representation is equal to h. Thus, for B_2 : $1^2+2\,(-1)^2+1^2+2\,(-1)^2+2\,(-1)^2=8$.
3. The vectors that compose the characters of different irreducible representation are orthogonal. Thus, under $A_1$and A_2: $(1\times 1)+2(1\times 1)+(1\times 1)+2[1\times(-1)]+2[1\times(-1)]=0$.
4. The characters of matrix representations, whether reducible or irreducible, of the operations in one and the same class are the same. This is evident in any character table.
5. The number of irreducible representations is equal to the number of classes. Thus, in C_{4v}, the number of irreducible representations is 5 and the number of symmetry classes is also 5.

At this point, it is convenient to summarize the meanings of each section of the character table, with reference to C above:

- The top row lists the *point group*, and then the *symmetry operators* in their *classes*; a digit before an operator indicates the *symmetry number k* of operators in the group.

- The first column, after the point-group heading, shows the labels of the *irreducible representations*, the *characters* of which are the digits in the row of each column under each operator heading; they indicate the *trace* of the symmetry matrix for that operation. The character 2 for the E label under the $\boldsymbol{E}$ heading indicates a *degeneracy* of 2.
- In the first of the final two columns, the functions x, y and z (also written as T_x, T_y, T_z) represent the symmetry for translational movements along Cartesian axes and/or p orbital functions; R_x, R_y and R_z represent the symmetry of rotational movements about the axes labelled by the subscripts. The final column lists the symmetry of specific quadratic functions that represent polarizability movements and/or d orbital functions.
- The meanings of superscripts and subscripts on the labels of irreducible representations have been discussed in Section 7.6.3.
- There are other functions that apply to f and g orbitals; they can be found in standard works on the chemical applications of group theory [7, 8].

7.7 The great orthogonality theorem

This theorem concerns the terms of those matrices that relate to the irreducible representations of point groups: *for any two irreducible representations* Γ_α and Γ_β, *the terms of the related matrices* $\mathbf{D}_\alpha(\boldsymbol{R})$ *and* $\mathbf{D}_\beta(\boldsymbol{R})$ *satisfy the following equation*:

$$\sum_{\boldsymbol{R}} \mathbf{D}^*_\alpha(\boldsymbol{R})_{i,j}\mathbf{D}_\beta(\boldsymbol{R})_{p,q} = \frac{h}{\sqrt{n_\alpha n_\beta}}\delta_{\alpha,\beta}\delta_{i,p}\delta_{j,q} \tag{7.37}$$

where $\boldsymbol{D}^*_\alpha(\boldsymbol{R})_{i,j}$ is the complex conjugate of $\mathbf{D}_\alpha(\boldsymbol{R})_{i,j}$, n_α and n_β are the dimensions of Γ_α and Γ_β, respectively, and the sum extends over all $\boldsymbol{R}$ symmetry operators of a group; each δ is Kronecker delta.[1] When all characters are real, $\mathbf{D}_\alpha(\boldsymbol{R})_{i,j}$ may be used instead of its conjugate. In applying this theorem, the $\mathbf{D}_\alpha$ and $\mathbf{D}_\beta$ matrices are assembled for particular values of i, j and p, q. For the i, j term in $\mathbf{D}_\alpha$ and the p,q term in $\mathbf{D}_\beta$ the sum $\mathbf{D}^*_\alpha(R)_{i,j}\,\mathbf{D}_\beta(\boldsymbol{R})_{p,q}$, or $\mathbf{D}_\alpha(\boldsymbol{R})_{i,j}\,\mathbf{D}_\beta(\boldsymbol{R})_{p,q}$ if the terms are real, is formed over all $\boldsymbol{R}$ operators in the group and compared with the right hand side of Eq. (7.37). Particular examples hereunder can be compared with the properties listed more qualitatively in Section 7.6.

[1] $\delta_{i,j} = 0$ for $i \neq j$; $\delta_{i,j} = 1$ for $i = j$.

(a) $\Gamma_\alpha \neq \Gamma_\beta; \alpha \neq \beta$.

The characters of the two representations are orthogonal, so that $\delta_{\alpha,\beta} = 0$: then

$$\sum_{\boldsymbol{R}} \mathbf{D}^*_\alpha(\boldsymbol{R})_{i,j}\,\mathbf{D}_\beta(\boldsymbol{R})_{p,q} = 0 \tag{7.38}$$

(b) $\Gamma_\alpha = \Gamma_\beta; \alpha = \beta(\delta_{\alpha,\beta} = 1)$.

If $i \neq p$ or/and $j \neq q$, $\delta_{i,p}$ or/and $\delta_{i,j} = 0$: then

$$\sum_{\boldsymbol{R}} \mathbf{D}^*_\alpha(\boldsymbol{R})_{i,j}\,\mathbf{D}_\beta(\boldsymbol{R})_{p,q} = 0 \tag{7.39}$$

(c) $\Gamma_\alpha = \Gamma_\beta(\delta_{\alpha,\beta} = 1)$;
$i = p$ and $j = q$ $(\delta_{i,p} = \delta_{i,j} = 1)$: then

$$\sum_{\boldsymbol{R}} |\mathbf{D}_\alpha(\boldsymbol{R})_{i,j}|^2 = h/n_\alpha \tag{7.40}$$

(d) Refer to the matrices developed in Eq. (7.15) for $C_{3v}, h = 6$. The implication of Eq. (7.40) is that the set of i, j matrix terms in an irreducible representation may be considered as a vector in h-dimensional space, and there are h^2 such vectors: for C_{3v}, there are 36 such i,j matrix terms (see Example 4.4). Do not worry about trying to visualize h-dimensional space: it is best regarded as a mathematical device for collating h related quantities in a vector like manner. Following Section 7.6.2, the vectors are mutually orthogonal, and from Eq. (7.40) the square of the magnitude of any vector is h/n_α. Because there are only h such vectors, it follows that

$$\sum_{\alpha=1}^{r} \boldsymbol{n}_\alpha^2 \leq h \tag{7.41}$$

where r is the total number of irreducible representations: a more detailed analysis [7] shows that $\sum_{\alpha=1}^{r} n_\alpha^2$ is exactly equal to h.

(e) $\Gamma_\alpha \neq \Gamma_\beta; i = j$ and $p = q$.
Here, the diagonals of the **D**-matrices will be used: then

$$\sum_{\boldsymbol{R}} \mathbf{D}_\alpha^*(\boldsymbol{R})_{i,j}\, \mathbf{D}_\beta(\boldsymbol{R})_{p,q} = \frac{h}{\sqrt{n_\alpha n_\beta}} \delta_{\alpha,\beta}\delta_{i,p}\ (\delta_{i,p} = \delta_{j,q}) \tag{7.42}$$

Consider the sums over all i and p:

$$\sum_{i=1}^{n_\alpha}\sum_{p=1}^{n_\beta} \mathbf{D}_\alpha^*(\boldsymbol{R})_{i,i}\, \mathbf{D}_\beta(\boldsymbol{R})_{p,p} = \frac{h}{\sqrt{n_\alpha n_\beta}}\delta_{\alpha,\beta}\sum_{i=1}^{n_\alpha}\sum_{p=1}^{n_\beta}\delta_{i,p} \tag{7.43}$$

The summations over i and p on the left hand side may written as

$$\sum_{i=1}^{n_\alpha}\sum_{p=1}^{n_\beta} \mathbf{D}_\alpha^*(R)_{i,i}\mathbf{D}_\beta(R)_{p,p} = \chi_\alpha^*\chi_\beta \tag{7.44}$$

since χ is the trace of a **D**-matrix. Summing now over all $\boldsymbol{R}$ operations:

$$\sum_{\boldsymbol{R}} \chi_\alpha^*(\boldsymbol{R})\chi_\beta(\boldsymbol{R}) = \sum_{i=1}^{k} g_i\chi_\alpha^*(\boldsymbol{R})_i\chi_\beta(\boldsymbol{R})_i \tag{7.45}$$

where k is the number of symmetry classes and g_i is the number of operations in the ith class. Strictly, $\chi_\alpha^*[\mathbf{D}(\boldsymbol{R})]$ terms should be used in Eq. (7.45), but the more usual notation will be followed here. The double summation on the right hand side of Eq. (7.43) is non-zero only when $p = i$ (and $n_\alpha = n_\beta$), when it becomes

$$\sum_{i=1}^{n_\alpha} 1 = n_\alpha \tag{7.46}$$

By combining results from Eqs. (7.43)–(7.46), assuming real characters.

$$\sum_{i=1}^{k} g_i \chi_\alpha(\boldsymbol{R})_i \chi_\beta(\boldsymbol{R})_i = h\delta_{\alpha,\beta} \tag{7.47}$$

treating the terms as real quantities, and the non-zero result, for $\alpha = \beta$, is

$$\sum_{i=1}^{k} g_i \left|\chi_\alpha(\boldsymbol{R})_i\right|^2 = h \tag{7.48}$$

(f) How many irreducible representations may be expected for any group? Consider the set of numbers $\sqrt{g_i}\chi_\alpha(\boldsymbol{R})$ as i ranges over k symmetry classes for a given value of α. The set forms a k-dimensional vector with components $\nu_{\alpha,i} = \sqrt{g_i}\chi_\alpha(\boldsymbol{R})_i$. If the group has r irreducible representations then there are r such vectors which, from Eq. (7.47) are orthogonal, namely, $\sum \nu^*_{\alpha,j}\nu_{\beta,j} = \boldsymbol{\nu}^*_{\alpha,j}\boldsymbol{\nu}_{\beta,j} = h\delta_{\alpha,\beta}$.

(g) There are only k orthogonal vectors in k-dimensional space (consider **i**, **j** and **k** in three dimensions) so that

$$r \le k \tag{7.49}$$

Accepting the equality in Eq. (7.41), r is exactly equal to k, that is, the number of non-equivalent irreducible representations in a point group is equal to the number of its symmetry classes. Some examples will show the use and value of the relationships developed above; C_{3v} is an example group.

Example 7.5

The character table for C_{3v} is shown here in expanded form; clearly, each entry in the normal table is the character of its **D**-matrix. The irreducible representations labelled here $\Gamma_1, \Gamma_2, \Gamma_3$, are written normally A_1, A_2 and E; all characters here are real.

C_{3v}	E	C_3	C_3^2	σ_v	σ'_v	σ''_v
Γ_1	(1)	(1)	(1)	(1)	(1)	(1)
Γ_2	(1)	(1)	(1)	(–1)	(–1)	(–1)
Γ_3	$\begin{pmatrix} 1 & 0 \\ 0 & 1 \end{pmatrix}$	$\begin{pmatrix} \overline{1/2} & \overline{\sqrt{3}/2} \\ \sqrt{3}/2 & \overline{1/2} \end{pmatrix}$	$\begin{pmatrix} \overline{1/2} & \sqrt{3}/2 \\ \overline{\sqrt{3}/2} & \overline{1/2} \end{pmatrix}$	$\begin{pmatrix} 1 & 0 \\ 0 & \overline{1} \end{pmatrix}$	$\begin{pmatrix} \overline{1/2} & \sqrt{3}/2 \\ \sqrt{3}/2 & 1/2 \end{pmatrix}$	$\begin{pmatrix} \overline{1/2} & \overline{\sqrt{3}/2} \\ \overline{\sqrt{3}/2} & 1/2 \end{pmatrix}$

If any of $\delta_{\alpha,\beta}, \delta_{i,p}$ or $\delta_{j,q} = 0$, then the right-hand side of Eq. (7.37) = 0. Two cases are considered:

1. $\alpha = \beta = 1; n_a = 1; i = p = 2; j = q = 1$ Then Eq. (7.37) reduces to Eq. (7.40), so that

$$\sum_{R} |\mathbf{D}_\alpha(\boldsymbol{R})_{i,j}|^2 = 1^2 + 1^2 + 1^2 + 1^2 + 1^2 + 1^2 = 6\,(\equiv h/n_\alpha).$$

This result exemplifies a vector in h-dimensional space. Similar results evolve for the other i, j terms, and for $\alpha = \beta = 2$.

2. $\alpha = \beta = 3; n_a = 2; i = p = 1; j = q = 2$. Then

$$\sum_R |\mathbf{D}_\alpha(R)_{i,j}|^2 = 0 + \left(\overline{\sqrt{3}/2}\right)^2 + \left(\sqrt{3}/2\right)^2 + 0 + \left(\sqrt{3}/2\right)^2 + \left(\overline{\sqrt{3}/2}\right)^2 = 3\,(\equiv h/n_\alpha).$$

Example 7.6
The usual form of the character table for C_{3v} is

C_{3v}	E	$2C_3$	$3\sigma_v$		
A_1	1	1	1	z	x^2+y^2, z^2
A_2	1	1	$\bar{1}$	R_z	
E	2	$\bar{1}$	0	(x, y) (R_x, R_y)	(x^2-y^2, xy) (xz, yz)

From the exact relationship of Eq. (7.41), there are three representations, and

$$n_{A_1}^2 + n_{A_2}^2 + n_E^2 = 6.$$

For $h = 6$, the integer values for n must be 1, 1 and 2, corresponding to the dimensionality of A_1, A_2 and E respectively. This equation holds for each column of the character table, taking into account the symmetry number for each class.

Example 7.7
Equations (7.45) and (7.47)–(7.48) refer directly to characters, the first of them being the more general. Using the irreducible representations A_1 and A_2 of C_{3v}, the right hand side of Eq. (7.45) sums to

$$(1 \times 1 \times 1) + (2 \times 1 \times 1) + [3 \times 1 \times (-1)] = 0$$

whereas for A_2 and A_2 the result is

$$(1 \times 1 \times 1) + (2 \times 1 \times 1) + [3 \times (-1) \times (-1)] = 6$$

which is the order h of the group. The same result follows from Eq. (7.48) because all characters in C_{3v} are real.

Example 7.8
The equations developed above can be used also to obtain the characters of a character table given those of others in the table, because they are not all independent. Let the total information on the characters for C_{3v} be as follows:

C_{3v}	E	$2C_3$	$3\sigma_v$
A_1	1	1	1
A_2	1	a	b
E	2	c	d

From Eq. (7.47), using A_1 and A_2 : $1 + 2a + 3b = 0$, so that the only possible answer is $a = 1$ and $b = -1$. In the $3\sigma_v$ column, $(3 \times 1^2) + [3 \times (-1)^2] + 3d^2 = 6$, so that $d = 0$. Finally, applying Eq. (7.47) to A_1 and E: $2 + 2c = 0$, so that $c = -1$, thus completing the characters of C_{3v}. Certain alternative procedures lead to the same results.

7.8 Reduction of reducible representations

A matrix representation Γ that is reducible to a set of irreducible representations can be block-factored, as exemplified by Eqs. (7.23) – (7.26). The set of blocked matrices, or *submatrices* $\mathbf{D}_i(\boldsymbol{R})$ comprises the set of irreducible representations of Γ. A given type of irreducible representation may appear among the blocked matrices once, more than once or not at all.

Consider a set of **D**-matrices $\mathbf{D}(W)$, $\mathbf{D}(X)$, $\mathbf{D}(Y)$ and $\mathbf{D}(Z)$ that exist in a block-factored matrix which, therefore, forms a reducible representation for a given group, and among which

$$\mathbf{D}(X)\mathbf{D}(Y) = \mathbf{D}(Z) \tag{7.50}$$

is assumed. Let these four matrices be replaced by another set $\mathbf{D}(W')$, $\mathbf{D}(X')$, $\mathbf{D}(Y')$ and $\mathbf{D}(Z')$, that are irreducible representations of the same group. Define a matrix **Q** such that

$$\left.\begin{aligned} \mathbf{Q}^{-1}\mathbf{D}(W)\mathbf{Q} &= \mathbf{D}(W') \\ \mathbf{Q}^{-1}\mathbf{D}(X)\mathbf{Q} &= \mathbf{D}(X') \end{aligned}\right\} \tag{7.51}$$

and so on. From Eq. (7.50) and Eq. (7.51)

$$\begin{aligned} \mathbf{D}(X')\mathbf{D}(Y') &= [\mathbf{Q}^{-1}\mathbf{D}(X)\,\mathbf{Q}]\,[\mathbf{Q}^{-1}\mathbf{D}(Y)\,\mathbf{Q}] = [\mathbf{Q}^{-1}\mathbf{D}(X)\,(\mathbf{Q}\mathbf{Q}^{-1})\,\mathbf{D}(Y)\mathbf{Q}] \\ &= [\mathbf{Q}^{-1}\mathbf{D}(X)\mathbf{D}(Y)\mathbf{Q}] = \mathbf{Q}^{-1}\mathbf{D}(Z)\mathbf{Q} = \mathbf{D}(Z') \end{aligned} \tag{7.52}$$

Thus, the two sets of matrices, related by similarity transformations, are isomorphic and their characters are equal.

The trace of the 9 × 9 matrix, Eq. (7.26), is 3, and it is evident from Eq. (7.26) that χ_{red} is the sum of the traces of its irreducible submatrices:

$$\chi_{\text{red}} = \chi_1 + \chi_2 + \chi_3 \tag{7.53}$$

and since the submatrices are equal,

$$\chi_{\text{red}} = 3\chi_1 \tag{7.54}$$

Generally, if $\chi_{\text{red}}(\boldsymbol{R})$ is the character of a reducible matrix Γ under the operation $\boldsymbol{R}$,

$$\chi_{\text{red}}(\boldsymbol{R}) = \sum_{\alpha=1}^{r} a_\alpha \chi_\alpha(\boldsymbol{R}) \tag{7.55}$$

where r is the number of irreducible representations, the same as the number of symmetry classes, $\chi_\alpha(\boldsymbol{R})$ is the character of the α irreducible representation for the operation $\boldsymbol{R}$, and a_α is the number of times that it appears in the reducible representation for the operation $\boldsymbol{R}$. From Eq. (7.45) and Eq. (7.47), assuming that all characters are real,

$$\sum_{\boldsymbol{R}} \chi_\alpha(\boldsymbol{R})\chi_\beta(\boldsymbol{R}) = h\delta_{\alpha,\beta} \tag{7.56}$$

For each operation $\boldsymbol{R}$, each side of Eq. (7.56) is multiplied by a_α and summed over α:

$$\sum_{\boldsymbol{R}} \chi_\beta(\boldsymbol{R}) \sum_{\alpha} a_\alpha \chi_\alpha(\boldsymbol{R}) = \sum_{\alpha} a_\alpha h\delta_{\alpha,\beta} \tag{7.57}$$

Substituting from Eq. (7.55), and since $\sum\limits_{\alpha} a_\alpha h\,\delta_{\alpha,\beta} = h\sum\limits_{\alpha} a_\alpha\,\delta_{\alpha,\beta}$

$$\sum_{\boldsymbol{R}} \chi_\beta(\boldsymbol{R})\chi_{\text{red}}(\boldsymbol{R}) = a_\alpha h \tag{7.58}$$

because $\sum\limits_{\alpha} a_\alpha h\,\delta_{\alpha,\beta}$ is non-zero only for $\alpha = \beta$, whence it follows that the number of times that the α irreducible representation appears in a given α irreducible representation Γ is

$$\begin{aligned} a_\alpha &= \tfrac{1}{h}\sum_{\boldsymbol{R}} \chi_\alpha(\boldsymbol{R})\chi_{\text{red}}(\boldsymbol{R}) \\ &= \tfrac{1}{h}\sum_{i=1}^{k} g_i\chi_\alpha(\boldsymbol{R})_i\chi_{\text{red}}(\boldsymbol{R})_i \end{aligned} \tag{7.59}$$

g_i is, as before, the number of operations in the i symmetry class; $\chi_{\text{red}}(\boldsymbol{R})$ is the character of the i symmetry class for the α reducible representation, and the sum is over the total of k symmetry classes.

Example 7.9

Let a representation in point group C_{3v} be

C_{3v}	$\boldsymbol{E}$	$2C_3$	$3\sigma_v$
Γ	3	0	1

Reference to the character table (see Example 7.6) shows that this representation is reducible: by inspection, $\Gamma = A_1 + E$, and confirmation arises through Eq. (7.59):

$$a_{A_1} = \frac{1}{6}\{(3\times1\times1)+(2\times1\times0)+(3\times1\times1)\} \quad = 1$$

$$a_{B_1} = \frac{1}{6}\{(3\times1\times1)+(2\times1\times0)+\,[3\times(-1)\times1)]\} = 0$$

$$a_E = \frac{1}{6}\{(3\times2\times1)+[2\times(-1)\times0]+(3\times0\times1)\} \;= 1$$

which leads to $\Gamma = A_1 + E$, as expected.

Example 7.10

In working in point group C_3, a representation may be given by the first three matrices of Eq. (7.15):

C_{3v}	E	$2C_3$	$3\sigma_v$
Γ	3	0	0

The character table for C_3 has been given in Section 7.6.4. By inspection, or from Eq. (7.59), $a_A = 1, a_\Gamma = 1$, $a_{\Gamma*} = 1$, which means that $\Gamma_{red} = A + \Gamma + \Gamma^*$. If the two components of the doubly degenerate E representation were first combined, then a_E would evaluate to 2: $a_E = \frac{1}{3}[(3 \times 2 \times 1) + 0 + 0] = 2$; thus, Γ_{red} would equal $A + 2E$. This is incorrect: the sum of the squares in the $\boldsymbol{E}$ column would equal 5, whereas $h = 3$.

7.9 Constructing a character table

Here, the properties of character tables are summarized, the character table for D_{3h} constructed on the basis of these properties, and the handling of complex characters examined.

7.9.1 Summary of the properties of character tables

1. The squares of the dimensions of the irreducible representations of a point group summed over the number r of irreducible representations is equal to the order h of the group:

$$\sum_{\alpha=1}^{r} n_\alpha^2 = h \tag{7.60}$$

2. The number of irreducible representations of a point group is equal to the number of symmetry classes:

$$r = k \tag{7.61}$$

3. The first row of A-type characters of any irreducible representation in a character table are each +1.
4. The squares of the characters of the identity matrix summed over the number of irreducible representations is equal to h:

$$\sum_{\alpha=1}^{r} |\chi_\alpha(\boldsymbol{E})|^2 = h \tag{7.62}$$

5. The rows of a character table summed over the number of symmetry classes obey the equation:

$$\sum_{i=1}^{k} g_i \chi_\alpha^*(\boldsymbol{R})_i \chi_\beta(\boldsymbol{R})_i = h\delta_{\alpha,\beta} \tag{7.63}$$

   If all characters are real, $\chi_\alpha(\boldsymbol{R})_i$ can be used.
6. The columns of a character table summed over the number of irreducible representations obey the equation:

$$\sum_{\alpha=1}^{r} \chi_\alpha^*(\boldsymbol{R})_i \chi_\beta(\boldsymbol{R})_j = \frac{h}{g_i}\delta_{i,j} \tag{7.64}$$

7. The dimension n_{red} of a reducible representation is equal to the sum of the dimensions n_a of the several irreducible representations into which it may be decomposed:

$$n_{\text{red}} = \sum_{\alpha=1}^{r} n_{\alpha} \tag{7.65}$$

7.9.2 Constructing the character table for point group D_{3h}

The character table for point group D_{3h} is constructed in this section. The group is written as

$$D_{3h}\{\boldsymbol{E}, \boldsymbol{C}_3, \boldsymbol{C}_3^2, \boldsymbol{C}'_2, \boldsymbol{C}'_{2'}, \boldsymbol{C}'_{2''}, \boldsymbol{\sigma}_h, \boldsymbol{S}_3, \boldsymbol{S}_3^2\boldsymbol{\sigma}_v, \boldsymbol{\sigma}'_v, \boldsymbol{\sigma}''_v\}$$

The designations $\boldsymbol{C}'_{2'}$ and $\boldsymbol{C}'_{2''}$ used here refer to different C'_2 operators, interrelated by the $\boldsymbol{C}_3$ operator. Evidently $h = 12$, and the operations are grouped in the symmetry classes $\boldsymbol{E}, 2\boldsymbol{C}_3, 3\boldsymbol{C}'_2, \boldsymbol{\sigma}_h, 2\boldsymbol{S}_3, 3\boldsymbol{\sigma}_v$. The structure of the table is indicated by the dotted line partitioning, and from (7.60), the only possible set of the n integers for six symmetry classes is 1, 1, 1, 1, 2, 2. Hence, a partial character table can be set up:

D_{3h}	$\boldsymbol{E}$	$2C_3$	$3C'_2$	$\boldsymbol{\sigma}_h$	$2S_3$	$3\boldsymbol{\sigma}_v$
A'_1	1	1	1	1	1	1
A'_2	1	a	b	1	a	b
E'	2	c	d	2	c	d
A''_1	1	1	1	$\bar{1}$	$\bar{1}$	$\bar{1}$
A''_2	1	a	b	$\bar{1}$	$-a$	$-b$
E''	2	c	d	$\bar{2}$	$-c$	$-d$

Applying Eq. (7.63) to the half E' row, with $\alpha = \beta = E'$, $2^2 + 2c^2 + 3d^2 = 6$ or $2c^2 + 3d^2 = 2$. The logical solution is $c = \pm 1$, $d = 0$. Applying Eq. (7.63) now to the half E' and half E'' rows together, $2 + 2c + (0\times 3d) = 0$, whence $c = -1$. From Eq. (7.64) with the entire $\boldsymbol{E}$ and $3\boldsymbol{C}'_2$ columns, $6 + 6b = 0$, so that $b = -1$.

Finally, applying Eq. (7.63) to the half A'_1 and half A'_2 rows together, $1 + 2a - 3 = 0$, whence $a = 1$. The character table can now be completed:

D_{3h}	E	$2C_3$	$3C'_2$	σ_h	$2S_3$	$3\sigma_v$
A'_1	1	1	1	1	1	1
A'_2	1	1	$\bar{1}$	1	1	$\bar{1}$
E'	2	$\bar{1}$	0	2	$\bar{1}$	0
A''_1	1	1	1	$\bar{1}$	$\bar{1}$	$\bar{1}$
A''_2	1	1	$\bar{1}$	$\bar{1}$	$\bar{1}$	1
E''	2	$\bar{1}$	0	$\bar{2}$	1	0

7.9.3 Handling complex characters

Point group S_4 is Abelian of order 4, and may be written as $S_4\{\boldsymbol{E}, \boldsymbol{S}_4, \boldsymbol{C}_2, \boldsymbol{S}_4^3\}$; $\boldsymbol{S}_4$ and $\boldsymbol{S}_4^3$ are not conjugated in this point group. The reader may wish to evaluate the similarity transformations $\boldsymbol{C}_2^{-1}\boldsymbol{S}_4\boldsymbol{C}_2$ and $\boldsymbol{C}_2^{-1}\boldsymbol{S}_4^3\boldsymbol{C}_2$, recalling that $\boldsymbol{C}_2$ is its own inverse, in order to confirm this statement.

There will be four one-dimensional irreducible representations, which need to be orthonormal. From Eq. (7.37) with $i = p$ and $j = q$, and since $n_a = n_\beta = 1$ for one-dimensional representations,

$$\sum_{j=1}^{4} \mathbf{D}_\alpha^*(\boldsymbol{R})_{i,j}\mathbf{D}_\beta(\boldsymbol{R})_{i,j} = h\delta_{\alpha,\beta} \tag{7.66}$$

and the representations $\Gamma_1 - \Gamma_4$ will obey this equation.

Let $\exp(\mathrm{i}2n\pi/4) = \varepsilon = \exp(\mathrm{i}n\pi/2)$, and the character table can be initiated by setting down the first column and then obtaining the remaining terms by appropriate multiplication. Point group S_4 is a cyclic group, so that the terms are multiples of $\mathbf{S}_4$.

Writing a first form of the character table

S_4	$\boldsymbol{S}_4$	$\boldsymbol{S}_4^2$	$\boldsymbol{S}_4^3$	$\boldsymbol{S}_4^4$
Γ_1	ε	ε^2	ε^3	ε^4
Γ_2	ε^2	ε^4	ε^6	ε^8
Γ_3	ε^3	ε^6	ε^9	ε^{12}
Γ_4	ε^4	ε^8	ε^{12}	ε^{16}

Applying Euler's theorem with de Moivre's extension, $\exp(\mathrm{i}n\pi/2) = \cos n\pi/2 + \sin n\pi/2$, the relationships between the powers can be obtained:

n	ε^n	n	ε^n
1, 9	i	3	–i
2, 6	–1	4, 8, 12, 16	1

Now the character table can be revised to

S_4	$\boldsymbol{S_4}$	$\boldsymbol{S_4^2}$	$\boldsymbol{S_4^3}$	$\boldsymbol{S_4^4}$
Γ_1	i	$\bar{1}$	$\bar{\mathrm{i}}$	1
Γ_2	$\bar{1}$	1	$\bar{1}$	1
Γ_3	$\bar{\mathrm{i}}$	$\bar{1}$	i	1
Γ_4	1	1	1	1

and is beginning to assume a recognizable form. The relationships $\boldsymbol{S_4^4} = \boldsymbol{E}$ and $\boldsymbol{S_4^2} = \boldsymbol{C_2}$ are now used, the table reordered and relabelled according to the rules described earlier, and the final columns added:

S_4	$\boldsymbol{E}$	S_4	C_2	S_4^3		
A	1	1	1	1	R_z	$z^2; x^2 + y^2$
B	1	$\bar{1}$	1	$\bar{1}$	z	$x^2 - y^2; xy$
E	$\left\{\begin{matrix}1\\1\end{matrix}\right.$	i $\bar{\mathrm{i}}$	$\bar{1}$ $\bar{1}$	$\bar{i}$ i	(x, y); (R_x, R_y)	(yz, zx)

In the two one-dimensional representations bracketed as E, the characters in the second of them are the complex conjugates of those in the first.

In order to complete the function columns of the table, a function operator (Appendix A11) is required. The p_x, p_y and p_z functions lie along the corresponding axes, and the $\boldsymbol{O_R}p$ table here lists the results:

	$\boldsymbol{O_E}$	$\boldsymbol{O_{S_4}}$	$\boldsymbol{O_{C_2}}$	$\boldsymbol{O_{S_4^3}}$
p_x	p_x	$\bar{p}_y$	$\bar{p}_x$	p_y
p_y	p_y	p_x	$\bar{p}_y$	$\bar{p}_x$
p_z	p_z	$\bar{p}_z$	p_z	$\bar{p}_z$

Thus, using now the appropriate projection operators (Appendix 11),

$$\begin{aligned} \boldsymbol{P}_A p_x &= p_x - p_y - p_x + p_y = 0 \\ \boldsymbol{P}_A p_y &= p_y + p_x - p_y - p_x = 0 \\ \boldsymbol{P}_A p_z &= p_z - p_z + p_z + p_z = 0 \end{aligned}$$

and it follows that the irreducible representation A is not spanned by x, y or z. A similar procedure with $\boldsymbol{P}_B$ shows that it is spanned by z. Finally, $\boldsymbol{P}_E$ is examined:

$$\begin{aligned} \boldsymbol{P}_E\, p_x &= p_x - p_y - p_x + p_y = 4p_x \\ \boldsymbol{P}_E\, p_y &= p_y + p_x - p_y - p_x = 4p_y \\ \boldsymbol{P}_E\, p_z &= p_z - p_z + p_z + p_z = 0 \end{aligned}$$

Thus, E is a doubly degenerate irreducible representation spanned by x and y. In this manner, the character table can be completed as shown

The majority of point groups have character tables with real characters. In some point groups, however, the two one-dimensional representations for the E characters are often added to produce real characters, as in the table above for S_4. In some applications, such as molecular orbital symmetries and selection rules, this formulation is satisfactory, but in reducing a representation by Eq. (7.59), an erroneous result can be obtained.

In the dihydrogenphosphate ion, $[H_2PO_4]^-$, a representation on the P–O vectors as a basis is

S_4	S_4	S_4^2	S_4^3	S_4^4
Γ_{red}	4	0	0	0

which reduces to $A + B + 2E$, which is erroneous as it does not obey Eqs. (7.60) – (7.65). In this case, the E irreducible representation must be set out fully, as in the S_4 character table above, where $\sum_R n_E^2 = 4$, which is h, for both parts of E. The correct reduction is obtained by taking E and E^* separately; then the irreducible representations are $A + B + \Gamma + \Gamma^*$, where Γ and Γ^* are the components of E before they are combined. Now, $\sum_R n^2_{\Gamma(\text{or } \Gamma^*)} = 4$, and the other requirements of the theory are also satisfied.

7.10 Direct products

Direct products have figured implicitly in the derivation and discussion of point groups in earlier chapters. For example, the combination of groups C_s and C_i, which was used in Section 3.12, given there in the Hermann–Mauguin form, leads immediately to $C2_h$ $(2/m)$. The direct product can also be used to develop character tables and point group representations.

7.10.1 Representations on direct product functions

The direct product of two matrices, indicated by the symbol $\otimes$, is very different from ordinary matrix multiplication. For example, given a 2×2 matrix **A** and a 3×3 matrix **B**, their direct product $\mathbf{A} \otimes \mathbf{B}$ is formed according to the equation

$$\mathbf{A} \otimes \mathbf{B} = \begin{pmatrix} a_{11} & a_{12} \\ a_{21} & a_{22} \end{pmatrix} \otimes \mathbf{B} \begin{pmatrix} a_{11}\mathbf{B} & \vdots & a_{12}\mathbf{B} \\ \cdots & \vdots & \cdots \\ a_{21}\mathbf{B} & \vdots & a_{22}\mathbf{B} \end{pmatrix} \tag{7.67}$$

that is, every element of each of the two matrices is individually and separately multiplied together.

This result can be used to generate character tables: in general, the elements of **A** and **B** are $a_{i,j}$ and $b_{k,l}$, respectively; then the general element of the product matrix **C** is $c_{i,j,k,l}$, but it can be allotted its own two dimensional label, if required. *In extenso*, Eq. (7.67) becomes

$$\mathbf{A} \otimes \mathbf{B} = \begin{pmatrix} a_{11} & a_{12} \\ a_{21} & a_{22} \end{pmatrix} \otimes \mathbf{B} \begin{pmatrix} a_{11}b_{11} & a_{11}b_{12} & a_{11}b_{13} & \vdots & a_{12}b_{11} & a_{12}b_{12} & a_{12}b_{13} \\ a_{11}b_{21} & a_{11}b_{22} & a_{11}b_{23} & \vdots & a_{12}b_{21} & a_{12}b_{22} & a_{12}b_{23} \\ a_{11}b_{31} & a_{11}b_{32} & a_{11}b_{33} & \vdots & a_{12}b_{31} & a_{11}b_{32} & a_{11}b_{33} \\ \cdots & \cdots & \cdots & \cdots & \cdots & \cdots & \cdots \\ a_{21}b_{11} & a_{21}b_{12} & a_{21}b_{13} & \vdots & a_{22}b_{11} & a_{22}b_{12} & a_{22}b_{13} \\ a_{21}b_{21} & a_{21}b_{22} & a_{21}b_{23} & \vdots & a_{22}b_{21} & a_{22}b_{22} & a_{22}b_{23} \\ a_{21}b_{31} & a_{21}b_{32} & a_{21}b_{33} & \vdots & a_{22}b_{31} & a_{22}b_{32} & a_{22}b_{33} \end{pmatrix} \tag{7.68}$$

A function can be formed by the direct product of irreducible representations, a process which is illustrated with the character table for point group T_d:

T_d	E	$8C_3$	$3C_2$	$6S_4$	$6\sigma_d$
A_1	1	1	1	1	1
A_2	1	1	1	−1	−1
E	2	−1	2	0	0
T_1	3	0	−1	1	−1
T_1	3	0	−1	−1	1

The product of any two of these representations requires the direct product of the characters for each symmetry class to be formed in turn and, in general, is itself not necessarily an irreducible representation. Thus,

$A_1 \otimes A_2$	1	1	1	-1	-1	$(= A_2)$
$A_2 \otimes E_2$	2	-1	2	0	0	$(= E)$
$E \otimes T_1$	6	0	-2	0	0	$(= T_1 + T_2)$
$T_1 \otimes T_2$	9	0	1	-1	-1	$(= A_2 + E + T_1 + T_2)$

These product functions are also representations of the given point group. The dimension of a product function is the product of the dimensions of the two functions forming the product. Thus, the dimension of $E \otimes T_1$ is $n_E n_{T_1}$, or 6; it is, of course, the character of $E \otimes T_1$ under the $\boldsymbol{E}$ symmetry class. The reducible representations of each of these direct product representations are shown above in parentheses. Applications of direct products and their ramifications to chemical bonding have been well covered in the literature [6–9], and two examples will be discussed shortly.

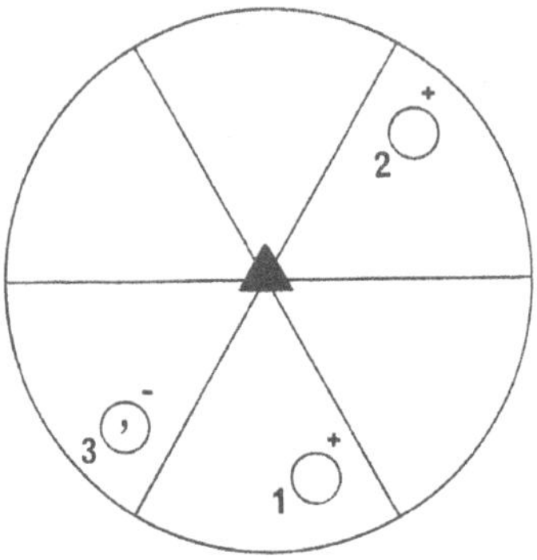

Fig. 7.8
Partial stereogram to show the operations C_3, C_i and $S_6^5(\equiv S_6^{-1})$, and the equivalence $C_iC_3 = S_6^5$: $(1)\xrightarrow{C_3}(2)\xrightarrow{C_i}(3)$; $(1)\xrightarrow{S_6^5}(3)$.

7.10.2 Formation of a character table by direct products

As an example of the application of Eq. (7.68) to the formation of a character table, consider the direct product of point groups C_i and C_{3v}. The symmetry operations expected from the products can be developed, for example, by considering the partial stereogram (Fig. 7.8), from which it is evident that $\boldsymbol{i}\boldsymbol{C}_3 = \boldsymbol{S}_6^5$. What is $\boldsymbol{R}$ in $\boldsymbol{i}\boldsymbol{R} = \boldsymbol{S}_6$?

The direct product $C_i \otimes C_{3v}$ will contain overall 12 columns and six rows, plus a title row and column. The component character tables generate the product table, and the complete table shown below is the expanded character table. The symbol C_2' is used for a twofold operation of a group that does not act along the principal axis direction, namely, the highest proper rotation direction and z reference axis; there are three such axes. For compactness, as before, a negative character has its minus sign as an overbar, as in $\overline{\sqrt{3}/2}$.

C_i	$\boldsymbol{E}$	$\boldsymbol{i}$
Γ_1	(1)	(1)
Γ_2	(1)	($\bar{1}$)

$\otimes$

C_{3v}	$\boldsymbol{E}$	$\boldsymbol{C_3}$	$\boldsymbol{C_3^2}$	$\boldsymbol{\sigma_v}$	$\boldsymbol{\sigma_v'}$	$\boldsymbol{\sigma_v''}$
Γ_1	(1)	(1)	(1)	(1)	(1)	(1)
Γ_2	(1)	(1)	(1)	($\bar{1}$)	($\bar{1}$)	($\bar{1}$)
Γ_3	$\begin{pmatrix} 1 & 1 \\ 0 & 1 \end{pmatrix}$	$\begin{pmatrix} \overline{1/2} & \overline{\sqrt{3}/2} \\ \sqrt{3}/2 & \overline{1/2} \end{pmatrix}$	$\begin{pmatrix} \overline{1/2} & \sqrt{3}/2 \\ \overline{\sqrt{3}/2} & \overline{1/2} \end{pmatrix}$	$\begin{pmatrix} 1 & 0 \\ 0 & \bar{1} \end{pmatrix}$	$\begin{pmatrix} \overline{1/2} & \sqrt{3}/2 \\ \sqrt{3}/2 & 1/2 \end{pmatrix}$	$\begin{pmatrix} \overline{1/2} & \overline{\sqrt{3}/2} \\ \overline{\sqrt{3}/2} & 1/2 \end{pmatrix}$

=

D_{3d}	$\boldsymbol{E}$	$\boldsymbol{C_3}$	$\boldsymbol{C_3^2}$	$\boldsymbol{C_2'}$	$\boldsymbol{C_{2'}'}$	$\boldsymbol{C_{2''}'}$	$\boldsymbol{i}$	$\boldsymbol{S_6^2}$	$\boldsymbol{S_6}$	$\boldsymbol{\sigma_v}$	$\boldsymbol{\sigma_v'}$	$\boldsymbol{\sigma_v''}$
A_{1g}	(1)	(1)	(1)	(1)	(1)	(1)	(1)	(1)	(1)	(1)	(1)	(1)
A_{2g}	(1)	(1)	(1)	($\bar{1}$)	($\bar{1}$)	($\bar{1}$)	(1)	(1)	(1)	($\bar{1}$)	($\bar{1}$)	($\bar{1}$)
E_g	$\begin{pmatrix} 1 & 0 \\ 0 & 1 \end{pmatrix}$	$\begin{pmatrix} \overline{1/2} & \overline{\sqrt{3}/2} \\ \sqrt{3}/2 & \overline{1/2} \end{pmatrix}$	$\begin{pmatrix} \overline{1/2} & \sqrt{3}/2 \\ \overline{\sqrt{3}/2} & \overline{1/2} \end{pmatrix}$	$\begin{pmatrix} 1 & 0 \\ 0 & \bar{1} \end{pmatrix}$	$\begin{pmatrix} \overline{1/2} & \sqrt{3}/2 \\ \sqrt{3}/2 & 1/2 \end{pmatrix}$	$\begin{pmatrix} \overline{1/2} & \overline{\sqrt{3}/2} \\ \overline{\sqrt{3}/2} & 1/2 \end{pmatrix}$	$\begin{pmatrix} 1 & 0 \\ 0 & 1 \end{pmatrix}$	$\begin{pmatrix} \overline{1/2} & \overline{\sqrt{3}/2} \\ \sqrt{3}/2 & \overline{1/2} \end{pmatrix}$	$\begin{pmatrix} \overline{1/2} & \sqrt{3}/2 \\ \overline{\sqrt{3}/2} & 1/2 \end{pmatrix}$	$\begin{pmatrix} 1 & 0 \\ 0 & \bar{1} \end{pmatrix}$	$\begin{pmatrix} \overline{1/2} & \sqrt{3}/2 \\ \sqrt{3}/2 & 1/2 \end{pmatrix}$	$\begin{pmatrix} \overline{1/2} & \overline{\sqrt{3}/2} \\ \overline{\sqrt{3}/2} & 1/2 \end{pmatrix}$
A_{1u}	(1)	(1)	(1)	(1)	(1)	(1)	($\bar{1}$)	($\bar{1}$)	($\bar{1}$)	($\bar{1}$)	($\bar{1}$)	($\bar{1}$)
A_{2u}	(1)	(1)	(1)	($\bar{1}$)	($\bar{1}$)	($\bar{1}$)	($\bar{1}$)	($\bar{1}$)	($\bar{1}$)	(1)	(1)	(1)
E_u	$\begin{pmatrix} 1 & 0 \\ 0 & 1 \end{pmatrix}$	$\begin{pmatrix} \overline{1/2} & \overline{\sqrt{3}/2} \\ \sqrt{3}/2 & \overline{1/2} \end{pmatrix}$	$\begin{pmatrix} \overline{1/2} & \sqrt{3}/2 \\ \overline{\sqrt{3}/2} & \overline{1/2} \end{pmatrix}$	$\begin{pmatrix} 1 & 0 \\ 0 & \bar{1} \end{pmatrix}$	$\begin{pmatrix} \overline{1/2} & \sqrt{3}/2 \\ \sqrt{3}/2 & 1/2 \end{pmatrix}$	$\begin{pmatrix} \overline{1/2} & \overline{\sqrt{3}/2} \\ \overline{\sqrt{3}/2} & 1/2 \end{pmatrix}$	$\begin{pmatrix} \bar{1} & 0 \\ 0 & \bar{1} \end{pmatrix}$	$\begin{pmatrix} 1/2 & \sqrt{3}/2 \\ \overline{\sqrt{3}/2} & 1/2 \end{pmatrix}$	$\begin{pmatrix} 1/2 & \overline{\sqrt{3}/2} \\ \sqrt{3}/2 & \overline{1/2} \end{pmatrix}$	$\begin{pmatrix} \bar{1} & 0 \\ 0 & 1 \end{pmatrix}$	$\begin{pmatrix} 1/2 & \overline{\sqrt{3}/2} \\ \overline{\sqrt{3}/2} & \overline{1/2} \end{pmatrix}$	$\begin{pmatrix} 1/2 & \sqrt{3}/2 \\ \sqrt{3}/2 & \overline{1/2} \end{pmatrix}$

The three C_2' operators are all of the same class.

The irreducible representations have been given the g or u subscript according as the characters of the matrices ($\chi_{\boldsymbol{R}}$) under $\boldsymbol{i}$ are $+1$ or -1, respectively. Finally, the symmetry operations are collected together in their classes and the traces of each matrix taken to form the characters of the conventional character table for point group D_{3d}, as shown below. The interpretation of the data in the final two columns will be discussed in Section 8.2.7 and Section 8.3.

D_{3d}	E	$2C_3$	$3C_2'$	i	$2S_6$	$3\sigma_d$	**Selection rules**	
A_{1g}	1	1	1	1	1	1		x^2+y^2, z^2
A_{2g}	1	1	–1	1	1	–1	R_z	
E_g	2	–1	0	2	–1	0	(R_x, R_y)	(x^2-y^2, xy) (xz, yz)
A_{1u}	1	1	1	–1	–1	–1		
A_{2u}	1	1	–1	–1	–1	1	z	
E_u	2	–1	0	–2	1	0	(x, y)	

The apparent anomaly $3\sigma_v$ (in C_{3v}) $\rightarrow$ $3\sigma_d$ (in D_{3d}) is simply a change of notation because the mirror symmetry planes now lie between the twofold axes.

7.10.3 How the direct product has been used

- The direct product of two point groups was used to derive another point group, as in the example: $C_{2v} \otimes C_i = D_{2h}$.
- Representations of point groups for product functions can be derived: if the wave functions ψ_1, ψ_2 and ψ_3 span A_2, E and E in point group C_{3v}, for example, then the product $\psi_1\,\psi_2\,\psi_3$ spans a representation given by the direct product $A_2 \otimes E \otimes E$, which may be evaluated as follows:

C_{3v}	E	$2C_3$	$3\sigma_v$
A_2	1	1	-1
E	2	-1	0
$A_2 \otimes E \otimes E$	4	1	0

This process is a multiplication of the characters of the matrices. If the multiplication of matrices be set out in full, say, for the C_3 symmetry class in C_{3v}, then

$$(1) \otimes \begin{pmatrix} \overline{\frac{1}{2}} & \overline{\frac{\sqrt{3}}{2}} \\ \frac{\sqrt{3}}{2} & \overline{\frac{1}{2}} \end{pmatrix} \otimes \begin{pmatrix} \overline{\frac{1}{2}} & \overline{\frac{\sqrt{3}}{2}} \\ \frac{\sqrt{3}}{2} & \overline{\frac{1}{2}} \end{pmatrix} = \begin{pmatrix} \frac{1}{4} & \frac{\sqrt{3}}{4} & \frac{\sqrt{3}}{4} & \frac{3}{4} \\ \overline{\frac{\sqrt{3}}{4}} & \frac{1}{4} & \overline{\frac{3}{4}} & \frac{\sqrt{3}}{4} \\ \overline{\frac{\sqrt{3}}{4}} & \overline{\frac{3}{4}} & \frac{1}{4} & \frac{\sqrt{3}}{4} \\ \frac{3}{4} & \overline{\frac{\sqrt{3}}{4}} & \overline{\frac{\sqrt{3}}{4}} & \frac{1}{4} \end{pmatrix} \qquad (7.69)$$

and the trace of the matrix is again 1.

- A character table can be formed also by the direct product of irreducible representations. Consider this process for the two component tables just used to form D_{3d}: Following Eq. (7.68),

C_i	E	i
A_g	1	1
A_u	1	−1

$\otimes$

C_{3v}	E	$2C_3$	C_3^2
A_1	1	1	1
A_2	1	1	−1
E	2	−1	0

$=$

$C_i \otimes C_{3v} = D_{3d}$	E	$2C_3$	$3C_2'$	i	$2S_6$	$3\sigma_d$
$A_1 \otimes A_g = A_{1g}$	1	1	1	1	1	1
$A_2 \otimes A_g = A_{2g}$	1	1	$\bar{1}$	1	1	$\bar{1}$
$E \otimes A_g = E_g$	2	$\bar{1}$	0	2	$\bar{1}$	0
$A_1 \otimes A_u = A_{1u}$	1	1	1	$\bar{1}$	$\bar{1}$	$\bar{1}$
$A_2 \otimes A_u = A_{2u}$	1	1	$\bar{1}$	$\bar{1}$	$\bar{1}$	1
$E \otimes A_u = E_u$	2	$\bar{1}$	0	$\bar{2}$	1	0

References 7

[1] Hahn T. (Ed.) *International tables for crystallography*, Vol. A. 5th ed. IUCr/Wiley, 2011.

[2] McCampbell TA, et al. *J. Chem. Crystallogr*. 2006; 36: 271.

[3] Pater W. *Appreciations*. Macmillan & Co, 1889.

[4] McWeeny R, *Symmetry*. Pergamon Press, 1963.

[5] Mulliken RS. *J. Chem. Phys*.1955; 23: 1977.

[6] Kettle SFA. *Symmetry and structure*. John Wiley & Sons, 1995.

[7] Bishop D. *Group Theory and Chemistry*, Clarendon Press, 1973.

[8] Cotton FA. *Chemical applications of group theory*. 3rd ed. John Wiley & Sons, 1990.

[9] Dresselhaus MS, et al. *Group theory: applications to the physics of condensed matter*. Springer, 2008.

Problems 7

7.1 Which pairs of these operations $\boldsymbol{C}_2$, $\boldsymbol{C}_3$, $\boldsymbol{\sigma}$, $\boldsymbol{E}$ commute?

7.2 Construct group multiplication tables for C_{2h} and D_{2d}. Is either of these groups Abelian?

7.3 Construct a group multiplication table for the cyclic group $G\{\boldsymbol{E}, \boldsymbol{A}, \boldsymbol{B}, \boldsymbol{C}\}$ of order 4, using the relations $\boldsymbol{A} = \boldsymbol{r}$, $\boldsymbol{B} = \boldsymbol{r}^2$, $\boldsymbol{C} = \boldsymbol{r}^3$. Which operation other than $\boldsymbol{E}$ is its own inverse?

7.4 Figure P7.1 shows a stereoview of the dithionate ion, $[S_2O_6]^{2-}$, as seen along the S–S bond. (a) Determine the point group of this species. (b) Construct a group multiplication table for this point group. (c) What are the subgroups of the point group? (d) Which, if any, of the subgroups is invariant to the point group of the species?

7.5 From a consideration of group multiplication tables, show that point group S_4 cannot be obtained from any combinations of point group operations, whereas S_6 is equivalent to a certain combination. Then, determine this combination. Are any subgroups of S_4 and S_6 invariant?

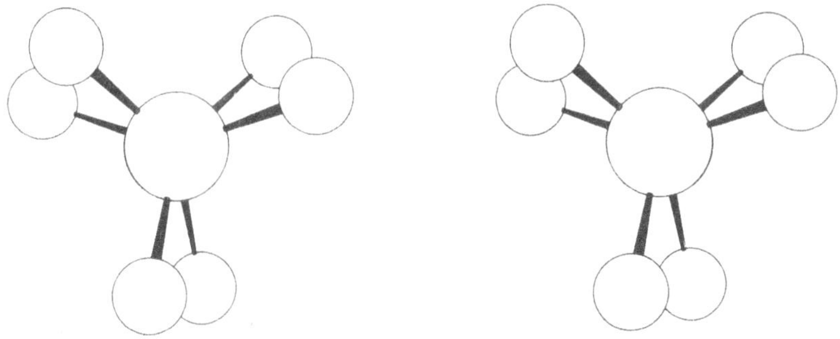

Fig. P7.1

7.6 Arrange the symmetry operators for the point groups of (a) the tetrachloroplatinate(II) ion, $[PtCl_4]^{2-}$, and (b) Z-1, 2-dichloroethene, CH_2Cl_2, in their symmetry classes; state the order of each group.

7.7 Prove that if a group member $\boldsymbol{B}$ is conjugate to $\boldsymbol{A}$, and $\boldsymbol{C}$ is conjugate to $\boldsymbol{B}$, then $\boldsymbol{C}$ is conjugate to $\boldsymbol{A}$.

7.8 Given that $H\{\boldsymbol{E}, \boldsymbol{C}_3, \boldsymbol{C}_3^2\}$ and $H'\{\boldsymbol{E}, \boldsymbol{\sigma}_v\}$ are subgroups of C_{3v}. To what extent are right and left cosets with H and with H' equal, and do they form the original point group with H or H'? Are any cosets invariant with respect to C_{3v}, and are there any other subgroups of C_{3v}? If so, list them.

7.9 What symmetry element or elements arise from the following combinations? (a) C_2 and σ_h; (b) S_4 and σ_v(S_4 lying in σ_v); (c) C_3 and C_i; (d) S_6 and $C_2(\perp S_6)$. In each case, what is the point group symbol in the Schönflies and Hermann–Mauguin notations?

7.10 Given matrix $\mathbf{A}$, identify matrix $\mathbf{B}$ in each of the following; $\varepsilon = \exp(i\alpha)$, where α is a constant.

$$\text{(a)}\quad \mathbf{A} = \begin{pmatrix} 1 & 0 & 1 \\ \bar{1} & 2 & 1 \\ 0 & 1 & 3 \end{pmatrix} \qquad \mathbf{B} = \begin{pmatrix} 1 & \bar{1} & 0 \\ 0 & 2 & 1 \\ 1 & 1 & 3 \end{pmatrix}$$

$$\text{(b)}\quad \mathbf{A} = \begin{pmatrix} 0 & 1 & \epsilon^* \\ \bar{1} & 3 & \epsilon \\ \epsilon & 2 & 3 \end{pmatrix} \qquad \mathbf{B} = \begin{pmatrix} 0 & 1 & \epsilon \\ \bar{1} & 3 & \epsilon^* \\ \epsilon^* & 2 & 3 \end{pmatrix}$$

$$\text{(c)}\quad \mathbf{A} = \begin{pmatrix} 1 & 0 & \bar{1} \\ 2 & 1 & 0 \\ \bar{1} & 1 & 2 \end{pmatrix} \qquad \mathbf{B} = \begin{pmatrix} 2 & \bar{4} & 3 \\ \bar{1} & 1 & \bar{1} \\ 1 & \bar{2} & 1 \end{pmatrix}$$

$$\text{(d)}\quad \mathbf{A} = \begin{pmatrix} 0 & \epsilon & 0 \\ \epsilon^* & 0 & 0 \\ 0 & \bar{1} & 1 \end{pmatrix} \qquad \mathbf{B} = \begin{pmatrix} 0 & \epsilon & 0 \\ \epsilon^* & 0 & \bar{1} \\ 0 & 0 & 1 \end{pmatrix}$$

What more can be said about the results for (c)?

7.11 Construct $\mathbf{D}$-matrices for the operations $\boldsymbol{C}_6(z)$ and $\boldsymbol{C}_2'(y)$ using basis vectors $\mathbf{i}$, $\mathbf{j}$ and $\mathbf{k}$ from the point group origin. Determine the nature and orientation of combinations formed by (a) $\mathbf{D}(\boldsymbol{C}_6)D(\boldsymbol{C}_2')$, and (b) $\mathbf{D}(\boldsymbol{C}_2')D(\boldsymbol{C}_6)$. [Appendix A3.7.1 may be helpful.]

7.12 The **D**-matrix below can be constructed in more than one way. Give two of them.

$$\mathbf{D}(\mathbf{C}_3^2) = \begin{pmatrix} \overline{1/2} & \sqrt{3}/2 & 0 \\ \overline{\sqrt{3}/2} & \overline{1/2} & 1 \\ 0 & 0 & 1 \end{pmatrix}$$

7.13 Use the 9 × 9 matrices in Eqs. (7.23) – (7.26) to obtain a reducible representation for the nitrogen dioxide molecule, and then reduce it.

7.14 How might the irreducible representations found for the previous problem be divided among the possible movements or displacements of the nitrogen dioxide molecule?

7.15 Given the following representations, find the irreducible representations in each case, using the appropriate character tables (a program is available for D_{6h}):

(a) C_{2v} :	3	−1	−3	1							
(b) C_{3v} :	5	2	−1								
(c) C_{3h} :	3	0	−1	−3	0	1					
(d) C_{4v} :	5	1	1	3	1						
(e) D_{6h} :	6	0	0	0	−2	0	0	0	−6	2	0
(f) O_h;	9	0	1	−1	3	3	0	3	1	1	

7.16 Show that the irreducible representations in C_3 are orthonormal.

7.17 Form the direct product representation $E_1 \otimes T_1 \otimes T_2$ in point group T_d, and find the irreducible representations of which it is composed.

7.18 Complete the following partial character table:

D_{3d}	E	$2C_3$	$3C_2'$	i	$2S_3$	$3\sigma_v$
A_{1g}	1	1	1	1	1	1
A_{2g}	1	1	−1	1	1	−1
E_g						
A_{1u}						
A_{2u}						
E_u						

7.19 Using the appropriate character tables, evaluate the direct product $C_i \otimes C_3$.

7.20 Show that Eq. (7.42) for the case that $\alpha = \beta$ may be written as: $\sum_R \chi^*(\boldsymbol{R})_i \chi(\boldsymbol{R})_p = \frac{h}{n}\delta_{i,p}$. What is the result of applying this equation in the following cases: (a) $C_{6v}, i = p = B_2$. (b) $C_{4h}, i = p = E_g$. (c) $D_{3h}, i = E'', j = A_2$?

7.21 What is the resultant of $\sum_R \chi_\alpha(\boldsymbol{R})\chi_\beta(\boldsymbol{R})$ in each of the following situations? (a) $C_{4v}: \alpha = \beta = B_1$. (b) D_{4v}; $\alpha = A_{2g}$, $\beta = A_{2u}$. (c) $D_{3h}: \alpha = \beta = E'$. (d) $T_d: \alpha = \beta = T_2$.

7.22 Justify the entries in last two columns of the character table for C_{2h}.

Applications of group theory

8

SYNOPSIS

- Structure and symmetry from models, experiment and theory
- Monte Carlo and molecular dynamics
- Symmetry adapted molecular orbitals
- Projection operators
- Crystal-field and ligand-field theory
- Vibrations of molecules: infrared and Raman activity
- Group theory derivation of point groups and space groups
- Factor group and site group analyses

8.1 Introduction

It will become evident, if not so already, that the descriptive, semi-analytical derivations used in the earlier chapters were implicitly applying some aspects of group theory to the matters under discussion. In this chapter, examples with both molecules and crystals will be examined in order to highlight applications of group theory to the symmetry and geometry of chemical species, making use of the theory developed in the previous chapter.

The many applications of group theory to chemical species include setting up wave functions that describe bonding in molecules and crystals, discussing the vibrations in molecular and crystalline materials, which is an important topic in the analysis of infrared and Raman spectra, and deriving point groups and space groups in a theoretical manner, to name but three. Without such information the observed symmetry and geometry of these species could not be explained.

'*But his cup of joy is full when the results of studies immediately find practical application*' [1]. In methane, as with many other substances, the inner electrons, $(1s)^2$ in this species, form a relatively inert atomic core, and bonding takes place here through the $(2s)^2$ and $(2p)^2$ electrons. The four C–H bonds in methane can be used as a basis set of functions for describing a representation in T_d, the point group of methane. By reducing this representation and comparing the result with the irreducible representations of T_d, a particular choice of atomic orbitals that lead to the symmetry of methane can be selected.

In boron trifluoride, more than one combination of atomic orbitals could produce its observed trigonal symmetry, and other chemical evidence must be brought to bear on the problem in order to choose the most suitable of the results.

In applying group theory to crystals, point groups and space groups can be derived in an analytical manner, as were their original deductions.

8.2 Structure and symmetry in molecules and ions

This section examines the symmetry and geometry of some chemical species, and the procedures involved in elucidating structure by models and by experimental and theoretical techniques, illustrated by straightforward examples; not all methods are necessarily applicable to all chosen examples.

8.2.1 Application of models

The literature contains numerous examples of the elucidation of structure by the help of models. In the early work on the X-ray analysis of crystal structures much attention was focussed on silicate minerals, and the structures of many types were determined in detail. In common with them all is an [SiO_4] tetrahedral 'structural unit'. Silicate structures exist in which the [SiO_4] unit appears as a discrete entity, as in $MgSiO_4$, for example, with the Si:O ratio of 1:4. Other structures evolve through differing methods of sharing of the oxygen atoms at the apices of the [SiO_4] unit: single chains and rings are formed where Si:O = 1:3, double chains with a ratio Si:O = 4:11, sheet structures where Si:O = 2:5, and silica itself, as in quartz with the ratio Si:O = 1:2; these modes of linking of [SiO_4] tetrahedra are illustrated in Fig. 8.1. The continued sharing of oxygen

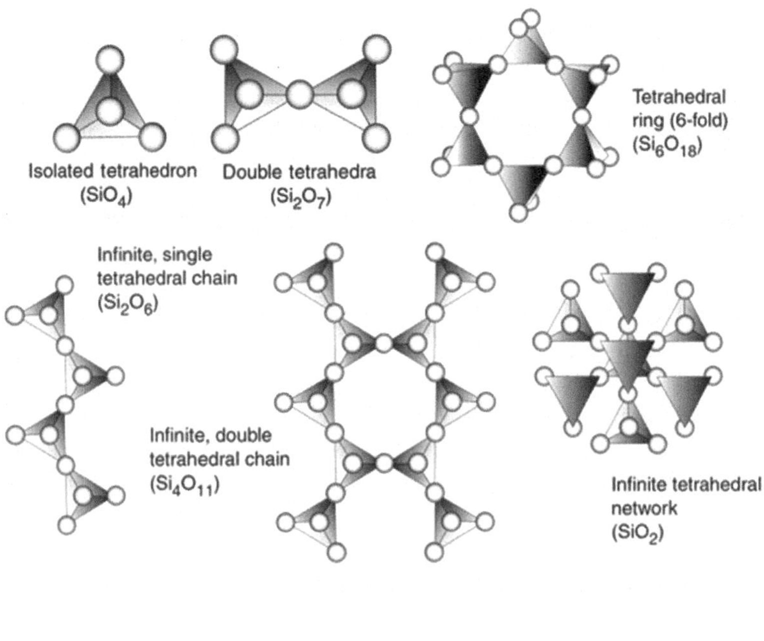

Fig. 8.1
The formation of silicate structures by sharing tetrahedral [SiO_4] structural units; the Si:O ratio is 1:4. As the extent of sharing increases, so the Si:O ratio decreases, ultimately to the 1:2 value in silica. The formation of a sheet silicate structure may be visualized by a side-to-side sharing of double chains to give Si:O = 2:5. [Reproduced by courtesy of Dr Phil Stoffer.]

atoms as side-to-side double chains gives the sheets structures with Si:O = 2:5; the ratio Si:O = 1:2 for silica itself, in the mineral quartz and its high temperature forms, involves a sharing of all oxygen atoms (Fig. 2.43).

It was from a disordered form of quartz that Bernal developed a model for water and other structures in the liquid state. In a practical example of the simulation of the structure of a monatomic liquid[1] he revealed its random nature of packing: a hard sphere model was obtained by packing identical, steel balls into an irregular shaped rubber ball. If the centres of the balls in the model are joined by straight lines which are then bisected perpendicularly by planes, the resulting closed figures are Voronoi polyhedra which fill the enclosing space completely (Fig. 8.2).

[1] 'monatomic liquid' implies the liquid state of a monatomic substance, such as argon.

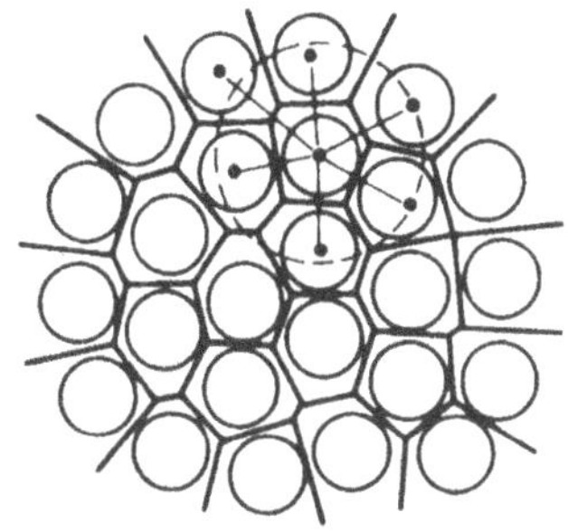

Fig. 8.2
Voronoi polyhedra in projection (also known as Dirichlet polygons) around randomly packed atoms in a monatomic liquid; the average coordination number is 5.5.

Voronoi polyhedra can be simulated if Plasticine balls are employed instead of steel in the aforementioned experiment. Individual balls from this practical simulation are found to have the shapes of Voronoi polyhedra, with average numbers of faces and coordination number of 5 and 13, respectively; the corresponding theoretically determined numbers are 5.1 and 13.6. The closest regular packing of equal spheres has a coordination number of 12, with an occupied volume per sphere of 1.351, whereas the close packing of irregular polyhedra has an occupied volume of approximately 1.53 per sphere, an increase of 13%. A model due to Bernal contrasts a randomly packed liquid, or 'heap', with the regularly packed array, or 'pile' (Fig. 8.3).

From his work with models, Bernal envisaged the water molecule as a regular tetrahedron with two charges of $+\frac{1}{2}$ and two of $-\frac{1}{2}$ at the four apices with an occupied volume of a sphere of radius approximately 1.38 Å, a value very close to the van der Waals radius of the oxygen atom. The positive apices link through hydrogen bonding to the oxygen atoms of neighbouring water molecules, and the negative apices are the two lone pairs similarly bonded, but to hydrogen atoms of neighbouring molecules. This model is still true in its essentials, and is useful in considering the role of water in ice and in hydrate structures.

8.2.2 Application of diffraction studies

Water has been examined in the liquid and solid states by X-ray, neutron and electron diffraction techniques. In the liquid state, a radial distribution curve displays evidence of nearest neighbour distances and of the average number

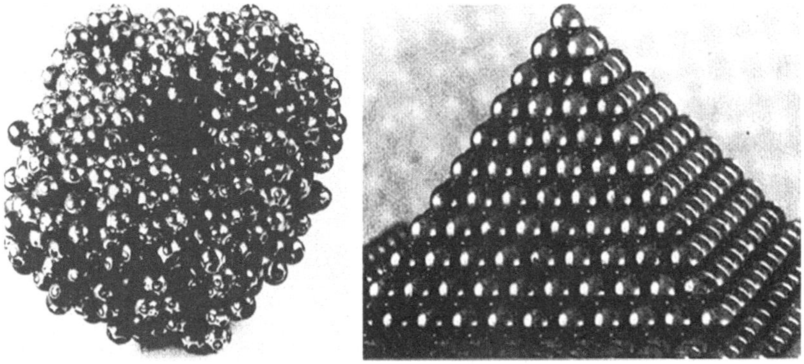

(a) (b)

Fig. 8.3
Bernal's stacking of identical spheres. (a) Random, close packed *heap* of average coordination number *ca.* 13. (b) Close packed *pile* of identical spheres of coordination number 12.

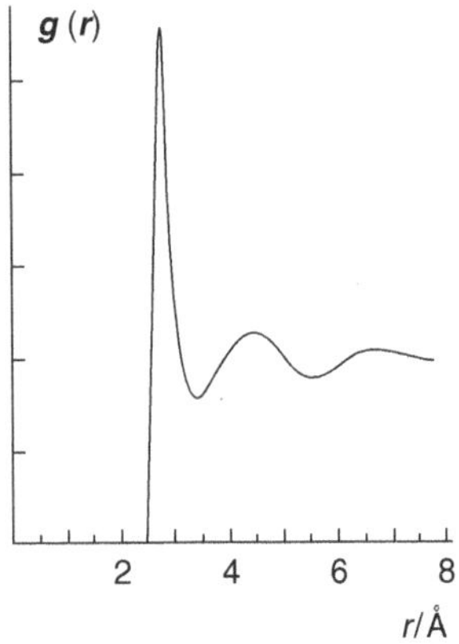

Fig. 8.4
Radial distribution curve for water, determined by X-ray diffraction. The strong peak at *ca.* 2.8 Å corresponds to the hydrogen-bonded O–H · · · H interaction.

of molecules around any given molecule of the liquid. Figure 8.4 shows an example of an X-ray radial distribution curve for water. The strong peak at approximately 2.8 Å represents the hydrogen-bonded H–O · · · H distance; the average coordination number found for water was 4.4.

In the solid state, sharp diffraction X-ray patterns are obtained from crystals of ice, from which the positions of the atoms in the unit have been determined. Similar results, but with greater precision, have been obtained by neutron diffraction, since the scattering of neutrons by hydrogen is much greater than that with X-rays. Figure 8.5a shows a stereoview of the structure of ice at 90 K. A tetrahedral disposition of bonds exists around each oxygen atom but, in any one molecule only two of the four tetrahedral directions carry hydrogen atoms (Fig. 8.5b). This structure is hexagonal, space group $P\frac{6_3}{m}mc$, with four molecules per unit cell. Below 95 K, ice transforms to a cubic structure with space group $Pn\bar{3}n$.

Boron trifluoride has been examined experimentally in both liquid and solid states. The trigonal-planar structure indicated by Fig. 7.4 is retained in these two states, but with some pairwise molecular association: the B–F distance found was 1.32 Å, which compares favourably with the sum of the covalent radii, but the intermolecular F · · · F distance found was 2.29Å is significantly less than the sum of the van der Waals radii, *ca.* 2.8 Å, which implies a degree of association between molecules.

The well-studied structure of benzene was first analysed as its hexamethyl derivative by Lonsdale in 1929, and found to be a planar hexagonal molecule within the limits experimental error; the C–C distance was given as 1.42 ± 0.03 Å. The position of the molecule in the triclinic unit cell accorded no higher

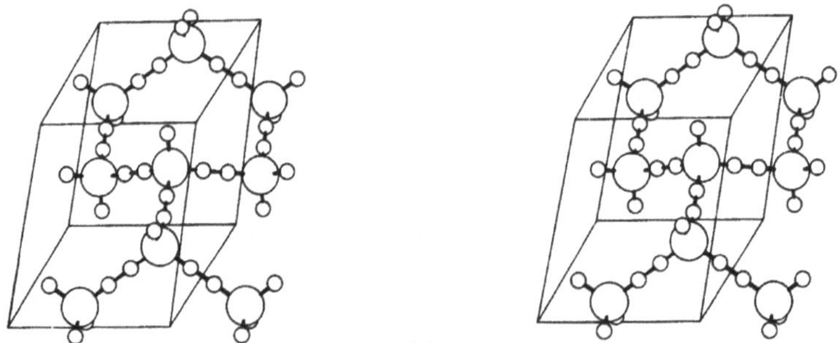
(a)

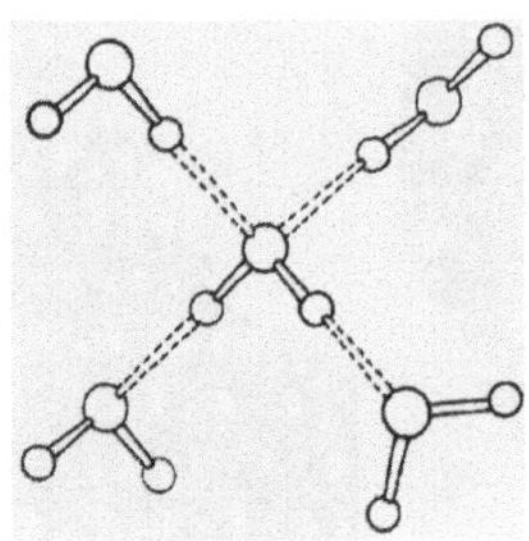
(b)

Fig. 8.5
(a) Stereoview of the unit cell and environs of the hydrogen bonded structure of ice at 90 K. The circles represent, in decreasing order of size, oxygen and statistical half-hydrogen atoms. (b) A tetrahedral array of bonds emanate from the oxygen atom, but from any one oxygen atom only two of these directions carry hydrogen atoms at any instant. [Reproduced by courtesy of Woodhead Publishing, UK.]

molecular site symmetry than $\bar{1}$. A more recent and precise determination by neutron diffraction on deuterated benzene found an average aromatic C–C bond length of 1.3972 ± 0.0007 Å, with an average C–D distance of 1.0864 ± 0.0009 Å. The latter distance is longer by neutron analysis than that found by X-ray diffraction because this analysis measures the distance between the carbon and hydrogen (deuterium) nuclei rather than between the maxima of their electron density distributions obtained from X-ray diffraction. The larger electronegativity of oxygen than that of hydrogen draws electron density away from hydrogen towards itself, thus decreasing the distance between the electron density maxima of the two atoms.

The many techniques in crystal structure analysis by X-ray diffraction and neutron diffraction, which determine the precise geometry and symmetry of chemical species, are well presented in the literature [2] and need not be discussed further here.

8.2.3 Application of theoretical studies

Molecular geometry may be determined by theoretical calculations based on wave mechanical principles. It was noted in an earlier chapter that the Schrödinger equation cannot be solved except in the very simplest of cases, and approximation methods must be applied to species more complex than the hydrogen molecule ion, H_2^+. As a start, the water molecule will be considered; on account its importance, it has been the subject of many detailed studies.

From experiment, the O–H bond length in the water molecule is approximately 0.1 nm, depending on the nature of the experimental probe, and the H–O–H bond angle *ca.* 104.4°. The electron configuration of oxygen, $(1s)^2\,(2s)^2\,(2p)^4$, implies bivalency, with bonding between O(2p) and H(1s) atomic orbitals. The directional character of p orbitals indicates a bond angle of 90°. In the simplest explanation, the hydrogen atoms would be said to repel each other, thus causing an opening out of the angle to a value greater than 90°.

Instead, consider two p orbitals, say p_u and p_v, making an angle θ_{uv} with each other, and let another similar orbital p_w in the plane of p_u and p_v be directed orthogonally to p_u (Fig. 8.6). Then

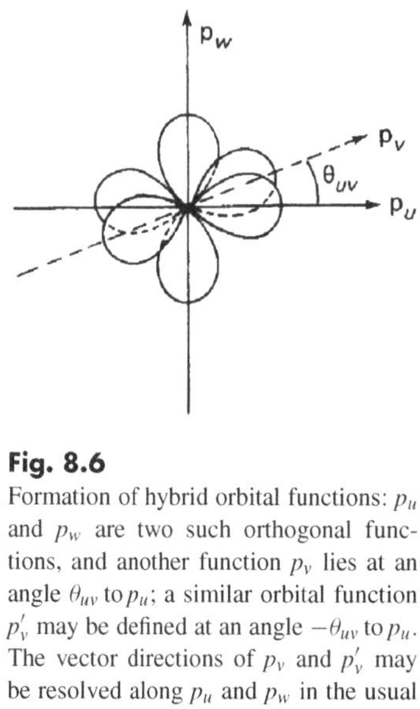

Fig. 8.6
Formation of hybrid orbital functions: p_u and p_w are two such orthogonal functions, and another function p_v lies at an angle θ_{uv} to p_u; a similar orbital function p'_v may be defined at an angle $-\theta_{uv}$ to p_u. The vector directions of p_v and p'_v may be resolved along p_u and p_w in the usual manner.

$$p_v = p_u \cos\theta_{uv} + p_w \sin\theta_{uv} \tag{8.1}$$

A similar orbital p'_v at an angle $-\theta_{uv}$ to p_u is given by

$$p'_v = p_u\, cos\, \theta_{uv} - p_w\, sin\, \theta_{uv} \tag{8.2}$$

Two hybrid orbitals h and h' may be postulated, using s and p atomic orbitals:

$$h = c_s s + c_p p_v$$

$$h' = c_s s + c_p p'_v \tag{8.3}$$

The normalization of both h and h' requires that $\int_\tau h^2 d\tau = \int_\tau h'^2 d\tau = c_s^2 + c_p^2 = 1$, and the orthogonality criterion gives $\int_\tau hh' d\tau = c_s^2 + c_p^2 \int_\tau p_v p'_v d\tau = c_s^2 + c_p^2 \cos 2\theta_{uv} = 0$. Hence, for the fractional characters c_s and c_p,

$$c_s^2 = \cos\Theta/(\cos\Theta - 1) \tag{8.4}$$

$$c_p^2 = 1/(1 - \cos\Theta) \tag{8.5}$$

where $\Theta = 2\theta_{uv}$ is the bond angle between the hybrids h and h', in the directions of p_v *and* p'_v. If the bond angle is taken as 104.4°, and the hybrid molecular wave function written as

$$\Psi = c_s\,\psi_s + c_p\,\psi_p \tag{8.6}$$

then for commonly occurring values of Θ, the following values for c_s and c_p may be derived:

Θ/ deg	90	109.4	120	180
c_p^2	1	3/4	2/3	1/2
c_s^2	0	1/4	1/3	1/2
Type	p	sp^3	sp^2	sp

A hybrid atomic orbital for oxygen, the central atom, may be written as a linear combination of atomic orbitals (LCAO), effectively the basis set:

$$\Psi = c_s\,\psi_s + c_p\,\psi_p \tag{8.7}$$

For the known H–O–H bond angle, it follows from Eq. (8.5) that $c_p^2 = 0.801$ whence $c_s^2 = 0.199$. Thus, the hybrid type character is approximately $p{:}s = 4{:}1$; had it been exactly tetrahedral the ratio would have been 3:1. The variation of c_s and c_p with bond angle Θ for this type of molecule is shown in Table 8.1.

In forming the molecule from its component atoms, energy is expended to unpair electrons in the 2s orbital, but it is more than compensated by the reduction in energy from the overlap of O(hybrid) with H(1s) compared with O(2p) and H(1s). The remaining 2s and 2p electrons occupy two other hybrid orbitals. They are equivalent to two lone pairs, and together with the hybrid orbitals are directed almost tetrahedrally. This configuration shows that the lone pairs make a considerable contribution to the polar nature of the water molecule, and the dipole moment of 1.85 D compared with 0.2 D for F_2O supports this view. Figure 8.7 illustrates some aspects of the discussion of this method of visualizing the water molecule.

Another way of looking at the water molecule is to assume an initial use of all 2s and 2p electrons to form four equivalent tetrahedral orbitals. Two of

Table 8.1 Contributions of s and p to hybrid sp^λ orbitals.

Orbital	Θ/°	c_s^2	c_p^2	$\lambda\ (= c_p^2/c_s^2)$
p	90	0	1	0
sp^4	104.4	0.2	0.8	4
sp^3	109.47	1/4	3/4	3
sp^2	120	1/3	2/3	2
sp	180	1/2	1/2	1
s	—	1	0	—

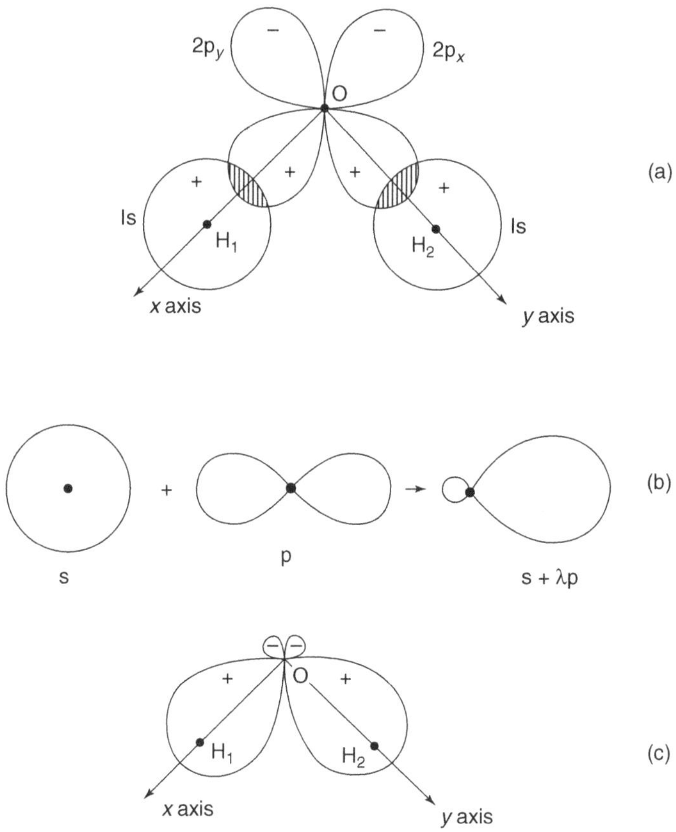

Fig. 8.7
Diagrammatic representation of the orbitals in the formation of a water molecule. (a) Bonding by overlap of O(2*p*) and H(1*s*) orbitals would lead to a bond angle of 90°. (b) Mixing of *s* and *p* orbitals to form hybrid orbitals $s + \lambda p$ on oxygen. (c) Bonding between the hybrid and hydrogen (1*s*) atoms leads to strong overlap, with a bond angle of *ca.* 104.4°. Orthogonality of the hybrids is supported by the fact that in the CH_3OH molecule the O—H bonding electrons are not significantly different from the configuration in the water molecule, thus providing a basis for characteristic bond lengths. [Reproduced by courtesy of Woodhead Publishing, UK.]

them then overlap the hydrogen atoms and the remaining non-bonding orbitals, containing the lone pairs, repel each other so as to modify the bond angle. In this way, the molecule is treated as a modified tetrahedron, similar to the Bernal model for water (compare Section 8.2.1 and Section 8.2.5.1).

A recent molecular dynamics study of water [3] revealed that many molecules did not have the expected tetrahedral configuration at any instant. Instead, two bonds were strong, and the other two disposed in an asymmetric manner. However, because the asymmetry is fluctuating rapidly it is averaged away over a time of *ca.* 10^2 fs, thus restoring the appearance of a tetrahedral coordination. This seems to be exactly what would be expected from both a chemical point of view, and the experimentally determined structure of ice, the symmetry of which exhibits an average of a random array of *ca.* 10^{16} frozen snapshots of a multitude of asymmetric structures (see Fig. 8.5b).

8.2.4 Monte Carlo and molecular dynamics techniques

Two theoretical methods of determining the structure of a liquid are Monte Carlo, which leads to equilibrium thermodynamic parameters and radial distribution functions $g(r)$, given a form for the potential $V(r)$, and molecular dynamics, which is time dependent and provides measures of transport properties.

8.2.4.1 Monte Carlo

In this technique, an atomic model is set up in a cubic unit cell, and an initial configuration of 200–300 atoms (see Table 8.4) is repeated by three-dimensional face-to-face stacking, so generating a macroscopic model by the translation group of a lattice. Figure 8.8 shows a projection of the model after elapsed simulation time; the cubic cell side a is chosen such that the desired number density N at a temperature T is achieved $\left(N = L/a^3\right)$. The initial configuration of atoms for a monatomic liquid is based on a face-centred cubic array.

The model in Fig. 8.8 contains a degree of translational symmetry that is not present in the liquid. It is assumed that, provided the range of the potential energy of interaction between atoms is less than $a/2$, the potential experienced by any given atom is not affected by the symmetry of the model. It is clear that motion of the species will take atoms out of the unit cell. However, as B, say, moves out from its cell, a translation image moves in to take its place; in this way, the atom density is conserved in each cell. A selection of results for liquid argon is listed in Table 8.2; the agreement with experimental data is of a high quality [4, 5].

8.2.4.2 Molecular dynamics

Molecular dynamics treats the evolution with time of systems of particles that interact through conservative forces[2] operating under the laws of classical mechanics. In effect, it tracks the motions of species in condensed phases by

[2] Forces that are derivable from the potential energy function through $F = -\mathrm{d}V/\mathrm{d}r$.

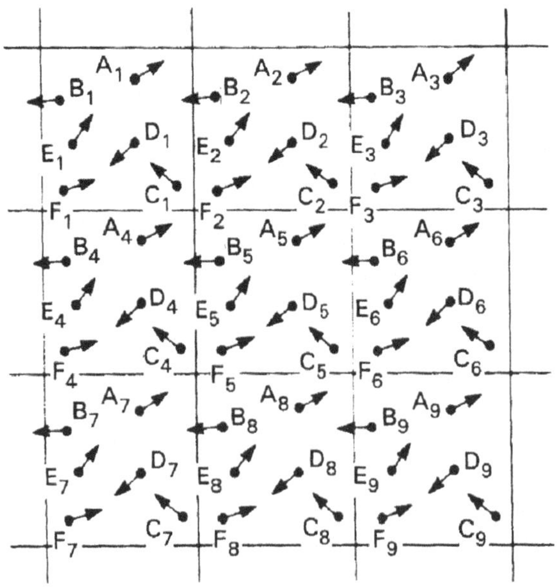

Fig. 8.8
Projection of a configuration of atoms in a Monte Carlo simulation of liquid argon. The positions of the atoms in any one cubic cell are random; the other cells are formed by the three-dimensional translation group. After sufficient simulation time, an equilibrium arrangement is attained: as one molecule moves out of a given cell so another moves in from an adjacent cell so as to maintain a constant number density. [Reproduced by courtesy of Professor A. J. C. Ladd.]

Table 8.2 Experimental and theoretical thermodynamic data for liquid argon.

		p/atm		$-U$/kJ mol^{-1}	
T/K	$10^6 V$/m^3 mol^{-1}	Calc	Expt	Calc	Expt
100.0	29.7	116	115	5.52	5.54
140.0	41.8	18	37	3.81	3.86
150.7	75.2	49	49	2.48	2.47

solving the Newtonian equations of motion, and has been particularly useful in elucidating transport and equilibrium parameters.

In applying molecular dynamics to a simulated liquid system, a set of initial coordinates is generated, usually in the form of a face centred unit cell of a cubic Bravais lattice, at a required density. Initial momenta configurations are assigned randomly, such that the system has the desired total energy, and boundary conditions are imposed, in the manner of the Monte Carlo method.

Many molecular dynamics calculations have been carried out with a hard sphere potential function. It is computationally straightforward, and shows that the structure of simple liquids is almost independent of their chemical nature, and may be approximated as an interaction of rigid spheres. This idea was present in the earlier work of Bernal with physical models (Section 8.2.1).

Figure 8.9 illustrates the results of molecular dynamics calculations of the interface of liquid argon, using 1500 atom sets. In Fig. 8.9a, the atoms are vibrating about their mean positions in the solid state, the sites of a face centred unit cell of a cubic lattice, and Fig. 8.9b shows the traces of the atoms, now in a typically liquid phase.

Many attempts have been made to simulate the properties of water, both by Monte Carlo and molecular dynamics techniques. Care is needed in specifying the pair potential between water molecules, because of the relatively long range effect of dipolar interactions. Table 8.3 lists the configurational energy,

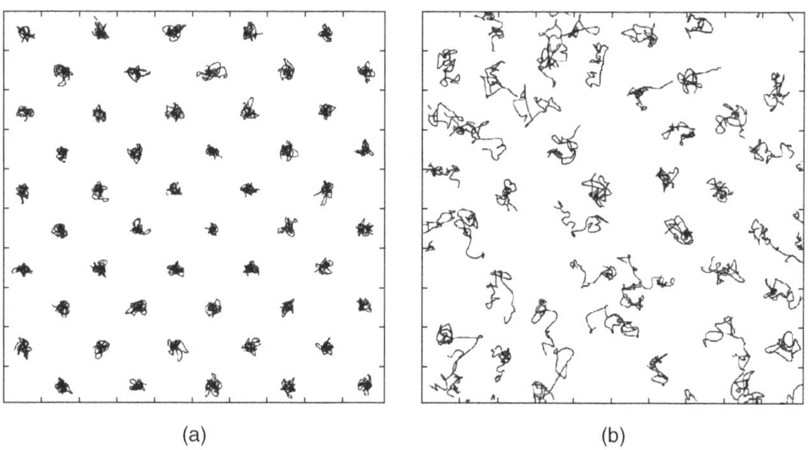

Fig. 8.9
Models of argon in the solid and liquid states from computer simulation of the Newtonian equations of motion. (a) Projection of a face centred unit cell of a cubic lattice, showing the initial configuration in the solid state. (b) Equilibrium liquid state, showing random configuration with only localized regions of order. [Reproduced by courtesy of Professor A. J. C. Ladd.]

Table 8.3 Thermodynamic properties of water by computer simulation.

System	$-E$/kJ mol^{-1}	pV/RT	C_V/J K^{-1}mol^{-1}
216[a]	43.1	0.05	100
256[b]	39.9 ± 0.3	0.6 ± 0.3	70
Experiment	41.1	0.05	75

The superscripts *a* and *b* link with Table 8.4.

an equation of state function and the constant volume heat capacity, compared with results from both molecular dynamics and experiment.

The structural properties of water were addressed by computing the radial distribution functions for O–O, O–H and H–H interactions, by sampling pair distributions after every 250 configurations. Table 8.4 lists the results for the positions and heights of the maxima for a 256 molecule system, and compares them with both molecular dynamics and experimental results, which are very satisfactory in representing thermodynamic and structural properties of the liquids studied [4, 5]. A recent study [6] includes a radial distribution function for water in cytosine solution (Fig. 8.10), with results in very good agreement with those in Table 8.4.

8.2.5 Symmetry adapted molecular orbitals

The molecular orbital method defines regions in a molecule where the probability of finding electrons, or electron density, is high. The molecular orbitals are formed by a combination of atomic orbitals, most commonly by the linear combination of atomic orbitals (LCAO) procedure. The molecular orbital method provides a straightforward model of molecular bonding, and most computational chemistry begins with the calculation of molecular orbitals of the system under consideration. In any set of molecular orbitals, the electrons supplied by

Table 8.4 Structural properties of water by computer simulation: positions r and heights M of maxima in radial distribution functions $g_{\text{I-J}}$.

		r_1/nm	M_1	r_2/nm	M_2
$g_{\text{O-O}}$	216[a]	0.285	3.09	0.470	1.13
	256[b]	0.285	3.11	0.530	1.06
	Experiment	0.283	2.31	0.425	1.08
$g_{\text{O-H}}$	216[a]	0.190	1.38	0.340	1.60
	256[b]	0.191	1.24	0.332	1.53
	Experiment	0.190	0.80	0.335	1.70
$g_{\text{H-H}}$	216[a]	0.250	1.50	0.390	1.20
	256[b]	0.250	1.15	0.375	1.07
	Experiment	0.235	1.04	0.400	1.08

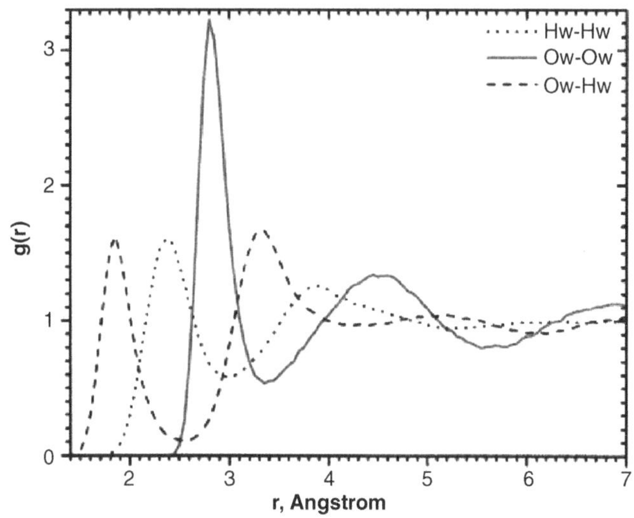

Fig. 8.10
Radial distribution for water as a function of distance, calculated by averaging over a 10 ps molecular dynamics simulation of cytosine solution. [Reproduced by permission The Royal Society of Chemistry.]

the atoms no longer belong to specific species but are distributed, or *delocalized*, over the whole molecule, and move under the influence of the atomic nuclei.

Each molecular orbital is represented by a wave function Ψ, compounded by the LCAO procedure. Generally, a molecular orbital contains two nuclei and up to two electrons: if two electrons are present in one and the same molecular orbital, the Pauli principle requires that their spins be opposed.

Bonding molecular orbitals are formed by in-phase interaction between atomic orbitals, antibonding molecular orbitals by out-of-phase interaction. In a simple example of a diatomic molecule, and adopting a common notation, bonding and antibonding molecular orbitals respectively may be specified as

$$\Psi = c_1\psi_1 + c_2\psi_2$$
$$\Psi^* = c_1\psi_1 - c_2\psi_2$$

where c_i represents the proportion of the ψ_i atomic orbital in Ψ. Clearly, there is no such formulation for a non-bonding molecular orbital; in other terminology, it forms a lone pair of electrons on the atom in question. The basic properties of molecular orbitals may be summarized as follows:

- A basis set of orbitals comprises the atomic orbitals that are available for molecular orbital interaction, both bonding and antibonding;
- The number of molecular orbitals is always equal to the number of atomic orbitals included in the LCAO or basis set;
- If the molecule under consideration has a degree of symmetry greater than unity, the consequent degenerate atomic orbitals are grouped by LCAO as symmetry adapted linear combinations of atomic orbitals. They belong to their representations of the molecular point group, and the wave functions that describe them are the *symmetry adapted linear combinations*, or SALCs;

- The number of molecular orbitals belonging to one representation is equal to the number of SALCs belonging to that representation;
- In any representation, the SALCs form more readily if their atomic orbitals are similar in energy.

Thus, if ψ_j is the jth basis function in a set of energy compatible molecular vectors and $c_{i,j}$ is the fractional contribution of this function to the ith SALC Ψ_i in a molecule, then

$$\Psi_i = \sum_j c_{i,j}\psi_j \tag{8.8}$$

The applications of the technique are expounded in detail in most books on quantum chemistry [7–10], and only a few examples will be considered in the remainder of this section.

8.2.5.1 The water molecule

The stretching modes of the water molecule can be used to establish a basis set for the molecule. The water molecule exhibits point group C_{2v}, and in Section 7.5.3 the representation Γ_{3n} was determined as 9 −1 1 3, which can be reduced to $3A_1 + A_2 + 2B_1 + 3B_2$. If the contributions from x, y, z (translational movements, $A_1 + B_1 + B_2$) and R_x, R_y, R_z (rotational movements, $A_2 + B_1 + B_2$) are subtracted, $2A_1$ and B_2 remains for vibrations. The A_1 stretch vibration is symmetric whereas the B_2 vibration is antisymmetric. The SALCs for water are then:

$$\Psi_{A_1} = \psi_1 + \psi_2$$

$$\Psi_{B_2} = \psi_1 - \psi_2 \tag{8.9}$$

Each SALC must be normalized. The previous section has shown that the sum of the squares of the similar LCAO coefficients c must be unity. Hence, the normalizing constant N is given by

$$N^2 \sum_{j=1}^{n} c_{i,j}^2 = 1 \tag{8.10}$$

Since $\sum_j c_{i,j}^2 = 1$ for the ith SALC, $N = 1\sqrt{2}$, and

$$\Psi_{A_1} = (\psi_1 + \psi_2)/\sqrt{2}$$

$$\Psi_{B_2} = (\psi_1 - \psi_2)/\sqrt{2} \tag{8.11}$$

From Section 7.7, the product of two irreducible representations is zero, and this condition is satisfied by SALCs, Eqs. (8.11). Figure 8.11 shows a molecular orbital energy level diagram for the water molecule. There are two bonding molecular orbitals from A_1 and B_1 symmetries. The O–H interactions are bonding in both symmetries $1a_1$ and $1b_1$, but both H–H interactions are antibonding (higher in energy), $3a_1$ and $2b_1$. Two molecular orbitals, of A_1 and B_2 symmetries are non-bonding, $2a_1$ and $1b_2$. The orbital occupancy may be described as $1a_1^2$, $1b_1^2$, $2a_1^2$, $2b_2^2$, where the superscript indicates the number of electrons in the molecular orbital, as indicated on Fig. 8.11. The same

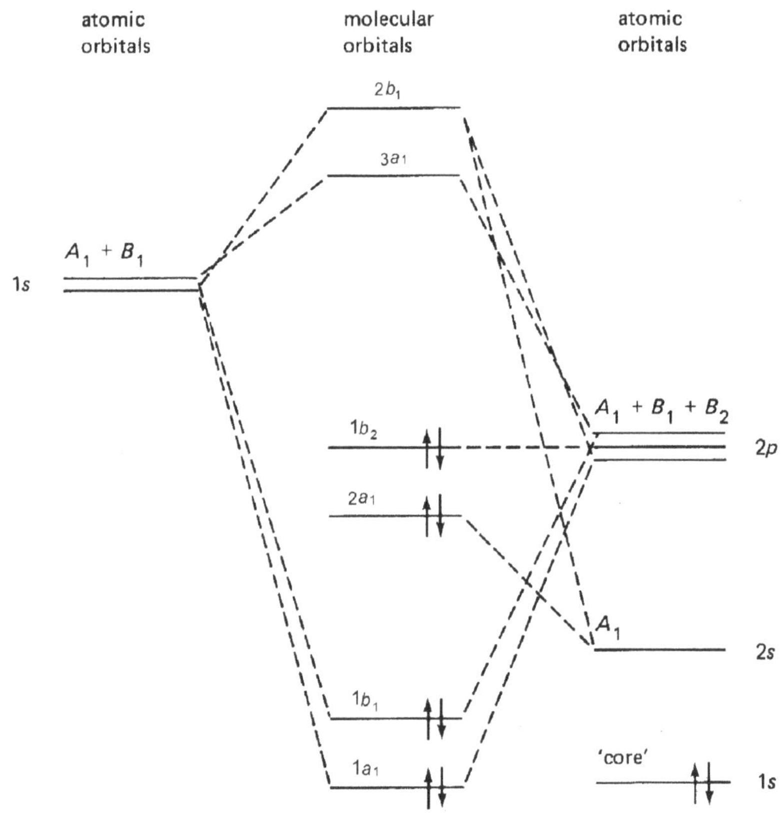

Fig. 8.11
Molecular orbital energy level diagram for the water molecule: $1a_1$ and $1b_2$ are bonding levels, $1a_1^\dagger$ and $1b_1^\dagger$ are non-bonding levels (lone pairs) and $1a_1^*$ and $1b_2^*$ are antibonding levels. [Reproduced by courtesy of Woodhead Publishing, UK.]

result may be obtained both by a representation based on internal coordinates, and by the use of the projection operator as will be exemplified in the next sections.

8.2.5.2 Methane

A projection operator may be said to map a vector space on to a subset of that space. It is a process akin to passing a light beam over a three-dimensional object on to a white screen; the result is a two-dimensional projection of the object. In other terms, it is like taking all points x, y, z of the object and setting one of them at zero. A similar process has already been encountered in the study of space groups, when a three-dimensional space group is projected on to a plane.

In the molecule of methane, there are four orbitals linking the atoms, so that four molecular orbitals are to be the expected result. Methane has the point group T_d, and the character table (Appendix A10.2) shows that the degenerate p orbitals span the T_2 irreducible representation. In order to obtain the desired representation for σ bonding in methane, it is convenient to detail the action of the symmetry operators on the C–H bond vectors:

$\boldsymbol{E}$ leaves all vectors unmoved $\quad \chi_E = 4$
$\boldsymbol{C}_3$ leaves one vector unmoved $\quad \chi_{C_3} = 1$

$\boldsymbol{C}_2$ moves all vectors	$\chi_{C_2} = 0$
$\boldsymbol{S}_4$ moves all vectors	$\chi_{S_4} = 0$
$\boldsymbol{\sigma}_d$ leaves two vectors unmoved	$\chi_{\sigma_d} = 2$

Hence, the representation

T_d	E	$8C_3$	$3C_2$	$6S_4$	$6\sigma_d$
Γ	4	1	0	0	2

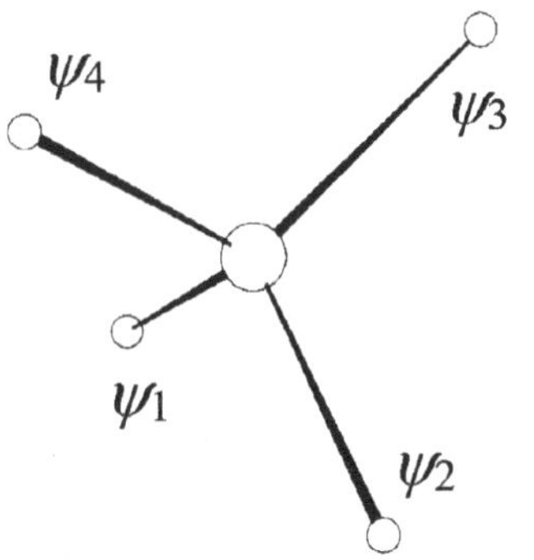

Fig. 8.12
Molecule of methane, CH_4; ψ_i, $i = 1-4$, represent hydrogen 1s atomic orbitals.

which reduces to $A_1 + T_2$.

Molecular orbitals, Ψ_i, $i = 1-4$, for methane can be generated from the hydrogen 1s atomic orbitals, ψ_i, $i = 1-4$, and the $(2s)^2$ and $(2p)^4$ orbitals of carbon. Each of the resulting orbitals consists of a combination of an s orbital spanning A_1 and three p orbitals spanning T_2. The projection operators (Appendix A11.4) $\boldsymbol{P}_{A_1}$ and $\boldsymbol{P}_{T_2}$ are applied to any three of the atomic orbitals; a fourth such orbital is not needed at this stage as it is not be linearly independent of the other three. Using Fig. 8.12, the action of all symmetry operations of T_d on the orbitals can be determined, as in the example in Appendix A11.4, and are listed below; the notation ${}_2\boldsymbol{C}_3^2$, for example, implies the rotation C_3^2 acting at ψ_2 on the other two orbitals, ψ_1 and ψ_3:

T_d	E	${}_1C_3$	${}_1C_3^2$	${}_2C_3$	${}_2C_3^2$	${}_3C_3$	${}_3C_3^2$	${}_4C_3$	${}_4C_3^2$	C_{2z}	C_{2y}	C_{2x}
ψ_1	ψ_1	ψ_1	ψ_1	ψ_3	ψ_4	ψ_4	ψ_2	ψ_2	ψ_3	ψ_2	ψ_4	ψ_3
ψ_2	ψ_2	ψ_4	ψ_3	ψ_2	ψ_2	ψ_1	ψ_4	ψ_3	ψ_1	ψ_1	ψ_3	ψ_4
ψ_3	ψ_3	ψ_2	ψ_4	ψ_4	ψ_1	ψ_3	ψ_3	ψ_1	ψ_2	ψ_4	ψ_2	ψ_1
	S_{4z}	S_{4z}^3	S_{4y}	S_{4y}^3	S_{4x}	S_{4x}^3	${}_1\sigma_d$	${}_2\sigma_d$	${}_3\sigma_d$	${}_4\sigma_d$	${}_5\sigma_d$	${}_6\sigma_d$
ψ_1	ψ_4	ψ_3	ψ_3	ψ_2	ψ_2	ψ_4	ψ_1	ψ_2	ψ_4	ψ_1	ψ_1	ψ_3
ψ_2	ψ_3	ψ_4	ψ_1	ψ_4	ψ_3	ψ_1	ψ_2	ψ_1	ψ_2	ψ_3	ψ_4	ψ_2
ψ_3	ψ_1	ψ_2	ψ_4	ψ_1	ψ_4	ψ_2	ψ_4	ψ_3	ψ_3	ψ_2	ψ_3	ψ_1

With some symmetry operations, Fig. 8.13 may be helpful as it aligns the symmetry elements of methane with the corresponding symmetry elements of a cube.

The projection operators $\boldsymbol{P}_{A_1}$ and $\boldsymbol{P}_{T_2}$ are now applied and, discounting the common factors,

$$\boldsymbol{P}_{A_1}\psi_1 = (1/2)(\psi_1 + \psi_2 + \psi_3 + \psi_4) \tag{8.12}$$

$$\boldsymbol{P}_{T_2}\psi_1 = \frac{1}{\sqrt{12}}(3\psi_1 - \psi_2 - \psi_3 - \psi_4) \tag{8.13}$$

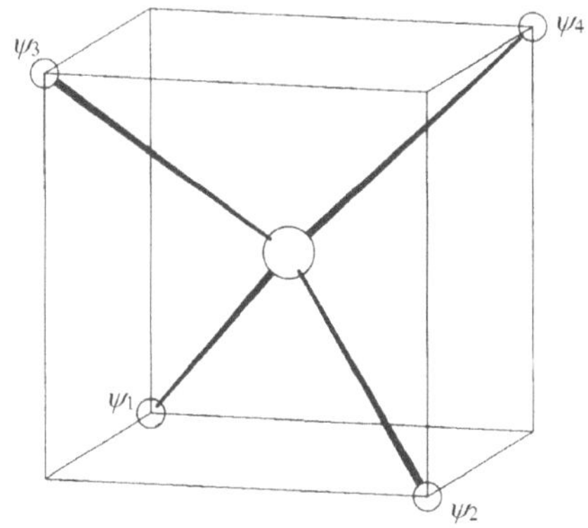

Fig. 8.13
The molecule of methane (T_d), Fig. 8.12, set in a cube (O_h) such that their common symmetry elements coincide; T_d is a subgroup of O_h.

$$\boldsymbol{P}_{T_2}\psi_2 = \frac{1}{\sqrt{12}}(-\psi_1 + 3\psi_2 - \psi_3 - \psi_4) \quad (8.14)$$

$$\boldsymbol{P}_{T_2}\psi_3 = \frac{1}{\sqrt{12}}(-\psi_1 - \psi_2 + 3\psi_3 - \psi_4) \quad (8.15)$$

These SALCs need now to be established in terms of normalized s, p_x. p_y and p_z functions. Eq. (8.12) is fully symmetric (A_1) and refers to a normalized s orbital: it is normalized, following Section 2.9.3, as

$$\frac{1}{2}(\psi_1 + \psi_2 + \psi_3 + \psi_4) \quad (8.16)$$

Writing Eq. (8.13) as a normalized, linear combination of p functions,

$$\frac{1}{\sqrt{12}}(3\psi_1 - \psi_2 - \psi_3 - \psi_4) = (c_x^2 + c_y^2 + c_z^2)^{1/2}(c_x p_x + c_y p_y + c_z p_z) \quad (8.17)$$

Using the tetrahedron model (Fig. 8.12), and also Fig. 8.13 as it aligns the symmetry elements of the tetrahedron with those of the cube, the effects of the $\boldsymbol{C}_3$ operation in the direction of ψ_1 are:

$$p_x \rightarrow p_y \qquad p_y \rightarrow p_z \qquad p_z \rightarrow p_x$$

so that $(c_x p_x + c_y p_y + c_z p_z)$ in Eq. (8.17) transforms to $(c_x p_y + c_y p_z + c_z p_x)$. By comparing coefficients, it follows that $c_x = c_z$, $c_y = c_x$ and $c_z = c_y$; then,

$$\frac{1}{\sqrt{12}}(3\psi_1 - \psi_2 - \psi_3 - \psi_4) = \pm\frac{1}{\sqrt{3}}(p_x + p_y + p_z) \quad (8.18)$$

Since resolving ψ_i $(i = 1 - 4)$ into its components shows that the positive square root is required in Eq. (8.10), the normalization factor is now $1/\sqrt{3}$.

Consider next the operation of $\boldsymbol{C}_3$ in the direction of ψ_2, then

$$p_x \rightarrow -p_y \qquad p_y \rightarrow p_z \qquad p_z \rightarrow -p_x$$

so that with Eq. (8.14) $(c_x p_x + c_y p_y + c_z p_z)$ transforms to $(-c_z p_x - c_x p_y + c_y p_z)$, and again the positive square applies:

$$\frac{1}{\sqrt{12}}(-\psi_1 + 3\psi_2 - \psi_3 - \psi_4) = \frac{1}{\sqrt{3}}(-p_x - p_y + p_z) \quad (8.19)$$

By similar reasoning with ψ_3 and Eq. (8.15),

$$\frac{1}{\sqrt{12}}(-\psi_1 - \psi_2 + 3\psi_3 - \psi_4) = \frac{1}{\sqrt{3}}(p_x - p_y - p_z) \qquad (8.20)$$

Solving Eqs. (8.18)–(8.20) for p_x, p_y and p_z, and including Eq. (8.16), leads to the result, expressed in matrix form,

$$\begin{pmatrix} 1/2 & 1/2 & 1/2 & 1/2 \\ 1/2 & \overline{1/2} & 1/2 & \overline{1/2} \\ 1/2 & \overline{1/2} & \overline{1/2} & 1/2 \\ 1/2 & 1/2 & \overline{1/2} & \overline{1/2} \end{pmatrix} \begin{pmatrix} \psi_1 \\ \psi_2 \\ \psi_3 \\ \psi_4 \end{pmatrix} = \begin{pmatrix} s \\ p_x \\ p_y \\ p_z \end{pmatrix} \qquad (8.21)$$

The 4×4 matrix is orthogonal, as is readily confirmed; hence, multiplication by its inverse, which here is the same as its transpose, gives

$$\begin{pmatrix} 1/2 & 1/2 & 1/2 & 1/2 \\ 1/2 & \overline{1/2} & \overline{1/2} & 1/2 \\ 1/2 & 1/2 & \overline{1/2} & \overline{1/2} \\ 1/2 & \overline{1/2} & 1/2 & \overline{1/2} \end{pmatrix} \begin{pmatrix} s \\ p_x \\ p_y \\ p_z \end{pmatrix} = \begin{pmatrix} \psi_1 \\ \psi_2 \\ \psi_3 \\ \psi_4 \end{pmatrix} \qquad (8.22)$$

and the final form of the molecular orbitals may be written as

$$\left.\begin{aligned} \Psi_1 &= 1/2(s + p_x + p_y + p_z) \\ \Psi_2 &= 1/2(s - p_x - p_y + p_z) \\ \Psi_3 &= 1/2(s + p_x - p_y - p_z) \\ \Psi_4 &= 1/2(s - p_x + p_y - p_z) \end{aligned}\right\} \qquad (8.23)$$

The character table for T_d shows that T_2 is spanned by three product terms that correspond to d_{xy}, d_{yz} and d_{zx}, as well as by the three p functions discussed above. It follow that symmetry arguments alone would predict the combinations either sp^3 or sd^3. However, chemical knowledge of energy levels leads to the adoption of the lower energy state, sp^3. A molecular orbital energy level diagram is shown in Fig. 8.14; the orbital occupancy may be written as $1a_1^2$, $1t_1^6$, indicating the threefold degeneracy of T_2 symmetry; as with the water molecule, the $(1s)^2$ electrons remain as a relatively inert core.

8.2.5.3 Benzene

Finally, here, the important π-electron system of benzene will be studied. Benzene exhibits point group D_{6h}, and the in-plane bonds involve the $2s, 2p_x$ and $2p_y$ functions of the carbon atoms. They combine with hydrogen $1s$ orbitals to form a hexagonal ring system of trigonal-planar σ bonds (Fig. 8.15a).

The six $2p_z$ orbitals on the carbon atoms directed normally to the plane of the ring (Fig. 8.15b) can be taken as basis vectors for a representation of π-bonding. The C_2' axes and σ_v planes pass through opposite carbon atoms and the C_2'' axes and σ_d planes pass through the mid-points of opposite C–C bonds. By considering the effects of the symmetry operations on the six vectors, as in previous examples, the following representation is obtained:

D_{6h}	E	$2C_6$	$2C_3$	C_2	$3C_2'$	$3C_2''$	i	$2S_3$	$2S_6$	σ_h	$3\sigma_d$	$3\sigma_v$
Γ_π	6	0	0	0	−2	0	0	0	0	−6	0	2

This representation is reduced easily with the program RD6H in the Web Program Suite, or otherwise, to give $B_{2g} + E_{1g} + A_{2u} + E_{2u}$. The π-bonding molecular orbitals are illustrated in Fig. 8.15c, which shows the positive and negative regions of the orbitals and their nodal planes, planes of zero electron density. The most stable (lowest energy) *bonding* molecular orbital π_1 corresponds to the positive combination of the six p_z orbitals, and transforms as A_{2u}; it may be noted, with reference to Section 7.6 that π_1 is symmetric to $\boldsymbol{C}_6$ and antisymmetric to $\boldsymbol{C}_i$ and $\boldsymbol{\sigma}_h$. The next energy level is represented by the degenerate pair π_2 and π_3; they transform as E_{1g}, and are symmetric to $\boldsymbol{C}_2$ and $\boldsymbol{C}_i$, as Fig. 8.15c shows. The orbitals are fully populated with two electrons each. The unpopulated orbitals $\pi_4 - \pi_6$ are *antibonding* molecular orbitals of higher energy; π_4 and π_5 have the same symmetry as π_2 and π_3, whereas the highest energy π_6 is B_{2g}, and is symmetric to $\boldsymbol{C}_i$ and antisymmetric to $\boldsymbol{C}_6$ and $\boldsymbol{\sigma}_h$; Fig. 8.16 shows another perspective of the π molecular orbitals of benzene.

A molecular π orbital energy level diagram is shown by Fig. 8.17. The stability of the benzene molecule is much greater than expected. The enthalpy of hydrogenation of cyclohexa-1,4-diene, C_6H_8, to cyclohexane, C_6H_{12}, is -239.2 kJ mol^{-1}: the similar quantity for benzene, C_6H_6, if containing three similar double bonds, might be expected as 1.5 times this value, or -358.8 kJ mol^{-1}, whereas the experimental value -208 kJ mol^{-1}; the difference of -150.8 kJ mol^{-1} is the delocalization energy, also called resonance energy, of benzene, the quantity indicated as D_π in Fig. 8.17.

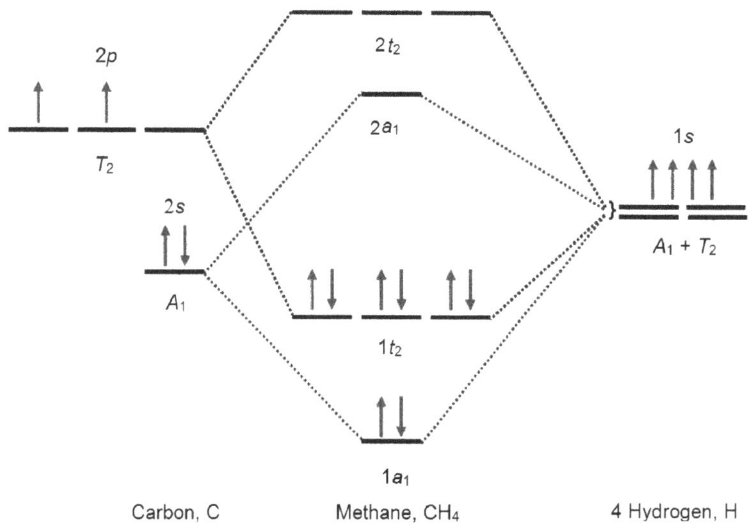

Fig. 8.14
Molecular orbital energy level diagram for methane; bonding takes place through the $1a_1$ and $1t_2$ molecular orbitals. [Reproduced by courtesy of Woodhead Publishing, UK.]

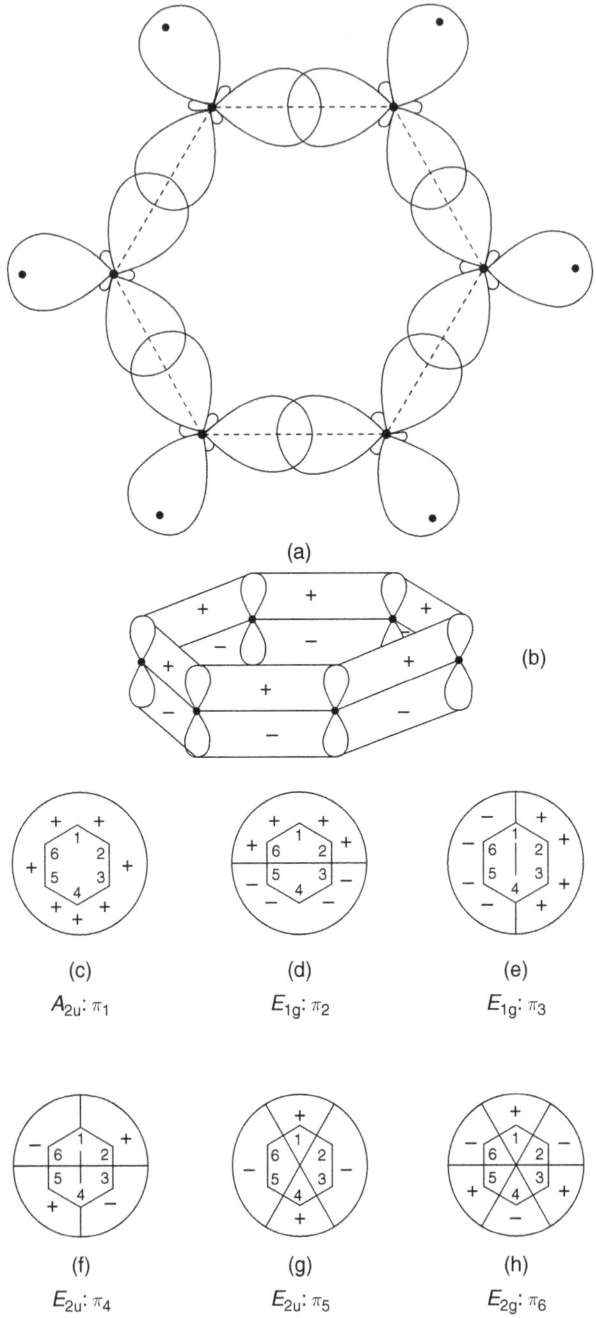

Fig. 8.15
Molecular orbital energy level diagram for benzene. (a) In-plane σ-bonding arises from the trigonal planar sp^2 molecular orbitals. (b) The out-of-plane π-bonding of lowest energy, equivalent to diagram (c), with no nodal planes (planes of zero electron density). Diagrams (d)–(h) indicate π-bonding levels of increasing energy: (d) and (e) are degenerate states with one nodal plane each; (f) and (g) are also degenerate states, but with two nodal planes each; (h) is the molecular orbital of highest energy, with three nodal planes. [Reproduced by courtesy of Woodhead Publishing, UK.]

The π-*bond order* in benzene is 0.667 and the σ-bond order is 1, so that the total bond order is 1.667, whereas that indicated by the Kekulé structure, 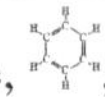, for benzene might be deduced as 1.5.

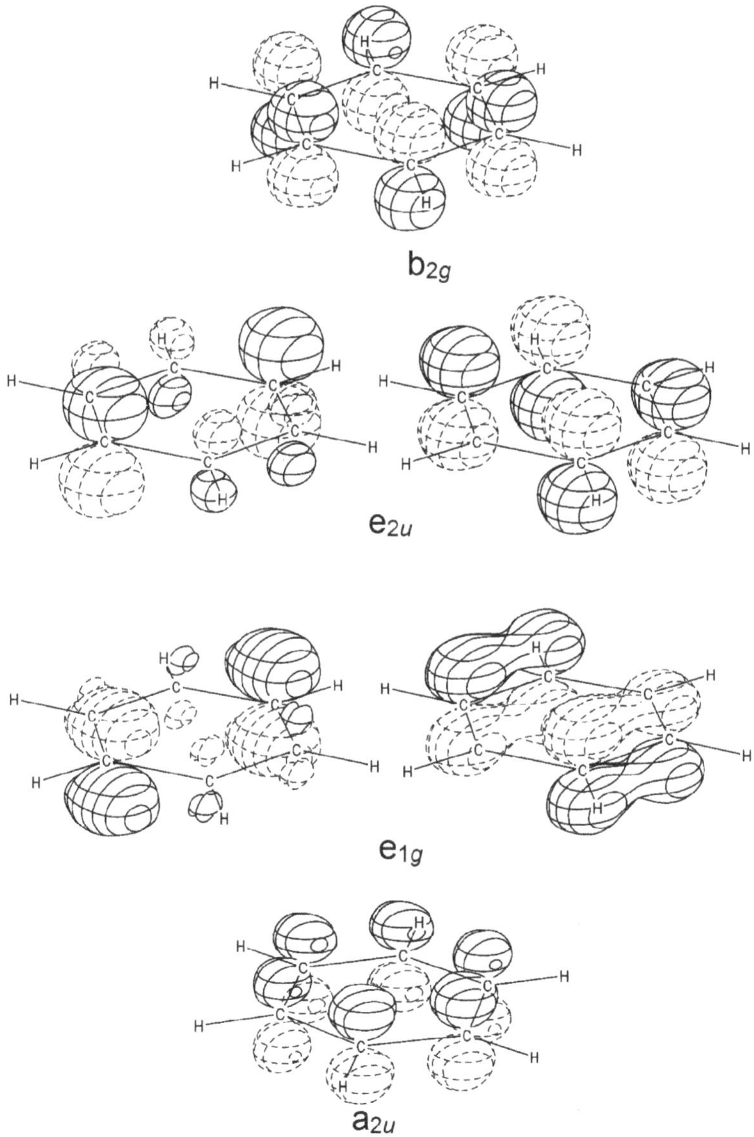

Fig. 8.16
Another view of π-molecular orbitals for benzene; the solid lines are positive regions and the dashed lines negative regions of the wave functions. [Jorgensen WL and Salem L. *The organic chemist's book of orbitals*. 1973; reproduced by courtesy of Elsevier.]

8.2.6 Transition metal compounds: crystal-field and ligand-field theories

The bonding of chemical groups to transition metals has been explained by two main theories. The first of them, *crystal-field theory* considers that the interaction between a transition metal and the bonding groups, or ligands, all acting as a point charges, arise through an electrostatic attraction between the positively charged central metal cation and the negatively charged surrounding ligands.

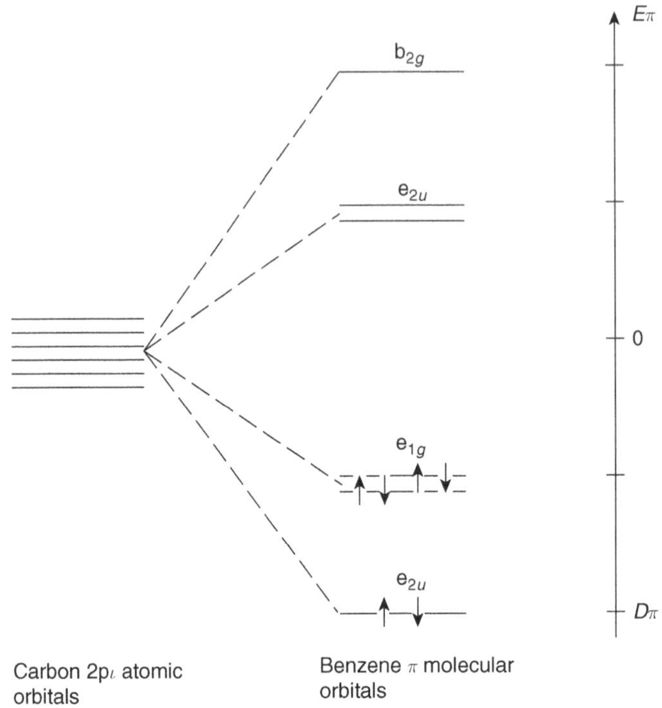

Fig. 8.17
Molecular orbital energy levels for π-bonding in benzene; D_π represents the π delocalization energy with respect to the zero level of *ca.* -150 kJ mol^{-1}.

In a complex of the type MX_6, for example, the focus is on the energy changes of the d levels of the metal cation as the ligands, originally effectively at infinity, approach it. Electron interaction would require the ligands to approach the cation along the lines of least repulsion, which can be seen from Fig. 2.39 to be the positive and negative directions of the x, y and z axes of the d orbital functions. Thus, the electrons from the ligands will be closer to some of the d orbitals than to others, and the fivefold degeneracy is split into two groups, with the d electrons closer to the ligands having the higher energy. The extent of splitting depends of several factors:

- the metal ion and, particularly, the number of its d electrons;
- the charge on the metal ion: the larger the charge, the greater the splitting;
- the coordination pattern of the ligand species around the cation;
- the electric field strength of the ligands: the stronger the field the greater the splitting.

In the MX_6 compounds, the commonest coordination is octahedral. The d orbitals are split into two sets with an energy difference Δ, with the d_{xy}, d_{zx} and d_{yz} functions being lower in energy than the d_z^2 and $d_{x^2-y^2}$: this situation arises because the former bonded group is further from the ligands, therefore experiencing less repulsion than the latter, which point directly to the corners of an octahedron. The three low energy levels are normally referred to as t_{2g}, whereas the levels of higher energy are the e_g.

The next most common coordination compounds are of type MX_4, in which four ligands form a tetrahedral coordination pattern around the metal ion. In a

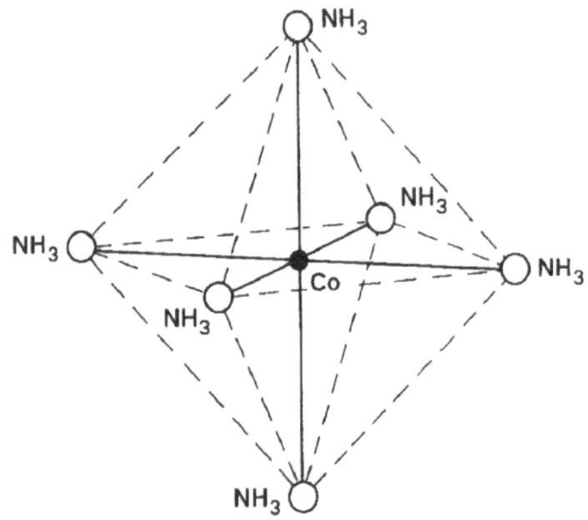

Fig. 8.18
The octahedral complex ion $[Co(NH_3)_6]^{3-}$, a typical transition metal coordination compound; its symmetry is $O_h(m\bar{3}m)$.

tetrahedral crystal field, the d orbitals are again split into two groups, with an energy difference of Δ_{tet} where the lower energy levels are d_z^2 and $d_{x^2-y^2}$; the higher energy levels d_{xy}, d_{zx} and d_{yz} are opposite to the octahedral case. Furthermore, since the ligand electrons in tetrahedral symmetry are not oriented directly towards the d orbitals, the energy splitting will be lower than in the octahedral case. Square planar and other complex geometries can also be described by crystal-field theory.

The crystal-field theory is developed solely in terms of electrostatic effects and fails to take account of overlap between the orbitals of the metal and the ligands. This implicit covalence leads to the molecular orbital description embraced by *ligand-field theory*, in which the electrons on the central metal ion are delocalized, a sort of expansion of the d orbitals. The repulsion is decreased by overlap, and the effect, which depends principally on the nature of the ligand, gives rise to the *nephelauxetic series* of ligands (Gk. *nephēle* = cloud, *auxēsis* = growth) which is related to the size of Δ:

$$\underline{I^-} < \underline{Br^-} < \underline{SCN^-} < \underline{Cl^-} < \underline{NO_3^-} < \underline{F^-} < \underline{OH^-} < \underline{H_2O} < \underline{NCS^-} < \underline{NH_3} < \underline{NO_2} < \underline{CN^-} < \underline{CO}$$

weak field (high spin) strong field (low spin)

A typical transition metal coordination compound is hexamminecobalt(III) chloride, $[Co(NH_3)_6]Cl_3$ (Fig. 8.18); this compound has octahedral symmetry, point group O_h ($m\bar{3}m$). It is evident from Fig. 2.39 that the electrons of the metal atom in the d_{z^2} and $d_{x^2-y^2}$ functions have less favourable energies for (greater repulsions to) the approach of lone-pair electrons than do the electrons in the other three d functions of the central atom, because they are directed towards the ligand positions. The result is that the degeneracy of the metal d orbitals is split into a group of two, the e_g^* molecular orbitals, and a group of three, the t_{2g} molecular orbitals of lower energy. The energy difference between the e_g^* and the t_{2g} levels is Δ_{oct}, is the ligand-field splitting energy parameter (LFSE).

The magnitude of Δ_{oct} is governed by both the nature and oxidation state of the metal, and the nature, number and geometry of the ligands. As an example, consider chromium(III) as a d^3 species. In the presence of weak field (high spin) ligands, such as SCN^- or Cl^- the d electrons are split in the manner of Fig. 8.19a, and Δ_{oct} is small in magnitude. However, if the ligand is strong field

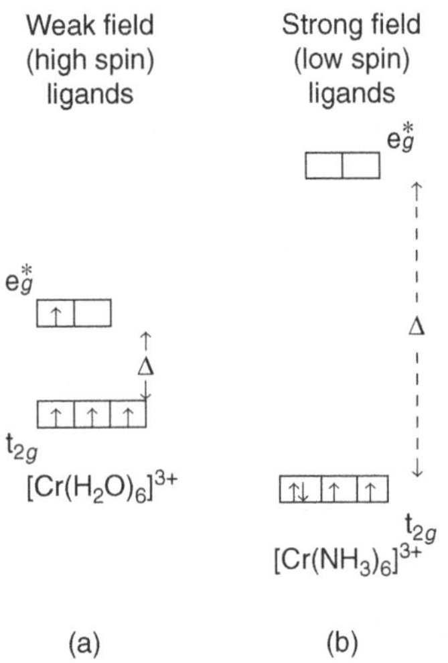

Fig. 8.19
Example of weak field and strong field electron configurations in chromium coordination compounds; the quantity Δ between the t_{2g} and e_g^* levels is the ligand-field stabilization energy.

(low spin), then the electron distribution of Fig. 8.19b is adopted and Δ_{oct} is large in magnitude. In both cases, the symmetry remains as O_h.

8.2.7 The hexacyanoferrate(II) ion

As a second and final example in this section, consider the octahedral hexacyanoferrate(II) ion, $\left[Fe(CN)_6\right]^{4-}$. The significant atomic orbital functions of this metal from the first transition series are again subscribed from the $3d, 4s$ and $4p$ energy levels of the atom.

The symmetry of the hexacyanoferrate(II) ion is $O_h(m\bar{3}m)$, and the six Fe–CN bond vectors can be used to define a representation for the Fe–CN σ-bonding. Since a centre of symmetry is present in this group, the g/u notation is needed with the irreducible representations in order to indicate the symmetry with respect to the $\boldsymbol{i}$ operator. The representation can be deduced readily by setting the octahedral ion within the framework of a cube, whereupon the corresponding symmetry directions can be visualized easily; hence the following reducible representation is derived:

O_h	E	$8C_3$	$6C_2$	$6C_4$	$3C_2\,(\equiv 3C_4^2)$	i	$6S_4$	$8S_6$	$3\sigma_h$	$6\sigma_d$
Γ	6	0	0	2	2	0	0	0	4	2

Following the method used previously, this representation reduces to $A_{1g} + E_g + T_{1u}$. Character tables are provided in Appendix A10, but that for point group O_h is repeated here for convenience:

O_h	E	$8C_3$	$6C_2$	$6C_4$	$3C_2(\equiv 3C_4^2)$	i	$6S_4$	$8S_6$	$3\sigma_h$	$6\sigma_d$		
A_{1g}	1	1	1	1	1	1	1	1	1	1		$x^2+y^2+z^2$
A_{2g}	1	1	–1	–1	1	1	–1	1	1	–1		
E_g	2	–1	0	0	2	2	0	–1	2	0		$(2z^2-x^2-y^2, x^2-y^2)$
T_{1g}	3	0	–1	1	–1	3	–1	0	–1	1	(R_x, R_y, R_z)	
T_{2g}	3	0	1	–1	–1	3	–1	0	–1	1		(xz, yz, xy)
A_{1u}	1	1	1	1	1	–1	–1	–1	–1	–1		
A_{2u}	1	1	–1	–1	1	–1	1	–1	–1	1		
E_u	2	–1	0	0	2	–2	0	1	–2	0		
T_{1u}	3	0	–1	1	–1	–3	–1	0	1	1	(x, y, z)	
T_{2u}	3	0	1	–1	–1	–3	1	0	1	–1		

From this table, the relevant data follow:

Atomic orbitals	Functions for *p*/*d* orbitals	Irreducible representations
3*d*	$z^2, x^2 - y^2$	E_g
	xy, zx, yz	T_{2g}
4*s*	$x^2 + y^2 + z^2$	A_{1g}
4*p*	x, y, z	T_{1u}

A molecular orbital energy level diagram can be drawn to illustrate the bonding of the Fe^{2+} cation with six coordinating CN^- anion ligands (Fig. 8.20). The six cyanide ligand symmetry orbitals are denoted ψ_i $(i = 1 - 6)$ for $+x$, $-x$, $+y$, $-y$, $+z$, $-z$ in order; then the orbital functions, as determined by projection operators, may be listed with their representations as follow:

s	$1/\sqrt{6}(\psi_1 + \psi_2 + \psi_3 + \psi_4 + \psi_5 + \psi_6)$	A_{1g}
p_x	$1\sqrt{2}(\psi_1 - \psi_2)$	T_{1u}
p_y	$1\sqrt{2}(\psi_3 - \psi_4)$	T_{1u}
p_z	$1\sqrt{2}(\psi_5 - \psi_6)$	T_{1u}
d_z^2	$1/\sqrt{12}(-\psi_1 - \psi_2 - \psi_3 - \psi_4 + 2\psi_5 + 2\psi_6)$	E_g
$d_{x^2-y^2}$	$1/2(\psi_1 + \psi_2 - \psi_3 - \psi_4)$	E_g

The d_{z^2} orbital has twice the probability along the z axis compared with the x and y directions. Three d functions on the metal atom, d_{xy}, d_{zx} and d_{yz}, of representation T_{2g} have no matching representations among the functions of the cyanide ligands, so that these d functions are non-bonding.

There are 18 electrons in the molecular orbitals of the complex ion, and each of the molecular orbitals a_{1g}, t_{1u} and t_{2g} contains a pair of electrons with opposed spins. Where symmetry functions of differing energy are combined, the resulting molecular orbitals are different in energy. The lower energy bonding molecular orbitals assume the character of the ligand, and the 12 electrons involved are effectively donated to the metal atom. The six electrons from the metal then occupy the t_{2g} non-bonding orbitals. This description is approximate, and a precise treatment would demand an approach of a more quantitative character.

The energy gap between the t_{2g} and e_g^* levels is of importance in the ligand-field theory of coordination compounds. In a free metal ion, the d orbitals have a degeneracy of five. In the centre of an octahedrally disposed field of negative centres (ligands) lying along the x, y and z axes, the d orbitals are split into a groups of two, e_g^*, and a group of three, t_{2g}, of lower energy. The d_{z^2} orbital can be resolved into a linear combination of $d_{z^2-x_2}$ and $d_{z^2-y_2}(2z^2 - x^2 - y^2)$, each of which is spatially equivalent to $d_{x^2-y^2}$, but such that d_{z^2} has a probability of twice that of $d_{x^2-y^2}$; the d_{xy}, d_{zx} and d_{yz} orbital functions are also equivalent.

Valence electrons from the metal atom engage with the t_{2g} and e_g^* molecular orbitals. The first three electrons occupy the t_{2g} levels singly. The fourth to sixth electrons can either pair with the t_{2g} or enter the higher e_g^* level. If the d

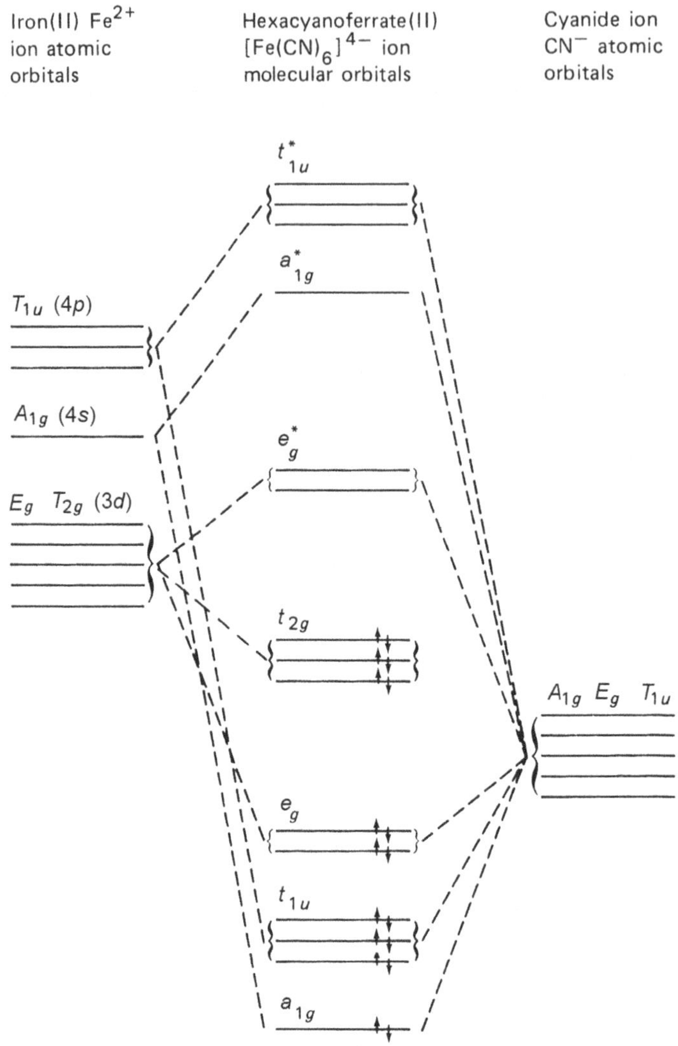

Fig. 8.20
Molecular orbital energy level diagram for the hexacyanoferrate(II) ion, $[Fe(CN)_6]^{4-}$. In the ion, molecular orbitals a_{1g}, t_{1u} and e_g are bonding, t_{2g} and e_g^* are non-bonding, and a_{1g}^* and t_{1u}^* are antibonding; the distance between the non-bonding t_{2g} and e_g^* is the ligand-field splitting energy, Δ. [Reproduced by courtesy of Woodhead Publishing, UK.]

orbital splitting energy is less than the pairing energy these electrons occupy the e_g^* level, whereas conversely, as in the case of the $[Fe\,(CN)_6]^{4-}$ ion, they pair in t_{2g}. The choice depends on the field strength of the ligand; the CN^- ion lies at the strong field (low spin) end of the nephelauxetic series of ligands, which favours pairing. The three d functions on the metal ion, d_{xy}, d_{zx} and d_{yz} of label t_{2g}, remain non-bonding.

The value and position of Δ relative to the t_{2g} and e_g^* energy levels varies with the strength of the ligand field. The e_g^* levels in this example lie at $\frac{3}{5}\Delta$ above the un-split d levels with the t_{2g} at $\frac{2}{5}\Delta$ below it. The variation of Δ with field strength is shown in Fig. 8.21. The fascinating field of coordination chemistry has a vast literature, of which [11–13] are significant examples.

8.3 Vibrational studies

A brief mention of vibration was given in Section 3.7.5, and it is desirable now to enlarge on this topic. The normal vibrational motions of a molecule result from a superimposition of a number of normal modes which are natural vibrations of the molecule, each characterized by a particular frequency.

The vibration of a normal mode may be likened to the vibration of a tuning fork. However, the tonal quality of a tuning fork is markedly different from that, say, of a violin string vibrating at the same frequency. The tuning fork emits an almost perfect sinusoidal vibration of a particular fundamental frequency, whereas the vibrating violin string contains a number of superimposed harmonics, which are integer multiples of the fundamental, as well as overtones.

In a similar manner the superimposition of the normal modes of a molecule leads to its characteristic vibrational frequency spectrum. A molecular frequency depends upon both the square root of the tension between the vibrating species, which arises from the interatomic forces between them, and the inverse of the square root of their effective mass. The halogen hydrides H*X* have a single vibration mode, and the frequencies are:

X	F	Cl	Br	I
$\overline{\nu}/\mathrm{cm}^{-1}$	3962	2886	2558	2233

It may be noted that spectroscopic 'frequencies' are traditionally quoted in the units of cm^{-1}; they are wave numbers: $\overline{\nu}/\mathrm{cm}^{-1} \times c/\mathrm{cm\ s}^{-1} = \nu/\mathrm{Hz}$, where c is the speed of light. For HCl, the vibration frequency is, thus, $2886\,\mathrm{cm}^{-1} \times 2.9979 \times 10^{10}\,\mathrm{cm\ s}^{-1} = 8.652 \times 10^{13}\,\mathrm{Hz}$.

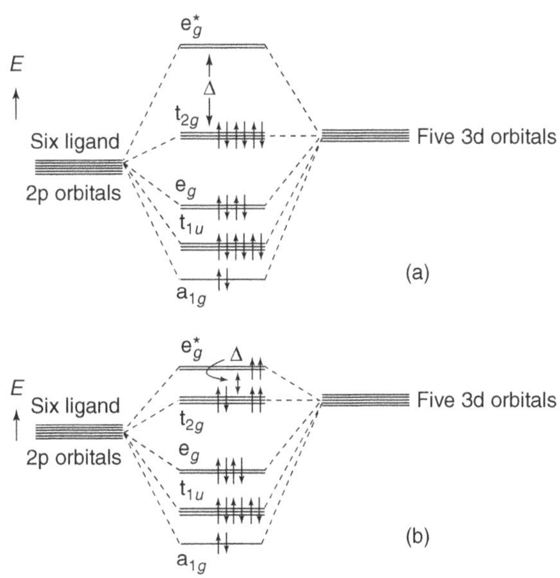

Fig. 8.21
Molecular orbital energy level diagrams for octahedral coordination. (a) Strong field (low spin), as with the $[\mathrm{Fe\,(CN)_6}]^{4-}$ ion, t_{2g}^6; the species has all spins coupled and is diamagnetic. (b) Weak field (high spin), as in the $[\mathrm{Co\,(SCN)_6}]^{3-}$ ion, $t_{2g}^6\ e_g^{*2}$; four unpaired electrons lead to significant paramagnetic properties. [Reproduced by courtesy of Woodhead Publishing, UK.]

8.3.1 Symmetry of normal modes

The number of normal vibrational modes for a molecule of N atoms is $3N - 6$, or $3N - 5$ for a diatomic species, as discussed in the previous chapter. The symmetry of the normal modes will be demonstrated with the water molecule as an example. The vibrations of symmetry related atoms are related one to the other, so that the normal modes are given symmetry labels, such as A_1 and E. In Section 7.5.3, the Γ_{3n} representation 9, $\bar{1}$, 1, 3 was derived for this species. It reduces to $3A_1 + A_2 + 2B_1 + 3B_2$. The rotational and translation motions span $A_1 + A_2 + 2B_1 + 2B_2$, which results in the symmetry $2A_1 + B_2$ for vibration, the expected number of three modes being the total dimensionality for vibration.

In order to associate the irreducible representations of the normal modes with the internal coordinates of the water molecule, refer again to Fig. 7.5; choose vectors $\mathbf{r}_a$ and $\mathbf{r}_b$ along O–H$_a$ and O–H$_b$, respectively, and let α be the H–O–H angle. With these internal coordinates as a basis, a representation can be generated by the unshifted atom method:

$C_{2}v$	E	C_2	σ_v	σ'_v
Γ_{vib}	3	1	1	3

Because no operation in C_{2v} interchanges $\mathbf{r}$ with α, $\Gamma_{\text{vib}} = \Gamma_{\mathbf{r}} + \Gamma_{\alpha}$. The four 3×3 matrices for Γ_{vib} based in internal coordinates can be block factored: the operation $\boldsymbol{C}_2$, for example, gives

$$\begin{pmatrix} 0 & 1 & 0 \\ 1 & 0 & 0 \\ 0 & 0 & 1 \end{pmatrix} \begin{pmatrix} \mathbf{r}_a \\ \mathbf{r}_b \\ \alpha \end{pmatrix} = \begin{pmatrix} \mathbf{r}'_a \\ \mathbf{r}'_b \\ \alpha' \end{pmatrix}$$

which leads to the character 0 for $\Gamma_{\mathbf{r}}$ and 1 for Γ_{α} under $\boldsymbol{C}_2$. Thus, the $\mathbf{r}$ vectors and the angle a form 2×2 and 1×1 bases, respectively:

C_{2v}	E	C_2	σ_v	σ'_v
$\Gamma_{\mathbf{r}}$	2	0	0	2
Γ_{α}	1	1	1	1

It is evident that $\Gamma_{\mathbf{r}}$ reduces to $A_1 + B_2$, and that Γ_{α} is the irreducible representation A_1.

The normal modes of the water molecule are illustrated in Fig. 8.22. The normal modes of A_1 symmetry must leave the symmetry of the molecule undisturbed, which can arise by the symmetric O–H bond stretch $\bar{\nu}_1$, and by the bending or 'breathing' H–O–H angle mode, $\bar{\nu}_2$. For B_1 symmetry, the character of –1 under C_2 symmetry in C_{2v} (Appendix 10) means that the hydrogen atoms must move in opposite sense, with consequent displacement of the oxygen atom, as shown by the asymmetric stretch $\bar{\nu}_3$.

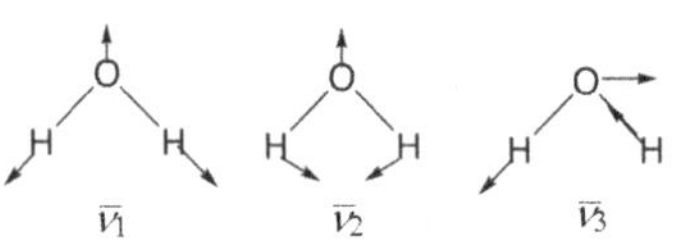

Fig. 8.22
Normal modes of the water molecule: $\bar{\nu}_1$, symmetric stretch, 3657 cm^{-1}; $\bar{\nu}_2$, symmetric bend, 1595 cm^{-1}; $\bar{\nu}_3$, asymmetric stretch, 3657 cm^{-1}.

8.3.2 Boron trifluoride

Boron trifluoride has point group D_{3h}, with the symmetry classes $\boldsymbol{E}$, $2\boldsymbol{C}_3$, $3\boldsymbol{C}_2'$, $\boldsymbol{\sigma}_h$, $2\boldsymbol{S}_3$, $3\boldsymbol{\sigma}_v$. A representation can be set up using orthogonal vectors set up at each of the three fluorine atoms (Fig. 8.23). Alternatively, a reducible representation can be generated in terms of the internal coordinates $\mathbf{r}_1, \mathbf{r}_2, \mathbf{r}_3$ directed along the three B–F bonds, and the F–B–F angle α.

The total number of degrees of freedom for boron trifluoride is twelve, of which six correspond to translational and rotational movements. The remaining six degrees correspond to the normal modes of vibration. Each normal mode forms a basis for an irreducible representation of the molecule, and each of the vectors that represent an instantaneous displacement of an atom may be compounded from the basis vectors.

The unshifted atom contributions (Section 7.5.3.1) lead to Γ_{3n} representation:

D_{3h}	$\boldsymbol{E}$	$2C_3$	$3C_2'$	σ_h	$2S_3$	$3\sigma_v$
Γ_{3n}	12	0	–2	4	–2	2

This representation reduces to $A_1' + A_2' + 3E' + 2A_2'' + E''$, which is the correct number of modes, 12. From the character table, it can be deduced that the translation function, Γ_{trans}, spans $E' + A_2''$ and the rotation function, Γ_{rot}, spans $A_2' + E'$; thus, Γ_{vib} corresponds to $A_1' + 2E' + A_2''$, which is correct for six vibrations because E' is doubly degenerate.

In terms of the internal coordinates $\mathbf{r}$, a representation may be determined on the B–F vectors $\mathbf{r}_n, n = 1 - 3$ (Fig. 8.23). The character of the matrix for each symmetry class in D_{3h} may be equated to the number of $\mathbf{r}$ vectors that are invariant under the class symmetry. Thus for $\boldsymbol{E}$,

$$\begin{pmatrix} 1 & 0 & 0 \\ 0 & 1 & 0 \\ 0 & 0 & 1 \end{pmatrix} \begin{pmatrix} \mathbf{r}_1 \\ \mathbf{r}_2 \\ \mathbf{r}_3 \end{pmatrix} = \begin{pmatrix} \mathbf{r}_1 \\ \mathbf{r}_2 \\ \mathbf{r}_3 \end{pmatrix},$$

and its character is 3, whereas for $\boldsymbol{C}_3$,

$$\begin{pmatrix} 0 & 0 & 1 \\ 1 & 0 & 0 \\ 0 & 1 & 0 \end{pmatrix} \begin{pmatrix} \mathbf{r}_1 \\ \mathbf{r}_2 \\ \mathbf{r}_3 \end{pmatrix} = \begin{pmatrix} \mathbf{r}_3 \\ \mathbf{r}_1 \\ \mathbf{r}_2 \end{pmatrix},$$

and its character is zero. Proceeding in this manner, the following reducible representation can be built, which will also represent Γ_α:

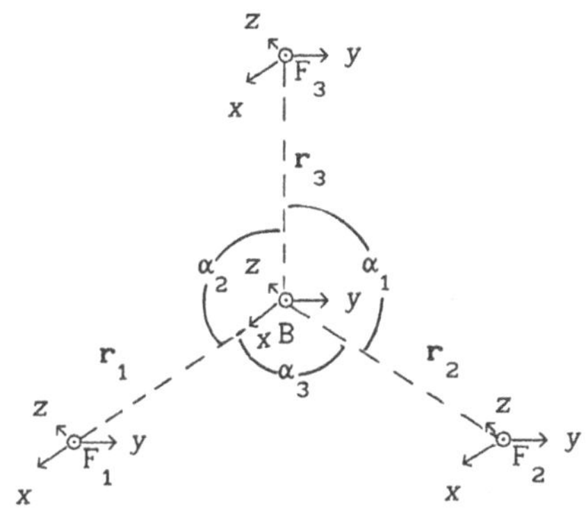

Fig. 8.23
Orthogonal x, y, z coordinate sets at the fluorine atoms, and $\mathbf{r}$, α internal coordinates for the boron trifluoride molecule, BF_3; $|\mathbf{r}_1| = |\mathbf{r}_2| = |\mathbf{r}_3|$ and $\alpha_1 = \alpha_2 = \alpha_3$. The z axis is normal to the plane, and the x axis is along $\mathbf{r}_1$; all operations of the point group, D_{3h}, act through the central, boron atom. [Reproduced by courtesy of Woodhead Publishing, UK.]

D_{3h}	E	$2C_3$	$3C_2'$	σ_h	$2S_3$	$3\sigma_v$
Γ_r	3	0	1	3	0	1
Γ_α	3	0	1	3	0	1

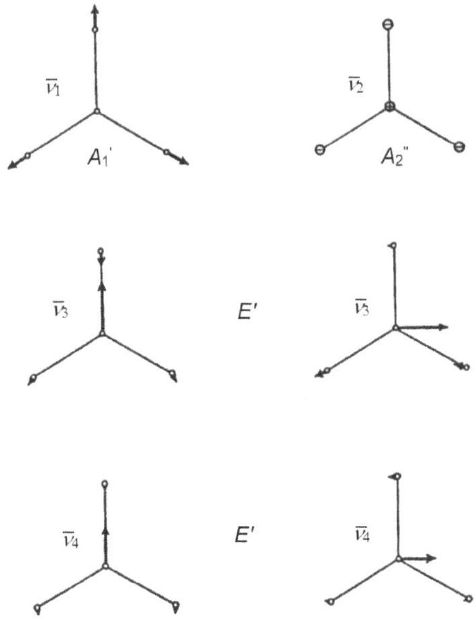

Fig. 8.24
Normal modes of vibration in boron trifluoride, and the corresponding irreducible representations: A_1', symmetric stretch; A_2'', symmetric bend (out-of-plane); E' ($\overline{\nu}_3$) and E' ($\overline{\nu}_4$) are the two doubly degenerate, asymmetric stretching and bending modes.

Reduction of these representations leads to

$$\Gamma_{\mathbf{r}} = A_1' + E'$$
$$\Gamma_\alpha = A_1' + E'$$

It is evident that $2A_1' + 2E'$ has been deduced, so that one result for A_1' is spurious. The A_1' representation in Γ_r is satisfied by the symmetric stretch $\overline{\nu}_1$, as shown in Fig. 8.24. For Γ_α, however, the representation A_1' would require simultaneous, equal increments or decrements in the angles α. Accordingly, one A_1' result is discarded, and the A_2'' irreducible representation assigned to the out-of-plane vibration $\overline{\nu}_2$. Such spurious cases arise with planar, tetrahedral and octahedral species, where the bond angles demonstrate closure (Section 7.2).

From the character table, the orbitals available for A_1' are d_{z^2} and s (s is always the totally symmetric A-type representation in any point group) and for E' the available functions are either p_x and p_y, or $d_{x^2-y^2}$ and d_{xy}. Group theory will not distinguish between these two possible sets of trigonal orbitals; from the point of view of symmetry, both are equally acceptable. However, from a chemical standpoint, since boron has no d orbitals that are energetically available for bonding, the s and p orbitals are selected to form molecular orbitals for the σ in-plane B–F bonds and π in-plane B–F bonds; they are of similar energy. The out-of-plane p_z orbital is π-bonded with the similar orbitals on the fluorine atoms.

Group theory determines symmetry based on the irreducible representations $A_2'' + E''$. Since there are no orbitals corresponding to E'', this representation corresponds to a degenerate pair of non-bonding orbitals: out-of-plane π- bonding is through A_2''; the full argument can be found in the literature [11] in terms of the $[CO_3]^{2-}$ ion, which also has D_{3h} symmetry.

8.3.3 Selection rules for infrared and Raman activity: dipole moment and polarizability

Vibrational frequency data on chemical species arise through infrared and Raman spectra; normally, both techniques are employed in a vibrational analysis. In infrared spectroscopy a radiation in the range 600 to 4000 cm^{-1} is passed through a sample of the species under investigation. The absorption at different wave numbers corresponds to excitation from the lowest vibrational energy level to higher levels. Not all molecules respond to infrared spectroscopy, and Raman spectroscopy forms a complementary procedure. In this method, the sample is irradiated with laser-generated (plane polarized) visible radiation. The molecules may then scatter radiation with a change in wavelength: on the one hand, when the irradiating photons collide with molecules and gives up some of their energy, the emergent less energetic *Stokes* scattered radiation is observed. On the other hand, when the photons collect energy from collision with the excited molecules, the higher frequency *anti-Stokes* radiation is produced. The actual emission of infrared and Raman spectra are governed by *selection rules* that are derived from a quantum mechanical analysis of molecular vibrations [14], and will be considered briefly here.

8.3.3.1 Infrared spectra

The general requirement for the production of infrared spectra by any species is that the molecule possesses an oscillating dipole. When interatomic bonds in a polar molecule stretch or bend in an applied electromagnetic field, the oscillating dipole reacts on the field of the source radiation, a part of which may be absorbed or emitted. Only those vibrations that can bring about a change in the dipole moment of the molecule can lead to the production of infrared spectra.

Selection rules for infrared spectra can be expressed in terms of vanishing integrals (Appendix A12). Let Ψ_i represent a wave function of a molecule in its initial (ground) state, and Ψ_f that of its final (excited) state. The change in energy is the fundamental transition $\Psi_i \rightarrow \Psi_f$, and the interaction between the molecule and the radiation is the *transition moment* $\mathcal{M}$, given by

$$\mathcal{M} = \int \Psi_i^* \boldsymbol{\mu}\, \Psi_f \mathrm{d}\tau \qquad (8.24)$$

where $\boldsymbol{\mu}$ is the dipole moment vector of the molecule. If this integral is non-zero, the transition will give rise to an infrared spectrum and the species is said to be *infrared active*. Conversely, if $\mathcal{M} = 0$, the particular transition is *infrared forbidden*.

Since $\boldsymbol{\mu}$ may be resolved into components, Eq. (8.24) may be broken down as

$$\mathcal{M}_i = \int \Psi_i \mu_i \Psi_f \mathrm{d}\tau \quad (i = x,\ y,\ z) \tag{8.25}$$

Ψ_i is used in place of its complex conjugate since the ground state wave function is real. Only one of Eqs. (8.25) need be non-zero for infrared activity to be possible and, as shown in Appendix A12, the necessary condition is that the direct product of the irreducible representations for the two vibrational states is or contains the irreducible representation to which μ_i belongs. This situation can be judged with the aid of the character table for the point group for the molecule under investigation.

The vibrational ground state wave function, Ψ_i, of the water molecule, for example, belongs to the A_1-type irreducible representation, and the direct product equations for dipole moment have been discussed in Appendix A12, where it was shown that the species is infrared active in both A_1 and B_2 symmetries.

The dipole moment variations can be related to the vibration modes, as shown diagrammatically below: (a) is the symmetric stretch; (b) is the symmetric bend; (c) is the asymmetric stretch of the H_2O molecule. It is clear that the dipole moment is changed by vibration in each of the modes (a)–(c) below, so that the species will give rise to infrared spectral lines.

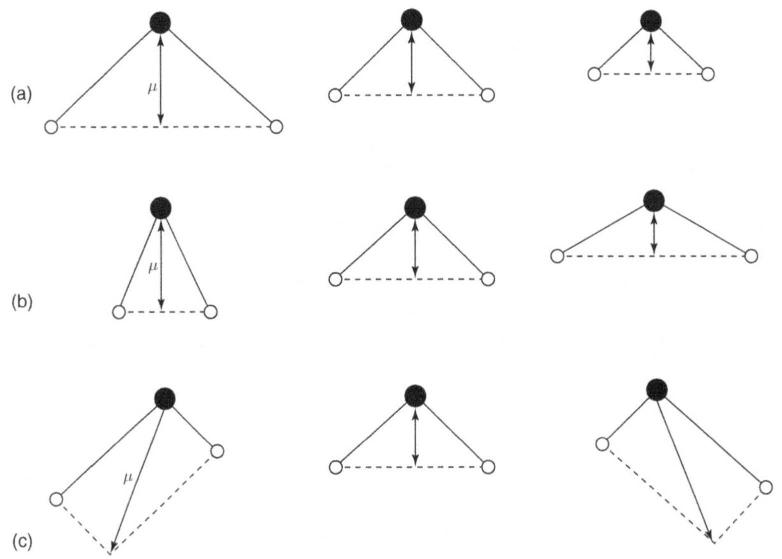

In the boron trifluoride vibration modes (Fig. 8.24), the A_1' symmetric stretch mode, along the B–F bonds, is totally symmetric and does not involve a change in the balance of bond moments; hence, this vibration is not infrared active. All other modes give rise to bands in the infrared spectrum.

8.3.3.2 Raman spectra

Raman activity is governed by the polarizability of the species. The electron density of a molecule is distorted in the presence of an applied electric field; electrons are attracted towards the positive pole of the field and the nuclei towards the negative pole. This separation gives rise to a polarizability P that is proportional to the strength of the electric field. Polarizability is a tensor property:

$$P_x = \alpha_{x,x} E_x + \alpha_{x,y} E_y + \alpha_{x,z} E_z$$
$$P_y = \alpha_{y,x} E_x + \alpha_{y,y} E_y + \alpha_{y,z} E_z$$
$$P_z = \alpha_{z,x} E_x + \alpha_{z,y} E_y + \alpha_{z,z} E_z$$

or concisely,

$$\mathbf{P} = \boldsymbol{\alpha} \mathbf{E} \tag{8.26}$$

where $\boldsymbol{\alpha}$ is the polarizability tensor. In Raman scattering, $\alpha_{i,j} = \alpha_{j,i}$ $(i, j = x, y, z)$, and if the vibration of the molecule changes the value of any $\alpha_{i,j}$ polarizability component, the vibration gives rise to Raman spectral lines. Polarizability can be represented by an ellipsoid, and its changes are shown by the diagram below for an AB_2 species as the electric field switches from positive to negative:

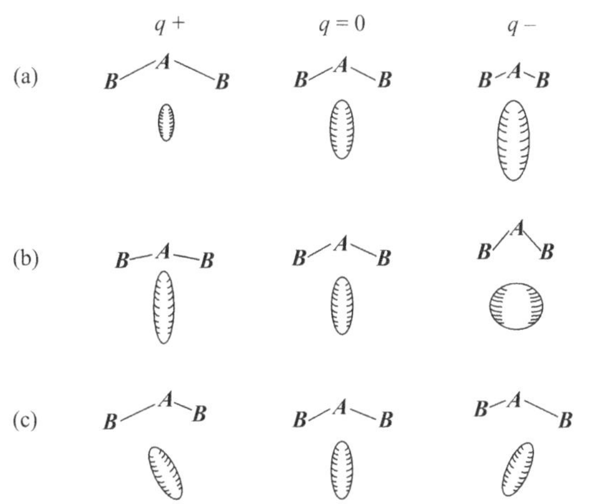

In the symmetric stretch (a), the polarizability ellipsoid changes *size*; in the symmetric bend (b), it changes *shape*; and in the asymmetric stretch (c), it is its *orientation* that alters. There exists also a *mutual exclusion rule*, namely, that a centrosymmetric species cannot exhibit both infrared and Raman activity on one and the same spectral band. In centrosymmetric species, infrared active modes, those for which the irreducible representations span translational movements x and/or y and/or z, have a character of -1 under operation $\boldsymbol{i}$, while those that span products xy and/or yz and/or zx have a character $+1$ under $\boldsymbol{i}$. In the character tables for centrosymmetric point groups, no one irreducible representation spans both the translational and product terms.

In group theory terms, a species is Raman active if any one of the six integrals $\mathscr{I}_{ij}$, $(i,j = 1-3)$ is non-zero:

$$\mathscr{I}_{ij} = \int \Psi_i \alpha_{ij} \Psi_f \, \mathrm{d}\tau \tag{8.27}$$

The Raman spectra arise from the inelastic scattering of radiation: if the frequency change $\Delta\nu$ is negative, that is, energy is transferred from the photon to the molecule, the Raman scattering is known as Stokes radiation; conversely, when $\Delta\nu$ is positive anti-Stokes spectra are produced.

Spectra can be examined for Raman activity by forming the direct products from the symbolic expression

$$\Psi_i \otimes \begin{pmatrix} \alpha_{x^2} \\ \alpha_{y^2} \\ \alpha_{z^2} \\ \alpha_{xy} \\ \alpha_{yz} \\ \alpha_{zx} \end{pmatrix} \otimes \Psi_f \tag{8.28}$$

where the irreducible representations spanned by the α_{ij} terms are determined from an appropriate character table.

Example 8.1

Determine the infrared and Raman activity for (a) ammonia (C_{3v}), and (b) benzene (D_{6h}) with $\mathbf{C}_2'$ and $\boldsymbol{\sigma}_v$ orientated so as to pass through opposite carbon atoms on the ring.

(a) Set up orthogonal axes at each atom and determine the number of unshifted atoms for each operation of the point group, remembering that all symmetry elements pass through the nitrogen atom. For NH_3,

C_{3v}	E	$2C_3$	$3\sigma_v$
Γ_{3n}	12	0	2

which reduces to $3A_1 + A_2 + 4E$. From the character table, translations span $A_1 + E$, and rotations $A_2 + E$; thus, vibrations are represented by $2A_1 + 2E$.

Infrared

For A_1: $A_1 \otimes \begin{pmatrix} A_1 \\ E \end{pmatrix} \otimes A_1 = A_1$ and E

For E: $A_1 \otimes \begin{pmatrix} A_1 \\ E \end{pmatrix} \otimes E = E$ and $A_1 + A_2 + E$

The molecule is infrared active, with two bands in its spectrum.

Raman

The product terms in C also span A_1 and E. Hence, there are two bands in the Raman spectrum, *coincident* with those in the infrared.

(b) In a similar manner with benzene, a representation may be deduced as

D_{6h}	E	$2C_6$	C_2	$3C_2'$	$3C_2''$	i	$2S_3$	$2S_6$	σ_h	$3\sigma_d$	$3\sigma_v$
Γ_{3n}	36	0	0	−4	0	0	0	0	12	0	4

which reduces to $2A_{1g} + 2A_{2g} + 2B_{2g} + 2E_{1g} + 4E_{2g} + 2A_{2u} + 2B_{1u} + 2B_{2u} + 4E_{1u} + 2E_{2u}$. The translational and rotational displacements together span $A_{2g} + E_{1g} + A_{2u} + E_{1u}$, so that vibrations span $2A_{1g} + A_{2g} + 2B_{2g} + E_{1g} + 4E_{2g} + A_{2u} + 2B_{1u} + 2B_{2u} + 3E_{1u} + 2E_{2u}$, corresponding to the required $3N - 6$, or 30 modes. These vibrations may be divided into in-plane and out-of-plane motions:

$$\Gamma_{\text{in}} = 2A_{1g} + A_{2g} + 4E_{2g} + 2B_{1u} + 2B_{2u} + 3E_{1u}$$
$$\Gamma_{\text{out}} = 2B_{2g} + E_{1g} + A_{2u} + 2E_{2u}$$

By forming the appropriate direct products, there results

Infrared active: $A_{2u} + 3E_{1u}$
Raman active: $2A_{1g} + E_{1g} + 4E_{2g}$

which the reader is invited to confirm, and to show that the other vibrations from the benzene molecule are both infrared and Raman forbidden.

Example 8.1 demonstrates the mutual exclusion rule: this result follows from the fact that the translational displacements x, y and z are antisymmetric (u) under $\boldsymbol{i}$ symmetry, whereas product terms such as z^2 and xy are symmetric (g) under $\boldsymbol{i}$ symmetry.

8.3.3.3 Polarization and Raman spectra

On examining laser-induced Raman scattered light with a second polarizer, it may be determined whether or not the scattered radiation is polarized or depolarized (the plane of polarization of the laser light rotated by 90°). A theoretical treatment [15–17] shows that polarized Raman spectra arise from transitions that involve the totally symmetric A_1-type vibrational mode, other transitions being depolarized. Typically, the depolarized radiation has an intensity 75% or more less than that of the polarized radiation. To the examples considered above, the following details may be added:

Species	Symmetry	Number of spectral lines	Nature
H_2O	A_1	2	Polarized
	B_2	1	Depolarized
NH_3	A_1	2	Polarized
	E	2	Depolarized
C_6H_6	A_{1g}	2	Polarized
	E_{1g}	1	Depolarized
	E_{2g}	4	Depolarized

8.3.4 Harmonics and combination vibrations

Just as a violin string can produce different tonal qualities according the manner and position in which the bow is applied, so under perturbations, thermal or radiant, spectra other than the first order type discussed so far can be excited and observed in a spectral record. Harmonics can be of the type $\simeq 2\overline{\nu}_2$, where $\overline{\nu}_2$ is the fundamental, whereas combinations are of the type $\simeq (\overline{\nu}_3 + \overline{\nu}_4)$. The approximate equality expresses the fact that the perturbations required to excite harmonics and overtones can also perturb the fundamental frequencies. An example that illustrates some of these properties is methane, which has been discussed elsewhere. Under the point group symmetry T_d of methane, a representation can be derived by the unshifted atom procedure:

T_d	E	$8C_3$	$3C_2$	$6S_4$	$6\sigma_d$
Atoms	5	2	1	1	3
$\chi_{\text{per atom}}$	3	0	$\overline{1}$	$\overline{1}$	1
Γ	15	0	$\overline{1}$	$\overline{1}$	3

which reduces to $A_1 + E + 2T_2$, a total of nine modes. The fundamental frequencies and examples of perturbation frequencies are listed in Table 8.5, where the extent of the applicability of the predictions can be judged. The symmetry of an overtone or combination term is the direct product (Appendix A10.2) of the symmetries of the contributing terms. Thus for $2\overline{\nu}_2$, as shown in Table 8.5, the symmetry of the overtone is $E \otimes E = A_1 + A_2 + E$. Further information on harmonic and combination frequencies can be gleaned from the literature [18].

8.4 Group theory and point groups

Point groups, both crystallographic and non-crystallographic were studied in an earlier chapter, and a derivation given by a descriptive, semi-analytical procedure. Here, they are examined more analytically, akin to their first derivations, among which those of Hessel [19] and Frankenheim [20] were probably the earliest. The concern here is with the crystallographic point groups, those that link with the 230 space groups. They may be defined as 3×3 orthogonal matrix groups that operate on a Bravais lattice so as to leave it unchanged. The orthogonal representations are either proper rotations or improper rotations. The matrix for a proper rotation has a determinant value of $+1$, whereas that for an improper rotation is -1.

The traces of all rotation matrices are integral. In fact, it is shown in Section 4.8 and Appendix A3.7 that in crystals, where the rotational degrees are $1, 2, 3, 4$ and 6, the trace of a rotation matrix is one of the integers $-1, 0, 1, 2$ or 3. This condition restricts the number of point groups in periodic crystals, although rotational symmetry *per se* can exist in any degree from 1 to infinity. The 32 point groups can be considered conveniently under four subheadings.

Table 8.5 Fundamental, overtone and combination frequencies for methane.

Assignment	Symmetry	Mode	Frequency/cm^{-1}
$\overline{\nu}_1$	A_1	Fundamental	2914
$\overline{\nu}_2$	E	Fundamental	1526
$\overline{\nu}_3$	T_2	Fundamental	3020
$\overline{\nu}_4$	T_2	Fundamental	1306
$2\overline{\nu}_2$	$A_1 + A_2 + E$	Overtone	3067[a]
$\overline{\nu}_3 - \overline{\nu}_4$	$A_1 + E + T_1 + T_2$	Combination	1720[b]

[a] Predicted $\simeq 2\overline{\nu}_2 = 3052$ cm^{-1}
[b] Predicted $\simeq (\overline{\nu}_3 - \overline{\nu}_4) = 1714$ cm^{-1}

8.4.1 Cyclic point groups

The cyclic groups of main crystallographic interest are C_n, $(n = 1, 2, 3, 4, 6)$. They are Abelian groups and comprise the operators $\boldsymbol{E}$ and $\mathbf{C}_n$. Thus, for $n = 4$, the group is $C_4\{\boldsymbol{E},\ C_4,\ \boldsymbol{C}_4^2(\equiv \boldsymbol{C}_2),\ \boldsymbol{C}_4^3\}$. The principal axis is C_4, and is the z reference axis for an example object under examination; the group order h is 4. Evidently, there are just five such cyclic groups.

8.4.2 Dihedral point groups

In order to generate dihedral point groups, other C_n axes are added to the cyclic groups. First, consider combining C_n with C_2. It is not difficult to see, particularly with the aid of a stereogram or a solid model, that C_2 must be either coincident with C_n. or normal to it, if indistinguishability is to be achieved after the symmetry operations. In addition, the matrix $\begin{pmatrix} \cos\theta & \overline{\sin\theta} & 0 \\ \sin\theta & \cos\theta & 0 \\ 0 & 0 & 1 \end{pmatrix}$ with a general angle θ would not otherwise lead to an integral trace -1 to $+3$. Four new, dihedral groups can be generated now by the direct products of C_2 with $C_n(n = 2, 3, 4, 6)$. For example,

$$C_4\left\{\boldsymbol{E},\ C_4,\ \boldsymbol{C}_4^2(\equiv \boldsymbol{C}_{2_{[001]}}),\ \boldsymbol{C}_4^3\right\} \otimes C_2\left\{\boldsymbol{E},\ \boldsymbol{C}_{2_{[100]}}\right\}$$

It is necessary to take note of the orientation of the $\boldsymbol{C}_2$ operators. For the direct product $\boldsymbol{C}_4 \otimes \boldsymbol{C}_{2_{[100]}}$, for example:

$$\begin{pmatrix} 0 & \overline{1} & 0 \\ 1 & 0 & 0 \\ 0 & 0 & 1 \end{pmatrix} \begin{pmatrix} 1 & 0 & 0 \\ 0 & \overline{1} & 0 \\ 0 & 0 & \overline{1} \end{pmatrix} = \begin{pmatrix} 0 & 1 & 0 \\ 1 & 0 & 0 \\ 0 & 0 & \overline{1} \end{pmatrix}$$

the result is the matrix for a twofold rotation along the direction [110] in a crystal with fourfold symmetry along z. Proceeding in this way builds up the group

$$\left\{\boldsymbol{E},\ C_4,\ \boldsymbol{C}_4^2(\equiv \boldsymbol{C}_{2_{[001]}}),\ \boldsymbol{C}_4^3,\ \boldsymbol{C}_{2_{[110]}},\ \boldsymbol{C}_{2_{[010]}},\ \boldsymbol{C}_{2_{[100]}},\ \boldsymbol{C}_{2_{[1\overline{1}0]}}\right\}$$

which is the dihedral group D_4. The other dihedral groups D_2, D_3 and D_6 can be deduced in this manner, with C_n, $(n = 2, 3, 6)$ remaining as the principal axis. Two dihedral groups belong to the cubic system, and they need special consideration.

8.4.3 Cubic rotation point groups

Another way of appreciating dihedral groups is through n-sided regular polygons containing rotations that leave the polygon invariant. They are, again, the rotations of the principal C_n axis along z, together with n twofold axes in the x, y plane; the diagram shows a tetrahedron set in a cube such that their common

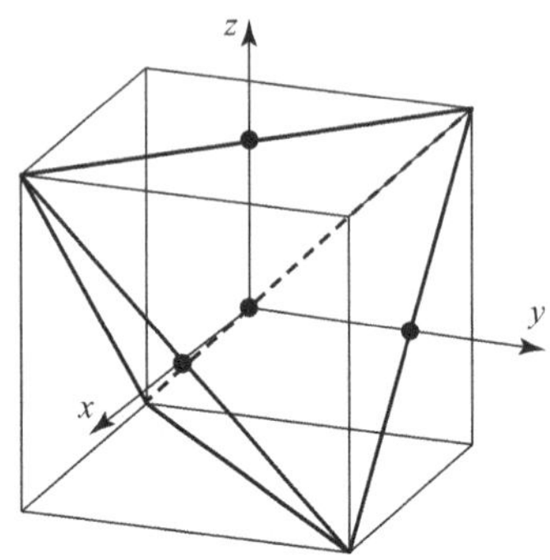

symmetry operators are coincident. An obvious feature of the regular tetrahedron is the presence of threefold proper rotations passing through the four apices and normal to the <111> form of planes. It may be noted here that a non-cyclic group of order greater than two must itself have an even order. Hence, a group $G\{\boldsymbol{E},\ 8\boldsymbol{C}_3,\ 8\boldsymbol{C}_3^2\}$ does not exist, one reason being that its order would be 17; thus, it is necessary to consider the combination of the $\boldsymbol{C}_3$ operators. Let the C_3 axes be along [111] and $[11\bar{1}]$. Then,

$$\begin{pmatrix} 0 & 1 & 0 \\ 0 & 0 & \bar{1} \\ \bar{1} & 0 & 0 \end{pmatrix} \begin{pmatrix} 0 & 0 & 1 \\ 1 & 0 & 0 \\ 0 & 1 & 0 \end{pmatrix} = \begin{pmatrix} 1 & 0 & 0 \\ 0 & \bar{1} & 0 \\ 0 & 0 & \bar{1} \end{pmatrix}$$

or $\boldsymbol{C}_{3_{[11\bar{1}]}}\boldsymbol{C}_{3_{[111]}} = \boldsymbol{C}_{2_{[100]}}$. Other combinations of $\boldsymbol{C}_3$ operators generate twofold axes along y and z, thus generating the cubic point group T.

Since a cube contains fourfold rotational symmetry, as well as the threefold symmetry just discussed, it is necessary to consider the combination of C_4 and C_3 symmetry operators. The C_4 axis may be set along the z axis, and the combination $\boldsymbol{C}_{4_{[001]}}\boldsymbol{C}_{3_{[111]}}$ determined in the usual manner:

$$\begin{pmatrix} 0 & \bar{1} & 0 \\ 1 & 0 & 0 \\ 0 & 0 & 1 \end{pmatrix} \begin{pmatrix} 0 & 0 & 1 \\ 1 & 0 & 0 \\ 0 & 1 & 0 \end{pmatrix} = \begin{pmatrix} \bar{1} & 0 & 0 \\ 0 & 0 & 1 \\ 0 & 1 & 0 \end{pmatrix}$$

The resulting matrix corresponds to a twofold rotation about [011] (Appendix A3.7.1). Since a product of two operations does not, in general, commute, the combination $\boldsymbol{C}_{3_{[111]}}\boldsymbol{C}_{4_{[001]}}$ could be different; it is, in fact, a twofold rotation about [101]. Building up a group in this way leads to the operations $\{\boldsymbol{E},\ 8\boldsymbol{C}_3,\ 6\boldsymbol{C}_2',\ 6\boldsymbol{C}_4,\ 3\boldsymbol{C}_2(\equiv 3\boldsymbol{C}_4^2)$, which represents the cubic point group O.

So far 11 rotation point groups, cyclic, dihedral and cubic, have been derived, as underlined:

$$\underline{C_1},\ \underline{C_2},\ \underline{C_3},\ \underline{C_4},\ \underline{C_6},\ \underline{D_2},\ \underline{D_3},\ \underline{D_4},\ \underline{D_6},\ \underline{T},\ \underline{O}$$

8.4.4 Point groups from combinations of operators

Since none of the eleven point groups deduced so far contains an improper rotation, or mirror plane, progress can be made by combining each of the above groups with a mirror plane operator. The group C_1 is the identity group, so that the product of the $\boldsymbol{\sigma}_h$ and $\boldsymbol{C}_1$ operators is $\sigma_h\boldsymbol{E}=\boldsymbol{\sigma}_h$, and this point group is usually written as C_s ($\equiv S_1$). The combination of $\boldsymbol{C}_2$ with $\boldsymbol{\sigma}_h$ may be formed as the direct product $\{\boldsymbol{E},\ \boldsymbol{C}_{s,(x,y)}\} \otimes \{\boldsymbol{E},\ \boldsymbol{C}_{2,[001]}\}$ and the result, determined with the appropriate matrices, is $\{\boldsymbol{E},\ \boldsymbol{C}_2,\ \boldsymbol{i},\ \boldsymbol{\sigma}_h\}$ and corresponds to point group C_{2h}. In general, the combination $\boldsymbol{\sigma}_h\boldsymbol{C}_n$ leads to the point groups of type C_{nh}.

Proceeding next with the D_n groups, and taking D_3 as an example: $D_3 \otimes \sigma_h = \{\boldsymbol{E},\ \boldsymbol{C}_3,\ \boldsymbol{C}_3^2,\ \boldsymbol{C}_2',\ \boldsymbol{C}_{2'}',\ \boldsymbol{C}_{2''}'\}\{\boldsymbol{E},\ \boldsymbol{\sigma}_h\}$. The appropriate group multiplication tables show that $\boldsymbol{C}_3\boldsymbol{\sigma}_h = \boldsymbol{S}_3$ and $\boldsymbol{C}_2'\boldsymbol{\sigma}_h = \boldsymbol{\sigma}_v$. Hence, the direct product evolves as $\{\boldsymbol{E},\ 2\boldsymbol{C}_3,\ 3\boldsymbol{C}_2',\ \boldsymbol{\sigma}_h,\ 2\boldsymbol{S}_3,\ 3\boldsymbol{\sigma}_v\}$, which is an expression of point group D_{3h}. Similar results are obtained with D_2, D_4 and D_6.

For the two cubic groups listed in Section 8.4.3, $\sigma_h \otimes T = T_h$, and $\sigma_h \otimes O = O_h$. It will be noted that with some of these σ_h combinations, specifically with $n = 2, 4$ and 6, a centre of symmetry has arisen. The eleven new groups determined are as follow:

$$C_{1h} \equiv \underline{C_s},\quad \underline{C_{2h}},\quad \underline{C_{3h}},\quad \underline{C_{4h}},\quad \underline{C_{6h}},\quad \underline{D_{2h}},\quad \underline{D_{3h}},\quad \underline{D_{4h}},\quad \underline{D_{6h}},\quad \underline{T_h},\quad \underline{O_h}$$

Since a centre of symmetry has been now identified as a group symmetry element, the products of the groups deduced so far that are lacking a centre of symmetry C_i may be studied. Just three new groups are derived from the combinations with C_i:

$$C_i \otimes C_1 = \underline{C_i},\quad C_i \otimes C_3 = \underline{S_6},\quad C_i \otimes D_3 = \underline{D_{3d}}$$

It is left as an exercise to the reader to show that no other products of the groups deduced so far with C_i lead to any other new groups.

The next combinations to investigate are the products of a σ_v plane containing the C_n axis and the point groups deduced above. In this case, the only the new point groups obtained are combinations with the cyclic groups, leading to

$$\underline{C_{2v}},\quad \underline{C_{3v}},\quad \underline{C_{4v}},\quad \underline{C_{6v}}.$$

One symmetry operation that has not appeared in any of the groups so far is $\boldsymbol{S}_4$. The *operator* $\boldsymbol{S}_4$ may be generated by the combination $\boldsymbol{C}_4^3\boldsymbol{i}$, which, in this case, gives the same result as $\boldsymbol{i}\boldsymbol{C}_4^3$:

$$\underset{\boldsymbol{C}_4^3}{\begin{pmatrix} 0 & 1 & 0 \\ \bar{1} & 0 & 0 \\ 0 & 0 & 1 \end{pmatrix}} \underset{\boldsymbol{i}}{\begin{pmatrix} \bar{1} & 0 & 0 \\ 0 & \bar{1} & 0 \\ 0 & 0 & \bar{1} \end{pmatrix}} = \underset{\boldsymbol{S}_4}{\begin{pmatrix} 0 & \bar{1} & 0 \\ 1 & 0 & 0 \\ 0 & 0 & \bar{1} \end{pmatrix}}$$

In the Hermann–Mauguin notation, $\bar{\boldsymbol{4}}$ is obtained by the combination $\boldsymbol{4}\bar{\boldsymbol{1}}$ or $\bar{\boldsymbol{1}}\boldsymbol{4}$. The *point group* $\underline{S_4}$, or $\bar{4}$ in the Hermann–Mauguin notation, is worthy of special mention: it is unique in that it cannot be determined by any other point group or combinations thereof; the point group product $C_4\,\sigma_h$ determines point group C_{4h}; for this reason, care is needed when defining the roto-reflection operation (see Section 3.10.1).

Although some of the point groups derived so far do include S_4 as a subgroup, C_{4h} and O_h for example, the two point groups yet undetected also involve S_4. Consider the combination of S_4, first with σ_v and then with T. Using the appropriate matrices, the operators in $\sigma_v \otimes S_4$ are: $\{\boldsymbol{E},\ \boldsymbol{\sigma}_v\}\{\boldsymbol{E},\ \boldsymbol{S}_4,\ \boldsymbol{C}_2,\ \boldsymbol{S}_4^3\} = \{\boldsymbol{E},\ \boldsymbol{S}_4,\ \boldsymbol{C}_2,\ \boldsymbol{S}_4^3,\ \boldsymbol{\sigma}_v,\ \boldsymbol{C}_2',\ \boldsymbol{\sigma}_v',\ \boldsymbol{C}_{2'}'\}$. Setting the resulting products into symmetry classes, in the standard orientation, it becomes $\{\boldsymbol{E},\ 2\boldsymbol{S}_4,\ \boldsymbol{C}_2,\ 2\boldsymbol{C}_2',\ 2\boldsymbol{\sigma}_d\}$, which corresponds to point group $\underline{D_{2d}}$; the change from σ_v to σ_d occurs because the vertical planes now lie between the twofold axes.

The remaining point group of the thirty-two is in the cubic system, and is obtained when point group S_4 is combined with T: $\{\boldsymbol{E},\ 4\boldsymbol{C}_3,\ 4\boldsymbol{C}_3^2,\ 3\boldsymbol{C}_2\}\{\boldsymbol{E},\ \boldsymbol{S}_4,\ \boldsymbol{C}_2,\ \boldsymbol{S}_4^3\} = \{\boldsymbol{E},\ 8\boldsymbol{C}_3,\ 3\boldsymbol{C}_2,\ 6\boldsymbol{S}_4,\ 6\boldsymbol{\sigma}_d\}$, and the result is an expression of point group $\underline{T_d}$. The reader should write these 32 point group symbols (underlined herein) now in the Hermann–Mauguin notation and note that they correspond exactly with those deduced in Section 3.6ff.

There are methods of deriving the 32 point groups, other than the two used in this work; they are group theory methods and guidance may be found in the literature [21–26].

8.5 Group theory and space groups

Space groups can be derived by the descriptive, semi-analytical operations that were applied in Chapter 5 and, perhaps more directly, by using group operations that may be expressed concisely by the Seitz operator:

$$\{\boldsymbol{R}\,|\,\mathbf{t}\}\mathbf{r} \tag{8.29}$$

This is a compact form of Eq. (5.2), and it implies the operator $\boldsymbol{R}$ acting on a vector $\mathbf{r}$, coupled with the translation vector $\mathbf{t}$; the diagram below is an illustration the Seitz operator:

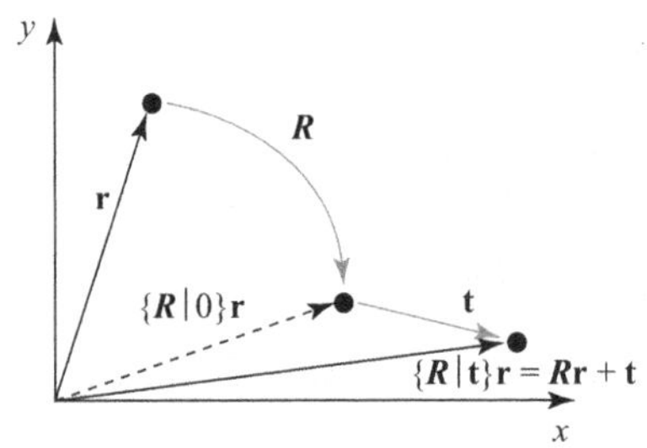

The operations of Eq. (5.3), for example, form a real affine group, of which space groups are subgroups; in no way is the definition of space group in Section 5.1 challenged by this approach (See also Appendix 13). The symmorphic and non-symmorphic groups need not be treated separately here; their difference has already been made clear.

The Seitz operators follow four important rules:

1. For two operators $\boldsymbol{R}_1$ and $\boldsymbol{R}_2$: $\{\boldsymbol{R}_2|\mathbf{t}_2\}\{\boldsymbol{R}_1|\mathbf{t}_1\} = \{\boldsymbol{R}_2\boldsymbol{R}_1|(\boldsymbol{R}_2\mathbf{t}_1 + \mathbf{t}_2)\}$.

For if $\{\boldsymbol{R}_1|\mathbf{t}_1\}\mathbf{r} = \mathbf{r}'$, then

$$\begin{aligned}\{\boldsymbol{R}_2|\mathbf{t}_2\}\{\boldsymbol{R}_1|\mathbf{t}_1\}\mathbf{r} &= \{\boldsymbol{R}_2|\mathbf{t}_2\}\mathbf{r}' = \boldsymbol{R}_2\mathbf{r}' + \mathbf{t}_2 \\ &= \boldsymbol{R}_2(\boldsymbol{R}_1\mathbf{r} + \mathbf{t}_1) + \mathbf{t}_2 = (\boldsymbol{R}_2\boldsymbol{R}_1)\mathbf{r} + \boldsymbol{R}_2\mathbf{t}_1 + \mathbf{t}_2 \\ &= \{\boldsymbol{R}_2\boldsymbol{R}_1|(\boldsymbol{R}_2\mathbf{t}_1 + \mathbf{t}_2)\}\mathbf{r}\end{aligned} \quad (8.30)$$

But since $(\boldsymbol{R}_2\mathbf{t}_1 + \mathbf{t}_2) = \mathbf{t}_3$, where $\mathbf{t}_3$ is another translation, it follows that the product of two Seitz operators is an operator in the same set as $\boldsymbol{R}_1$ and $\boldsymbol{R}_2$.

2. The identity operator is $\{\boldsymbol{E}|\mathbf{0}\}$.
3. The inverse operator $\{\boldsymbol{R}|\mathbf{t}\}^{-1}$ is $\{\boldsymbol{R}^{-1}|-\boldsymbol{R}^{-1}\mathbf{t}\}$.
   For $\{\boldsymbol{R}^{-1}|-\boldsymbol{R}^{-1}\mathbf{t}\}\{\boldsymbol{R}|\mathbf{t}\} = \{\boldsymbol{R}^{-1}\boldsymbol{R}|(\boldsymbol{R}^{-1}\mathbf{t} - \boldsymbol{R}^{-1}\mathbf{t})$ from (8.30), which is $\{\boldsymbol{E}|\mathbf{0}\}$; hence,

$$\{\boldsymbol{R}|\mathbf{t}\}^{-1} = \{\boldsymbol{R}^{-1}|-\boldsymbol{R}^{-1}\mathbf{t}\} \quad (8.31)$$

4. The associative law (Section 7.2) holds for Seitz operators.

8.5.1 Triclinic and monoclinic space groups

Since there is only one triclinic lattice and two point groups consistent with this lattice, it transpires that the triclinic space groups are soon determined as C_1^1 and C_i^1. The Schönflies notation for space groups is the point group notation augmented by a numerical superscript to distinguish between different space groups belonging to one and the same point group. The derivation of space groups will be considered here only in sufficient detail to show the application of the Seitz operator in deriving space groups. The Hermann–Mauguin symbol provides all the information needed to define the space group, once the meanings of the three positions in the symbol are clearly understood.

The argument in Section 4.4.3 will be followed, but for the monoclinic system. In this system, there are three point groups, C_2, C_s and C_{2h} and two lattices: the lattice types are not distinguished by the space group symbol; thus, in the Schönflies notation, all monoclinic space groups symbols are based on the cyclic symbol, C. The twofold axis is set along z: this is known as the crystallographic *first setting*, whereas the conventional crystallographic choice places the twofold axis along y, known as the *second setting*.

In the monoclinic system, the orientation of a symmetry direction must not conflict with the conditions of the monoclinic lattice. For example, the translation vector $\mathbf{b}$ is related to $-\mathbf{b}$ by a twofold rotation about z, which means that the b direction is perpendicular to that of c. If it were not so, then the twofold rotation would be no longer compatible with the monoclinic lattice. Thus, any symmetry additional to C_2 must occupy the same direction, in this case that normal to z, as in C_{2h}.

Let a twofold axis lie along z and a glide plane normal to it with a translational component of $b/2$. The point group is C_{2h} with the operators

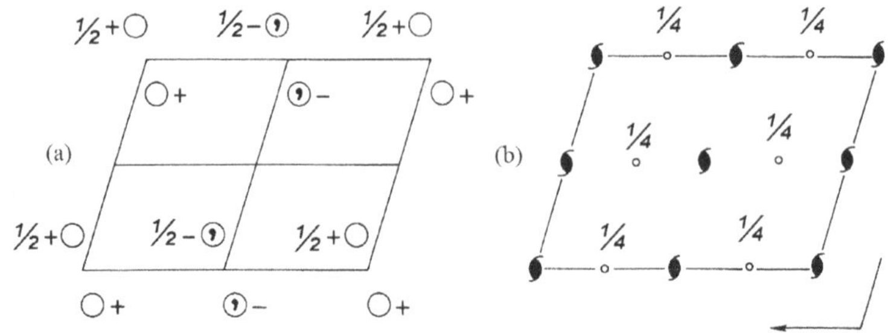

Fig 8.25
Space group $C_{2h}^5(P2_1/b)$ illustrated on the a, b plane. (a) General equivalent positions. (b) Symmetry elements; the arrow indicates the direction of the b glide translation. The standard setting of this space group may be achieved by a shift of origin to 0, ¼, ¼ .

$\{\boldsymbol{E},\ \boldsymbol{C}_2,\ \boldsymbol{i},\ \boldsymbol{\sigma}_h\}$, as with any space group derived from it; the differences with space groups arise from the values and positions of $\mathbf{t}$. Consider the following symmetry operators in the Seitz notation and their actions:

$$\{\boldsymbol{E}|\mathbf{0}\}\begin{pmatrix}x\\y\\z\end{pmatrix}=\begin{pmatrix}x\\y\\z\end{pmatrix}\qquad \{\boldsymbol{C}_{2[001]}|\mathbf{t}_{0,0,{}^1/_2}\}\begin{pmatrix}x\\y\\z\end{pmatrix}=\begin{pmatrix}\bar{x}\\\bar{y}\\{}^1/_2+z\end{pmatrix}$$

$$\{\boldsymbol{\sigma}_{h(x,y,0)}|\mathbf{t}_{0,{}^1/_2,0}\}\begin{pmatrix}x\\y\\z\end{pmatrix}=\begin{pmatrix}x\\{}^1/_2+y\\\bar{z}\end{pmatrix}\quad \{\boldsymbol{i}|\mathbf{t}_{p,q,r}\}\begin{pmatrix}x\\y\\z\end{pmatrix}=\begin{pmatrix}2p-x\\2q-y\\2r-z\end{pmatrix}$$

The expected four general equivalent positions are revealed and, since the centre of symmetry arises from the product $\{\boldsymbol{\sigma}_h|\mathbf{t}_{0,{}^1/_2,0}\}\{\boldsymbol{C}_{2[001]}|\mathbf{t}_{0,0,{}^1/_2}\}=\{\boldsymbol{\sigma}_h\boldsymbol{C}_{2[001]}|\boldsymbol{\sigma}_h\mathbf{t}_{0,0,{}^1/_2}+\mathbf{t}_{0,{}^1/_2,0}\}$, from Eq. (8.30), it follows that it lies at 0, ¼, ¼, *vide* Fig. 5.18; $p=0, q=r={}^1/_4$. This space group is C_{2h}^5 or $P2_1/b$, which is equivalent to $P2_1/c$ in the second setting.

The standard orientation places the centre of symmetry at the origin; thus, a shift to the position 0, ¼, ¼, is required in order to achieve this presentation. Do not confuse the $\boldsymbol{C}$ symbol here (***bold italic***) with that for C centring (*italic*) in the Hermann–Mauguin notation. Fig. 8.25 shows the general equivalent positions and symmetry elements for space group C_{2h}^5.

8.5.2 Orthorhombic space groups

If a second twofold axis is added to the group C_2, it has been shown already that it can only be coincident with it, which is trivial in this context, or perpendicular to it, in which case the orthorhombic system is invoked. The orthorhombic point groups are D_2, C_{2v} and D_{2h}. Consider the space group already examined in Section 5.4.9.2.

A mirror plane is normal to the x axis and intersects it at p, a glide plane is normal to the y axis intersecting it at q with a translation of $a/2$, and a twofold axis or twofold screw axis lies along at $[0,0,z]$. The point group is C_{2v}, and has the operators $\{\boldsymbol{E},\ \boldsymbol{C}_2,\ \boldsymbol{\sigma}_v,\ \boldsymbol{\sigma}'_v\}$; four general equivalent positions are to be expected. Proceeding in a more general way:

$$\{\boldsymbol{E}|\mathbf{0}\}\begin{pmatrix}x\\y\\z\end{pmatrix}=\begin{pmatrix}x\\y\\z\end{pmatrix}$$

$$\{\boldsymbol{\sigma}_{[100]}|\mathbf{t}_{0,0,0}\}\begin{pmatrix}x\\y\\z\end{pmatrix}=\begin{pmatrix}\bar{x}\\y\\z\end{pmatrix}+\begin{pmatrix}0\\0\\0\end{pmatrix}$$ From the setting, the translation 2 $(p, 0, 0)$ must be added $\rightarrow \begin{pmatrix}\bar{x}\\y\\z\end{pmatrix}+\begin{pmatrix}2p\\0\\0\end{pmatrix}$

$$\{\boldsymbol{\sigma}_{[010]}|\mathbf{t}_{1/2,0,0}\}\begin{pmatrix}\bar{x}\\y\\z\end{pmatrix}+\begin{pmatrix}2p\\0\\0\end{pmatrix}=\begin{pmatrix}\bar{x}\\\bar{y}\\z\end{pmatrix}+\begin{pmatrix}2p+1/2\\0\\0\end{pmatrix}$$ The translation 2$(0, q, 0)$ must be added $\rightarrow \begin{pmatrix}\bar{x}\\\bar{y}\\z\end{pmatrix}+\begin{pmatrix}2p+1/2\\2q\\0\end{pmatrix}$

$$\{\boldsymbol{C}_{2_{[001]}}|\mathbf{t}_{0,0,0}\}\begin{pmatrix}\bar{x}\\\bar{y}\\z\end{pmatrix}+\begin{pmatrix}2p+1/2\\2q\\0\end{pmatrix}=\begin{pmatrix}x\\y\\z\end{pmatrix}+\begin{pmatrix}2p+1/2\\2q\\0\end{pmatrix}$$ Since the **t**-translation is $0, 0, 0$, and $\sum_{\boldsymbol{R}} \mathbf{t} = 0$, it follows that $p = 1/4$ and $q = 0$, giving the correct orientation for C_{2v}^4 or $Pma2$; the diagram for this space group is that of Fig. 5.23.

8.5.3 Tetragonal space groups

The unique axis in the tetragonal system is chosen as z, and is S_4 in the space group of symbol D_{2d}^4 ($P\bar{4}2_1c$). The S_4 axis is at $0, 0, z; 0, 0, 0$, which, in ITA notation, means that the S_4 axis is the line $0, 0, z$ and the inversion point on that axis is 0, 0, 0, the origin. The point group operators are $\{\boldsymbol{E},\ 2\boldsymbol{S}_4,\ \boldsymbol{C}_2,\ 2\boldsymbol{C}_2'(x),\ 2\boldsymbol{\sigma}_d\}$; the space group Seitz operator parallel to x is $\{C_{2_{[100]}}|t_{1/2,0,0}\}$ at (x, q, r), and the c glide operator is $\{\boldsymbol{\sigma}_{[110]}|t_{0,0,1/2}\}$ at (p, p, z). Proceeding as before,

$$\{\boldsymbol{E}|\mathbf{0}\}\begin{pmatrix}x\\y\\z\end{pmatrix}=\begin{pmatrix}x\\y\\z\end{pmatrix}$$

$$\{\boldsymbol{\sigma}_{[110]}|\mathbf{t}_{0,0,1/2}\}\begin{pmatrix}x\\y\\z\end{pmatrix}+\begin{pmatrix}0\\0\\1/2\end{pmatrix}=\begin{pmatrix}\bar{y}\\\bar{x}\\z\end{pmatrix}+\begin{pmatrix}0\\0\\1/2\end{pmatrix}$$ From the setting, the translation $(p, p, 0)$ must be added $\rightarrow \begin{pmatrix}\bar{y}\\\bar{x}\\z\end{pmatrix}+\begin{pmatrix}p\\p\\1/2\end{pmatrix}$

$$\{\boldsymbol{C}_{2_{[100]}}|\mathbf{t}_{1/2,0,0}\}\begin{pmatrix}x\\y\\z\end{pmatrix}=\begin{pmatrix}x\\\bar{y}\\\bar{z}\end{pmatrix}+\begin{pmatrix}1/2\\0\\0\end{pmatrix}$$ From the setting, the translation $(0, 2q, 2r)$ must be added $\rightarrow \begin{pmatrix}x\\\bar{y}\\\bar{z}\end{pmatrix}+\begin{pmatrix}1/2\\2q\\2r\end{pmatrix}$

From Eq. (8.30),

$$\{C_{2[100]}|\mathbf{t}_{1/2,0,0}\}\{\sigma_{[110]}|\mathbf{t}_{0,0,1/2}\}\begin{pmatrix}x\\ \bar{y}\\ \bar{z}\end{pmatrix}+\begin{pmatrix}1/2\\ 2q\\ 2r\end{pmatrix}$$

$$=\{C_{2[100]}\sigma_{[110]}|(C_{2[100]}\mathbf{t}_{0,0,1/2}+\mathbf{t}_{1/2,0,0}\}=\begin{pmatrix}0&\bar{1}&0\\1&0&0\\0&0&\bar{1}\end{pmatrix}+\begin{pmatrix}1/2+p\\ p+2q\\ 1/2+2r\end{pmatrix}$$

Operating finally with S_4^3:

$$\{S^3_{4[00z]}|\mathbf{t}_{0,0,0}\}\begin{pmatrix}x\\ y\\ z\end{pmatrix}=\begin{pmatrix}\bar{y}\\ x\\ \bar{z}\end{pmatrix}+\begin{pmatrix}0\\0\\0\end{pmatrix}$$

Since $\sum_{R}\mathbf{t}=0, p=1/2,\ q=1/4,$ $r=1/4$. Strictly, the equations give $p=-1/2,\ q=-1/4,\ r=-1/4$; but since it is crystallographically equivalent to add or subtract to a coordinate, positive values for p, q and r are obtained directly.

8.5.4 Cubic space groups

As was seen in Chapter 5, cubic space groups present a little more complexity because of the inclined threefold axes. In Section 5.4.14.1 space group $T^4(P2_13)$ was discussed. From the group $\{E,\ 4C_3,\ 4C_3^2,\ 3C_2\}$, there will be twelve general positions for this space group (Table 5.5), with the addition of the appropriate screw axes components. An interesting feature of this and, indeed, the other cubic space groups is the presence of threefold screw axes, which are not obvious from an examination of the general equivalent positions. The translation component of these axes is $1/3$ (or $2/3$) of the repeat distance in the direction of the axes; the actual locations of the screw axes depend on the space group (See also Section 5.4.14).

Of the space groups $I23$ and $I2_13$, there are two situations to note:

- The twofold axes intersect one another, as do the twofold screw axes, in space group $I23$, and the highest point symmetry is 23.
- No such intersections occur for any twofold axes in $I2_13$, and its highest point symmetry is only 3. The 3_1 axes intersect the plane $z=0$ at $(1/6, 1/6; 1/3, 1/6)$ and the 3_2 axes at $(1/6, 2/3; 1/3, 2/3)$.

Referring back to Fig. 5.42a,b, it is prudent to consider the effect on the threefold axis at A of the twofold screw axis at $\left[1/4, 0, z\right]$. From Eq. (7.4), and remembering that C_2 is its own inverse, a similarity transformation on the C_3 operator along [111] gives the operator O :

$$
\begin{aligned}
\boldsymbol{O} &= \{\boldsymbol{C}_{2_{[001]}}|\mathbf{t}_{1/2,0,1/2}\}^{-1}\{\boldsymbol{C}_{3_{[111]}}|\mathbf{t}_{0,0,0}\}\{\boldsymbol{C}_{2_{[001]}}|\mathbf{t}_{1/2,0,1/2}\} \\
&= \{\boldsymbol{C}_{2_{[001]}}|\mathbf{t}_{1/2,0,1/2}\}^{-1}\{\boldsymbol{C}_{3_{[111]}}\boldsymbol{C}_{2_{[001]}}|\boldsymbol{C}_{3_{[111]}}\mathbf{t}_{1/2,0,1/2}+\mathbf{t}_{0,0,0}\} \\
&= \{\boldsymbol{C}_{2_{[001]}}|\mathbf{t}_{1/2,0,1/2}\}^{-1}\{\boldsymbol{C}^2_{3_{[1\bar{1}1]}}|\mathbf{t}_{1/2,1/2,0}\} \\
&= \{\boldsymbol{C}_{2_{[001]}}|\mathbf{t}_{1/2,0,\overline{1/2}}\}\{\boldsymbol{C}^2_{3_{[1\bar{1}1]}}|\mathbf{t}_{1/2,1/2,0}\} \\
&= \{\boldsymbol{C}_{2_{[001]}}\boldsymbol{C}^2_{3_{[1\bar{1}1]}}|(\boldsymbol{C}_{2_{[001]}}\mathbf{t}_{1/2,1/2,0}+\mathbf{t}_{1/2,0,\overline{1/2}}\} \\
&= \{\boldsymbol{C}_{2_{[001]}}\boldsymbol{C}^2_{3_{[1\bar{1}1]}}|\mathbf{t}_{0,\overline{1/2},1/2}\} \\
&= \begin{pmatrix} 0 & 0 & \bar{1} \\ 1 & 0 & 0 \\ 0 & \bar{1} & 0 \end{pmatrix} + \begin{pmatrix} 0 \\ \overline{1/2} \\ 1/2 \end{pmatrix}
\end{aligned}
\tag{8.32}
$$

Applying this operator: $\boldsymbol{O}(x, y, z) \rightarrow -z,\ -1/2+x,\ 1/2-y$. Adding 1 to the first and second coordinates leads to the crystallographically equivalent position $1-z,\ 1/2+x,\ 1/2-y$, which represents the operator $\boldsymbol{C}^2_{3_{[11\bar{1}]}}$, a threefold symmetry axis through $1-x,\ 1/2-x,\ x$ [27], not in the same class as $\boldsymbol{C}_{3_{[111]}}$. It intersects the plane of projection ($z=0$) at $x=1, y=-1/2$ ($\equiv 1/2$), which is position C in Fig. 5.42a. The coordinates $\bar{z},\ 1/2+x,\ 1/2-y$ correspond to general equivalent position (8) in the international tables [21]; $\boldsymbol{C}_3$ and $\boldsymbol{C}_3^2$ are different symmetry classes in point group T, and hence, also in this space group. This analysis is a confirmation of the related work in Section 5.4.14.1 (see also references [18]–[20] of Chapter 5).

Most probably, it will have been noticed that some of the work in this section was implicit in Section 5.5. However, the Seitz operator provides an elegant way of manipulating space group (and point group) operations, applying matrices *in extenso* only when arriving at the final answer.

8.6 Factor groups

Any space group G may be represented by the Seitz operators $\{\boldsymbol{E}|\mathbf{t}_n\}$ and $\{\boldsymbol{R}_j|\mathbf{t}_j\}$, where t_n refers to the translational components **a**, **b** and **c** of the translation subgroup T, and $\{\boldsymbol{R}_j|\mathbf{t}_j\}$ refers to the other symmetry operators and their translations. From the definition of lattice, the group T is an infinite group, whereas $\{\boldsymbol{R}_j|\mathbf{t}_j\}$ refers to a finite group of order h. That T is an invariant subgroup of G may be shown by finding the conjugate of any member of the translation group. If a typical member of T is $\{\boldsymbol{E}|\mathbf{t}_n\}$ and that of the space group $\{\boldsymbol{R}|\mathbf{t}_n+\mathbf{t}'\}$, where $\mathbf{t}'$ is added to preserve generality, then the conjugate of $\{\boldsymbol{E}|\mathbf{t}_n\}$ is given by the similarity transformation of Eq. (7.4):

$$
\begin{aligned}
\{\boldsymbol{R}|\mathbf{t}_n+\mathbf{t}'\}^{-1}\{\boldsymbol{E}|\mathbf{t}_n\}\{\boldsymbol{R}|\mathbf{t}_n+\mathbf{t}'\} &= \{\boldsymbol{R}|\mathbf{t}_n+\mathbf{t}'\}^{-1}\{\boldsymbol{ER}|\}\{\boldsymbol{E}|\mathbf{t}_n+\mathbf{t}'+\mathbf{t}_n\} \\
&= \{\boldsymbol{R}|\mathbf{t}_n+\mathbf{t}'\}^{-1}\{\boldsymbol{R}|2\mathbf{t}_n+\mathbf{t}'\} \\
&= \{\boldsymbol{R}^{-1}|-\boldsymbol{R}^{-1}(\mathbf{t}_n+\mathbf{t}')\}^{-1}\{\boldsymbol{R}|2\mathbf{t}_n+\mathbf{t}'\} \\
&= \{\boldsymbol{R}^{-1}\boldsymbol{R}|\boldsymbol{R}^{-1}(2\mathbf{t}_n+\mathbf{t}')-\boldsymbol{R}^{-1}(\mathbf{t}_n+\mathbf{t}')\} \\
&= \{\boldsymbol{E}|\boldsymbol{R}^{-1}\mathbf{t}_n\}
\end{aligned}
$$

The group T is its own conjugate, so that it is invariant, and the member $\{\boldsymbol{E}|\boldsymbol{R}^{-1}\ \mathbf{t}_n\}$ is itself a translation.

Since T is invariant, there is only one way in which it can be formed. Thus, if T be removed from G, there remains the *factor group* of the space group, of order h. Consider space group $P2_1/m$ (C_{2h}^2). If the translation group T is ignored, the isomorphism, or one-to-one relationship, of the symmetry operations of the point group and space group is clear:

$$\begin{array}{llcccc} \text{Point group} & C_{2h}\ (2/m): & \boldsymbol{E} & C_2 & \boldsymbol{i} & \sigma_h \\ \text{Space group} & C_{2h}^2\ (P2_1/m): & \boldsymbol{E} & C_2 & \boldsymbol{i} & \sigma_h \end{array}$$

Hence, a character table for $P2_1/m$ may be derived from that for $2/m$; it is the C_{2h} character table discussed earlier. In centred space groups, it is desirable to transform a centred cell to a primitive-type unit cell. Such a cell may be obtained by an appropriate transformation, for example, $C \rightarrow P$,

$$\begin{pmatrix} \overline{1/2} & 1/2 & 0 \\ 1/2 & 1/2 & 0 \\ 0 & 0 & 1 \end{pmatrix} \begin{pmatrix} \mathbf{a}_C \\ \mathbf{b}_C \\ \mathbf{c}_C \end{pmatrix} = \begin{pmatrix} \mathbf{a}_P \\ \mathbf{b}_P \\ \mathbf{c}_P \end{pmatrix}$$

or by defining a Wigner–Seitz cell (Section 4.9.8).

8.6.1 Factor group analysis of iron(II) sulphide

Iron(II) sulphide has been stated to crystallize in space group $P6_3mc$ with two formula entities in the unit cell. Each iron atom is coordinated by a slightly distorted octahedron of six sulphur atoms, and each sulphur atom by six iron atoms at the corners of a trigonal prism; the relevant special equivalent positions in the space group are listed below:

$$\begin{array}{llll} 6 & c & m & x,\ \bar{x},\ z;\quad x,\ 2x,\ z;\quad 2\bar{x},\ \bar{x},\ z; \\ & & & \bar{x},\ x,\ 1/2+z;\quad \bar{x},\ 2\bar{x},\ 1/2+z;\quad 2x,\ x,\ 1/2+z. \\ 2 & b & 3m & 1/3,\ 2/3,\ z;\quad 2/3,\ 1/3,\ 1/2+z. \\ 2 & a & 3m & 0,\ 0,\ z;\quad 0,\ 0,\ 1/2+z \end{array}$$

It is clear that the twofold positions are occupied, and the iron atoms may be placed in Wyckoff a positions with the sulphur atoms in b.

The factor group is C_{6v}, and a Γ_{3n} representation may be determined on the basis of the unshifted atoms, as in working with a point group:

C_{6v}	$\boldsymbol{E}$	$2C_6$	$2C_3$	C_2	$3\sigma_v$	$3\sigma_d$
Γ_{3n} (Fe)	2	0	2	0	2	0
Γ_{3n}	6	0	0	0	2	0

which reduces to $A_1 + B_1 + E_1 + E_2$; a similar analysis applies to the sulphur atoms. The overall dimensionality is twelve, which is correct for the contents of the unit cell.

A primitive unit cell is normally preferred in any vibration analysis since it involves half the number of vibrations of even the simplest centred cell. Care must be exercised when dealing with non-symmorphic space groups because it is possible that atoms in equivalent positions in neighbouring unit cells may not vibrate in phase with those in the reference unit cell. Further study of this topic lies outside the remit of this book but is discussed in the literature [22, 23].

8.6.2 Symmetry ascent and correlation

The scattering of incident radiation by atoms in a crystal, often termed *lattice vibrations*, is of two types: that in which the atoms in unit cell of the crystal move in phase with one another produces *acoustic phonons*, whereas that in which the atoms vibrate with respect to one another leads to *optic phonons*. The acoustic phonons can be likened to translations and rotations in a free molecule situation, so it is the optic phonons that give rise to infrared and Raman spectra in the crystalline state.

In iron(II) sulphide, the acoustic modes belong to the translations that span A_1 and E_1 in C_{6v}, so that infrared activity arises for these modes, whereas Raman spectra arise from A_1, E_1 and E_2, with coincidence in A_1 and E_1.

These results may be reached by a *symmetry ascent* technique. The iron and sulphur atoms lie on sites of symmetry C_{3v}, and the translational movements for that group span A_1 (z) and E_1 (x, y). The ascent of symmetry from C_{3v} to C_{6v} can be determined most readily from a C_{6v} *correlation table* [28]:

C_{3v}	$\longrightarrow$	C_{6v}
A_1	$\longrightarrow$	A_1
A_1	$\longrightarrow$	B_1
E	$\longrightarrow$	E_1
E	$\longrightarrow$	E_2

The vibrations of an atom may be resolved along x, y and z directions. Thus, two iron atoms on C_{3v} sites in irreducible representations $A_1 + E$ ascend to C_{6v} as $A_1 + B_1 + E_1 + E_2$, so that for both atomic species together $\Gamma_{3n} = 2A_1 + 2B_1 + 2E_1 + 2E_2$, and extracting the vibrational symmetries gives $\Gamma_{\text{vib}} = A_1 + 2B_1 + E_1 + 2E_2$, as determined above (remember *coincidence*).

8.6.3 Site group and factor group analyses

An alternative technique for studying vibrations in crystals is *site group analysis*. It makes the approximation that vibrational coupling between the individual entities is negligible, so that the modes and activities may be

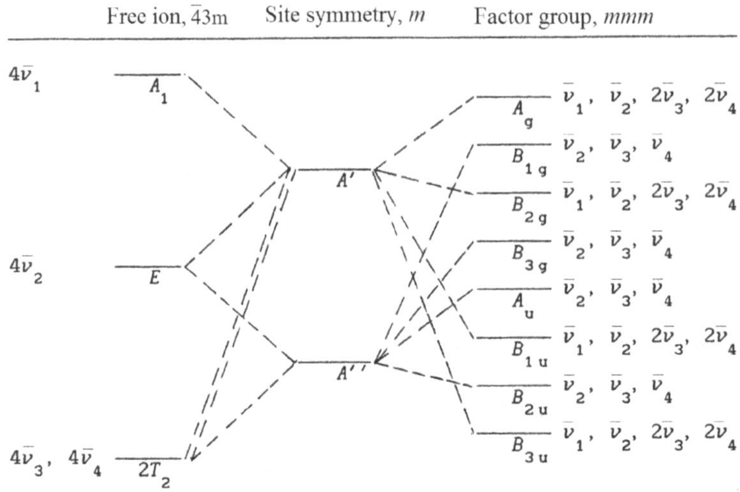

Fig. 8.26
Correlation diagram for the $[CrO_4]^{2-}$ ion in potassium chromate. The free ion modes $A_1 + E + 2T_2$, the nine vibrational degrees of freedom for the ion, correspond to frequencies ν_1–ν_4 in the left hand column; those in the right hand column are the 36 frequencies for the factor group. [Reproduced by courtesy of Woodhead Publishing, UK.]

obtained as already discussed, but using the site symmetry of the species in its space group. Where the site symmetry is of higher degree than twofold but lower than that of the crystal environment, site group analysis splits the degeneracies of the normal modes.

Potassium chromate crystallizes in space group *Pnma* (D_{2h}^{16}), with $Z = 4$. The symmetry of the chromate ion, $[CrO_4]^{2-}$, is $\bar{4}3m$ (T_d), but occupies sites of symmetry m in the unit cell:

$$4 \quad c \quad m \pm \left\{x, {}^1\!/\!_4, z; {}^1\!/\!_2 - x, {}^3\!/\!_4, {}^1\!/\!_2 + z\right\}$$

The T_d symmetry of the chromate ion leads to four normal vibrational modes; the activity in the infrared for T_2, and in the Raman for $A_1 + E + T_2$ account for the total of six modes. The Raman and infrared T_2 modes are coincident, and the nine vibration modes together with the six for translation and rotation make up the fifteen degrees of freedom expected for a single chromate ion.

The spectra recorded for potassium chromate are more complex than this analysis suggests: single spectra are replaced by multiplets, and weak infrared activity is observed for the A_1 and E type modes. It is a weakness of the site group method that coupling is ignored, particularly with ionic crystals.

The factor group for potassium chromate is D_{2h}: there are eight K^+ ions on two Wyckoff *c* sites, symmetry *m*, and four $[CrO_4]^{2-}$ ions on another such set of special positions. There are thirty-six vibrational modes, twelve rotational and twelve translational, making sixty modes in all for the four chromate anions.

A correlation table can be set up (Fig. 8.26) to show first the symmetry descent from $\bar{4}3m$ in the free ion, to m, the site symmetry in the crystal, and then ascent to mmm,(D_{2h}) which is the factor group for this example. The 16 frequencies $4\bar{\nu}_1(A_1)$, $4\bar{\nu}_2(E)$, $4\bar{\nu}_3(T_2)$ and $4\bar{\nu}_4(T_2)$ correspond to the 36 internal vibrational modes of the chromate ions. Together with the six rotation and six translation modes for the chromate ion, the total of its 60 degrees of freedom are accounted for. Twenty-four translational modes arise for the eight potassium ions; there are no rotational modes for a monatomic species.

On account of the centrosymmetry of the crystal, no coincidences can exist between the infrared and Raman spectra. Comparison with the experimental data shows that the predicted Raman spectra are present, but fewer than expected infrared bands appear [29, 30]. Further details on lattice dynamics and correlation procedures are recorded in the literature [31–34].

References 8

[1] Dubos RJ. *Louis Pasteur, free lance of science*. Da Capo Press, 1986.
[2] Ladd M and Palmer R. *Structure determination by X-ray crystallography*. Springer, 2013, and references therein.
[3] Kühne T and Khaliullin R. *Nat. Commun.* 2013; 4: 1450.
[4] Ladd AJC. *Computer simulation of liquids*. PhD Thesis, University of Cambridge, UK, 1977.
[5] Ladd AJC. *Mol. Phys.* 1977; 33: 1039.
[6] Furmanchuk A, et al. *Phys. Chem. Chem. Phys.* 2010; 12: 3363.
[7] Simons J. *An introduction to theoretical chemistry*. Cambridge University Press, 2003.
[8] Atkins PW. *Molecular quantum mechanics*. Oxford University Press, 2010.
[9] McQuarrie DA. *Quantum chemistry*. University Science Books, 2007.
[10] Bishop D. *Group theory and chemistry*. Clarendon Press, 1973.
[11] Cotton FA. *Chemical applications of group theory*. 3rd ed. John Wiley & Sons, 1990.
[12] Figgis BN and Hitchman MA. *Ligand field theory and its applications*. Wiley-VCH, 2000.
[13] Kettle SFA. *Physical inorganic chemistry: a coordination chemistry approach*. Oxford University Press, 2000.
[14] Hollas JM. *Modern spectroscopy*. 4th ed. John Wiley, 2004.
[15] Wilson EB, Decius JC and Cross PC *Molecular vibrations*. McGraw-Hill, 1955.
[16] Woodward LA. *Introduction to the theory of molecular vibrations and vibrational spectroscopy*. Oxford University Press, 1972.
[17] Gilson TR and Hendra PJ. *Laser Raman spectroscopy*. Wiley, 1970.
[18] Herzber, G. *Molecular spectra and molecular structure*, Vol. II. Van Nostrand, 1945.
[19] Hessel JFC. *Kristall. Gehlers Physikalische Wörterbuch*, Leipzig: 1830.
[20] Frankenheim ML. *Nova Acta Acad.*, Vol. 19. Breslau: von Grass, 1842.
[21] Hahn T (Ed.). *International tables for crystallography*, Vol. A. 5th ed. IUCr/Wiley, 2011.
[22] Burns G and Glazer AM. *Space groups for solid state scientists*. 2nd ed. Academic Press, 1978, and references therein.
[23] Burns G and Glazer AM. *Space groups for solid state scientists*. 3rd ed. Elsevier, 2013.
[24] Streitwolf H-W. *Gruppentheorie in der Festkörperphysik*. Akad. Verlags. Leipzig, 1967. [Eng. trans. Sykes JB. *Group theory in solid state physics*. London: Macdonalds, 1971].
[25] Hall LW. *Group theory and symmetry in chemistry*. McGraw-Hill, 1969.

[26] Hilton H. *Mathematical crystallography and the theory of groups of movements.* Clarendon, 1903; reprinted New York: Dover, 1963.
[27] Geometric interpretation of matrix column representation of symmetry operation. <http://www.cryst.ehu.es/cryst/matrices.html>.
[28] Atkins PW, Child, MS, Phillips CSG. *Tables for group theory.* Oxford University Press, 2006; online at <http://www.cinam.univ-mrs.fr/klein/teach/inorganic/tables_for_group_theory.pdf>
[29] Couture L. *J. Chem. Phys.* 1947; 15: 153.
[30] Carter RL. *J. Chem. Educ.* 1971; 48: 297.
[31] Dove MT. *Introduction to lattice dynamics.* Cambridge University Press, 2005.
[32] Decius JC and Hexter RM. *Molecular vibrations in crystals.* Mc-Graw-Hill, 1977.
[33] Vedder W and Hornig DF. In: Thompson HW (ed) *Advances in inorganic chemistry*, Vol. II. Interscience Publishers, 1961.
[34] Buttery HJ. *J. Chem. Soc.* (A) 1970; 471.

Problems 8

8.1 The angular orbital functions of two of the five d orbitals have the following forms:
1. $f(\theta) = 3\cos^2\theta - 1$;
2. $f(\theta) = \sin^2\theta\cos 2\phi$.
(a) Determine the polynomial name for each of these functions; (b) Obtain the normalizing constant N for each function.

8.2 Ammonia, NH_3, has the point group C_{3v}. Set up a representation and reduce it.

8.3 Continuing from Problem 2: suggest the most likely combination of atomic orbitals for ammonia if the H–N–H bond angle is *ca.* 107°.

8.4 Determine the irreducible representations of the possible orbital functions for the σ in-plane bonds in the $[NiF_4]^{2-}$ ion, point group D_{4h}. What orbitals are most probable for this anion?

8.5 Determine the transformation properties for the p orbital functions of nitrogen in ammonia. How does the result affect the degeneracy of the p orbitals ion this molecule?

8.6 Reduce the following representations: (a) C_{4v}: 4 0 0 0–2. (b) D_{4d}: 6 0–2 0–2 0 0. (c) D_{6h}: 3 0–4 0 2–2 2 0 0 0 12 0 4. [A program RD6H is available in the Web Program Suite for reducing a representation in D_{6h}. *The source code for this program is also supplied. If you are familiar with FORTRAN, you may wish to modify RD6H for another point groups. Please do not refer errors in any modified program to the author.*]

8.7 The species $[PtCl_6]^{2-}$ undergoes a symmetrical extension (Jahn–Teller effect)[3] of the two axial bond lengths. What is the symmetry of the resulting species? Form the irreducible representations using the new bond vectors as a basis. Let C_2 lie along the x axis, with σ_v in the x, z plane.

[3] Distortion of electron density around the cation when the electrons are asymmetrically disposed in its d orbitals.

8.8 (a) Obtain a Γ_{3n} representation for the dinitrogen monoxide molecule and reduce it. (b) Allocate the irreducible representations to the molecular movements. (c) Determine the irreducible representations for vibration and allocate them to the modes of vibration for this molecule. Sketch the normal modes and decide which are infrared active and which Raman active.

8.9 Show how is it that in point group C_4 the operators $\boldsymbol{C}_4$ and $\boldsymbol{C}_4^3$ are in different symmetry classes, yet in the same class in point group C_{4v}?

8.10 Show that the combinations of the operator $\boldsymbol{S}_4$ with (a) $\boldsymbol{i}$ and (b) $\boldsymbol{\sigma}_v$ do not lead to point groups other than those already derived in the text.

8.11 A 4_1 screw axis lies along the line $[x, 0, 0]$. Formulate a Seitz operator for this symmetry and list the coordinates derivable from it.

8.12 Find the nature and orientation of the symmetry element obtained by the combination of $\bar{4}$ along z preceded by a c glide plane at $y = 0$. What is the space group for this combination?

8.13 The Seitz symbol for a general space group rotation is $\{\boldsymbol{R}|\mathbf{t}\}$, where $\mathbf{t}$ need not be zero. Evaluate the effect of $\{\boldsymbol{R}|\mathbf{t}\}^n\mathbf{r}$, where n is an integer. Hence, determine the total number of crystallographic screw axes.

8.14 (a) What symmetry element evolves from the product of operations $\boldsymbol{4}_{[00z]}\,\boldsymbol{b}_{(1/4,y,z)}$? (b) What is the result if the product the product is formed as $\boldsymbol{b}_{(1/4,y,z)}\,\boldsymbol{4}_{[00z]}$? (c) How do the results relate to the scheme in Section 5.4.6?

8.15 A carbonyl compound $Cr(CO)_3(C_6\,H_6)$ crystallizes in space group $P2_1/m$, with two molecules in the unit cell. (a) What can be deduced about the molecular symmetry and its site symmetry in the crystal? (b) What is the factor group for this crystal? (c) What is the symmetry of the $Cr(CO)_3$ moiety? (d) Use the six C—O vibrators to generate a representation for the factor group, and reduce it. (e) How many infrared and Raman bands are predicted? (f) Will they be coincident?

8.16 Perovskite (named after the Russian mineralogist Count Lev Alexsevich von Perovski) exhibits several crystal forms. One of them is cubic, space group $Pm\bar{3}m(O_h)$, No. 221, with one formula entity in the unit cell; the unit cell dimension has been reported as *ca.* 4.04 Å. (a) How would the atoms be placed in the unit cell? Ionic radii: $r_{Ba^{2+}} = 1.49$Å, $r_{O^{2-}} = 1.25$Å. (You may need to refer to ITA, or the Bilbao Crystallographic Server, reference [27], link WYCKPOS). Sketch your resulting structure. (b) The crystal is compressed along [001]. What is the space group of the resulting structure? (c) If an atom at ½, ½, ½ is displaced along the z axis by and amount Δz, what then is the space group? (d) Can the crystal from either (a) or (b) exhibit a pyroelectric effect? If so, how so, and if not, why not.

8.17 Sketch the Wigner–Seitz unit cell for a lattice described by a cubic F unit cell. What figure is produced, and what is its point group symmetry?

8.18 Benzene belongs to point group D_{6h}. The symmetric C–C stretch mode is allocated to A_{1g}. Is it infrared and/or Raman active in this mode?

8.19 Spectral data for the BF_3 molecule are listed below:

$\bar{\nu}$ / cm^{-1}		
Infrared		**Raman**
–	$\bar{\nu}_1$	888
692	$\bar{\nu}_2$	–
1454	$\bar{\nu}_3$	1454
480	$\bar{\nu}_4$	480

Boron trifluoride was studied in Section 8.3.2. Assign the given frequencies to the vibrational modes.

8.20 If the infrared and Raman active modes for benzene are
Infrared active: $A_{2u}, 3E_{1u}$
Raman active: $2A_{1g}, E_{1g}, 4E_{2g}$,
how would they be affected by contamination with monochlorobenzene?

8.21 (a) Sketch the molecule of E-dichloroethene and write its point group? (b) How are the reference axes set for this molecule? (c) What are the characters of the reducible representations with the bases (i) hydrogen atoms $1s$, and (ii) chlorine atoms $2p$ orbitals of this species? (d) Reduce the representations that you devise. Give your answers in the Hermann-Mauguin notation.

8.22 What are the direct products of the irreducible representations in point group $2/m$. What states will interact under an operator that transforms like the A_u irreducible representation?

8.23 Construct SALCs for the water molecule by the use of projection operators.

Computer-assisted studies

9

SYNOPSIS

- Derivation of point groups
- Recognition of point groups
- Internal and Cartesian coordinates
- Molecular geometry
- Best-fit plane
- Reduction of a representation in point group D_{6h}
- Unit cell reduction
- Matrix operations
- Zone symbols and Miller indices
- Linear least squares

9.1 Introduction

From time to time throughout the text, reference has been made to computer programs designed to assist in the study of the textual material. Computers do not teach; rather, they assist in handling problems that are lengthy, so that the reader need not be deterred from attempting them. The Web Program Suite has been prepared to accompany the work in this book; it can be accessed via the publisher's web site, http://www.oup.co.uk/companion/ladd.

Computing is an essential feature of any modern scientific work, and the programs described here should be of help in understanding aspects of the text and in solving problems. Programs on the topics featured in the synopsis above are supplied as IBM-compatible .EXE files, and it is suggested that the complete program suite be loaded from the appropriate web reference into a personal program folder. Each program may be executed by a double click on the program name. Most data files carry the suffix .TXT but some require the suffix .DAT. The operation of each program is mostly self-explanatory, but the following notes may help with the use of the programs.

9.2 Derivation of point groups

In Section 3.6ff, the derivation of point groups is discussed and the program EULR, which follows the steps of the derivation, there described. The program is not interactive, but it shows how the various restrictions listed in the text influence the number of calculations that are needed. Finally, six sets of axes are given, on which the point groups can be built. The program has an extension to three commonly occurring non-crystallographic point groups. The results appear on the screen and also in a file ANGOUT.TXT. Finally, the user is invited to tackle the problems presented on the monitor at the end of the program run.

9.3 Recognition of point groups

There are several ways in which the recognition of a point group can be approached systematically. The program SYMM is based on a method [1] described in Section 3.11, and the flow diagram of the program is shown by Fig. 3.28. It is necessary to supply a model number between 1 and 102: appropriate numbers for models are given in bold type in the legends for Figs. 3.19–3.26b, as well as for the models described in Appendix A1.2. The program is interactive and all instructions appear on the monitor screen. If an incorrect response to a question is given, the user is directed back to the point of error. Two such returns are allowed before the program rejects that particular application for further preliminary study. Note that for convenience of output, the program refers to $m\bar{3}m$ and $m\bar{3}$ as $m3m$ and $m\bar{3}$, respectively; it lists also the Schönflies symbol in parentheses.

9.4 Internal and Cartesian coordinates

This program, INTXYZ, converts the geometry of a molecule in terms of its internal coordinates, that is, bond lengths, bond angles and torsion angles, into a set of Cartesian coordinates. The data must be supplied in a file named CART.TXT. Section 2.8.3 gives details of the input data, and a sample data set is included in the program folder. The format of the data must be followed exactly. For those not familiar with FORTRAN, the input format (A4,I6,3F10.5) has the following meaning, assuming that each input line is divided into a number of equal 'cells':

- A4: atom identification of up to 4 characters *beginning* in cell 1;
- I6: an integer of up to 6 digits *ending* in cell 10;
- 3F10.5: three real numbers of length 10 each, ending in cells 20, 30 and 40, respectively.

Problem 2.13 is based around this program, and invokes the next program in this set. The results are sent to the file METRIC.TXT

9.5 Molecular geometry

The program MOLGOM calculates bond lengths, bond angles and torsion angles. It requires unit cell dimensions input at the keyboard and crystallographic x, y and z coordinates from a file named MOLDAT.TXT. If Cartesian coordinates are used in the data file, then $a = b = c = 1$ is the input for the cell dimensions. The results are output to the file GEOM.TXT; it contains *inter alia* the molecular geometry listed in the CART.TXT file. A sample MOLDAT file is given in the program folder.

9.6 Best-fit plane

The program PLANE obtains the best-fit plane to a set of X, Y, Z coordinates. The equation of a plane, usually given in the form involving four unknowns, is

$$A'X + B'Y + C'Z + D = 0$$

but it is convenient to reset it as

$$AX + BY + C = Z$$

which involves only three unknown quantities. It is then treated by the usual technique of least squares. The input is at the keyboard and comprises the number of data and their X, Y and Z coordinates. The equation of the plane and the deviations of the input data points from the plane are output to the monitor screen, indicating the end of the program run.

9.7 Reduction of a representation in point group D_{6h}

The rather lengthy reduction of a representation in point group D_{6h} can be carried out with the program RD6H. On entering the program, the procedure is self-explanatory. The source code for this program is supplied, so that those readers who are familiar with FORTRAN may wish to modify it for other point groups. The irreducible representations are output to the monitor screen.

9.8 Unit cell reduction

The program LEPAGE was kindly made available by courtesy of Professor A. L. Spek, University of Utrecht. For reduction, choose the *D* option and enter the unit cell parameters, one to a line. It may be desirable to vary the precision by means of the *C* option. All instructions appear on the monitor screen.

9.9 Matrix operations

The program MATOPS performs operations on one 3×3 matrix or on two, according to choice. The output is in the file MATOUT.TXT. If operations are required on only one matrix **A**, neglect all reference to **B** in the output; the **B** matrix will have been set to unity by the program. Thus, meaningful results will then be the determinant, transpose, cofactor and inverse of **A**. There will be no inverse of any matrix if the determinant is less than or equal to zero. For operations on two matrices, as well as the results above, there will be the sum, difference and product of **A** and **B**.

9.10 Zone symbol or Miller indices

This program carries out the calculations of a zone symbol from two triplets of Miller indices, or Miller indices from two zone symbols. The two triplets are input on one line, with space between each entry, and the output is to the monitor screen.

No problems have been prepared specifically for this chapter since the applications of these programs arise naturally through the previous eight chapters and their problems sections.

9.11 Linear least squares

The program LSLI determines the best-fit straight line to a series of data points that must number at least 3 and, in this program, cannot exceed 100. Data must be entered from the keyboard; an example data set is LSSQ.TXT. Program execution follows input of the program name, LSLI; unit weights for each observation are assumed.

It is implicit that errors in x_i are significantly smaller than those in y_i. The goodness-of-fit is reflected by the values of $\sigma(A)$, $\sigma(B)$ and Pearson's r coefficient. If the errors in the parameters A and B are to be propagated to another quantity, then they follow the law given in Section 2.8.4.2.

Reference 9

[1] Ladd, MFC. *Int. J. Math. Educ. Sci. Tech.* 1976; 7: 395.

Stereoviews and crystal models

A1

A1.1 Stereoviews and stereoviewing

Stereoviews have been used to illustrate the three-dimensional character of crystal structures since 1926, and the technique is now commonplace. Computer programs are available [1,2] for preparing the two views needed for producing the desired three-dimensional image.

Two diagrams of a given object are necessary, set approximately 63 mm apart, in order to form a three-dimensional visual image. In viewing stereoscopic diagrams, each eye should see only the appropriate half of the complete illustration.

The simplest procedure is direct viewing with a *stereoviewer*, whereupon the three-dimensional image appears centrally between the stereoscopic pair of diagrams. A supplier of a relatively inexpensive stereoviewer is 3Dstereo.com, Inc, 1930 Village Center Circle, #3-333, Las Vegas, NV 89134, USA.

Alternatively, one can train the unaided eyes to defocus, and allow each eye to see only the appropriate diagram. The eyes must be relaxed and look straight ahead; it may help to close the eyes for a moment, then open them wide and allow them to relax without consciously focusing on the diagram. The viewing process may be aided by holding a white card edgeways between the two drawings.

Finally, a simple stereoviewer can be constructed. A pair of plano-convex or biconvex lenses, each of focal length approximately 100 mm and diameter approximately 30 mm, is mounted between two opaque cards, with the centres of the lenses approximately 63 mm apart. The card frame must be so shaped that the lenses may be brought close to the eyes. Figure A1.1 illustrates the construction of the stereoviewer. It may be helpful to obscure a segment S of each lens, closest to the nose region N, of maximum depth approximately 25% of the lens diameter.

A1.2 Crystal models

In this section, instructions are given for making crystal models that may be helpful in studying crystal symmetry. Other sets of instructions can be found in the literature [3–5]. (If plano-convex lenses are used, set the convex side of the lens away from the observer.)

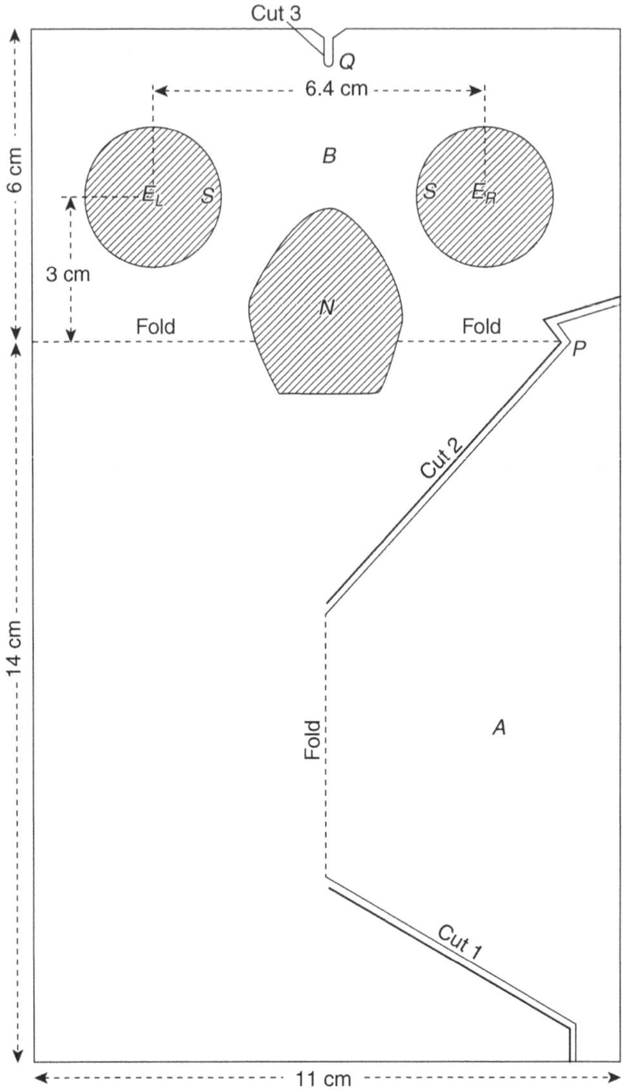

Fig. A1.1
Prepare two pieces of thin card as shown; cut out and discard the shaded portions. Make cuts along the double lines, Cut 1 and Cut 2. Glue the two cards together with the lenses E_L and E_R in position. To use, fold the portions *A* and *B* backwards, and set the projection *P* into the cut at *Q*; strengthen the fold with Sellotape, if necessary. View the stereo diagrams from the side marked *B*.

A1.2.1 Cube

On a thin card, draw a square of side, say 40 mm. On each side of this square draw another identical square. Lightly score the edges of the first square and fold the other four to form five faces of a cube, and fasten the edges with Sellotape. There is an advantage in leaving the sixth face of the cube open, as indicated hereunder, and the presence of the sixth face can be imagined, or the model inverted.

A1.2.2 Tetrahedron

On another similar card, draw an equilateral triangle of side *ca.* $39.5\sqrt{2}$ mm; this measure is just less than the length of the face diagonal of the cube above.

On each side of the triangle, draw another identical triangle. Lightly score the edges of the first triangle, fold the other three triangles in the same sense to meet at an apex, and seal the apex with glue or with Sellotape. Note that on placing the tetrahedron inside the cube so that an edge of the tetrahedron represents a face diagonal of the cube, the assembly aligns the symmetry elements common to both models.

A1.2.3 Model with $\bar{4}$ symmetry

A model similar to one of the tetrahedra shown in Fig. 3.1 has been chosen because it exhibits a fourfold inversion axis, which is one of the more difficult symmetry elements to appreciate from drawings.

Mark out a thin card in accordance with the dimensions listed in the legend to Fig. A1.2; then cut along the solid lines, discarding the shaded portions. Make folds in the same direction along the dotted lines; the sections *ADNP* and *CFLM* are glued internally, and *EFHJ* is glued externally.

If these three models are used with the point group recognition program SYMM, allocate model number **7** for the cube, **19** for the tetrahedron and

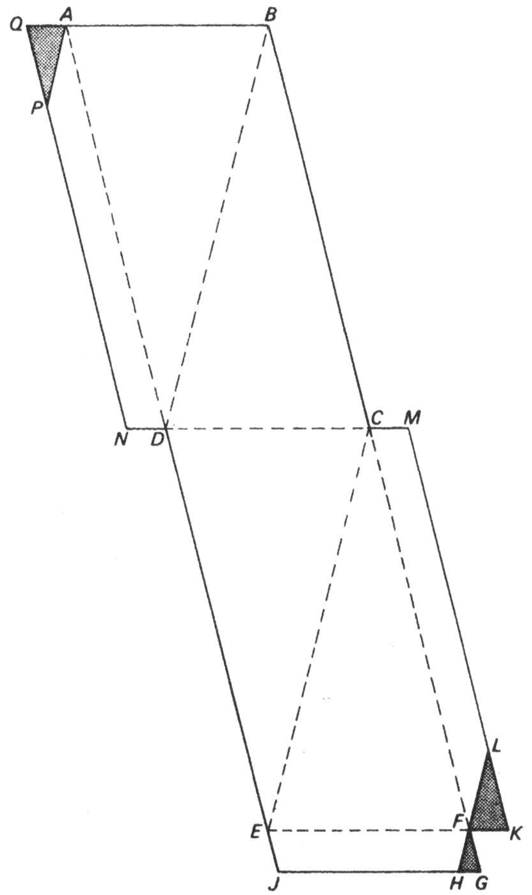

Fig. A1.2
Data for the construction of a tetragonal crystal with a $\bar{4}$ axis: $NQ = AD = BD = BC = DE = CE = CF = KM = 100$ mm; $AB = CD = EF = GJ = 50$ mm; $AP = PQ = FL = KL = 20$ mm; $AQ = DN = CM = FK = FG = FH = EJ = 10$ mm.

86 for the third (tetragonal) model. The identification of the point group of each model then proceeds along the lines indicated by the block diagram of Fig. 3.28, on which the program SYMM is based.

References

[1] ORTEP-3 for Windows. <http://www.chem.gla.ac.uk/~louis/software/ortep3/>.
[2] Mercury—crystal structure visualization, exploration and analysis made easy. <http://www.ccdc.cam.ac.uk/Solutions/FreeSoftware/Pages/FreeMercury.aspx>.
[3] Introduction to the crystal class models. <http://crystallography.org.uk/wp-content/uploads/2011/07/crystal_classes.pdf>.
[4] Crystal forms. <http://webmineral.com/help/Forms.shtml>.
[5] Models of crystal shapes. <http://learn-science.20m.com/student_crystals.html>.

Analytical geometry of direction cosines

A2

A2.1 Direction cosines of a line

In Fig. A2.1, P_1 is the point x_1, y_1, z_1 referred to x, y and z orthogonal axes. Lines from P_1 perpendicular to the x, y and z axes cut them at A, B and C, respectively. Thus, $OA = x_1$, $OB = y_1$, and $OC = z_1$. The direction cosines of OP_1 are given by $\cos \chi_1 = x_1/OP_1$, $\cos \psi_1 = y_1/OP_1$, and $\cos \omega_1 = z_1/OP_1$. Hence,

$$\cos^2 \chi_1 + \cos^2 \psi_1 + \cos^2 \omega_1 = \left(x_1^2 + y_1^2 + z_1^2\right)/OP_1^2 \qquad \text{(A2.1)}$$

Since x_1, y_1 and z_1 are the projections of OP_1 on to the x, y and z axes, it follows that

$$x_1^2 + y_1^2 + z_1^2 = OP_1^2 \qquad \text{(A2.2)}$$

Hence,

$$\cos^2 \chi_1 + \cos^2 \psi_1 + \cos^2 \omega_1 = 1 \qquad \text{(A2.3)}$$

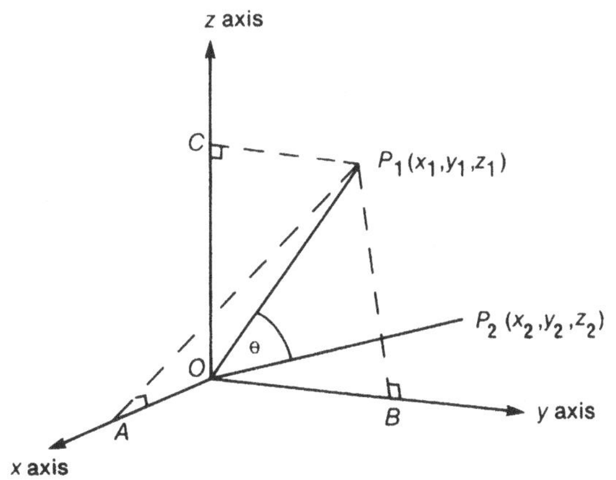

Fig. A2.1
The direction cosines of OP_1 and OP_2, referred to the rectangular axes x, y and z: $\chi_n = \angle AOP_n$, $\psi_n = \angle BOP_n$, $\omega_n = \angle COP_n$ $(n = 1, 2)$.

A2.2 Angle between two lines

The point $P_2(x_2, y_2, z_2)$ is constructed such that $OP_2 = OP_1 = r$. For OP_1, from the foregoing,

$$x_1 = r\cos\chi_1, \qquad y_1 = r\cos\psi_1, \qquad \text{and} \qquad z_1 = r\cos\omega_1$$

and for OP_2,

$$x_2 = r\cos\chi_2, \qquad y_2 = r\cos\psi_2, \qquad \text{and} \qquad z_2 = r\cos\omega_2$$

where $\cos\chi_2$, $\cos\psi_2$ and $\cos\omega_2$ are the direction cosines of the line OP_2. If the origin is shifted from O to P_1, then the coordinates of P_2 become

$$x_2' = x_2 - x_1, \qquad y_2' = y_2 - y_1, \qquad z_2' = z_2 - z_1 \tag{A2.4}$$

so that the length P_1P_2 is given by

$$\begin{aligned}(P_1P_2)^2 &= (x_2 - x_1)^2 + (y_2 - y_1)^2 + (z_2 + z_1)^2 \\ &= r^2 \left(\cos^2\chi_1 + \cos^2\chi_2 - 2\cos\chi_1\cos\chi_2 \right. \\ &\quad + \cos^2\psi_1 + \cos^2\psi_2 - 2\cos\psi_1\cos\psi_2 \\ &\quad \left. + \cos^2\omega_1 + \cos^2\omega_2 - 2\cos\omega_1\cos\omega_2\right)\end{aligned} \tag{A2.5}$$

Using Eq. (A2.3),

$$(P_1P_2)^2 = 2r^2\left[1 - (\cos\chi_1\cos\chi_2 + \cos\psi_1\cos\psi_2 + \cos\omega_1\cos\omega_2)\right] \tag{A2.6}$$

In the isosceles triangle OP_1P_2

$$P_1P_2/2 = r\sin(\theta/2) \tag{A2.7}$$

Therefore,

$$(P_1P_2)^2/4r^2 = \sin^2(\theta/2) = (1 - \cos\theta)/2$$

or

$$(P_1P_2)^2/2r^2 = 1 - \cos\theta \tag{A2.8}$$

Comparing Eq. (A2.6) and Eq. (A2.8),

$$\cos\theta = \cos\chi_1\cos\chi_2 + \cos\psi_1\cos\psi_2 + \cos\omega_1\cos\omega_2 \tag{A2.9}$$

Vectors and matrices

A3

A3.1 Introduction

This appendix is but a brief introduction to vector algebra, mainly for the purpose of defining the use of vectors and matrices in the study of the geometry of crystals and molecules. The reader who wishes for more mathematical detail on vector algebra should refer to a standard text on this subject.

A3.2 Vectors

Ordinary numbers and their representations are *scalar* quantities; they have the magnitude expressed by the quantity itself. A vector has both magnitude and direction. Thus, the number 30, and the variable x written in *italics*, are scalars whereas 30° N, and the variable **X** written in bold type, represent vectors; the *unit vector* may be defined as **i**, where $|\mathbf{i}| = 1$.

A3.2.1 Sum, difference and scalar product of two vectors

In Fig. A3.1, **a** and **b** are two vectors drawn from the origin O; QS is perpendicular to OP. From the Pythagorean extension theorem,

$$PQ^2 = OQ^2 + OP^2 - 2OP\,OS$$

or

$$PQ^2 = OQ^2 + OP^2 - 2OP\,OQ\cos\angle QOP \qquad \text{(A3.1)}$$

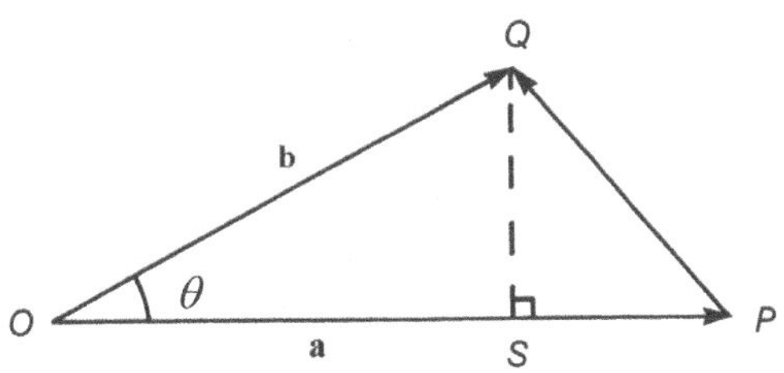

Fig. A3.1
Vectors **a** and **b** at an angle θ to each other; the vector **PQ** is **b** − **a**.

which can be written as

$$PQ^2 = a^2 + b^2 - 2ab\cos\angle\mathbf{a}\,\mathbf{b} \tag{A3.2}$$

In vector notation

$$\mathbf{PQ} = \mathbf{b} - \mathbf{a} \tag{A3.3}$$

Since the algebraic manipulations of addition and subtraction must take place along a line, with vectors resolved as necessary, a product term is modified by the cosine of the angle between the forward directions of the vectors; thus, the scalar product is

$$\mathbf{PQ}\cdot\mathbf{PQ} = (\mathbf{b} - \mathbf{a})\cdot(\mathbf{b} - \mathbf{a})$$

or

$$PQ^2 = a^2 + b^2 - 2\mathbf{a}\cdot\mathbf{b} \tag{A3.4}$$

Comparing Eq. (A3.4) and Eq. (A3.2),

$$\mathbf{a}\cdot\mathbf{b} = a\,b\cos\angle\mathbf{a}\,\mathbf{b} \tag{A3.5}$$

which defines the *scalar product* of two vectors, and may be written for Fig. A3.1 as

$$\mathbf{a}\cdot\mathbf{b} = a\,b\cos\theta \tag{A3.6}$$

where θ is the angle between the *forward directions* of **a** and **b**, and a and b are the magnitudes of the vectors, which may be written also as $|\mathbf{a}|$ and $|\mathbf{b}|$. It follows that the scalar product, also called the *dot product*, of a vector with itself is the square of the magnitude of the vector:

$$\mathbf{a}\cdot\mathbf{a} = |\mathbf{a}|^2 = a^2 \tag{A3.7}$$

since the angle $\angle\mathbf{a}\,\mathbf{a}$ is then zero. Further, easily verified, results with dot products are:

$$\left.\begin{aligned} &\mathbf{a}\cdot\mathbf{b} = \mathbf{b}\cdot\mathbf{a} \\ &p(\mathbf{a}\cdot\mathbf{b}) = (p\mathbf{a})\cdot\mathbf{b} = \mathbf{a}\cdot(p\mathbf{b}) \\ &\mathbf{c}\cdot(\mathbf{a}\cdot\mathbf{b}) = \mathbf{c}\cdot\mathbf{a} + \mathbf{c}\cdot\mathbf{b} \end{aligned}\right\} \tag{A3.8}$$

where p is a scalar constant; the operations of addition and subtraction of vectors involve a scalar product where the vector components are not along one and the same line.

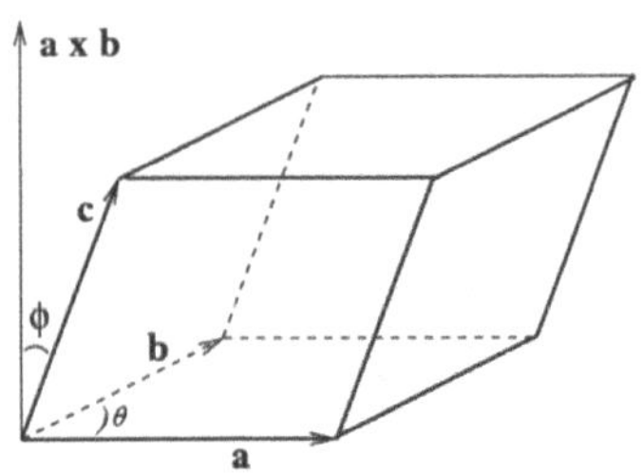

Fig. A3.2
General parallelepiped $\mathbf{a}, \mathbf{b}, \mathbf{c}$. Vectors **a** and **b** are at an angle θ to each other, and the product $\mathbf{a}\times\mathbf{b}$ is a vector of magnitude $ab\sin\theta$, normal to the plane of a, b and directed according to the right-hand rule. The volume of the parallelepiped is $|\mathbf{c}\cdot(\mathbf{a}\times\mathbf{b})| = abc\cos\phi$, where ϕ is the angle between $\mathbf{a}\times\mathbf{b}$ and **c**.

A3.2.2 Vector product of two vectors

A3.2.2.1 Definition of vector product

Consider first the two vectors **a** and **b** from the origin O in Fig. 3.2. The *vector product*, also called *cross product*, is defined as

$$\mathbf{a}\times\mathbf{b} = \mathbf{i}\,a\,b\sin\theta \tag{A3.9}$$

where $a\,b\sin\theta$ is the area of the parallelogram formed by the vectors **a** and **b**, and **i** is a unit vector normal to the plane containing **a** and **b**. The direction of **i** is such that the three vectors **a**, **b** and **i** follow the right-hand rule: if the thumb and first and middle finger of the right hand are held mutually perpendicular, then with the first finger as **a**, the middle finger as **b**, then the thumb gives the direction of **i**. Note that the sequence **a**, **b**, **i**, is the same as that for the reference axes x, y, z. It follows that

$$\mathbf{b} \times \mathbf{a} = -\mathbf{a} \times \mathbf{b} \tag{A3.10}$$

A3.2.2.2 Coordinate notation

Let **c** be a vector with components c_1, c_2, c_3 with respect to the orthogonal directions of **a**, **b** and **c**. Then,

$$\begin{aligned} \mathbf{a} \times \mathbf{b} &= (\mathbf{i}a_1 + \mathbf{j}a_2 + \mathbf{k}a_3) \times (\mathbf{i}b_1 + \mathbf{j}b_2 + \mathbf{k}b_3) \\ &= a_1b_1\mathbf{i} \times \mathbf{i} \ + \ a_1b_2\mathbf{i} \times \mathbf{j} \ + a_1b_3\mathbf{i} \times \mathbf{k} + \\ &\quad a_2b_1\mathbf{j} \times \mathbf{i} \ + a_2b_2\mathbf{j} \times \mathbf{j} \ + a_2b_3\mathbf{j} \times \mathbf{k} + \\ &\quad a_3b_1\mathbf{k} \times \mathbf{i} + a_3b_2\mathbf{k} \times \mathbf{j} + a_3b_3\mathbf{k} \times \mathbf{k} \\ &= (a_2b_3 - a_3b_2)\mathbf{i} + (a_3b_1 - a_1b_3)\mathbf{j} + (a_1b_2 - a_2b_1)\mathbf{k} \end{aligned} \tag{3.11}$$

where **i**, **j** and **k** are unit vectors along **a**, **b** and **c**, and $\mathbf{i} \times \mathbf{j} = \mathbf{k}$, $\mathbf{j} \times \mathbf{i} = -\mathbf{k}$, with cyclic permutations. Thus, if $\mathbf{a} \times \mathbf{b} = \mathbf{c}$, the scalar components of the vector **c** are $c_1 = (a_2b_3 - a_3b_2)$, $c_2 = (a_3b_1 - a_1b_3)$ and $c_3 = (a_1b_2 - a_2b_1)$.

A3.3 Volume of a parallelepiped

An important application of the vector product arises in determining the volume of a general parallelepiped formed by three sides a, b, c, parallel to the x, y and z axes, respectively, and the three interaxial angles α, β and γ, defined as in Section 2.2.1.

Referring again to Fig. A3.2, the volume of a parallelepiped is the area of the base multiplied by the perpendicular height, that is, the magnitude $|\mathbf{a} \times \mathbf{b}|$ multiplied by $|\mathbf{c}|\ |\cos\phi|$: note that if **a** and **b** are interchanged, then $\mathbf{a} \times \mathbf{b}$ would point downward: the angle ϕ would be larger than $\pi/2$, and $\cos\phi$ would be negative. Thus, the volume V is written as

$$V = |\mathbf{c} \cdot (\mathbf{a} \times \mathbf{b})| \tag{A3.12}$$

or $|\mathbf{b} \cdot (\mathbf{c} \times \mathbf{a})|$ or $|\mathbf{c} \cdot (\mathbf{a} \times \mathbf{b})|$. In order to evaluate V, **a**, **b**, and **c** are expressed in terms of a set of orthogonal unit vectors **i**, **j** and **k**:

$$\begin{aligned} \mathbf{a} &= a_1\mathbf{i} + a_2\mathbf{j} + a_3\mathbf{k} \\ \mathbf{b} &= b_1\mathbf{i} + b_2\mathbf{j} + b_3\mathbf{k} \\ \mathbf{c} &= c_1\mathbf{i} + c_2\mathbf{j} + c_3\mathbf{k} \end{aligned} \tag{A3.13}$$

Then, using Eq. (A3.12) with expansion of the vectors in terms of unit vectors **i**, **j** and **k** in the directions of **a**, **b** and **c**, respectively, and recalling that products such as $\mathbf{i} \times \mathbf{i} = 0$, $\mathbf{i} \times \mathbf{j} = \mathbf{k}$ and $\mathbf{j} \times \mathbf{i} = -\mathbf{k}$,

$$V = (c_1\mathbf{i} + c_2\mathbf{j} + c_3\mathbf{k}) \cdot (a_1b_2\mathbf{k} - a_1b_3\mathbf{j} - a_2b_1\mathbf{k} + a_2b_3\mathbf{i} + a_3b_1\mathbf{j} - a_3b_2\mathbf{i}) \tag{A3.14}$$

which, after simplification, may be expressed as the determinant

$$\begin{vmatrix} a_1 & a_2 & a_3 \\ b_1 & b_2 & b_3 \\ c_1 & c_2 & c_3 \end{vmatrix} \tag{A3.15}$$

Since rows and columns of a determinant can be interchanged without altering its value,

$$V^2 = \begin{vmatrix} a_1 & a_2 & a_3 \\ b_1 & b_2 & b_3 \\ c_1 & c_2 & c_3 \end{vmatrix} \begin{vmatrix} a_1 & b_1 & c_1 \\ a_2 & b_2 & c_2 \\ a_3 & b_3 & c_3 \end{vmatrix} \tag{A3.16}$$

Multiplying the determinants, according to the rules for matrices (Appendix A3.4.7 and Appendix A3.4.9), leads to

$$V^2 = \begin{vmatrix} a_1a_1 + a_2a_2 + a_3a_3 & a_1b_1 + a_2b_2 + a_3b_3 & a_1c_1 + a_2c_2 + a_3c_3 \\ b_1a_1 + b_2a_2 + b_3a_3 & b_1b_1 + b_2b_2 + b_3b_3 & b_1c_1 + b_2c_2 + b_3c_3 \\ c_1a_1 + c_2a_2 + c_3a_3 & c_1b_1 + c_2b_2 + c_3b_3 & c_1c_1 + c_2c_2 + c_3c_3 \end{vmatrix} \tag{A3.17}$$

which may be expressed in vector notation as the determinant

$$V^2 = \begin{vmatrix} \mathbf{a}\cdot\mathbf{a} & \mathbf{a}\cdot\mathbf{b} & \mathbf{a}\cdot\mathbf{c} \\ \mathbf{b}\cdot\mathbf{a} & \mathbf{b}\cdot\mathbf{b} & \mathbf{b}\cdot\mathbf{c} \\ \mathbf{c}\cdot\mathbf{a} & \mathbf{c}\cdot\mathbf{b} & \mathbf{c}\cdot\mathbf{c} \end{vmatrix} \tag{A3.18}$$

Evaluating:

$$V^2 = a^2b^2c^2 + ab\ \cos\gamma\ bc\cos\alpha\ ca\cos\beta + ac\cos\beta\ ba\cos\gamma\ bc\cos\alpha \\ -ca\cos\beta\ b^2ca\ \cos\beta - bc\cos\alpha\ a^2bc\ \cos\alpha\ - ab\cos\gamma\ c^2ab\cos\gamma$$

which simplifies to

$$V = abc(1 - \cos^2\alpha - \cos^2\beta - \cos^2\gamma + 2\cos\alpha\cos\beta\cos\gamma)^{1/2} \tag{A3.19}$$

A3.4 Matrices

A matrix is an ordered rectangular array of numbers or variables. The matrix may be symbolized by **A** and its size as $m \times n$. Its elements a_{ij} run from $i = 1$ to m down the columns and from $j = 1$ to n along the rows.

A3.4.1 General matrix

Thus, in the 4×3 matrix **A**,

$$\mathbf{A} = \begin{pmatrix} a_{11} & a_{12} & a_{13} \\ a_{21} & a_{22} & a_{23} \\ a_{31} & a_{32} & a_{33} \\ a_{41} & a_{42} & a_{43} \end{pmatrix} \tag{A3.20}$$

where i runs from 1 to 4 and j runs from 1 to 3. In most cases, the matrices of interest herein will be 3×3 size matrices; a matrix in which $m = n$ is a *square* matrix.

A3.4.2 Row matrix

A row matrix consists of a single line array, dimensions $1 \times n$. The array

$$\mathbf{R} = (a_{11} \ \ a_{12} \ \ a_{13}) \qquad \text{(A3.21)}$$

is a row matrix, or *row vector*, of size 1×3.

A3.4.3 Column matrix

The matrix

$$\mathbf{C} = \begin{pmatrix} a_{11} \\ a_{21} \\ a_{31} \\ a_{41} \end{pmatrix} \qquad \text{(A3.22)}$$

is a column matrix, or *column vector*, of dimension 4×1.

A3.4.4 Symmetric, skew-symmetric, equal and identity matrices

A *symmetric* matrix $\mathbf{A}$ has elements $a_{ij} = a_{ji}$ for all i and j; otherwise it is *skew-symmetric*; the matrices $\mathbf{A}$ and $\mathbf{B}$ below are both skew-symmetric. Note that a symmetric matrix is equal to its transpose (Appendix A3.4.6). *Equal* matrices have both the same dimensions and equal corresponding elements; such matrices need not be square. The *identity* matrix $\mathbf{I}$ has diagonal elements of unity and zero elements otherwise.

$$\mathbf{A} = \begin{pmatrix} 1 & 0 & 1 \\ 0 & 1 & 2 \\ \bar{2} & 1 & 1 \end{pmatrix} \quad \mathbf{B} = \begin{pmatrix} 1 & 1 & 2 \\ 0 & \bar{1} & 0 \\ 1 & 0 & 0 \end{pmatrix} \quad \mathbf{I} = \begin{pmatrix} 1 & 0 & 0 \\ 0 & 1 & 0 \\ 0 & 0 & 1 \end{pmatrix} \qquad \text{(A3.23)}$$

For neatness, the negative sign here is placed above the digit to which it refers, as with Miller indices.

A3.4.5 Addition and subtraction of matrices

Two matrices $\mathbf{A}$ and $\mathbf{B}$ may be added or subtracted if $\mathbf{A}$ has both the same number of columns and same number of rows as $\mathbf{B}$. Thus, in $\mathbf{C} = \mathbf{A} \pm \mathbf{B}$, $c_{ij} = a_{ij} \pm b_{ij}$:

$$\begin{pmatrix} 1 & 0 & 1 \\ 0 & 1 & 2 \\ \bar{2} & 1 & 1 \end{pmatrix} + \begin{pmatrix} 1 & 1 & 2 \\ 0 & \bar{1} & 0 \\ 1 & 0 & 0 \end{pmatrix} = \begin{pmatrix} 2 & 1 & 3 \\ 0 & 0 & 2 \\ \bar{1} & 1 & 1 \end{pmatrix} \quad \text{and} \quad \begin{pmatrix} 1 & 0 & 1 \\ 0 & 1 & 2 \\ \bar{2} & 1 & 1 \end{pmatrix} - \begin{pmatrix} 1 & 1 & 2 \\ 0 & \bar{1} & 0 \\ 1 & 0 & 0 \end{pmatrix} = \begin{pmatrix} 0 & \bar{1} & \bar{1} \\ 0 & 2 & 2 \\ \bar{3} & 1 & 1 \end{pmatrix}$$

A3.4.6 Transposition

A transposed matrix has a_{ij} replaced by a_{ji} for all i and j. Thus, if a matrix $\mathbf{A}$ is

$$\mathbf{A} = \begin{pmatrix} 1 & 0 & 1 \\ 0 & 1 & 2 \\ \bar{2} & 1 & 1 \end{pmatrix} \qquad \text{(A3.24)}$$

its transpose $\mathbf{A}^{\mathrm{T}}$ is given by

$$\mathbf{A}^{\mathrm{T}} = \begin{pmatrix} 1 & 0 & \bar{2} \\ 0 & 1 & 1 \\ 1 & 2 & 1 \end{pmatrix} \qquad \text{(A3.25)}$$

A3.4.7 Multiplication with matrices

If a matrix is multiplied by a scalar quantity s, then all elements a_{ij} of the matrix are transformed to sa_{ij}. Two matrices **A** and **B** can be multiplied if and only if **A** has the dimensions $m \times p$ and **B** has the dimensions $p \times n$, so that their product **C** has the dimensions $m \times n$. The element c_{ij} of **C** is given by multiplying the ith row of **A** and the jth column of **B** element by element. In general,

$$c_{ij} = \sum_{k=1}^{p} \mathbf{A}_{ik}\mathbf{B}_{kj} \qquad \text{(A3.26)}$$

For a 3×3 matrix, $c_{12} = a_{11}b_{12} + a_{12}b_{21} + a_{13}b_{23}$. Thus, if c_{12} is formed from matrices **A** and **B** in Eq. (A3.23), $c_{12} = (1 \times 1) + (0 \times \bar{1}) + (1 \times 0) = 1$. Note that the position of c_{ij} in **C** is the junction of the ith row and j^{th} column.

The multiplication of **A** and **B** from Eq. (A3.23) to give **C** is written concisely as

$$\mathbf{C} = \mathbf{A}\,\mathbf{B} \qquad \text{(A3.27)}$$

It is important to note that, in general, $\mathbf{B}\,\mathbf{A} \neq \mathbf{A}\,\mathbf{B}$. Thus, using the matrices from Eq. (A3.23),

$$\mathbf{A}\,\mathbf{B} = \begin{pmatrix} 2 & 1 & 2 \\ 2 & \bar{1} & 0 \\ \bar{1} & \bar{3} & \bar{4} \end{pmatrix} \quad \text{and} \quad \mathbf{B}\,\mathbf{A} = \begin{pmatrix} \bar{3} & 3 & 5 \\ 0 & \bar{1} & \bar{2} \\ 1 & 0 & 1 \end{pmatrix} \qquad \text{(A3.28)}$$

A3.4.8 Some multiplicative properties of matrices

The following properties of matrices are useful from time to time; the list is not exhaustive.

$$\left.\begin{array}{c} \mathbf{A}(\mathbf{BC}) = (\mathbf{AB})\mathbf{C} \\ (\mathbf{A}+\mathbf{B})\mathbf{C} = \mathbf{AC} + \mathbf{BC} \\ s(\mathbf{A}+\mathbf{B}) = s\mathbf{A} + s\mathbf{B} \\ \text{where } s \text{ is a scalar constant} \\ (\mathbf{A}+\mathbf{B})^{\mathrm{T}} = \mathbf{A}^{\mathrm{T}} + \mathbf{B}^{\mathrm{T}} \\ (\mathbf{AB})^{\mathrm{T}} = \mathbf{B}^{\mathrm{T}}\mathbf{A}^{\mathrm{T}} \end{array}\right\} \qquad \text{(A3.29)}$$

A3.4.9 Determinant values of matrices

The determinant of a matrix is a mathematical tool of importance in many applications; here the concern is with its use in evaluating the inverse of matrix. Since an inverse exists only for a square matrix, the same condition applies to determinants.

A3.4.9.1 2 x 2 Matrix

Consider the two-dimensional matrix

$$\mathbf{a} = \begin{pmatrix} a_{11} & a_{12} \\ a_{21} & a_{22} \end{pmatrix}$$

The determinant, written variously as det(**a**), | **a** | or Δ, is given by

$$\det(\mathbf{a}) = \begin{vmatrix} a_{11} & a_{21} \\ a_{12} & a_{22} \end{vmatrix} = a_{11}a_{22} - a_{12}a_{21} \qquad \text{(A3.30)}$$

This is straightforward; a 3 × 3 matrix is a little more involved.

A3.4.9.2 3 x 3 Matrix

Let the general 3 × 3 matrix be

$$\mathbf{A} = \begin{pmatrix} a_{11} & a_{12} & a_{13} \\ a_{21} & a_{22} & a_{23} \\ a_{31} & a_{32} & a_{33} \end{pmatrix} \qquad \text{(A3.31)}$$

One could repeat the matrix side-to-side, delete the final column and cross-multiply in a manner similar to that in evaluating zone symbols, as in Example 2.1 of Section 2.5. A more general procedure is given by

$$\det(\mathbf{A}) = \begin{vmatrix} a_{11} & a_{12} & a_{13} \\ a_{21} & a_{22} & a_{23} \\ a_{31} & a_{32} & a_{33} \end{vmatrix} = a_{11}\begin{vmatrix} a_{22} & a_{23} \\ a_{32} & a_{33} \end{vmatrix} - a_{12}\begin{vmatrix} a_{21} & a_{23} \\ a_{31} & a_{33} \end{vmatrix} + a_{13}\begin{vmatrix} a_{21} & a_{22} \\ a_{31} & a_{32} \end{vmatrix} \qquad \text{(A3.32)}$$

If either procedure is applied to matrix **A** from Eq. (A3.23), then det(**A**) is formed as

$$\det(\mathbf{A}) = 1\begin{vmatrix} 1 & 2 \\ 1 & 1 \end{vmatrix} - 0\begin{vmatrix} 0 & 2 \\ \bar{2} & 1 \end{vmatrix} + 1\begin{vmatrix} 0 & 1 \\ \bar{2} & 1 \end{vmatrix} = -1 + 2 = 1 \qquad \text{(A3.33)}$$

Some useful properties of determinants are:

- A matrix and its transpose have the same determinant value; *vide* Eq. (A3.15).
- Interchanging two columns of a determinant multiplies its value by −1.
- If any row or column is zero, the determinant value is zero.

A3.4.10 Inverse of a matrix

The inverse of a *square* matrix may be obtained through its *cofactor* matrix. Using matrix **A** from Eq. (A3.23) as an example, the *cofactor* matrix **D**, is obtained with elements d_{ij} given by

$$d_{ij} = (-1)^{i+j} M_{ij} \qquad \text{(A3.34)}$$

where M_{ij} is the *minor* determinant of **A**, obtained by striking out the row and column containing the *ij* element; thus,

$$d_{12} = -\begin{vmatrix} 0 & 2 \\ \bar{2} & 1 \end{vmatrix} = 4$$

Proceeding in this manner,

$$\mathbf{D} = \begin{pmatrix} \bar{1} & \bar{4} & 2 \\ 1 & 3 & \bar{1} \\ \bar{1} & \bar{2} & 1 \end{pmatrix} \qquad \text{(A3.35)}$$

The transpose of the cofactor matrix is the *adjoint* matrix $\mathbf{A}^\dagger$, where $a^\dagger_{i,j} = d_{j,i}$, but there is no need to use it explicitly here. Finally, the inverse of **A**, written as $\mathbf{A}^{-1}$, is given by

$$\mathbf{A}^{-1} = \frac{1}{\det(\mathbf{A})}\mathbf{D}^{\mathrm{T}} = \frac{1}{1}\begin{pmatrix} \bar{1} & 1 & \bar{1} \\ \bar{4} & 3 & \bar{2} \\ 2 & \bar{1} & 1 \end{pmatrix} = \begin{pmatrix} \bar{1} & 1 & \bar{1} \\ \bar{4} & 3 & \bar{2} \\ 2 & \bar{1} & 1 \end{pmatrix} \qquad \text{(A3.36)}$$

A simpler, equivalent procedure is to form the elements of $\mathbf{A}^{-1}$ directly as

$$a^{-1}_{ij} = (-1)^{i+j} M_{ji} \qquad \text{(A3.37)}$$

where M_{ji} is the minor determinant formed by deleting the row and column containing the *ji* element of the matrix **A**. Thus, the a^{-1}_{23} element of the inverse matrix is given by Eq. (A3.34), but with M_{ji} obtained by striking out the row and column of its 3,2 element:

$$a^{-1}_{23} = \frac{1}{\det(\mathbf{A})}(-1)^5 \begin{vmatrix} 1 & 1 \\ 0 & 2 \end{vmatrix} = -2\,.$$

A3.4.11 Orthogonal and unitary matrices

A matrix that fulfils the condition

$$\mathbf{A}^{-1} = \mathbf{A}^{\mathrm{T}} \qquad \text{(A3.38)}$$

is said to be an *orthogonal* matrix. All orthogonal matrices are square but not necessarily symmetrical, as in the following example:

$$\mathbf{A} = \begin{pmatrix} \cos\theta & \overline{\sin\theta} & 0 \\ \sin\theta & \cos\theta & 0 \\ 0 & 0 & 1 \end{pmatrix} \qquad \mathbf{A}^{-1} = \begin{pmatrix} \cos\theta & \sin\theta & 0 \\ \overline{\sin\theta} & \cos\theta & 0 \\ 0 & 0 & 1 \end{pmatrix} \qquad \mathbf{A}^{\mathrm{T}} = \begin{pmatrix} \cos\theta & \sin\theta & 0 \\ \overline{\sin\theta} & \cos\theta & 0 \\ 0 & 0 & 1 \end{pmatrix}$$

A matrix **A** is *unitary* if its adjoint is equal to its inverse

$$\mathbf{A}^\dagger = \mathbf{A}^{-1}$$

that is, $a^\dagger_{ij} = a^{-1}_{ij}$ for all *i* and *j*: matrix **A** below is unitary:

$$\mathbf{A} = \begin{pmatrix} 1 & 0 & 0 \\ 0 & 0 & \mathrm{e}^{\mathrm{i}} \\ 0 & \mathrm{e}^{-2\mathrm{i}} & 0 \end{pmatrix} \qquad \mathbf{A}^\dagger = \begin{pmatrix} 1 & 0 & 0 \\ 0 & 0 & \mathrm{e}^{2\mathrm{i}} \\ 0 & \mathrm{e}^{-\mathrm{i}} & 0 \end{pmatrix} \qquad \mathbf{A}^{-1} = \begin{pmatrix} 1 & 0 & 0 \\ 0 & 0 & \mathrm{e}^{2\mathrm{i}} \\ 0 & \mathrm{e}^{-\mathrm{i}} & 0 \end{pmatrix}$$

If **A** is a complex matrix, the adjoint matrix $\mathbf{A}^\dagger$ is complex conjugate of the transpose, $(\mathbf{A}^{\mathrm{T}})^*$, as can be seen in the matrices above.

A3.4.12 Matrices, rows and columns

In certain transformations, row and column matrices are involved. Thus, the multiplication of a row $\mathbf{x}$ by a matrix $\mathbf{A}$ with the result $\mathbf{x}'$ would be written as

$$\mathbf{xA} = \mathbf{x}'$$

For example:

$$\underset{\mathbf{x}}{(1\ \ 2\ \ 3)} \underset{\mathbf{A}}{\begin{pmatrix} 1 & 0 & 1 \\ 0 & 2 & 1 \\ 1 & 3 & \bar{1} \end{pmatrix}} = \underset{\mathbf{x}'}{(4\ 13\ 0)}$$

whereas if $\mathbf{x}$ is treated as a column,

$$\underset{\mathbf{A}}{\begin{pmatrix} 1 & 0 & 1 \\ 0 & 2 & 1 \\ 1 & 3 & \bar{1} \end{pmatrix}} \underset{\mathbf{x}}{\begin{pmatrix} 1 \\ 2 \\ 3 \end{pmatrix}} = \underset{\mathbf{x}''}{\begin{pmatrix} 4 \\ 7 \\ 4 \end{pmatrix}}$$

Note also that

(i) $\mathbf{A}^{\mathrm{T}}\mathbf{x} = \mathbf{x}'$
(ii) $(\mathbf{A}^{\mathrm{T}})^{-1} = (\mathbf{A}^{-1})^{\mathrm{T}}$

A3.5 Normal to a plane (*hkl*)

In Fig. A3.3, (*hkl*) is any plane in a lattice, making intercepts a/h, b/k and c/l with the x, y and z axes, respectively. The normal $\mathbf{n}$ to the plane is, from Appendix A3.2.2, given by any cross product such as $\mathbf{p}_1 \times \mathbf{p}_2$. Thus, remembering that $\mathbf{a} \times \mathbf{b} = -\mathbf{b} \times \mathbf{a}$, and $\mathbf{b} \times \mathbf{b} = 0$, $\mathbf{n}$ may be written as

$$\begin{aligned} \mathbf{n} &= \left(\frac{\mathbf{a}}{h} - \frac{\mathbf{b}}{k}\right) \times \left(\frac{\mathbf{b}}{k} - \frac{\mathbf{c}}{l}\right) \\ &= \frac{1}{hk}\mathbf{a} \times \mathbf{b} + \frac{1}{kl}\mathbf{b} \times \mathbf{c} + \frac{1}{lh}\mathbf{c} \times \mathbf{a} \end{aligned} \qquad \text{(A3.39)}$$

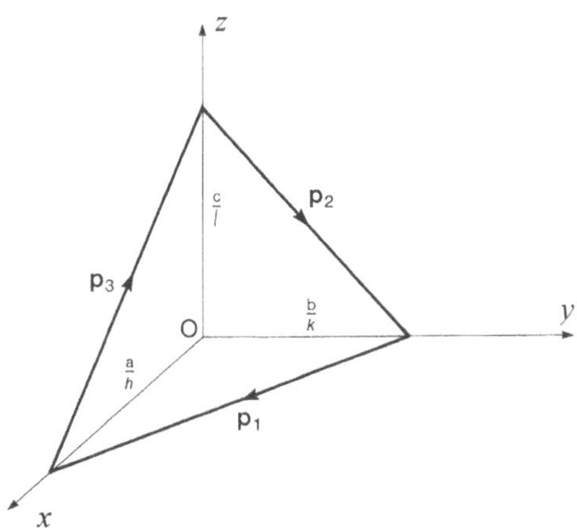

Fig. A3.3
A plane (*hkl*); the vectors $\mathbf{p}_1 = (\mathbf{a}/h) - (\mathbf{b}/k)$, $\mathbf{p}_2 = (\mathbf{b}/k) - (\mathbf{c}/l)$ and $\mathbf{p}_3 = (\mathbf{c}/l) - (\mathbf{a}/h)$.

Multiplying by hkl yields

$$\mathbf{n} = h(\mathbf{b} \times \mathbf{c}) + k(\mathbf{c} \times \mathbf{a}) + l(\mathbf{a} \times \mathbf{b}) \tag{A3.40}$$

A3.6 Solution of linear simultaneous equations

Matrices form the basis of Cramer's method for solving systems of simultaneous equations; the following example illustrates the procedure:

$$\left.\begin{aligned} x_1 + x_2 - x_3 &= 6 \\ x_1 - x_2 + x_3 &= 2 \\ x_1 \qquad - 2x_3 &= 4 \end{aligned}\right\} \tag{A3.41}$$

Writing Eq. (A3.41) in the following matrix form:

$$\underset{\mathbf{S}}{\begin{pmatrix} 1 & 1 & \bar{1} \\ 1 & \bar{1} & 1 \\ 1 & 0 & \bar{2} \end{pmatrix}} \underset{\mathbf{x}}{\begin{pmatrix} x_1 \\ x_2 \\ x_3 \end{pmatrix}} = \underset{\mathbf{x}'}{\begin{pmatrix} 6 \\ 2 \\ 4 \end{pmatrix}} \tag{A3.42}$$

In general, $x_i = \frac{\det(\mathbf{S}_i)}{\det(\mathbf{S})}$, where $\mathbf{S}_i$ is the appropriate determinant, in which the column $\mathbf{x}'$ replaces the ith column of the matrix $\mathbf{S}$. Thus, in our example

$$\det(\mathbf{S}) = \begin{vmatrix} 1 & 1 & \bar{1} \\ 1 & \bar{1} & 1 \\ 1 & 0 & \bar{2} \end{vmatrix} = 4 \qquad \det(\mathbf{S}_1) = \begin{vmatrix} 6 & 1 & \bar{1} \\ 2 & \bar{1} & 1 \\ 4 & 0 & \bar{2} \end{vmatrix} = 16$$

$$\det(\mathbf{S}_2) = \begin{vmatrix} 1 & 6 & \bar{1} \\ 1 & 2 & 1 \\ 1 & 4 & \bar{2} \end{vmatrix} = 8 \qquad \det(\mathbf{S}_3) = \begin{vmatrix} 1 & 1 & 6 \\ 1 & \bar{1} & 2 \\ 1 & 0 & 4 \end{vmatrix} = 0$$

Hence, $x_1 = 4$, $x_2 = 2$, $x_3 = 0$.

A3.7 Useful matrices

It is noted first that from Appendix A6 the *trace* of the rotation matrix, Eq. (A6.6), that is, the sum of the diagonal elements, is $(1 + 2\cos\phi)$, for either orthogonal or non-orthogonal axes, where ϕ is the angle of rotation for a symmetry axis in a crystal. Since $|\cos\phi| \leq 1$, an *integral* value of the trace can be only –1, 0, 1, 2 or 3. Hence, the following results obtain:

$(1 + 2\cos\phi)$	ϕ	*R*-fold axis
−1	180°	2
0	120°	3
1	90°	4
2	60°	6
3	360°	1

This result may be compared with that from Section 4.8.

The set of matrices that follows is useful in working with point groups and space groups. Some of them will be self-evident from the work that has already been studied. Others may be prepared by suitable matrix combinations. For example, consider the stereogram in Fig. A3.4. Rotation of a point x, y, z, about [111] has the following actions:

$$\underset{(1)}{x,\, y,\, z} \xrightarrow{3_{[111]}} \underset{(2)}{z,\, x,\, y} \xrightarrow{3_{[111]}} \underset{(3)}{y,\, z,\, x}$$

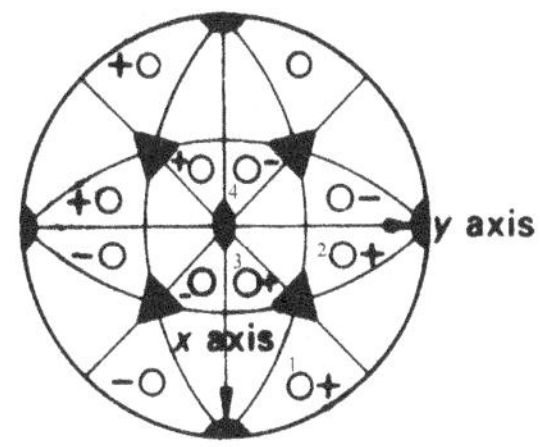

Fig. A3.4
Cubic stereogram, point group 23 : refer to Appendix A3.7 for the implication of points 1–4 on the diagram.

But what of a threefold rotation of x, y, z about $[11\bar{1}]$? From the stereogram, the sequence of symmetry operations

$$\underset{(1)}{x,\, y,\, z} \xrightarrow{3^2_{[111]}} \underset{(3)}{y,\, z,\, x} \xrightarrow{2_{[0y0]}} \underset{(4)}{\bar{y},\, z,\, \bar{x}}$$

describes a rotation of the point x, y, z about $[11\bar{1}]$. In matrix notation:

$$\underset{\mathbf{2}_{[0y0]}}{\begin{pmatrix} \bar{1} & 0 & 0 \\ 0 & 1 & 0 \\ 0 & 0 & \bar{1} \end{pmatrix}} \underset{\mathbf{3}_{[111]}}{\begin{pmatrix} 0 & 0 & 1 \\ 1 & 0 & 0 \\ 0 & 1 & 0 \end{pmatrix}} \underset{\mathbf{3}_{[111]}}{\begin{pmatrix} 0 & 0 & 1 \\ 1 & 0 & 0 \\ 0 & 1 & 0 \end{pmatrix}} = \underset{\mathbf{3}_{[11\bar{1}]}}{\begin{pmatrix} 0 & \bar{1} & 0 \\ 0 & 0 & 1 \\ \bar{1} & 0 & 0 \end{pmatrix}}.$$

The trace of the matrix $\mathbf{3}_{[11\bar{1}]}$ is zero; clearly it corresponds to threefold rotation, in this case, about the direction $[11\bar{1}]$.

The set of matrices given below are designated mostly by a $[UVW]$ symbol, the direction of the corresponding symmetry operator; where $\mathbf{m}_{[UVW]}$ is listed, the plane normal to $[UVW]$ is implied. Any rotation matrix $\mathbf{R}$ may be converted to a $\bar{\mathbf{R}}$ matrix by negating all non-zero elements in $\mathbf{R}$. Thus, from the first example, $\bar{\mathbf{2}}_{[100]}$, which is equivalent to $\mathbf{m}_{[100]}$, that is, the plane $(0, y, z)$, is obtained by negating all non-zero elements of the matrix for $\mathbf{2}_{[100]}$. The symbolism $\mathbf{r}(x', y', z') = \mathbf{R}\mathbf{r}(x, y, z)$ means that the point $\mathbf{r}(x', y', z')$ results from multiplying the point $\mathbf{r}(x, y, z)$ by the matrix $\mathbf{R}$, in the order indicated.

A3.7.1 Rotation and reflection matrices on orthogonal axes

A3.7.1.1 Twofold rotation and mirror ($\bar{2}$) reflection

$$\mathbf{2}_{[100]} = \begin{pmatrix} 1 & 0 & 0 \\ 0 & \bar{1} & 0 \\ 0 & 0 & \bar{1} \end{pmatrix} \quad \mathbf{2}_{[010]} = \begin{pmatrix} \bar{1} & 0 & 0 \\ 0 & 1 & 0 \\ 0 & 0 & \bar{1} \end{pmatrix} \quad \mathbf{2}_{[001]} = \begin{pmatrix} \bar{1} & 0 & 0 \\ 0 & \bar{1} & 0 \\ 0 & 0 & 1 \end{pmatrix}$$

$$\mathbf{2}_{[011]} = \begin{pmatrix} \bar{1} & 0 & 0 \\ 0 & 0 & 1 \\ 0 & 1 & 0 \end{pmatrix} \quad \mathbf{2}_{[101]} = \begin{pmatrix} 0 & 0 & 1 \\ 0 & \bar{1} & 0 \\ 1 & 0 & 0 \end{pmatrix} \quad \mathbf{2}_{[110]} = \begin{pmatrix} 0 & 1 & 0 \\ 1 & 0 & 0 \\ 0 & 0 & \bar{1} \end{pmatrix}$$

$$\mathbf{2}_{[01\bar{1}]} = \begin{pmatrix} \bar{1} & 0 & 0 \\ 0 & 0 & \bar{1} \\ 0 & \bar{1} & 0 \end{pmatrix} \quad \mathbf{2}_{[\bar{1}01]} = \begin{pmatrix} 0 & 0 & \bar{1} \\ 0 & \bar{1} & 0 \\ \bar{1} & 0 & 0 \end{pmatrix} \quad \mathbf{2}_{[1\bar{1}0]} = \begin{pmatrix} 0 & \bar{1} & 0 \\ \bar{1} & 0 & 0 \\ 0 & 0 & \bar{1} \end{pmatrix}$$

From the foregoing, the matrices for mirror reflection operation $m(\equiv \bar{2}_{\perp m})$ across the planes normal to [100], [010], [001], [011], [101], [110], $(01\bar{1})$, $(\bar{1}01)$ and $(1\bar{1}0)$ may be obtained by negating the non-zero elements of the above six matrices.

A3.7.1.2 Threefold rotation inclined to x, y, z at $\cos^{-1}(1/\sqrt{3})$, approximately 54.74°

$$\mathbf{3}_{[111]} = \begin{pmatrix} 0 & 0 & 1 \\ 1 & 0 & 0 \\ 0 & 1 & 0 \end{pmatrix} \quad \mathbf{3}_{[\bar{1}11]} = \begin{pmatrix} 0 & \bar{1} & 0 \\ 0 & 0 & 1 \\ \bar{1} & 0 & 0 \end{pmatrix} \quad \mathbf{3}_{[1\bar{1}1]} = \begin{pmatrix} 0 & \bar{1} & 0 \\ 0 & 0 & \bar{1} \\ 1 & 0 & 0 \end{pmatrix} \quad \mathbf{3}_{[11\bar{1}]} = \begin{pmatrix} 0 & 1 & 0 \\ 0 & 0 & \bar{1} \\ \bar{1} & 0 & 0 \end{pmatrix}$$

A3.7.1.3 Fourfold rotation

$$\mathbf{4}_{[100]} = \begin{pmatrix} 1 & 0 & 0 \\ 0 & 0 & \bar{1} \\ 0 & 1 & 0 \end{pmatrix} \quad \mathbf{4}_{[010]} = \begin{pmatrix} 0 & 0 & 1 \\ 0 & 1 & 0 \\ \bar{1} & 0 & 0 \end{pmatrix} \quad \mathbf{4}_{[001]} = \begin{pmatrix} 0 & \bar{1} & 0 \\ 1 & 0 & 0 \\ 0 & 0 & 1 \end{pmatrix} \quad \bar{\mathbf{4}}_{[001]} = (-1)\,\mathbf{4}_{[001]} = \begin{pmatrix} 0 & 1 & 0 \\ \bar{1} & 0 & 0 \\ 0 & 0 & \bar{1} \end{pmatrix}$$

A3.7.1.4 Sixfold rotation

$$\mathbf{6}_{[001]} = \mathbf{2}_{[001]}\mathbf{3}^2_{[001]} = \begin{pmatrix} \bar{1} & 0 & 0 \\ 0 & \bar{1} & 0 \\ 0 & 0 & 1 \end{pmatrix} \begin{pmatrix} -1/2 & -\sqrt{3}/2 & 0 \\ \sqrt{3}/2 & -1/2 & 0 \\ 0 & 0 & 1 \end{pmatrix} \begin{pmatrix} -1/2 & -\sqrt{3}/2 & 0 \\ \sqrt{3}/2 & -1/2 & 0 \\ 0 & 0 & 1 \end{pmatrix}$$

$$= \begin{pmatrix} 1/2 & -\sqrt{3}/2 & 0 \\ \sqrt{3}/2 & 1/2 & 0 \\ 0 & 0 & 1 \end{pmatrix}$$

A3.7.2 Rotation and reflection matrices on rhombohedral axes

$$\mathbf{3}_{[111]} = \begin{pmatrix} 0 & 0 & 1 \\ 1 & 0 & 0 \\ 0 & 1 & 0 \end{pmatrix} \quad \mathbf{m}_{(0yz)} = \begin{pmatrix} 1 & 0 & 0 \\ 0 & 0 & 1 \\ 0 & 1 & 0 \end{pmatrix}$$

(i) The matrix $\mathbf{3}_{[111]}$ here is identical with that for $\mathbf{3}_{[111]}$ on orthogonal axes.
(ii) Other points related by m symmetry may be obtained by the matrix combinations $\mathbf{3}_{[111]}\,\mathbf{m}_{(0yz)}$.

A3.7.3 Rotation and reflection matrices on hexagonal axes, *x*, *y*, *u*, *z*

$$\mathbf{2}_{[120]} = \begin{pmatrix} \bar{1} & 1 & 0 \\ 0 & 1 & 0 \\ 0 & 0 & \bar{1} \end{pmatrix} \quad \mathbf{m}_{[120]} = \begin{pmatrix} 1 & \bar{1} & 0 \\ 0 & \bar{1} & 0 \\ 0 & 0 & 1 \end{pmatrix} \; [\equiv \mathbf{m} \,||\, (x,\ 0,\ z)]$$

$$\mathbf{2}_{[210]} = \begin{pmatrix} 1 & 0 & 0 \\ 1 & \bar{1} & 0 \\ 0 & 0 & \bar{1} \end{pmatrix} \quad \mathbf{m}_{[210]} = \begin{pmatrix} \bar{1} & 0 & 0 \\ \bar{1} & 1 & 0 \\ 0 & 0 & 1 \end{pmatrix} \; [\equiv \mathbf{m} \,||\, (0,\ y,\ z)]$$

$$\mathbf{3}_{[0001]} = \begin{pmatrix} 0 & \bar{1} & 0 \\ 1 & \bar{1} & 0 \\ 0 & 0 & 1 \end{pmatrix} \quad \mathbf{6}_{[001]} = \begin{pmatrix} 1 & \bar{1} & 0 \\ 1 & 0 & 0 \\ 0 & 0 & 1 \end{pmatrix}$$

A4 Stereographic projection of a circle is a circle

In Fig. A4.1, AB is the trace of a small circle on a vertical section of a sphere, centre O, of arbitrary radius. The lines PA and PX are equal in length and are generators of a cone about PQ as axis; the points where the cone touches the sphere form the circle. The trace AX of its right section is one diameter of an ellipse; AB is the trace of one circular section of this ellipse, and XY is the trace of the section symmetrically inclined to the axis PQ, the conjugate circular section. Let BZ be drawn parallel to the diameter ST of the primitive. Then,

$$\angle ZBP = \angle PAB \tag{A4.1}$$

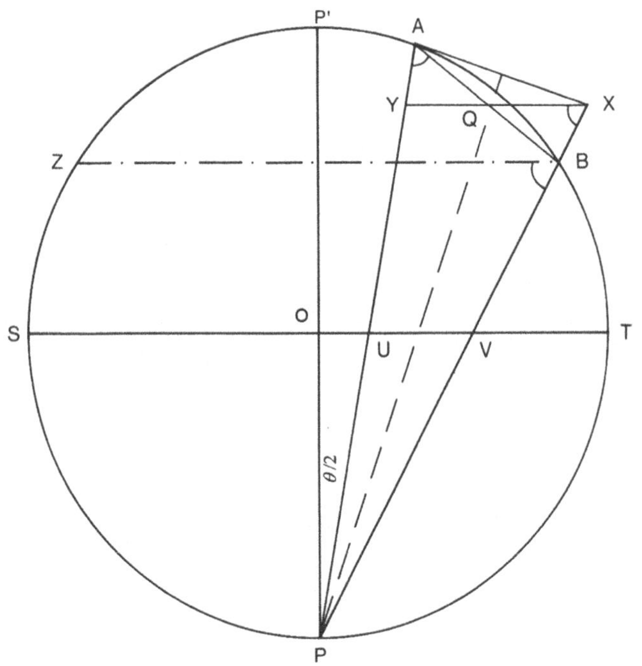

Fig. A4.1
The stereographic projection of a circle: vertical section of the sphere, diameter ST, AB is the trace of the plane of a small circle on the surface of the sphere, and UV is the diameter of the projection of the small circle on to the primitive; U is the stereographic projection of the point A.

since they stand on the equal arcs, *PZ* and *PB*. Triangles *PAB* and *PXY* are congruent, so that

$$\angle PAB = \angle PXY = \angle ZBP \tag{A4.2}$$

Hence,

$$XY \parallel BZ \parallel ST \tag{A4.3}$$

and, therefore, *UV* is a diameter of a circle on the primitive. It follows that all angular relationships on a crystal are reproduced truly on a stereogram: it is an *angle-true* and *symmetry-true* representation of a crystal.

A5 Best-fit plane

The equation of a plane may be written as $A'X + B'Y + C'Z = D'$. When there are more than three pairs of experimentally determined X, Y and Z coordinates, the best plane through them may be obtained by the method of least squares. For this process it is convenient to rearrange the equation to the form $AX + BY + C = Z$. If the error of fit ε_i for the ith measurement is given by

$$\varepsilon_i = AX_i + BY_i + C_i - Z_i \tag{A5.1}$$

then the least squares estimates of A, B and C are obtained by minimizing the sum

$$\sum_i \varepsilon_i^2 = \sum_i (AX_i + BY_i + C - Z_i)^2 \tag{A5.2}$$

Thus, the partial differentials of the function with respect A, B and C are set equal to zero. There result the normal equations, from which A, B and C can be solved by the procedure described in Appendix A3.6. The program PLANE in the Web Program Suite has been constructed by this procedure, and can be used to determine a best-fit plane and also the distances of the data points from that plane. The method of least squares is discussed in detail in the literature [1].

Reference

[1] Whittaker ET and Robinson G. *The calculus of observations*. Blackie & Son, 1924.

General rotation matrices

A6

In this appendix, a matrix **S** is derived for the anticlockwise rotation of a point X, Y, Z by an angle ϕ about an axis normal to the plane of X and Y, with an angle γ between the X and Y axes. Since the rotation axis is normal to the plane, the Z coordinate of the point remains unchanged.

In Fig. A6.1, $\angle PQN = \pi - \gamma$, and $\angle QPN = \gamma - \pi/2$. Then,

$$\begin{aligned} X &= r\cos\theta - y\cos\gamma \\ Y &= r\sin\theta/\sin\gamma \end{aligned} \tag{A6.1}$$

It follows that

$$\begin{aligned} X' &= r\cos(\theta+\phi) - Y'\cos\gamma \\ Y' &= r\sin^{-1}\gamma\sin(\theta+\phi) \end{aligned} \tag{A6.2}$$

Expanding Eq. (A6.2), substituting for $r\sin\theta$ and $r\cos\theta$ from Eq. (A6.1) and rearranging leads to

$$X' = X(\cos\phi - \cos\gamma\sin\phi/\sin\gamma) - Y\left(\sin\gamma\sin\phi - \cos^2\gamma\sin\phi/\sin\gamma\right) \tag{A6.3}$$

$$Y' = X(\sin\phi/\sin\gamma) + Y(\cos\phi) + \cos\gamma\sin\phi/\sin\gamma) \tag{A6.4}$$

Written concisely:

$$\mathbf{X}' = \mathbf{SX} \tag{A6.5}$$

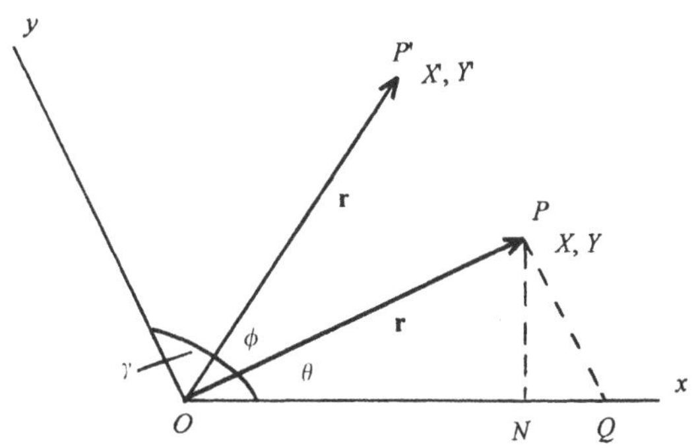

Fig. A6.1
Vector ***OP*** of length $|\mathbf{r}|$ at a general angle θ to the x axis; ***OP'*** is the same vector after being rotated anticlockwise by an angle ϕ from the direction of ***OP***. The general angle between the x and y axes is γ; for threefold and sixfold symmetry, γ has the value 120°.

where the matrix $\mathbf{S}$ is given by

$$\mathbf{S} = \begin{pmatrix} (\cos\phi - \cos\gamma \sin\phi/\sin\gamma) & -(\sin\gamma\sin\phi - \cos^2\gamma\sin\phi/\sin\gamma) & 0 \\ (\sin\phi/\sin\gamma) & (\cos\phi + \cos\gamma\sin\phi/\sin\gamma) & 0 \\ 0 & 0 & 1 \end{pmatrix} \tag{A6.6}$$

The matrix in Eq. (A6.6) will suffice for all rotational operations that are encountered in studying point groups where the rotation axis, z in this discussion, is normal to the x, y plane. For threefold rotation in the cubic system, the stereogram for point group 432 (Fig. 3.12c) shows that that a fourfold anticlockwise rotation about the x axis followed by a fourfold anticlockwise rotation about the z axis is equivalent to a threefold anticlockwise rotation about the direction [111].

A general equation for the rotation of $\mathbf{X}\,(X, Y, Z)$ is $\mathbf{X}' = \mathbf{RX}$. Thus, from Eq. (A6.6) for a fourfold rotation $\mathbf{R}_x$ about the x axis in the cubic system ($\gamma = 90°$), by interchange of axes,

$$\underset{\mathbf{R}_x}{\begin{pmatrix} 1 & 0 & 0 \\ 0 & 0 & \bar{1} \\ 0 & 1 & 0 \end{pmatrix}} \underset{\mathbf{X}}{\begin{pmatrix} X \\ Y \\ Z \end{pmatrix}} = \underset{\mathbf{X}'}{\begin{pmatrix} X \\ \bar{Z} \\ Y \end{pmatrix}} \tag{A6.7}$$

and from a successive rotation $\mathbf{R}_x$ about the z axis,

$$\underset{\mathbf{R}_z}{\begin{pmatrix} 0 & \bar{1} & 0 \\ 1 & 0 & 0 \\ 0 & 0 & 1 \end{pmatrix}} \underset{\mathbf{X}'}{\begin{pmatrix} X \\ \bar{Z} \\ Y \end{pmatrix}} = \underset{\mathbf{X}''}{\begin{pmatrix} Z \\ X \\ Y \end{pmatrix}} \tag{A6.8}$$

Hence, the matrix for the rotation $\mathbf{3}_{[111]}$ in the cubic system is obtained by the product $\mathbf{R}_Z\mathbf{R}_x$:

$$\underset{\mathbf{R}_x\mathbf{R}_z = \mathbf{3}_{[111]}}{\begin{pmatrix} 0 & 0 & 1 \\ 1 & 0 & 0 \\ 0 & 0 & 1 \end{pmatrix}} \underset{\mathbf{X}}{\begin{pmatrix} X \\ Y \\ Z \end{pmatrix}} = \underset{\mathbf{X}''}{\begin{pmatrix} Z \\ X \\ Y \end{pmatrix}} \tag{A6.9}$$

Trigonometric identities

A7

The following relationships are often useful in the trigonometrical manipulations of the geometrical structure factor equations (Section 6.6ff).

$$\cos A + \cos B = 2\cos\frac{A+B}{2}\cos\frac{A-B}{2} \tag{A7.1}$$

$$\cos A - \cos B = -2\sin\frac{A+B}{2}\sin\frac{A-B}{2} \tag{A7.2}$$

$$\sin A + \sin B = 2\sin\frac{A+B}{2}\cos\frac{A-B}{2} \tag{A7.3}$$

$$\sin A - \sin B = 2\cos\frac{A+B}{2}\sin\frac{A-B}{2} \tag{A7.4}$$

$$\cos(A+B) = \cos A\cos B - \sin A\sin B \tag{A7.5}$$

$$\cos(A-B) = \cos A\cos B + \sin A\sin B \tag{A7.6}$$

$$\sin(A+B) = \sin A\cos B + \cos A\sin B \tag{A7.7}$$

$$\sin(A-B) = \sin A\cos B - \cos A\sin B \tag{A7.8}$$

$$\sin 2A = 2\sin A\cos A \tag{A7.9}$$

$$\cos 2A = \cos^2 A - \sin^2 A = 2\cos^2 A - 1 = 1 - 2\sin^2 A \tag{A7.10}$$

$$\sin^2(2\pi n/4) = 0 \text{ for } n \text{ even or } 1 \text{ for } n \text{ odd} \tag{A7.11}$$

$$\cos^2(2\pi n/4) = 1 \text{ for } n \text{ even or } 0 \text{ for } n \text{ odd} \tag{A7.12}$$

$\cos 2\pi(\theta + n/2)$ is equivalent to $\cos 2\pi(\theta - n/2)$, where n is an integer. (A7.13)

In manipulating geometrical structure factor equations, it is sometimes desirable to modify a coordinate by adding ± 1; it results in a crystallographically equivalent value for the coordinate: $\frac{1}{2} + x \equiv -\frac{1}{2} + x$.

A8 Spherical polar coordinates

A8.1 Polar coordinates

In the diagram in Fig. A8.1, the *polar coordinates* r, θ and ϕ are defined by

$$\left.\begin{aligned} x &= r\sin\theta\cos\phi \\ y &= r\sin\theta\sin\phi \\ z &= r\cos\theta \end{aligned}\right\} \qquad \text{(A8.1)}$$

where

$$r^2 = x^2 + y^2 + z^2 \qquad \text{(A8.2)}$$

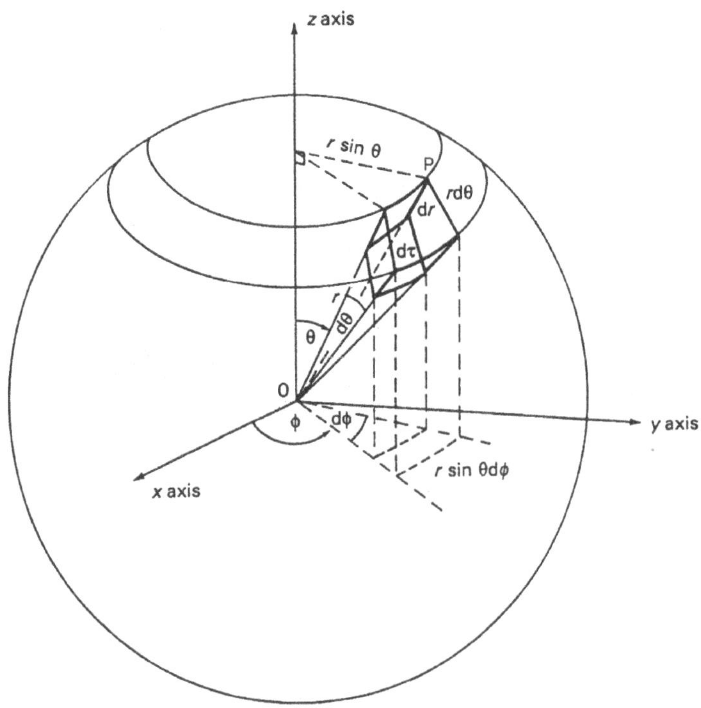

Fig. A8.1
Conversion of Cartesian coordinates to spherical polar coordinates. The distance OP is $|\mathbf{r}|$, where $\mathbf{r}$ is the vector from the origin O to the point $P(x, y, z)$. [Reproduced by courtesy of Woodhead Publishing, UK.]

A8.2 Volume element

The volume element $d\tau$ in Fig. A8.1 corresponds to the infinitesimal quantity $dx\,dy\,dz$ in Cartesian coordinates; it is straightforward to show the relationship

$$d\tau = r^2 \sin\theta \, dr \, d\theta \, d\phi \qquad \text{(A8.3)}$$

The limits of the polar variables that correspond to x, y and z each between $-\infty$ and $+\infty$ are

$$\left.\begin{array}{l} 0 \le r \le +\infty \\ 0 \le \theta \le \pi \\ 0 \le \phi \le 2\pi \end{array}\right\} \qquad \text{(A8.4)}$$

A9 The gamma function, $\Gamma(n)$

The gamma function [1,2] may be defined by the equation

$$\Gamma(n) = \int_0^{\infty} x^{n-1} \exp\left(-ax^2\right) \mathrm{d}x \tag{A9.1}$$

where a is a constant and n is a positive number; the integrals occur in studying *inter alia* quantum chemistry and atomic scattering factors. The following results are useful:

1. For $n > 0$ and integral

$$\Gamma\,(n) = (n-1)! \tag{A9.2}$$

2. For $n > 0$

$$\Gamma\,(n+1) = n\Gamma\,(n) \tag{A9.3}$$

and if n is also integral

$$\Gamma(n+1) = n! \tag{A9.4}$$

3.

$$\Gamma\,(1/2) = \sqrt{\pi} \tag{A9.5}$$

Example A9.1
Consider the solution of the integral

$$\mathrm{I} = \int_0^{\infty} x^4 \exp\left(-x^2/2\right) \mathrm{d}x$$

Let $x^2/2 = t$, so that $x = (2t)^{1/2}$ and $\mathrm{d}x = (2t)^{-1/2}\mathrm{d}t$. Then

$$\mathrm{I} = 2\sqrt{2} \int_0^{\infty} t^{3/2} \exp\,(-t)\,\mathrm{d}t = 2\sqrt{2}\,\Gamma\,(5/2) = 3\,(\pi/2)^{1/2}$$

Occasionally, the reduction formula hereunder is useful:

$$\int x^n \exp\,(ax)\,\mathrm{d}x = \frac{x^n \exp\,(ax)}{a} - \frac{n}{a} \int x^{n-1} \exp\,(ax)\,\mathrm{d}x \tag{A9.6}$$

References

[1] Sebah P and Gourdon X. Introduction to the gamma function. <http://numbers.computation.free.fr/Constants/Miscellaneous/gammaFunction.html> (2002).
[2] eFunda Inc. Gamma function finder. <http://www.efunda.com/math/gamma/findgamma.cfm> (2013).

A10 Point group character tables and related data

A10.1 Introduction

In this appendix, the character tables listed are those likely to be encountered in studying the textual matter and problems in this book. A full list of character tables and other relevant data are given in various text books to which reference has been made in earlier sections.

The penultimate column of each character table refers to *rotational and translational* movements, and is needed in vibrational analyses, and the final column to *product functions*, needed for other spectroscopic analyses and for some aspects of molecular bonding. It may be noted that x^2, y^2 and z^2, singly or in additive terms, transform as the totally symmetric irreducible representation in all point groups.

In the final two columns of certain character tables, parentheses are used: for example, $(R_x,\ R_y)$ means that they are the same and are counted once, implying twofold degeneracy of the irreducible representation to which they span, whereas $x^2 + y^2$, z^2 implies that they are different and can be counted twice. In cubic point groups, twofold and threefold degenerate irreducible representations occur, and entries such as $(x,\ y,\ z)$ appear in cubic point groups. Note that the symbol '*E*', herein distinguished as *E* (label of the twofold degenerate irreducible representation) and ***E*** (the identity symmetry operator); also in the cubic groups the symbol *T* is used for both the point group *T* and the threefold degenerate irreducible representation.

A10.2 Character tables

C_s Group

$C_s\ (\equiv S_1)$	$\boldsymbol{E}$	i		
A'	1	1	$x,\ y,\ R_z$	$z^2,\ x^2,\ y^2,\ xy$
A''	1	-1	$z,\ R_x,\ R_y$	$yz,\ zx$

C_i Group

$C_i (\equiv S_2)$	E	i		
A_g	1	1	R_z, R_x, R_y	$z^2, x^2, y^2, xy, yz, zx$
A_u	1	−1	z, x, y	

C_n Groups

C_1	E
A	1

C_2	E	C_2		
A	1	1	z, R_z	x^2, y^2, z^2, xy
B	1	−1	x, y, R_x, R_y	yz, zx

C_3	E	C_3	C_3^2		$\varepsilon = \exp(i2\pi/3)$
A	1	1	1	z, R_z	$x^2 + y^2, z^2$
E	$\left\{\begin{matrix}1\\1\end{matrix}\right.$	$\begin{matrix}\varepsilon\\\varepsilon^*\end{matrix}$	$\left.\begin{matrix}\varepsilon^*\\\varepsilon\end{matrix}\right\}$	$(x, y), (R_x, R_y)$	$(x^2 - y^2, xy), (yz, zx)$

C_4	E	C_4	C_2	C_4^3		
A	1	1	1	1	z, R_z	$x^2 + y^2, z^2$
B	1	−1	1	−1		$x^2 - y^2, xy$
E	$\left\{\begin{matrix}1\\1\end{matrix}\right.$	$\begin{matrix}i\\-i\end{matrix}$	$\begin{matrix}-1\\-1\end{matrix}$	$\left.\begin{matrix}-i\\i\end{matrix}\right\}$	$(x, y), (R_x, R_y)$	(yz, zx)

C_5	E	C_5	C_5^2	C_5^3	C_5^4		$\varepsilon = \exp(i2\pi/5)$
A	1	1	1	1	1	z, R_z	$x^2 + y^2, z^2$
E_1	$\left\{\begin{matrix}1\\1\end{matrix}\right.$	$\begin{matrix}\varepsilon\\\varepsilon^*\end{matrix}$	$\begin{matrix}\varepsilon^2\\\varepsilon^{*2}\end{matrix}$	$\begin{matrix}\varepsilon^{*2}\\\varepsilon^2\end{matrix}$	$\left.\begin{matrix}\varepsilon^*\\\varepsilon\end{matrix}\right\}$	$(x, y), (R_x, R_y)$	(yz, zx)
E_2	$\left\{\begin{matrix}1\\1\end{matrix}\right.$	$\begin{matrix}\varepsilon^2\\\varepsilon^{*2}\end{matrix}$	$\begin{matrix}\varepsilon^*\\\varepsilon\end{matrix}$	$\begin{matrix}\varepsilon\\\varepsilon^*\end{matrix}$	$\left.\begin{matrix}\varepsilon^{*2}\\\varepsilon^2\end{matrix}\right\}$		$(x^2 - y^2, xy)$

C_6	E	C_6	C_3	C_2	C_3^2	C_6^5		$\varepsilon = \exp(i2\pi/6)$
A	1	1	1	1	1	1	z, R_z	x^2+y^2, z^2
B	1	−1	1	−1	1	−1		
E_1	$\begin{Bmatrix}1\\1\end{Bmatrix}$	ε ε^*	$-\varepsilon^*$ $-\varepsilon$	−1 −1	$-\varepsilon$ $-\varepsilon^*$	ε^* ε	$(x, y), (R_x, R_y)$	(xz, yz)
E_2	$\begin{Bmatrix}1\\1\end{Bmatrix}$	$-\varepsilon^*$ $-\varepsilon$	$-\varepsilon$ $-\varepsilon^*$	1 1	$-\varepsilon^*$ $-\varepsilon$	$-\varepsilon$ $-\varepsilon^*$		(x^2-y^2, xy)

D_n Groups

D_2	E	$C_2(z)$	$C_2(y)$	$C_2(x)$		
A	1	1	1	1		x^2, y^2, z^2
B_1	1	1	−1	−1	z, R_z	xy
B_2	1	−1	1	−1	y, R_y	zx
B_3	1	−1	−1	1	x, R_x	yz

D_3	E	$2C_3$	$3C_2$		
A_1	1	1	1		x^2+y^2, z^2
A_2	1	1	−1	z, R_z	
E	2	−1	0	$(x, y), (R_x, R_y)$	$(x^2-y^2, xy), (zx, xy)$

D_4	E	$2C_4$	$C_2(=C_4^2)$	$2C_2'$	$2C_2''$		
A_1	1	1	1	1	1		x^2+y^2, z^2
A_2	1	1	1	−1	−1	z, R_z	
B_1	1	−1	1	1	−1		x^2-y^2
B_2	1	−1	1	−1	1		xy
E	2	0	−2	0	0	$(x, y), (R_x, R_y)$	(zx, yz)

D_5	E	$2C_5$	$2C_5^2$	$5C_2$		
A_1	1	1	1	1		x^2+y^2, z^2
A_2	1	1	1	−1	z, R_z	
E_1	2	2 cos 72°	2 cos 144°	0	$(x, y), (R_x, R_y)$	(zx, yz)
E_2	2	2 cos 144°	2 cos 72°	0		(x^2-y^2, xy)

D_6	E	$2C_6$	$2C_3$	C_2	$3C_2'$	$3C_2''$		
A_1	1	1	1	1	1	1		x^2+y^2, z^2
A_2	1	1	1	1	−1	−1	z, R_z	
B_1	1	−1	1	−1	1	−1		
B_2	1	−1	1	−1	−1	1		
E_1	2	1	−1	−2	0	0	$(x, y), (R_x, R_y)$	(zx, yz)
E_2	2	−1	−1	2	0	0		(x^2-y^2, xy)

C_{nv} Groups

C_{2v}	E	C_2	$\sigma(xz)$	$\sigma(yz)$		
A_1	1	1	1	1	z	x^2, y^2, z^2
A_2	1	1	−1	−1	R_z	xy
B_1	1	−1	1	−1	x, R_y	zx
B_2	1	−1	−1	1	y, R_x	yz

C_{3v}	E	$2C_3$	$3\sigma_v$		
A_1	1	1	1	z	x^2+y^2, z^2
A_2	1	1	−1	R_z	
E	2	−1	0	$(x, y), (R_x, R_y)$	$(x^2-y^2, xy), (zx, xy)$

C_{4v}	E	$2C_4$	C_2	$2\sigma_v$	$2\sigma_d$		
A_1	1	1	1	1	1	z	x^2+y^2, z^2
A_2	1	1	1	−1	−1	R_z	x^2-y^2
B_1	1	−1	1	1	−1		
B_3	1	−1	1	−1	1		xy
E	2	0	−2	0	0	$(x, y), (R_x, R_y)$	$(zx, yz), (x^2-y^2, xy)$

C_{6v}	E	$2C_6$	$2C_3$	C_2	$3\sigma_v$	$3\sigma_d$		
A_1	1	1	1	1	1	1	z	x^2+y^2, z^2
A_2	1	1	1	1	−1	−1	R_z	
B_1	1	−1	1	−1	1	−1		
B_2	1	−1	1	−1	−1	1		
E_1	2	1	−1	−2	0	0	$(x, y), (R_x, R_y)$	(zx, yz)
E_2	2	−1	−1	2	0	0		(x^2-y^2, xy)

C_{nh} Groups

C_{2h}	E	C_2	i	σ_h		
A_g	1	1	1	1	R_z	x^2, y^2, z^2, xy
B_g	1	−1	1	−1	R_x, R_y	zx, yz
A_u	1	1	−1	−1	z	
B_u	1	−1	−1	1	x, y	

$C_{3h}(\equiv S_3)$	E	C_3	C_3^2	σ_h	S_3	S_3^5		$\varepsilon = \exp(i2\pi/3)$
A'	1	1	1	1	1	1	R_z	x^2+y^2, z^2
E'	{1 1	ε ε^*	ε^* ε	1 1	ε ε^*	ε^* ε}	(x, y)	(x^2-y^2, xy)
A''	1	1	1	−1	−1	−1	z	
E''	{1 1	ε ε^*	ε^* ε	−1 −1	$-\varepsilon$ $-\varepsilon^*$	$-\varepsilon^*$ $-\varepsilon$}	(R_x, R_y)	(zx, yz)

C_{4h}	E	C_4	C_2	C_4^3	i	S_4^3	σ_h	S_4		
A_g	1	1	1	1	1	1	1	1	R_z	x^2+y^2, z^2
B_g	1	−1	1	−1	1	−1	1	−1		x^2-y^2, xy
E_g	1 1	i −i	−1 −1	−i i	1 1	i −i	−1 −1	−i i	(R_x, R_y)	(zx, yz)
A_u	1	1	1	1	−1	−1	−1	−1	z	
B_u	1	−1	1	−1	−1	1	−1	1		
E_u	1 1	i −i	−1 −1	−i i	−1 −1	−i i	1 1	i −i	(x, y)	

C_{6h}	E	C_6	C_3	C_2	C_3^2	C_6^5	i	S_3^5	S_6^5	σ_h	S_6	S_3		$\varepsilon = \exp(i2\pi/6)$
A_g	1	1	1	1	1	1	1	1	1	1	1	1	R_z	x^2+y^2, z
B_g	1	−1	1	−1	1	−1	1	−1	1	−1	1	−1		
E_{1g}	1 1	ε ε^*	$-\varepsilon^*$ $-\varepsilon$	−1 −1	$-\varepsilon$ $-\varepsilon^*$	ε^* ε	1 1	ε ε^*	$-\varepsilon^*$ $-\varepsilon$	−1 −1	$-\varepsilon$ $-\varepsilon^*$	ε^* ε	(R_x, R_y)	(zx, yz)
E_{2g}	1 1	$-\varepsilon^*$ $-\varepsilon$	$-\varepsilon$ $-\varepsilon^*$	1 1	$-\varepsilon^*$ $-\varepsilon$	$-\varepsilon$ $-\varepsilon^*$	1 1	$-\varepsilon^*$ $-\varepsilon$	$-\varepsilon$ $-\varepsilon^*$	1 1	$-\varepsilon^*$ $-\varepsilon$	$-\varepsilon$ $-\varepsilon^*$		(x^2-y^2, xy)
A_u	1	1	1	1	1	1	−1	1	−1	−1	−1	−1	z	
B_u	1	−1	1	−1	1	−1	−1	1	−1	1	−1	1		
E_{1u}	1 1	ε ε^*	$-\varepsilon^*$ $-\varepsilon$	−1 −1	$-\varepsilon$ $-\varepsilon^*$	ε^* ε	−1 −1	$-\varepsilon$ $-\varepsilon^*$	ε^* ε	1 1	ε ε^*	$-\varepsilon^*$ $-\varepsilon$	(x, y)	
E_{2u}	1 1	$-\varepsilon^*$ $-\varepsilon$	$-\varepsilon$ $-\varepsilon^*$	−1 −1	$-\varepsilon^*$ $-\varepsilon$	$-\varepsilon$ $-\varepsilon^*$	−1 −1	ε^* ε	ε ε^*	−1 −1	ε^* ε	ε ε^*		

D_{nh} Groups

D_{2h}	E	$C_2(z)$	$C_2(y)$	$C_2(x)$	i	$\sigma(xy)$	$\sigma(zx)$	$\sigma(yz)$		
A_g	1	1	1	1	1	1	1	1		x^2, y^2, z^2
B_{1g}	1	1	−1	−1	1	1	−1	−1	R_z	xy
B_{2g}	1	−1	1	−1	1	−1	1	−1	R_y	zx
B_{3g}	1	−1	−1	1	1	−1	−1	1	R_x	yz
A_u	1	1	1	1	−1	−1	−1	−1		
B_{1u}	1	1	−1	−1	−1	−1	1	1	z	
B_{2u}	1	−1	1	−1	−1	1	−1	1	y	
B_{3u}	1	−1	−1	1	−1	1	1	−1	x	

D_{3h}	E	$2C_3$	$3C_2$	σ_h	$2S_3$	$3\sigma_v$		
A_1'	1	1	1	1	1	1		x^2+y^2, z^2
A_2'	1	1	−1	1	1	−1	R_z	
E'	2	−1	0	2	−1	0	(x, y)	(x^2-y^2, xy)
A_1''	1	1	1	−1	−1	−1		
A_2''	1	1	−1	−1	−1	1	z	
E''	2	−1	0	−2	1	0	(R_x, R_y)	(zx, yz)

D_{4h}	E	$2C_4$	C_2	$2C_2'$	$2C_2''$	i	$2S_4$	σ_h	$2\sigma_v$	$2\sigma_d$		
A_{1g}	1	1	1	1	1	1	1	1	1	1		x^2+y^2, z^2
A_{2g}	1	1	1	−1	−1	1	1	1	−1	−1	R_z	
B_{1g}	1	−1	1	1	−1	1	−1	1	1	−1		x^2-y^2
B_{2g}	1	−1	1	−1	1	1	−1	1	−1	1		xy
E_g	2	0	−2	0	0	2	0	−2	0	0	(R_x, R_y)	(zx, yz)
A_{1u}	1	1	1	1	1	−1	−1	−1	−1	−1		
A_{2u}	1	1	1	−1	−1	−1	−1	−1	1	1	z	
B_{1u}	1	−1	1	1	−1	−1	1	−1	−1	1		
B_{2u}	1	−1	1	−1	1	−1	1	−1	1	−1		
E_u	2	0	−2	0	0	−2	0	2	0	0	(x, y)	

D_{6h}	E	$2C_6$	$2C_3$	C_2	$3C_2'$	$3C_2''$	i	$2S_3$	$2S_6$	σ_h	$3\sigma_d$	$3\sigma_v$		
A_{1g}	1	1	1	1	1	1	1	1	1	1	1	1		x^2+y^2, z^2
A_{2g}	1	1	1	1	−1	−1	1	1	1	1	−1	−1	R_z	
B_{1g}	1	−1	1	−1	1	−1	1	−1	1	−1	1	−1		
B_{2g}	1	−1	1	−1	−1	1	1	−1	1	−1	−1	1		
E_{1g}	2	1	−1	−2	0	0	2	1	−1	−2	0	0	(R_x, R_y)	
E_{2g}	2	−1	−1	2	0	0	2	−1	−1	2	0	0		(x^2-y^2, xy)
A_{1u}	1	1	1	1	1	1	−1	−1	−1	−1	−1	−1		
A_{2u}	1	1	1	1	−1	−1	−1	−1	−1	−1	1	1	z	
B_{1u}	1	−1	1	−1	1	−1	−1	1	−1	1	−1	1		
B_{2u}	1	−1	1	−1	−1	1	−1	1	−1	1	1	−1		
E_{2g}	2	1	−1	−2	0	0	−2	−1	1	2	0	0	(x, y)	
E_{2u}	2	−1	−1	2	0	0	−2	1	1	−2	0	0		

D_{nd} Groups

D_{2d}	E	$2S_4$	C_2	$2C_2'$	$2\sigma_d$		
A_1	1	1	1	1	1		x^2+y^2, z^2
A_2	1	1	1	−1	−1	R_z	
B_1	1	−1	1	1	−1		x^2-y^2
B_2	1	−1	1	−1	1	z	xy
E	2	0	−2	0	0	$(x, y), (R_x, R_y)$	(zx, yz)

D_{3d}	E	$2C_3$	$3C_2$	i	$2S_6$	$3\sigma_d$		
A_{1g}	1	1	1	1	1	1		x^2+y^2, z^2
A_{2g}	1	1	−1	1	1	−1	R_z	
E_g	2	−1	0	2	−1	0	(R_x, R_y)	$(x^2-y^2, xy), (zx, yz)$
A_{1u}	1	1	1	−1	−1	−1		
A_{2u}	1	1	−1	−1	−1	1	z	
E_u	2	−1	0	−2	1	0	(x, y)	

S_n Groups

S_4	E	S_4	C_2	S_4^3		
A	1	1	1	1	R_z	x^2+y^2, z^2
B	1	−1	1	−1	z	x^2-y^2, xy
E	$\begin{Bmatrix}1\\1\end{Bmatrix}$	$\begin{matrix}i\\-i\end{matrix}$	$\begin{matrix}-1\\-1\end{matrix}$	$\begin{matrix}-i\\i\end{matrix}$	$(x, y), (R_x, R_y)$	(zx, yz)

S_6	E	C_3	C_3^2	i	S_6^5	S_6		$\varepsilon = \exp(i2\pi/3)$
A_g	1	1	1	1	1	1	R_z	x^2+y^2, z^2
E_g	$\begin{Bmatrix}1\\1\end{Bmatrix}$	$\begin{matrix}\varepsilon\\\varepsilon^*\end{matrix}$	$\begin{matrix}\varepsilon^*\\\varepsilon\end{matrix}$	$\begin{matrix}1\\1\end{matrix}$	$\begin{matrix}\varepsilon\\\varepsilon^*\end{matrix}$	$\begin{matrix}\varepsilon^*\\\varepsilon\end{matrix}$	(R_x, R_y)	$(x^2-y^2, xy), (zx, yz)$
A_u	1	1	1	−1	−1	−1	z	
E_u	$\begin{Bmatrix}1\\1\end{Bmatrix}$	$\begin{matrix}\varepsilon\\\varepsilon^*\end{matrix}$	$\begin{matrix}\varepsilon^*\\\varepsilon\end{matrix}$	$\begin{matrix}-1\\-1\end{matrix}$	$\begin{matrix}-\varepsilon\\-\varepsilon^*\end{matrix}$	$\begin{matrix}-\varepsilon^*\\-\varepsilon\end{matrix}$	(x, y)	

For S_1, S_2 and S_3, see C_s, C_i and C_{3h}, respectively.

Cubic groups

T	E	$4C_3$	$4C_3^2$	$3C_2$		$\varepsilon = \exp(i2\pi/3)$
A	1	1	1	1		$x^2 + y^2 + z^2$
E	$\begin{cases} 1 \\ 1 \end{cases}$	ε ε^*	ε^* ε	1 1		$(2z^2 - x^2 - y^2, x^2 - y^2)$
T	3	0	0	−1	$(R_x, R_y, R_z), (x, y, z)$	(xy, zx, yz)

T_h	E	$4C_3$	$4C_3^2$	$3C_2$	i	$4S_6$	$4S_6^5$	$3\sigma_h$		$\varepsilon = \exp(i2\pi/3)$
A_g	1	1	1	1	1	1	1	1		$x^2 + y^2 + z^2$
E_g	$\begin{cases} 1 \\ 1 \end{cases}$	ε ε^*	ε^* ε	1 1	1 1	ε ε^*	ε^* ε	$\begin{rcases} 1 \\ 1 \end{rcases}$		$(2z^2 - x^2 - y^2,$ $x^2 - y^2)$
T	3	0	0	−1	3	0	0	−1	$(R_x, R_y, R_z), (x, y, z)$	(xy, zx, yz)
A_u	1	1	1	1	−1	−1	−1	−1		
E_u	$\begin{cases} 1 \\ 1 \end{cases}$	ε ε^*	ε^* ε	1 1	−1 −1	$-\varepsilon$ $-\varepsilon^*$	$-\varepsilon^*$ $-\varepsilon$	$\begin{rcases} -1 \\ -1 \end{rcases}$		
T_u	3	0	0	−1	−3	0	0	1	(x, y, z)	

T_d	E	$8C_3$	$3C_2$	$6S_4$	$6\sigma_d$		
A_1	1	1	1	1	1		$x^2 + y^2 + z^2$
A_2	1	1	1	−1	−1		
E	2	−1	2	0	0		$(2z^2 - x^2 - y^2, x^2 - y^2)$
T_1	3	0	−1	1	−1	(R_x, R_y, R_z)	
T_2	3	0	−1	−1	1	(x, y, z)	(xy, zx, yz)

O	E	$8C_3$	$3C_2\ (=3C_4^2)$	$6C_4$	$6C_2$		
A_1	1	1	1	1	1		$x^2+y^2+z^2$
A_2	1	1	1	−1	−1		
E	2	−1	2	0	0		$(2z^2-x^2-y^2,\ x^2-y^2)$
T_1	3	0	−1	1	−1	(x, y, z), (R_x, R_y, R_z)	
T_2	3	0	−1	−1	1		(xy, zx, yz)

O_h	E	$8C_3$	$6C_2'$	$6C_4$	$3C_2(=3C_4^2)$	i	$6S_4$	$8S_6$	$3\sigma_h$	$6\sigma_d$		
A_{1g}	1	1	1	1	1	1	1	1	1	1		$x^2+y^2+z^2$
A_{2g}	1	1	−1	−1	1	1	−1	1	1	−1		
E_g	2	−1	0	0	2	2	0	−1	2	0		$(2z^2-x^2-y^2,\ x^2-y^2)$
T_{1g}	3	0	−1	1	−1	3	1	0	−1	−1	(R_x, R_y, R_z)	
T_{2g}	3	0	1	−1	−1	3	−1	0	−1	1		(xy, zx, yz)
A_{1u}	1	1	1	1	1	−1	−1	−1	−1	−1		
A_{2u}	1	1	−1	−1	1	−1	1	−1	−1	1		
E_u	2	−1	0	0	2	−2	0	1	−2	0		
T_{1u}	3	0	−1	1	−1	−3	−1	0	1	1	(x, y, z)	
T_{2u}	3	0	1	−1	−1	−3	1	0	1	−1		

Linear groups

$C_{\infty v}$	E	$2C_\infty^\phi$	...	$\infty\sigma_v$		
$A_1 \equiv \Sigma^+$	1	1	...	1	z	$x^2+y^2,\ z^2$
$A_2 \equiv \Sigma^-$	1	1	...	−1	R_z	
$E_1 \equiv \Pi$	2	$2\cos\phi$	...	0	(x, y), (R_x, R_y)	(zx, yz)
$E_2 \equiv \Delta$	2	$2\cos 2\phi$	...	0		$(x^2-y^2,\ xy)$
$E_3 \equiv \Phi$	2	$2\cos 3\phi$	...	0		
...	...	...	...	...		

$D_{\infty h}$	E	$2C_{\infty}^{\phi}$	...	$\infty\sigma_v$	i	$2S_{\infty}^{\phi}$	...	∞C_2		
Σ_g^+	1	1	...	1	1	1	...	1		x^2+y^2, z^2
Σ_g^-	1	1	...	−1	1	1	...	−1	R_z	
Π_g	2	$2\cos\phi$	...	0	2	$-2\cos\phi$	...	0	(R_x, R_y)	(zx, yz)
Δ_g	2	$2\cos 2\phi$	...	0	2	$2\cos 2\phi$	...	0		(x^2-y^2, xy)
...	...	...	...	...	...	...	...	...		
Σ_u^+	1	1	...	1	−1	−1	...	−1	Z	
Σ_u^-	1	1	...	−1	−1	−1	...	1		
Π_u	2	$2\cos\phi$	...	0	−2	$2\cos\phi$	...	0	(x, y)	
Δ_u ...	2	$2\cos 2\phi$	...	0	−2	$-2\cos 2\phi$	...	0		

Icosahedral groups

I	E	$12C_5$	$12C_5^2$	$20C_3$	$15C_2$		$\varphi^{\pm} = (1 \pm \sqrt{5})/2$
A	1	1	1	1	1		$x^2+y^2+z^2$
T_1	3	φ^+	φ^-	0	−1	$(x, y, z), (R_x, R_y, R_z)$	
T_2	3	φ^-	φ^+	0	−1		
G	4	−1	−1	1	0		
H	5	0	0	−1	1		$(2z^2-x^2-y^2, x^2-y^2, xy, yz, zx)$

I_h	E	$12C_5$	$12C_5^2$	$20C_3$	$15C_2$	i	$12S_{10}$	$12S_{10}^3$	$20S_6$	15σ		$\varphi^{\pm} = (1 \pm \sqrt{5})/2$
A_g	1	1	1	1	1	1	1	1	1	1		$x^2 + y^2 + z^2$
T_{1g}	3	φ^+	φ^-	0	−1	3	φ^-	φ^+	0	−1	(R_x, R_y, R_z)	
T_{2g}	3	φ^-	φ^+	0	−1	3	φ^+	φ^-	0	−1		
G_g	4	−1	−1	1	0	4	−1	−1	1	0		
H_g	5	0	0	−1	1	5	0	0	−1	1		$(2z^2 - x^2 - y^2, x^2 - y^2, xy, yz, zx)$
A_u	1	1	1	1	1	−1	−1	−1	−1	−1		
T_{1u}	3	φ^+	φ^-	0	−1	−3	$-\varphi^-$	$-\varphi^+$	0	1	(x, y, z)	
T_{2u}	3	φ^-	φ^+	0	−1	−3	$-\varphi^+$	$-\varphi^-$	0	1		
G_u	4	−1	−1	1	0	−4	1	1	−1	−1		
H_u	5	0	0	−1	1	−5	0	0	1	−1		

A10.3 Direct products of irreducible representations and other related data

A10.3.1 Non-degenerate irreducible representations

$$A \otimes A = B \otimes B = A$$
$$A \otimes B = B$$
$$A \otimes E = B \otimes E = E$$
$$A \otimes T = B \otimes T = T$$
$$A \otimes E_1 = B \otimes E_2 = E_1$$
$$A \otimes E_2 = B \otimes E_1 = E_2$$

A10.3.2 Gerade/ungerade

$$g \otimes g = g$$
$$u \otimes u = g$$
$$g \otimes u = u$$

A10.3.3 Prime/double prime

$$' \otimes ' = '' \otimes '' = '$$
$$' \otimes '' = ''$$

A10.3.4 Subscripts to irreducible representations[1]

$$1 \otimes 1 = 2 \otimes 2 = 1$$
$$1 \otimes 2 = 2$$

[1] Where there are subscripts 1, 2 and 3, $1 \otimes 2 = 3$, $2 \otimes 3 = 1$ and $3 \otimes 1 = 2$ (Examples: D_2 and D_{2h}).

A10.3.5 Twofold degenerate irreducible representations[2]

$$E_1 \otimes E_1 = A_1 + A_2 + E_2$$
$$E_2 \otimes E_2 = A_1 + A_2 + E_1$$
$$E_1 \otimes E_2 = B_1 + B_2 + E_1$$
$$E \otimes E = A_1 + A_2 + E$$

[2] In point groups where the principal axis is C_4 or C_2, $E \otimes E = A_1 + A_2 + B_1 + B_2$.

A10.3.6 Threefold degenerate irreducible representations

$$E \otimes T_1 = E \otimes T_2 = T_1 + T_2$$
$$T_1 \otimes T_1 = T_2 \otimes T_2 = A_1 + E + T_1 + T_2$$
$$T_1 \otimes T_2 = A_2 + E + T_1 + T_2$$

A10.3.7 Infinitely degenerate irreducible representations

$$\Sigma^+ \otimes \Sigma^+ = \Sigma^- \otimes \Sigma^- = \Sigma^+$$
$$\Sigma^+ \otimes \Sigma^- = \Sigma^-$$
$$\Sigma \otimes \Delta = \Delta$$
$$\Sigma \otimes \Pi = \Pi$$
$$\Pi \otimes \Pi = \Sigma^+ + \Sigma^- + \Delta$$
$$\Pi \otimes \Delta = \Pi + \Phi$$
$$\Delta \otimes \Delta = \Sigma^+ + \Sigma^- + \Gamma$$

A10.4 Other useful relationships

$$\varepsilon = \exp(i2\pi/n)$$
$$\varepsilon^* = \exp(-i2\pi/n)$$
$$\varepsilon\varepsilon^* = 1$$
$$\varepsilon + \varepsilon^* = 2\cos(2\pi/n)$$
$$\varepsilon - \varepsilon^* = i2\sin(2\pi/n)$$

where ε^* is the complex conjugate of ε.
For the particular case of $n = 3$:

$$\varepsilon^2 = \varepsilon^*$$
$$\varepsilon + \varepsilon^* = -1$$
$$\varepsilon - \varepsilon^* = i\sqrt{3}$$

$$\varphi^+ = \tfrac{1}{2}(1 + \sqrt{5}) = 1.61803\ldots = -2\cos 144°$$
(φ^+ is the 'golden number', Section 3.13.1)
$$\varphi^- = \tfrac{1}{2}(1 - \sqrt{5}) = -0.61803\ldots = -2\cos 72°$$
$$\varphi^+\varphi^+ = 1 + \varphi^+$$
$$\varphi^-\varphi^- = 1 + \varphi^-$$
$$\varphi^+\varphi^- = -1$$

Linear, unitary and projection operators

A11

A11.1 Linear operators

An *operator* has the property of changing one function into another function. Thus, if an operator is specified by $\boldsymbol{O}$, and

$$\boldsymbol{O}(2x^2 + x) = 4x + 1, \qquad \text{(A11.1)}$$

then the operator $\boldsymbol{O}$ is clearly the differential operator $\frac{d}{dx}$. There are many such operators, and a particular case is the *linear* operator. An operator is linear if, for any function f

$$\boldsymbol{O}kf = k(\boldsymbol{O}f) \qquad \text{(A11.2)}$$

and

$$\boldsymbol{O}(f_1 + f_2) = \boldsymbol{O}f_1 + \boldsymbol{O}f_2 \qquad \text{(A11.3)}$$

where k is a constant. Linear operators possess several important properties:

For two linear operators with a functions f:

- $(\boldsymbol{O}_1 + \boldsymbol{O}_2)f = \boldsymbol{O}_1 f + \boldsymbol{O}_2 f$ (A11.4)
- $\boldsymbol{O}_1 \boldsymbol{O}_2 f = \boldsymbol{O}_1(\boldsymbol{O}_2 f)$ (A11.5)
- $\boldsymbol{O}_1(\boldsymbol{O}_2 + \boldsymbol{O}_3) = \boldsymbol{O}_1\boldsymbol{O}_2 + \boldsymbol{O}_1\boldsymbol{O}_3$ (A11.6)
- $\boldsymbol{O}_1(\boldsymbol{O}_2\boldsymbol{O}_3) = (\boldsymbol{O}_1\boldsymbol{O}_2)\boldsymbol{O}_3$ (A11.7)

Except in special cases, $\boldsymbol{O}_i\boldsymbol{O}_j \neq \boldsymbol{O}_j\boldsymbol{O}_i$: this result is easily demonstrated, and may be compared with the product of two rotations $\boldsymbol{R}_i\boldsymbol{R}_j$ where $\boldsymbol{R}$ is greater than 2.

Example A11.1

Let $\boldsymbol{O}_1 = \frac{d}{dx}$, $\boldsymbol{O}_2 = x^2$, $\boldsymbol{O}_3 = \frac{d^2}{dx^2}$, $\boldsymbol{O}_4 = k$, and $f_1 = x^3 - 2x + 1$, $f_2 = 2x^2 - 3$. Then,

(a) $\boldsymbol{O}_1 f_1 = 3x^2 - 2$
(b) $\boldsymbol{O}_1 kf_1 = 6x^2 - 4$
(c) $\boldsymbol{O}_1(f_1 + f_2) = 3x^2 + 4x - 2$
(d) $\boldsymbol{O}_1\ \boldsymbol{O}_2 f_1 = 5x^4 - 6x^2 + 2x$
(e) $\boldsymbol{O}_2\ \boldsymbol{O}_1 f_1 = 3x^4 - 2x^2$ ($\neq\ \boldsymbol{O}_1\ \boldsymbol{O}_2 f$)

A11.2 Operators in function space

Let a symmetry operation $\boldsymbol{R}$ be represented by a *transformation operator* $\boldsymbol{O_R}$. A transformation operator acts on the functions of function space and follows the rules for linear operators. A set of such operators for all symmetry operations $\boldsymbol{R}$ in a point group is homomorphic (Section 7.3) with the set of symmetry operations themselves. If an operation $\boldsymbol{R}$ moves a vector from a position $\mathbf{r}$ to a new position $\mathbf{r}'$, a transformation operator $\boldsymbol{O_R}$ is associated with $\boldsymbol{R}$ according to

$$\boldsymbol{O_R} f(\mathbf{r}') = f(\mathbf{r}) \tag{A11.8}$$

that is, the new function $\boldsymbol{O_R} f$ assigns the value of the original f to the new location $\mathbf{r}'$; thus, $\mathbf{r}'$ is related to $\mathbf{r}$ by an equation similar to Eq. (7.10).

Let $\boldsymbol{R}$ and $\boldsymbol{R}'$ be two symmetry operators in a point group, such that $\boldsymbol{R}$ moves a vector $\mathbf{r}$ to $\mathbf{r}'$, and then $\boldsymbol{R}'$ moves $\mathbf{r}'$ to $\mathbf{r}''$. From the definition of a group, there will be another symmetry operation $\boldsymbol{R}''$ in the group, given by

$$\boldsymbol{R}'' = \boldsymbol{R}'\boldsymbol{R} \tag{A11.9}$$

that moves $\mathbf{r}$ directly to $\mathbf{r}''$.

The associated transformation operator moves a function f defined at $\mathbf{r}$ to the new location $\mathbf{r}''$

$$\boldsymbol{O_R} f(\mathbf{r}'') = f(\mathbf{r}) \tag{A11.10}$$

For another function g in the same space as f, Eq. (A11.8) leads to

$$\boldsymbol{O_R} g(\mathbf{r}'') = g(\mathbf{r}') \tag{A11.11}$$

Since g is any function, let it be equal to $\boldsymbol{O_R} f$; then with Eq. (A11.8)

$$g(\mathbf{r}') = \boldsymbol{O_R} f(\mathbf{r}') = f(\mathbf{r}) \tag{A11.12}$$

Applying this result to Eq. (A11.11),

$$[\boldsymbol{O_{R'}}(\boldsymbol{O_R} f)](\mathbf{r}'') = g(\mathbf{r}') = (\boldsymbol{O_R} f)\mathbf{r}' = f(r) \tag{A11.13}$$

which, from Eq. (A11.10), shows that

$$\boldsymbol{O_{R''}} = \boldsymbol{O_{R'}}\boldsymbol{O_R} \tag{A11.14}$$

Thus, the relationship, Eq. (A11.9), for symmetry operators is paralleled by Eq. (A11.14) for the corresponding transformation operators.

Example A11.2

Consider a function space spanned by the three p orbital functions. From the matrix $\mathbf{A}$ in Appendix A3.4.11, an anticlockwise rotation $\boldsymbol{C}_\theta$ moves $\mathbf{r}\,(x, y, z)$ to $\mathbf{r}'(x', y', z')$ where $|\mathbf{r}| = |\mathbf{r}'|$, and

$$\begin{aligned} x' &= x\cos\theta - y\sin\theta \\ y' &= x\sin\theta + y\cos\theta \\ z' &= z \end{aligned}$$

and inverting these equations gives

$$\begin{aligned} x &= x'\cos\theta + y'\sin\theta \\ y &= -x'\sin\theta + y'\cos\theta \\ z &= z' \end{aligned}$$

Considering p_x and using a result from Appendix A8, the operator $\boldsymbol{O}_\theta$ that corresponds to the symmetry operation $\boldsymbol{C}_\theta$ has the property

$$\boldsymbol{O}_\theta p_x(x', y', z') = p_x(x, y, z) = f(\mathbf{r})x/r$$

Substituting for x and r

$$\boldsymbol{O}_\theta p_x(x', y', z') = f(\mathbf{r}')(x' \cos\theta + y' \sin\theta)/r'$$

and since there are primes on both sides, neatness suggests the form

$$\boldsymbol{O}_\theta p_x(x, y, z) = f(\mathbf{r})(x \cos\theta + y \sin\theta)/r$$

with similar equations for p and p_z.

A11.3 Unitary operators

Again, from Appendix A8, the right hand side of Eq. (A11.14) may be recast as

$$\boldsymbol{O}_\theta p_x = p_x \cos\theta + p_y \sin\theta + (p_z \times 0)$$

Similarly,

$$\begin{aligned} \boldsymbol{O}_\theta p_y &= -p_x \sin\theta + p_y \cos\theta + (p_z \times 0) \\ \boldsymbol{O}_\theta p_z &= p_x \times 0 \quad + p_y \times 0 \quad + (p_z \times 1) \end{aligned}$$

Since $\boldsymbol{O}_\theta p_x$ is a row vector, these equations form the matrix $\mathbf{D}(\mathbf{C}_\theta)$ which shows a formal equivalence with Eq. (7.9)

$$\mathbf{D}(\boldsymbol{C}_\theta) = \begin{pmatrix} \cos\theta & -\sin\theta & 0 \\ \sin\theta & \cos\theta & 0 \\ 0 & 0 & 1 \end{pmatrix} \tag{A11.15}$$

For convenience, p_x, p_y and p_z will be renamed here as p_i, $i = 1$–3, the $\mathbf{D}$-matrix elements as $\mathbf{D}(\boldsymbol{R})_{ij}$ and the unit vectors **i, j, k** as $\mathbf{e}_i$, $i = 1$–3. The operations discussed in Section 7.5.2, in real space, may be now summarized as

$$\boldsymbol{R}\mathbf{e}_i = \sum_{j=1}^{3} \mathbf{r}_{ji}\,\mathbf{e}_i \tag{A11.16}$$

Similarly, the p function transformation in function space is given by

$$\boldsymbol{O_R} p_i = \sum_{j=1}^{3} \mathbf{D}(\boldsymbol{R})_{ji} p_j \tag{A11.17}$$

which is a formalism of the equation in Example A11.2. Thus, by analogy with Eq. (A11.14),

$$\mathbf{D}(\boldsymbol{R}'') = \mathbf{D}(\boldsymbol{R}')\mathbf{D}(\boldsymbol{R}) \tag{A11.18}$$

From the nature of a scalar, the transformation operators leave the scalar product of two functions unchanged:

$$\boldsymbol{O_R} f_1 \cdot \boldsymbol{O_R} f_2 = f_1 \cdot f_2 \tag{A11.19}$$

These operators are *unitary operators* and can be represented by unitary matrices (Appendix A3.4.11) which lead to unitary representations for point groups; thus, $\mathbf{D}(\boldsymbol{C}_3)$ in unitary form is $\begin{pmatrix} \varepsilon^* & 0 & 0 \\ 0 & \varepsilon & 0 \\ 0 & 0 & 1 \end{pmatrix}$. Representations are unitary if the basis functions are orthogonal and normalized to one and the same constant, usually unity. If orthonormal functions are employed, this condition holds implicitly.

A11.4 Projection operators

Consider a set of orthonormal functions ψ_n that form a basis for the βth n_β-dimensional irreducible representation Γ_β in a given point group of order h. Then, from Eq. (A11.17)

$$\sum_{\boldsymbol{R}} \mathbf{D}_\alpha(\boldsymbol{R})^*_{pq} \boldsymbol{O}_{\mathbf{R}} \psi_{\beta,q} = \sum_{p=1}^{n_\beta} \sum_{\boldsymbol{R}} \mathbf{D}_\alpha(\boldsymbol{R})^*_{pq} \mathbf{D}_\beta(\boldsymbol{R})_{pq} \psi_{\beta,q} \tag{A11.20}$$

Since the basis functions are orthonormal, the **D**-matrices are unitary. Pre-multiplication by $D^*_\alpha(\mathbf{R})_{ij}$ from a representation of the same symmetry $\boldsymbol{R}$, followed by summation over all $\boldsymbol{R}$ and reversal of the order of summation on the right hand side, leads to

$$\sum_{\boldsymbol{R}} \mathbf{D}_\alpha(\boldsymbol{R})^*_{ij} \boldsymbol{O}_{\boldsymbol{R}} \psi_{\beta,q} = \sum_{p=1}^{n_\beta} \sum_{\boldsymbol{R}} \mathbf{D}_\alpha(\boldsymbol{R})^*_{ij} \mathbf{D}_\beta(\boldsymbol{R})_{pq} \boldsymbol{\psi}_{\beta,q} \tag{A11.21}$$

From the great orthogonality theorem, Eq. (7.37), the right hand side of Eq. (A11.21) may be equated to

$$\sum_{p=1}^{n_\beta} \frac{h}{\sqrt{n_\alpha n_\beta}} \delta_{\alpha\beta}\, \delta_{jq}\, \psi_{\beta,p}$$

which is zero if $\alpha \neq \beta$, but which otherwise gives $(h/n)\,\delta_{jq}\psi_{\alpha,i}$ because the sum over p is zero except for $p = i$.

Thus, Eq. (A11.21) becomes

$$\boldsymbol{P}_{\alpha,ij}\, \psi_{\beta,q} = \frac{h}{n_\alpha} \delta_{\alpha\beta}\, \delta_{jq}\, \psi_{\beta,i} \tag{A11.22}$$

whereupon the *projection operator* $\mathbf{P}_{\alpha,ij}$ is given by

$$\boldsymbol{P}_{\alpha,ij} = \sum_{\boldsymbol{R}} \mathbf{D}_\alpha(\boldsymbol{R})^*_{ij} \boldsymbol{O}_{\boldsymbol{R}} \tag{A11.23}$$

and is a linear combination of operators $\boldsymbol{O}_{\boldsymbol{R}}$ with coefficients from the **D**-matrices of the Γ_α representation. For the non-zero case of Eq. (A11.22), with $\alpha = \beta$ and $j = q$,

$$\boldsymbol{P}_{\alpha,ij}\psi_{\alpha,j} = \frac{h}{n_\alpha} \psi_{\alpha,i} \tag{A11.24}$$

so that $P_{\alpha,ij}$ operating on the function ψ_α at location j reproduces that function multiplied by h/n_a at a new location i.

Another projection operator P_α may be defined that is a linear combination, having properties similar to those just described, by using the matrix elements from Eq. (A11.23) for the special case of $i = j$:

$$P_\alpha = \sum_{i=1}^{n_\alpha} P_{\alpha,ii} = \sum_{i=1}^{n_\alpha} \sum_{R} \mathbf{D}_\alpha(\boldsymbol{R})^*_{ii} \boldsymbol{O_R} \tag{A11.25}$$

Recalling that $\sum_i \mathbf{D}_\alpha(\boldsymbol{R})^*_{ii} = \chi^*_\alpha$, and summing first over i,

$$P_\alpha = \sum_R \chi^*_\alpha \boldsymbol{O_R} \tag{A11.26}$$

where the coefficient of $\boldsymbol{O_R}$ is the complex conjugate of the characters of $\boldsymbol{R}$ in the representation Γ_α. Where the characters are real, $\chi^*_\alpha(\boldsymbol{R})$ is replaced by $\chi_\alpha(\boldsymbol{R})$. Whereas Eq. (A11.23) needs the complete **D**-matrix in order to define $P_{\alpha,ij}$, P_α may be obtained readily from the character table of the appropriate point group. From Eqs. (A11.21)–(A11.22), and with $i = j$,

$$P_\alpha\, \psi_{\beta,q} = \sum_{i=1}^{n_\alpha} \frac{h}{n_\alpha} \delta_{\alpha\beta}\, \delta_{iq}\, \psi_{\beta,\,i}$$

Hence, for $\alpha = \beta$ (and $i = q$)

$$P_\alpha\, \psi_{\alpha,i} = \frac{h}{n_\alpha}\, \psi_{\alpha,\,i} \tag{A11.27}$$

The projection operator P_α acting on a function that is a member of the Γ_α function space reproduces that function multiplied by h/n, but any function not belonging to the Γ_α function space is annihilated. The operator P_α may be applied to any combination of basis functions in the space Γ_α. Thus, from the linear combination

$$\Phi_\alpha = \sum_{i=1}^{n} c_i\, \psi_{\alpha,i} \tag{A11.28}$$

and

$$P_\alpha\, \Phi_\alpha = \frac{h}{n_\alpha} \sum_{R} \chi^*_\alpha(\boldsymbol{R}) \boldsymbol{O_R}\, \Phi_{\alpha i}. \tag{A11.29}$$

Thus, P_α generates a sum of members of a basis set spanning the irreducible representation Γ_α.

Example A11.3

Apply the projection operator to point group C_{3v}. The basis set of functions is α (spanning $A1$), β (spanning $A2$) and μ, v (spanning the two-dimensional E representation). Referring to the *extended* character table for C_{3v}, the effects of $\boldsymbol{O_R}$ on each function follows they behave like row vectors:

C_{3v}	O_E	O_{C_3}	$O_{C_3^2}$	O_{σ_v}	$O_{\sigma_v'}$	$O_{\sigma_v''}$
α	α	α	α	α	α	α
β	β	β	β	β	β	β
μ	μ	$\left(\mu\overline{1/2}\ \ \nu\overline{\sqrt{3}/2}\right)$	$\left(\overline{\mu/2}\ \ \nu\sqrt{3}/2\right)$	μ	$\left(\mu\overline{1/2}\ \ \nu\sqrt{3}/2\right)$	$\left(\mu\overline{1/2}\ \ \nu\overline{\sqrt{3}/2}\right)$
ν	ν	$\left(\mu\sqrt{3}/2\ \ \overline{\nu/2}\right)$	$\left(\mu\overline{\sqrt{3}/2}\ \ \overline{\nu/2}\right)$	$-\nu$	$\left(\mu\sqrt{3}/2\ \ \nu/2\right)$	$\left(\mu\overline{\sqrt{3}/2}\ \ \nu/2\right)$

Apply the projection operators to the four functions $\alpha,\ \beta,\ \mu$ and ν, using the above table:

$$\begin{aligned} \boldsymbol{P}_{A_1}\alpha &= \alpha+\alpha+\alpha+\alpha+\alpha+\alpha = 6\alpha \\ \boldsymbol{P}_{A_2}\alpha &= \alpha+\alpha+\alpha-\alpha-\alpha-\alpha = 0 \\ \boldsymbol{P}_{E}\alpha &= 2\alpha-\alpha-\alpha = 0 \end{aligned}$$

so that $\boldsymbol{P}_{A_1}$ operating in α multiplies it by h/n, but with A_2 and E α is annihilated. With $\beta,\ \mu$ and ν, the corresponding results are 6β, 3μ and 3ν, respectively. The two parts of E together satisfy the h/n_α requirement; multi-dimensional representations must be treated in this manner in order to achieve the correct result.

Vanishing integrals

A12

A12.1 Introduction

If the sine function

$$\int_{-p}^{p} \sin 2\pi kx \, dx \tag{A12.1}$$

is integrated, the result is zero for any limit $\pm p$, because of the antisymmetry of the sine function across the origin. This type of function could not span the totally symmetric A_1 irreducible representation. Consider next the integral

$$\mathscr{I} = \int \psi_1 \psi_2 \, d\tau \tag{A12.2}$$

where ψ_1 and ψ_2 are wave functions and the integration is over the space of τ. Since the value of the integral $\mathscr{I}$ is independent of molecular orientation, any symmetry operation acting on $\mathscr{I}$ is equivalent to the transformation $\mathscr{I} \rightarrow \mathscr{I}$, that is, a product $\psi_1\psi_2$ can be the basis for the A_1 irreducible representation. For example, in point group C_{3v}, suppose that ψ_1 and ψ_2 span A_1 and E respectively, then the table hereunder follows:

	E	$2C_3$	$3\sigma_v$
$\psi_1(A_1)$	1	1	1
$\psi_2(E)$	2	-1	0

The product $\psi_1\psi_2$ is 2 -1 0, which has the symmetry of E: A_1 is not present ($A_1 \otimes E = E$, Appendix A10.2.1) and the integral, Eq. (A12.2), vanishes because ψ_1 and ψ_2 do not transform as the same irreducible representation of the point group. Had ψ_2 also spanned A_1, then $\psi_1\psi_2$ would have had the symmetry of A_1, and the integral would not necessarily have vanished; it could however, vanish for reasons unrelated to symmetry [1]. The arguments may be extended to products of more than two functions.

Example A12.1
Can the integrals, Eq. (A12.2), of the functions (a) $x^2 - y^2$, and (b) $x^2 - y^2 + z^2$ be non-vanishing when integrated over a square centred at the origin?

A square corresponds with point group D_{4h}. Hence, from the character table for this point group: (a) The function $x^2 - y^2$ spans B_{1g} and the integral vanishes. (b) The function $x^2 - y^2 + z^2$ spans $A_{1g} + B_{1g}$, and the integral can be now non-vanishing.

A12.2 Spectroscopic applications

In applying quantum mechanical principles to spectroscopic transitions, integrals of the form

$$\int \psi_i^* \boldsymbol{\mu}\ \psi_f\, \mathrm{d}\tau \qquad \text{(A12.3)}$$

are encountered, where i and f are initial and final states of the system, and $\boldsymbol{\mu}$ is a transitional moment operator, the dipole moment operator, for example, which may be written as

$$\boldsymbol{\mu} = \sum_i e_i \mathbf{r}_i = \sum_i e_i(\mathbf{i}x_i + \mathbf{j}y_i + \mathbf{k}z_i)_i \qquad \text{(A12.4)}$$

where e_i is the charge of the ith particle, and x_i, y_i and z_i are the coordinates of its position vector $\mathbf{r}_i$, where $\mathbf{i}$, $\mathbf{j}$ and $\mathbf{k}$ are unit vectors along x, y and z, respectively. Now, Eq. (A12.3) can be resolved into three integrals, that along the x axis, for example, being

$$\int \psi_i^* \mu_x\ \psi_f\, \mathrm{d}\tau \qquad \text{(A12.5)}$$

The integral, Eq. (A12.3), will be non-zero if any one of its component integrals like Eq. (A12.5) is non-zero: for this condition to exist, the direct product of ψ_i, μ_x and ψ_f, or its counterpart in y or z, should contain the fully symmetric type A_1 representation [2].

The ground state vibrational wave function ψ_i of an atom has the same mathematical form as that of an s atomic orbital, so it is spherically (totally) symmetric, which means that there are no symmetry restrictions attached to it.

Consider the water molecule, point group C_{2v}. The character table shows that x, y and z have the symmetries B_1, B_2 and A_1, respectively. Following Section 8.3.3.2, it can be shown that this molecule has vibrational (final) states ψ_f corresponding to A_1 and B_2 symmetries. Hence, the integral Eq. (A12.3) may be investigated by the direct products

$$\Gamma_{\psi_1} \otimes \begin{pmatrix} \Gamma_x \\ \Gamma_y \\ \Gamma_z \end{pmatrix} \otimes \Gamma_{\psi_f} \qquad \text{(A12.6)}$$

With $\Gamma_{\psi_1} = A_1$ and $\Gamma_{\psi_f} = A_1$,

$$\begin{aligned} A_1 \otimes B_1 \otimes A_1 &= B_1 \\ A_1 \otimes B_2 \otimes A_1 &= B_2 \\ A_1 \otimes A_1 \otimes A_1 &= A_1 \end{aligned}$$

and with $\Gamma_{\psi_1} = A_1$ and $\Gamma_{\psi_f} = B_2$,

$$A_1 \otimes B_1 \otimes B_2 = A_2$$
$$A_1 \otimes B_2 \otimes B_2 = A_1$$
$$A_1 \otimes A_1 \otimes B_2 = B_2$$

Thus, Eq. (A12.3) can be non-zero in both A_1 and B_2 symmetries because the direct products contain the fully symmetric A_1 representation in each case.

References

[1] Bishop DM. *Group theory and chemistry*. Clarendon Press, 1973.
[2] McWeeny R. *Symmetry*. Pergamon Press, 1963.

A13 Affine groups

A13.1 Introduction

The application of affine group theory (Lat. *affinis* = related) to crystal symmetry stems from Klein's renowned Erlangen program [1], to which Professor Boeyens referred in his Foreword to this book. A crystal that can be represented by a pattern of points P^n such that translations leave it invariant forms a lattice; examples exist implicitly in Chapter 4, wherein the important value of n is 3. Symmetry can be said to refer to the set of isometric mappings that leave the entire crystal invariant; thus, a three-dimensional space group is a group of isometries, an isometry being a one-to-one mapping of one metric space into another such space while preserving the distance between each pair of points. Hence, two crystal patterns are equivalent if their groups of isometries are equal.

A13.2 Linear mappings

A linear mapping M on a three-dimensional space P^3 follows the rules

$$\begin{aligned} M(Q+R) &= M(Q)+M(R) \\ M(\kappa \cdot R) &= \kappa \cdot M(R) \end{aligned} \tag{A13.1}$$

where Q and R are elements in the three-dimensional vector space and κ is a scalar quantity. In a hexagonal lattice (Fig. 4.16b), for example, M could be a threefold rotation having the matrix

$$\mathbf{M} = \begin{pmatrix} \overline{1/2} & \overline{\sqrt{3}/2} & 0 \\ \sqrt{3}/2 & \overline{1/2} & 0 \\ 0 & 0 & 1 \end{pmatrix} \tag{A13.2}$$

with respect to orthogonal reference axes, and the inverse transformation matrix $\mathbf{M}^{-1}$ exists, such that $\mathbf{M}^{-1}\mathbf{M} = \mathbf{E}$.

A13.3 Affine mappings and affine groups

In a crystal, the symmetry operations of translation, rotation and reflection, or combinations thereof, constitute affine mappings. The Seitz notation (Section 8.5) represents an affine mapping: the Seitz operator was discussed in the context of the representation of symmetry operations and the derivation of space groups, and given there as $\{\boldsymbol{R}|\mathbf{t}\}$ where $\boldsymbol{R}$ is the linear part and $\boldsymbol{t}$ its translation part. It was shown in that section that the product of two mappings follows the rule

$$\{\boldsymbol{R}_2|\mathbf{t}_2\}\{\boldsymbol{R}_1|\mathbf{t}_1\} = \{\boldsymbol{R}_2\boldsymbol{R}_1|\boldsymbol{R}_2\mathbf{t}_1 + \mathbf{t}_2\} \tag{A13.3}$$

which is a general result; the inverse mapping was also described, and is it straightforward to show, from Eq. (A13.3), that

$$\{\boldsymbol{R}|\mathrm{t}\}\left\{\boldsymbol{R}^{-1}| -\boldsymbol{R}^{-1}\mathbf{t}\right\} = E \tag{A13.4}$$

These mappings form affine groups, examples of which occur implicitly in Chapters 5 and 8. All members of a three-dimensional space group lie in an affine group A_3, with the additional property that the linear part of the group is an isometry. Since

$$(\boldsymbol{R}_1\boldsymbol{R}_2)^{\mathrm{T}} = \boldsymbol{R}_2{}^{\mathrm{T}}\boldsymbol{R}_1{}^{\mathrm{T}} = \boldsymbol{R}_2{}^{-1}\boldsymbol{R}_1{}^{-1} = (\boldsymbol{R}_1\boldsymbol{R}_2)^{-1} \tag{A13.5}$$

the product of two orthogonal matrices is orthogonal, so that affine mappings with orthogonal linear parts form subgroups of the affine group known as Euclidean groups, of which every space group is a subgroup.

A13.4 Space groups and space group types

A *space group* is characterized by both the symmetry expressed in its symbol and its lattice translations. A *space group type* is also characterized by its symbol, but its lattice dimensions are arbitrary. The number of space groups is theoretically infinite, but the number of space group types is 230.

The material $FeTa_2O_6$ crystallizes in the tetragonal system: one form is a rutile (TiO_2) structure type and another, so-called tri-rutile type, is similar but with a c dimension three times that of the rutile type structure. Both are designated by the symmetry symbol $P\frac{4_2}{m}nm$, but constitute two space groups, albeit one space group type. Expressed in another way, a space group is the set of symmetry operators for a specific crystal structure, whereas a space group type is one of 230 ways of arranging the symmetry operators in space. It follows that most of the descriptive material of Chapter 5 is about space group types, except that in Section 5.4.15.1 to Section 5.4.15.4 it refers to space groups.

Two space groups have the same space group type if they are conjugate under affine transformation [2]. On account of this condition, 22 space group types form 11 enantiomorphic pairs: only one group in each of the following pairs counts as an affine space group, so that there are 219 such groups in all.

$$P4_1/P4_3;\ \ P4_122/P4_322;\ \ P4_12_12/P4_32_12;$$
$$P3_1/P3_2;\ \ P3_121/P3_221;\ \ P3_112/P3_212;$$
$$P6_1/P6_5;\ \ P6_2/P6_4;\ \ P6_122/P6_522;\ \ P6_222/P6_422;\ \ P4_132/P4_332$$

From a chemical point of view, however, it may become necessary to determine the absolute configuration of a molecule, that is, to differentiate between the two groups in an enantiomorphic pair, and procedures for achieving it are discussed in the literature [3].

A13.5 Conclusion

For those who wish to explore further this esoteric but fundamental material, some particular references of interest are given [2, 4–6]. *"Choose one definite objective and drive ahead toward it. You may never reach your goal, but you will find something of interest on the way."* (Felix Klein; 1849–1925).

References

[1] Klein, F. (1872) *Vergleichende Betrachtungen über neuere geometrische Forschungen*, Deichert, Erlangen. Available in translation at http://math.ucr.edu/home/baez/erlangen/erlangen_tex.pdf

[2] Hahn, T. [Editor] (2011) *International Tables for Crystallography*, Vol. A, Fifth edition, reprinted with corrections, IUCr. Oxford: Wiley-Blackwell.

[3] Ladd, M. and Palmer, R. (2013) *Structure Determination by X-ray Crystallography*, Fifth edition, Springer Science+Business Media.

[4] Müller, U. (2013) *Symmetry Relationships between Crystal Structures*, IUCr. Oxford: Oxford University Press.

[5] Igodt, P. (2009) *Crystallographic Groups and Their Generalizations*, Amer. Math. Soc.

[6] Fried, D. and Goldman, W. M. (1983). <http://www2.math.umd.edu/~wmg/ThreeDimAffCryst.pdf>

Tutorial solutions

Solutions 1

1.1 (a) One vertical *m* plane, passing through the handle of the teacup. (b) Two vertical *m* planes intersecting at 90°. [Their line of intersection is a twofold rotation axis.] (c) Three mutually perpendicular *m* planes. [There are also three twofold rotation axes perpendicular to the *m* planes, intersecting at the point of intersection of the *m* planes.] (d) As for (c). (e) As for (b). (f) No symmetry; all faces are different.

(a) (b)

Fig. S1.1
(a) Unit cell of an ideal, square lattice. (b) Unit cell of an ideal, centred, rectangular lattice.

1.2 Refer to Fig. S1.1. (a) A square of points. (b) A centred rectangle of points.

1.3 Following the argument in Section 1.5, and using the symmetry of the octahedron, which is the same as that of the cube, the resistance to a current across opposite apices is $\frac{1}{2}\ \Omega$.

1.4 (a) Replacing Cl by F, or F by Cl, would double the number of *m* planes. (b) The molecular plane and the vertical plane passing through C(2) and C(5).

1.5 The letters F G J P Q R do not exhibit *m* symmetry.

1.6 (a) Odd. (b) Even. (c) Odd. (d) Even. (e) Odd. (f) Odd.

1.7 It's not hard to get full marks for this one.

1.8 (a) Numbers of *m* planes on the figures are, in order from left to right: 3, 4, 5 and 6. (b) Equilateral triangle; square; pentagon; hexagon. (c) The number of sides of each figure is equal to the number of its *m* lines. (c) Unlike the polygons in Fig. 1.2, the rectangle is not a *regular* polygon.

1.9 None; they each possess three *m* planes. [They differ in other symmetry elements, as will be realized later.]

1.10 The vertical symmetry line of a parabola passes through a minimum [or a maximum in the case of an inverted parabola]. From the equation, $Y = f(X) = X^2 + 6X + 5$, so that $f'(X) = 2X + 6$. At the minimum, $f'(X) = 0$, so that $X = -3$. Hence the *m* line is parallel to the *Y* axis and passes through the point $X = -3$.

1.11 Reflection symmetry exists between bars 9 and 10. This extract would sound the same if played backwards.

1.12 Near-reflection symmetry in the whole structure and in its parts (columns, domes, wall), cylindrical symmetry in the columns and domes.

Solutions 2

2.1 (a) $(3\bar{1}0)$. (b) $(\bar{3}212)$. (c) $(40\bar{3})$. (d) $(1\bar{4}2)$. (e) $(08\bar{1})$. (f) $(1\bar{1}6)$.

2.2 (a) $[\bar{1}10]$. (b) $[031]$. (c) $[0\bar{1}1]$. (d) $[\bar{3}0\bar{1}]$. (e) $[231]$. (f) $[11\bar{1}]$. For each value of $[UVW]$, $[\bar{U}\bar{V}\bar{W}]$ is also a solution.

2.3 (a) $[\bar{1}10]$. (b) $[031]$. (c) $[012]$. (d) $[031]$, $[231]$. (e) $[231]$. (f) $[11\bar{1}]$.

2.4 (a) $(\bar{1}10)$. (b) $(2\bar{1}2)$. (c) $(12\bar{1})$. (d) $(\bar{1}0\bar{1})$. (e) (010). (f) $(\bar{1}\bar{1}\bar{1})$.

2.5 The planes must satisfy the Weiss zone equation, in this case $h - 2k + 3l = 0$; two possible planes are (210) and (121). Check by the cross-multiplication rule.

2.6 By the cross-multiplication rule, the zone symbol is $[k_1l_2 - k_2l_1, l_1h_2 - l_2h_1, h_1k_2 - h_2k_1]$. From the Weiss zone equation, $(k_1l_2 - k_2l_1)(mh_1 + nh_2) + (l_1h_2 - l_2h_1)(mk_1 + nk_2) + (h_1k_2 - h_2k_1)(ml_1 + nl_2) = 0$. Expansion of this equation shows it to be true.

2.7 Refer to Fig. 2.9. Let d_1 be the normal to $(h_1k_1l_1)$ from the origin O. Its direction cosines are $\cos\chi_1$, $\cos\psi_1$ and $\cos\omega_1$, and equal to d_1h_1/a, d_1k_1/b and d_1l_1/c, respectively. Hence, from Appendix A2, Eq. (A2.3), $\frac{h_1^2}{a^2} + \frac{k_1^2}{b^2} + \frac{l_1^2}{c^2} = \frac{1}{d_1^2}$, and $\cos\chi_1 = \frac{h_1}{a}/r_1$, where $r_1 = \left(\frac{h_1^2}{a^2} + \frac{k_1^2}{b^2} + \frac{l_1^2}{c^2}\right)^{1/2}$. Similar expressions are obtained for $\cos\psi_1$ and $\cos\omega_1$, for the direction cosines of the normal d_2 to the plane $(h_2k_2l_2)$ and for r_2. The angle between the normals now follows from Appendix A2, Eq. (A2.9):

$$\cos\phi = \frac{\frac{h_1h_2}{a^2} + \frac{k_1k_2}{b^2} + \frac{l_1l_2}{c^2}}{\left(\frac{h_1^2}{a^2} + \frac{k_1^2}{b^2} + \frac{l_1^2}{c^2}\right)^{1/2}\left(\frac{h_2^2}{a^2} + \frac{k_2^2}{b^2} + \frac{l_2^2}{c^2}\right)^{1/2}}.$$

2.8 From the data, $c/a = 0.9539/0.5287 = 1.8042$. Thus, following Fig. 2.31 ($\sin\beta = 1$, and 001 replacing 100), $\angle 001 - 101 = \tan^{-1} 1.8042 = 61.003°$. Similarly, $\frac{c/2}{a} = \tan^{-1} \angle 001 - 102 = 42.054° = p$. Thus, $q = 61.003° - 42.054° = 18.949°$. The angle between q and r is the angle between zones [010] and $[\bar{1}\,01]$, which is 90°, *vide* Solution 2.7. In the right-angled triangle $(p + q)rt$, $\cos 63.90°$ (given) $= \cos 61.003° \cos r$, so that $r = 24.833°$. Finally, in the right-angled triangle qrs, $\cos\angle 102 - 111 = \cos 18.949° \cos 24.833°$. Thus, $\angle 102 - 111 = 30.87°$.

2.9 First solve triangle 010–111–110 for ϕ_4 and ψ. From Eq. (2.23), $\cos\phi_4 = \frac{\cos 49.15° - (\cos 71.90 \cos 55.75)°}{(\sin 71.90 \sin 55.75)°}$, so that $\phi_4 = 52.41°$. Again, but with Eq. (2.26), $\sin\psi = \frac{\sin 55.75° \sin 52.40°}{\sin 49.15°}$, so that $\psi = 59.99°$. Now use triangle 001–101–111. From Eq. (2.26), $\sin\phi_1 = \frac{\sin 90° \sin(90-71.90)°}{\sin(90-49.15)°} = \frac{\cos 71.90°}{\cos 49.15°}$, so that $\phi_1 = 28.358°$. Then, $180 - \beta = \phi_4 + \psi = 52.41° + 28.36° = 80.77°$, so that $\beta = 99.23°$. From Fig. 2.30, $\frac{a\sin\beta}{b} = \tan\angle 100 - 110 = \tan 34.25°$, so that $\frac{a}{b} = 0.6809$. From Fig. 2.26, $c/a = \frac{\sin\phi_3}{\sin\phi_4}$, so that $\frac{c}{b} = \frac{a}{b}\frac{c}{a} = 0.5999 \times 0.6809 = 0.408$. Summarizing for gypsum: $a : b : c = 0.681 : 1 : 0.4084$, $\beta = 99.23°$.

2.10 From the cross-multiplication rule, the zone axis symbol is $[\bar{1}\,1\,\bar{1}]$ or $[1\,\bar{1}\,1]$. From either of these zone symbols, $(0 \times U) + (-1 \times V) + (1 \times W) = 0$. Using the planes (121) and (231) for $(h_1k_1l_1)$ and $(h_2k_2l_2)$, $p + 2q = 0$, $2p + 3q = 1$ and $p + q = 1$. Hence, from the first two equations $q = -1$. Thus, $p = 2$.

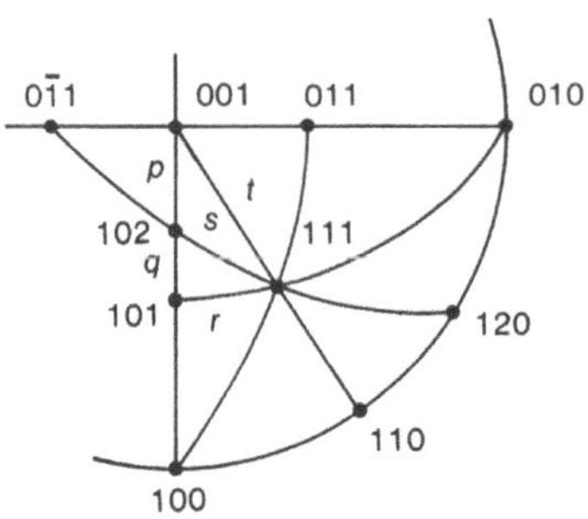

Fig. S2.1
Partial stereogram of Fig. P2.1, with convenience letters added. [McKie D and McKie C. *Essentials of crystallography*. 1986; reproduced by courtesy of Blackwell Scientific Publications.]

2.11 Begin with pole c, at the intersection of zones [(100)–(001] and [(111)–(010)], that is, $[0\,\bar{1}\,0]$ and $[10\bar{1}]$, so that by the cross multiplication rule, c is (101). Since pole a is given as 210, the zone axis symbol for zone a–g is $[\bar{1}21]$. Pole b lies in zones $[0\bar{1}\,1]$ and $[\bar{1}21]$, and is 311. Pole d is in zones $[\bar{1}21]$ and $[(001)–(1\bar{1}0)]$, which gives $1\,\bar{1}3$ for pole d. For pole e, the zones are [100] and $[\bar{1}21]$, which makes pole e $0\bar{1}2$. Only one zone circle is defined through f, namely, $[\bar{1}21]$. From symmetry, the intersection between $[(010) - (0\bar{1}0)]$ and $[(001) - (\bar{1}00)]$ is either $\bar{1}01$ or $\bar{1}0\bar{1}$. Assume the former as it is shown as • on the stereogram. Now, f lies in $[\bar{1}21]$ and $[(010) - (\bar{1}01)]$; thus, pole f evaluates to $\bar{1}\,\bar{1}\,1$. Pole g, lying at the intersection of [001] and $\bar{1}\,21]$ is clearly $\bar{2}\,\bar{1}\,0$.
Summarizing (Fig. S2.1):

a	*b*	*c*	*d*	*e*	*f*	*g*	Zone axis
210	311	101	$1\bar{1}3$	$0\bar{1}2$	$\bar{1}\bar{1}1$	210	$[\bar{1}21]$

2.12 The best plane from program PLANE is $-1.0563X - 1.7324Y + 4.3803 = Z$; the rms deviation calculates as 0.777 with standard deviation 0.850.

2.13 The internal coordinates will be listed in the file GEOM.TXT. Atom C6 lies above the C1–C5 plane.

2.14 A plot of the atoms in the unit cell and its environs reveals that the shortest Cl- - -Cl interatomic distance is between atoms at $^1/_4$, y, z, and $^3/_3$, $\bar{y}$, z. Hence, d^2(Cl- - -Cl) $= a^2/4 + 4y^2b^2$, so that d(Cl . . . Cl) = 4.634 Å. The superposition of errors (see Section 2.6.6) shows that the variance of d(Cl . . . Cl) is given by

$$[2d\sigma\,(d)]^2 = \left[2a\sigma\,(a)/4\right]^2 + \left[8y^2b\sigma\,(b)\right]^2 + \left[8b^2y\sigma\,(y)\right]^2$$

so that $\sigma(d) = 0026$ Å. It may be noted that this answer calculates as 0.02640 Å to four significant figures. Using just the third term, that in $\sigma(y)$, the result is 0.02629 Å. Thus, the error in a distance between atoms arises, normally, mostly from the errors in the corresponding atomic coordinates.

2.15 In Fig. S2.2, $OA = a/h$ and $OB = b/k$. The plane ($hkil$) intercepts the u axis at p; draw DE parallel to AO. Since OD bisects $\angle AOB$, $\angle AOD = 60°$, so that $\Delta\, ODE$ is equilateral; hence, $OD = DE = OE = p$. Triangles EBD and OBA are similar, so that $EB/DE = OB/OA = (b/k)/(a/h)$. Since $EB = b/k - p$, $p = ab/(ak + bh)$. From symmetry. $a = b = u$, so that $u/p = h + k$. Writing u/p as $-i$, since p lies on the negative side of the u axis ($OD = -u/p$), it follows that $i = -(h + k)$.

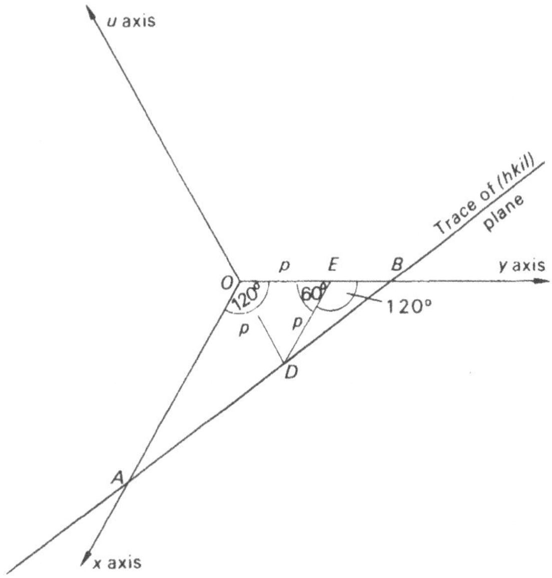

Fig. S2.2
Construction to show that the Miller–Bravais index i is equal to $-(h + k)$.

2.16 Set up $\int \psi_{2,1,1}\psi^*_{2,1,1}\,\mathrm{d}\tau$. Consider the integral

$$\mathscr{I} = \frac{1}{64\pi}(Z/a_0)^3 \int_0^\infty \rho\exp(-\rho)\,r^2\mathrm{d}r \int_0^\pi \sin^2\theta\,\mathrm{d}\theta \int_0^{2\pi} \exp(\mathrm{i}\phi)\exp(-\mathrm{i}\phi)\,\mathrm{d}\phi$$

with its separate parts $\mathscr{I}_r$ $\mathscr{I}_\theta$ and $\mathscr{I}_\phi$. Then:

$$\mathscr{I}_\phi = \int_0^{2\pi} \exp(\mathrm{i}\phi)\exp(-\mathrm{i}\phi)\,\mathrm{d}\phi = \int_0^{2\pi} \mathrm{d}\phi = 2\pi$$

$$\mathscr{I}_\theta = \int_0^{\pi} \sin^2\theta\,\sin\theta\,\mathrm{d}\theta = \int_0^{\pi} \sin^3\theta\,\mathrm{d}\theta = \int_0^{\pi} \sin^3\theta\,\mathrm{d}\theta$$

$$= -(1/3)\cos\theta(\sin 2\theta + 2)\,|_0^\pi = (4/3)$$

$$\mathscr{I}_r = (1/64\pi)(Z/a_0)^3 \int_0^\infty \rho^2 \exp(-\rho)\, r^2 \mathrm{d}r$$
$$= (1/64\pi)(Z/a_0)^4 \int_0^\infty r^4 \exp(-Zr/a_0)\, \mathrm{d}r$$

Let $Zr/a_0 = t$, so that $r = (a_0/Z)t$ and $\mathrm{d}r = (a_0/Z)\,\mathrm{d}t$. Then,

$$\mathscr{I}_r = (1/64\pi)(Z/a_0)^3 (a_0/Z)^5 \int_0^\infty t^4 \exp(-t)\mathrm{d}t = (1/64\pi)\,\Gamma(5) = (1/64\pi)4!$$

Combining the three integrals,

$$\int \psi_{2,1,1}\psi^*_{2,1,1}\, \mathrm{d}\tau = 2\pi\,(4/3)(1/64\pi)4! = 1$$

so that $\psi_{2,1,1}$ is a normalized function.

Solutions 3

3.1 (a) Refer to Figs. 3.3c, 3.4b and 3.4c. (b) All parts of a body must respond correctly to all symmetry elements of the body; *m* symmetry would be achieved only in the special case of a 60° parallelogram, or rhombus.

3.2 Refer, in order, to Figs. 3.12a, 3.12b and 3.12c.
Subgroups. $mm2$: 1, 2, m. $\bar{4}2m$: 1, 2, m, 222, $mm2$, $\bar{4}$. 23: 1, 2, 222, 3.
$mm2$ and a centre $= mmm$. $\bar{4}2m$ and a centre $= \frac{4}{m}mm$. 23 and a centre $= m\bar{3}$.

3.3 m: Monoclinic (y axis unique, by convention); $m \perp y$.
422: Tetragonal; 4 along z; 2 along x and y; 2 along $[1\bar{1}0]$ and $[110]$.
$\bar{6}m2$: Hexagonal; $\bar{6}$ along z; $m \perp x,\ y,\ u$ ($\equiv\ \bar{2}$ along $x,\ y,\ u$); $2 \perp x, y$ and u.
$\bar{4}3m$: Cubic; $\bar{4}$ along x, y, z; 3 along $<111>$; $m \perp\ <1\bar{1}0>$.

3.4 Using the equations of Section 3.6.3:
522: $\angle 52 = 90°, \angle 22 = 36°$; 5 along z; 2 along x, y, u, v, w; 2 at 36° to x, y, u, v, w (v and w are used here for the additional reference axes that are equivalent under fivefold rotation)
$\bar{8}2m$: $\angle\bar{8}2 = 90°, \angle 2\,2 = 45°, \angle 2\bar{2} = 22\tfrac{1}{2}°$; $\bar{8}$ along z; 2 along x, y, u, v at 45° to one another; $\bar{2}$ (m) at $22\tfrac{1}{2}°$ to x, y, u, v and normal to z.

3.5 $432, 43\bar{2}, 4\bar{3}2, \bar{4}32, 4\bar{3}\bar{2}, \bar{4}3\bar{2}, \bar{4}\bar{3}2, \bar{4}\bar{3}\bar{2}$. The symbols involving one or three $\bar{R}$ elements are untenable (Section 3.6.3). Hence, only $432, 4\bar{3}\bar{2}, \bar{4}3\bar{2}$ and $\bar{4}\bar{3}2$ are unique point group symbols. Furthermore, $\bar{4}\bar{3}2$ and $\bar{4}\bar{3}2$ are equivalent, so that the three symbols in standard form are $432, \bar{4}3m$ and $m\bar{3}m$.

3.6 (a) mmm. (b) $2/m$.

3.7

	{010}	{110}	{123}
$mm2$	2(s)	4(g)	4(g)
$4mm$	4(s)	4(s)	8(g)
432	6(s)	12(s)	24(g)

3.8 (a) $1, \bar{1}$. (b) $1, 2/m$. (c) $2mm$, mmm. (d) $m, 4/m$. (e) $2mm, \frac{4}{m}mm$. (f) $m, \frac{4}{m}mm$. (g) $1, \bar{3}$. (h) $m, \bar{3}m$. (i) $3, 6/m$. (j) $3m1, \frac{6}{m}mm$. (k) $3, m\bar{3}$. (l) $2, m\bar{3}m$.

3.9 (a) 3. (b) 2 or $2/m$ (y axis unique). (c) $\bar{4}3m$. (d) 32 or 6.

3.10 (a) $\frac{6}{m}mm$. (b) $mm2$. (c) $mm2$. (d) $mm2$. (e) mmm. (f) $mm2$. (g) m. (h) m. (i) $\bar{6}m2$. (j) $mm2$. (k) $mm2$. (l) mmm. (m) $mm2$. (n) $\frac{6}{m}mm$.

3.11 $\bar{4}3m$, $3m$, $mm2$, m, 1.

3.12 (a) 2 or m. (b) 2.

3.13 $m \perp x$ 4 along z

$$\underset{\mathbf{R}_2}{\begin{pmatrix} \bar{1} & 0 & 0 \\ 0 & 1 & 0 \\ 0 & 0 & 1 \end{pmatrix}} \underset{\mathbf{R}_1}{\begin{pmatrix} 0 & \bar{1} & 0 \\ 1 & 0 & 0 \\ 0 & 0 & 1 \end{pmatrix}} = \underset{\mathbf{R}_3}{\begin{pmatrix} 0 & 1 & 0 \\ 1 & 0 & 0 \\ 0 & 0 & 1 \end{pmatrix}}$$

Hence, $x, y, z \longrightarrow y, x, z$, or $\mathbf{x}' = \mathbf{R}_2\,\mathbf{R}_1\,\mathbf{x}$. The matrix $\mathbf{R}_3$ refers to a mirror plane at 45° to the x and y axes, containing the z axis, namely, the m plane normal to $[1\,\bar{1}0]$.

3.14 $m \perp y$ 3 along z

$$\underset{\mathbf{R}_2}{\begin{pmatrix} 1 & 0 & 0 \\ 1 & \bar{1} & 0 \\ 0 & 0 & 1 \end{pmatrix}} \underset{\mathbf{R}_1}{\begin{pmatrix} 0 & \bar{1} & 0 \\ 1 & \bar{1} & 0 \\ 0 & 0 & 1 \end{pmatrix}} = \underset{\mathbf{R}_3}{\begin{pmatrix} 0 & \bar{1} & 0 \\ \bar{1} & 0 & 0 \\ 0 & 0 & 1 \end{pmatrix}}$$

Matrix $\mathbf{R}_3$ corresponds to an m plane normal to the u axis, but it is of the *same form* as $\mathbf{R}_2$. Thus, the m generated by $\mathbf{R}_2\,\mathbf{R}_1$ is coincident with the m form given initially, and so there is no non-trivial meaning for a third position in this, or any other, trigonal *point group* symbol.

3.15 Subgroups of $I_h(m\bar{3}5)$

Schönflies notation			**Hermann–Mauguin notation**		
I	T_h	T	235	$\bar{4}3m$	23
D_{5d}		D_{3d}	$m\bar{5}$		$\bar{3}m$
	D_{2h}			mmm	
D_5	D_3	D_2	25	32	222
	C_{2h}			$\frac{2}{m}$	
C_{5v}	C_{3v}	C_{2v}	$m5$	$3m$	$mm2$
C_5	C_3	C_2	5	3	2
S_{10}		S_6	$\bar{5}$		$\bar{3}$
	C_s			m	
	i			$\bar{1}$	
	E			1	

Subgroups of I (235)

Schönflies notation			**Hermann–Mauguin notation**		
	T			23	
D_5	D_3	D_2	25	32	222
C_5	C_3	C_2	5	3	2
	E			1	

3.16

	Octahedron	**Icosahedron**	**Truncated icosahedron**
Vertices	6	12	60
Faces	8	20	32
Edges	12	30	90

Vertices + Faces = Edges + 2. (First deduced by Euler.)

Solutions 4

4.1

	(i)	(ii)
(a)	4*mm*	6*mm*
(b)	Square	Hexagonal
(c)	Another square may be outlined as *p* unit cell	The symmetry at each point is degraded to 2*mm*. A rectangular net is produced, and a *p* cell outlined.

In both cases, the transformation equations are:

$$\mathbf{a}' = \frac{\mathbf{a}}{2} - \frac{\mathbf{b}}{2} \text{ and } \mathbf{b}' = \frac{\mathbf{a}}{2} + \frac{\mathbf{b}}{2}$$

(There are other similar choices that give the same area.)

4.2 Monoclinic; 1.77 nm.

4.3 (i) Refer to Section 4.1. (i) Orthorhombic *B* is a lattice, equivalent to orthorhombic *C*.

(ii) Orthorhombic *C*, but a special case with $a = b$. [Don't forget the meaning of the symbol ¢.]

(iii) Triclinic, equivalent to triclinic *P*. (iv) Not a lattice.

4.4 (i) $\mathbf{a}' = -\mathbf{a} + \mathbf{c}$; $\mathbf{b}' = -\mathbf{b}$; $\mathbf{c}' = \mathbf{a}$. (**b** is negated in order to preserve right-handed axes.)

(ii) $\mathbf{a}' = \frac{\mathbf{b}}{2} + \frac{\mathbf{c}}{2}$; $\mathbf{b}' = \frac{\mathbf{a}}{2} + \frac{\mathbf{c}}{2}$; $\mathbf{c}' = \frac{\mathbf{a}}{2} + \frac{\mathbf{b}}{2}$.

(iii) $\mathbf{a}' = \frac{\mathbf{a}}{2} - \frac{\mathbf{b}}{2}$; $\mathbf{b}' = \frac{\mathbf{a}}{2} + \frac{\mathbf{b}}{2}$; $\mathbf{c}' = \mathbf{c}$.

(Other choices are possible, provided **a**, **b** and c form a right-handed set.)

4.5 Rotation about *c* does not lead to indistinguishability, and the symmetry at each point is no more than $\bar{1}$; the cell is a special case of the triclinic system with $\gamma = 90°$.

4.6 Refer to Fig. 4.24. From the figure or Eq. (4.43) or Eq. (A3.19), $V_F/V_R = 4$.

4.7 $\mathbf{a}' = -\mathbf{a}/2 + \mathbf{b}/2 + \mathbf{c}/2$. It is straightforward to show, by reference to Appendix A3 as necessary, that $a' = 4.582_8$ Å, and $a' = b' = c'$. Since $\cos\alpha' = \frac{\mathbf{b}'\cdot\mathbf{c}'}{b'c'}$, $\alpha' = \cos^{-1}\left\{\frac{(\mathbf{a}/2-\mathbf{b}/2+\mathbf{c}/2)\cdot(\mathbf{a}/2+\mathbf{b}/2-\mathbf{c}/2)}{a'^2}\right\} = \cos^{-1}\left\{\frac{a^2}{4}(2\cos\alpha - 1)\right\}/a'^2 = 80.41°$ and $\alpha' = \beta' = \gamma'$.

4.8 The determinant of the matrix is 2; hence, $V_{II} = 2V_I$. The inverse matrix is $\begin{pmatrix} 1/2 & 1/2 & 0 \\ -1/2 & 1/2 & 0 \\ 0 & 0 & 1 \end{pmatrix}$ and its transpose is $\begin{pmatrix} 1/2 & -1/2 & 0 \\ 1/2 & 1/2 & 0 \\ 0 & 0 & 1 \end{pmatrix}$. Hence,

$$\left.\begin{aligned} x' &= 0.5(0.123 + 0.671) = 0.397 \\ y' &= 0.5(0.123 - 0.671) = -0.274 \\ z' &= 0.314 \end{aligned}\right\}$$

4.9 (100) : (110) : (1$\bar{3}$0) : (042) = 1 : 0.71 : 0.32 : 0.22. The likely boundary faces are (100) and (110).

4.10 (a) Refer to Fig. S4.1.

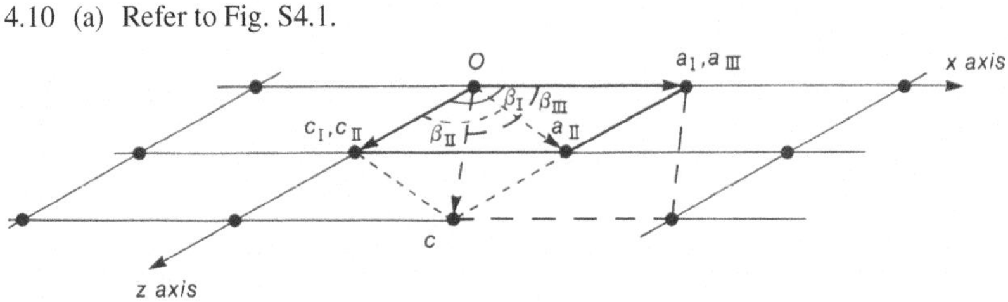

Fig. S4.1
Monoclinic unit cells I, II and III, drawn from one and the same origin, in projection on to (010).

(b) Cell I, P; Cell II, P; Cell III, B or I. Cell III would not be chosen as it twice the volume of standard monoclinic cells I and II. The choice of cell II would generally be preferred because it has the same size as cell I and a less obtuse, more convenient β angle.

(c)

$$\mathbf{a}_{\mathrm{II}} = \mathbf{S}_1\mathbf{a}_{\mathrm{I}} \quad \mathbf{S}_1 = \begin{pmatrix} 1 & 0 & 1 \\ 0 & 1 & 0 \\ 0 & 0 & 1 \end{pmatrix}$$

$$\mathbf{a}_{\mathrm{III}} = \mathbf{S}_2\mathbf{a}_{\mathrm{I}} \quad \mathbf{S}_2 = \begin{pmatrix} 1 & 0 & 0 \\ 0 & 1 & 0 \\ 1 & 0 & 2 \end{pmatrix}$$

$$\mathbf{a}_{\mathrm{III}} = \mathbf{S}_3\mathbf{a}_2 \quad \mathbf{S}_3 = \begin{pmatrix} 1 & 0 & \bar{1} \\ 0 & 1 & 0 \\ 1 & 0 & 1 \end{pmatrix}$$

(d) Cell II: (chosen): $a^* = 0.201$, $b^* = 0.0660$, $c^* = 0.174$ Å^{-1}, $\beta^* = 61.42^{\circ}$

(e) (i) $(hkl)_{\mathrm{II}} = \mathbf{S}_1(hkl)_{\mathrm{I}}$ $(hkl)_{\mathrm{II}} = \begin{pmatrix} 1 & 0 & 1 \\ 0 & 1 & 0 \\ 0 & 0 & 1 \end{pmatrix}\begin{pmatrix} 1 \\ 3 \\ 2 \end{pmatrix} = (332)$

(ii) $[UVW]_{\mathrm{III}} = \mathbf{S}_2^{\mathrm{T}}[UVW]\mathbf{a}_{\mathrm{II}}$ $[UVW]_{\mathrm{III}} = \begin{pmatrix} 1/2 & 0 & \overline{1/2} \\ 0 & 1 & 0 \\ 1/2 & 0 & 1/2 \end{pmatrix}\begin{pmatrix} 2 \\ 1 \\ \bar{3} \end{pmatrix} = [52\bar{1}]$

(iii) $(x,\ y,\ z)_{\mathrm{I}} = (\mathbf{S}_1^{-1})^{\mathrm{T}}\mathbf{x}_{\mathrm{III}}$ $(x, y, z)_{\mathrm{I}} \begin{pmatrix} 1 & 0 & 1 \\ 0 & 1 & 0 \\ 0 & 0 & 2 \end{pmatrix}\begin{pmatrix} 0.6 \\ 0.5 \\ \overline{0.3} \end{pmatrix}$

$$= (0.300, 0.500, -0.600)$$

4.11 In the close-packed cubic structure, the spheres are in contact across a face diagonal of the cube. Hence, $a\sqrt{2} = 4r$, where a is the edge length of the cube and r the radius of the spherical atom. Since the unit cell contains four atoms, the volume occupied per atom is $(4r^3/\sqrt{2})^3/4$, $4r^3\sqrt{2}$. The volume of the atom is $4\pi r^3/3$, so that the packing fraction is $\frac{4\pi r^3}{3} \div 4\sqrt{2}r^3$, or $\pi/\sqrt{18}$. Thus, the packing fraction is 0.74. The same value obtains for the close packed hexagonal, but the body centred cubic packing fraction is 0.68. The trick with the hexagonal close packed structure is to realize where contact occurs; Fig. S4.2 may help.

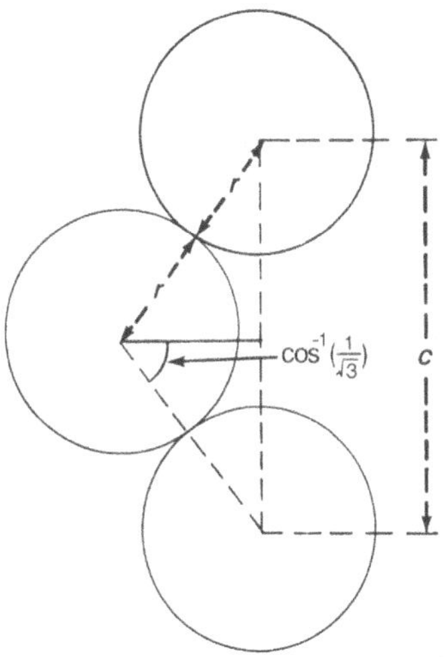

Fig. S4.2
Vertical section through a close packed hexagonal unit cell.

4.12 A $\bar{4}$ axis produces the following succession of points (Appendix A3.7.1):

$$x, y, z \rightarrow \bar{y}, x, \bar{z} \rightarrow \bar{x}, \bar{y}, z \rightarrow y, \bar{x}, \bar{z}$$

The negative signs indicate perpendicularity between x, y and z. Using, in addition, the scalar products $x\mathbf{a} \cdot y\mathbf{b}$ and $(-y)\mathbf{a} \cdot x\mathbf{b}$, then to satisfy indistinguishability $xyab\cos\gamma = -xyab\cos\gamma$, so that $\gamma = 90^{\circ}$, and by similar arguments, $\alpha = \beta = 90^{\circ}$. Since x and y are interchanged, indistinguishability requires that $a = b$, but no such restriction applies to z. Hence, $a = b \not\subset c$, and $a = \beta = \gamma = 90^{\circ}$.

Solutions 5

5.1 (a) (i) $4mm$. (ii) $6mm$. (b) (i) Square. (ii) Hexagonal. (c) (i) if the unit cell is centered, then another square can be drawn to form a conventional unit cell of half the area of the centered unit cell. (ii) If the unit cell is centered it is no longer hexagonal; each point is degraded to the $2mm$ symmetry of the rectangular system, and may be described by a conventional p unit cell. The transformation equations in each example are:

$$\mathbf{a}' = \mathbf{a}/2 + \mathbf{b}/2; \qquad \mathbf{b}' = -\mathbf{a}/2 + \mathbf{b}/2$$

Note. A regular hexagon of points with another point placed at its centre does not represent a centred hexagonal unit cell, but three adjacent p hexagonal unit cells in different relative orientations. Without the point at the centre, the hexagon of points does not represent a lattice.

5.2 A C unit cell may be obtained by the transformations:

$$\mathbf{a}_C = \mathbf{a}_F; \quad \mathbf{b}_C = \mathbf{b}_F; \quad \mathbf{c}_C = -\mathbf{a}_F/2 + \mathbf{c}_F/2.$$

The new c dimension is obtained from evaluating the dot product:

$$(-\mathbf{a}/2 + \mathbf{c}/2) \cdot (-\mathbf{a}/2 + \mathbf{c}/2)$$

so that $c_C = 5.7627$ Å; a and b are unchanged. The angle β_C in the transformed unit cell is obtained by evaluating

$$\cos\beta' = \mathbf{a} \cdot (-\mathbf{a}/2 + \mathbf{c}/2)/a'c' = (-a + c\cos\beta)/(2c')$$

thus, $\beta_C = 139.29°$. $V_C(C \text{ cell})/V_F(F \text{ cell}) = ½$. (Count the number of unique lattice points in each cell; each lattice point is associated with a unique portion of the volume.)

5.3 (a) The symmetry is no longer tetragonal, although the lattice is true; it is now orthorhombic. (b) The tetragonal symmetry is apparently restored, but the lattice is no longer true; the lattice points are not all in the same environment in the same orientation. (c) A tetragonal F unit cell is formed and represents a true tetragonal lattice. However, tetragonal F is equivalent to tetragonal I cell, of smaller volume, under the transformation and is, therefore, the standard setting.

$$\mathbf{a}_I = \mathbf{a}_F/2 + \mathbf{b}_F/2; \quad \mathbf{b}_I = -\mathbf{a}_F/2 + \mathbf{b}_F/2; \quad \mathbf{c}'_I = \mathbf{c}_F$$

5.4 For the F unit cell, $r^2_{[31\bar{2}]}/\text{Å}^2 = \mathbf{r}_{[31\bar{2}]} \cdot \mathbf{r}_{[31\bar{2}]} = 3^2 a^2 + 1^2 b^2 + 2^2 c^2 + 2 \times 3 \times (-2) \times 6 \times 8 \times \cos 110°$, so that $r = 28.74$ Å. To obtain the value in the C unit cell, this calculation could be repeated with the dimensions of the C unit cell, leading to 28.74 Å. Alternatively, the transformation matrix could be used to obtain the F equivalent of $[31\bar{2}]_C$, and then the original F cell dimensions used on it. The matrix for this C unit cell length in terms of the F cell is:

$$\mathbf{S} = \begin{pmatrix} 1 & 0 & 0 \\ 0 & 1 & 0 \\ -½ & 0 & ½ \end{pmatrix} \quad \text{so that} \quad (\mathbf{S}^{-1})^T = \begin{pmatrix} 1 & 0 & 1 \\ 0 & 1 & 0 \\ 0 & 0 & 2 \end{pmatrix}$$

Then, $[UVW]_F = (\mathbf{S}^{-1})^T [UVW]_C = [11\bar{4}]_F$, which confirms $r_{[11\bar{4}]}$ as 28.74 Å.

5.5 The a, b plane is a rectangle. Unless c, intersecting a and b, lies at 90° to the plane, a twofold rotation would neither map **a** on to –**a**, nor **b** on to –**b**: still less would any higher degree of rotation suffice. Thus, 'diclinic' is not an eighth crystal system; it has $\bar{1}$ symmetry at each point, and is a special case of the triclinic system in which the γ angle is 90°.

5.6 (a) Plane group $c2mm$ is shown in Fig. S5.1, with the coordinates listed below it.

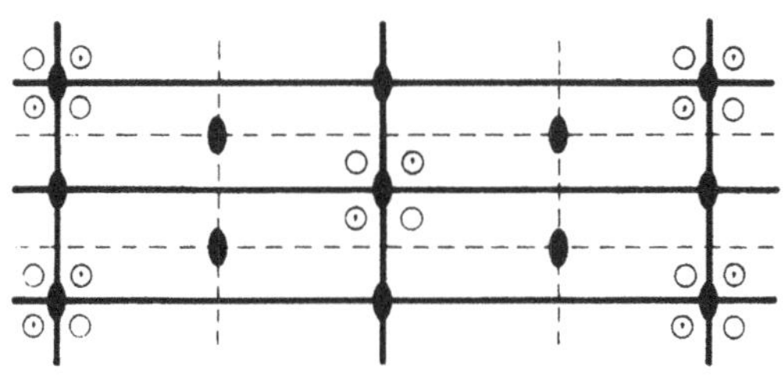

Fig. S5.1
Plane group $c2mm$, origin on $2mm$.

			Origin on **2*mm***				
			$\left(0, 0; \frac{1}{2}, \frac{1}{2}\right)+$				**Limiting conditions**
8	(f)	1	$x, y;$	$x, \bar{y};$	$\bar{x}, y;$	$\bar{x}, \bar{y}$	$hk: h+k=2n$
4	(e)	m	$0, y;$	$0, \bar{y};$			—
4	(d)	m	$x, 0;$	$\bar{x}, 0$			—
4	(c)	2	$\frac{1}{4}, \frac{1}{4};$	$\frac{1}{4}, \frac{3}{4}$			As above + $hk: h=2n, (k=2n)$
2	(b)	$2mm$	$0, \frac{1}{2}$				—
2	(a)	$2mm$	$0, 0$				—

(b) Plane group $p2mg$ is shown in Fig. S5.2; this diagram also shows the minimum number of motifs P, V and Z.

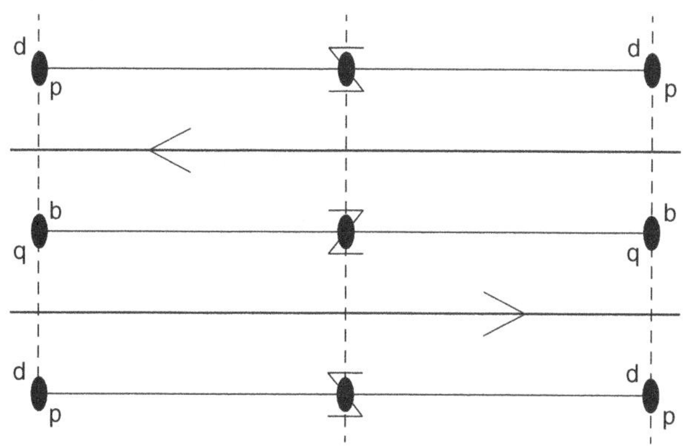

Fig. S5.2
The minimum necessary number of letters P, V and Z in plane group $p2mg$.

(c) If the symmetry elements are arranged with 2 at the intersection of m and g, they do not form a group. Attempts to draw such an arrangement lead to continued halving of the repeat distance parallel to the g line. (d) Refer to Fig. 5.2. (e) $p6mm$ is illustrated by Fig. S5.3.

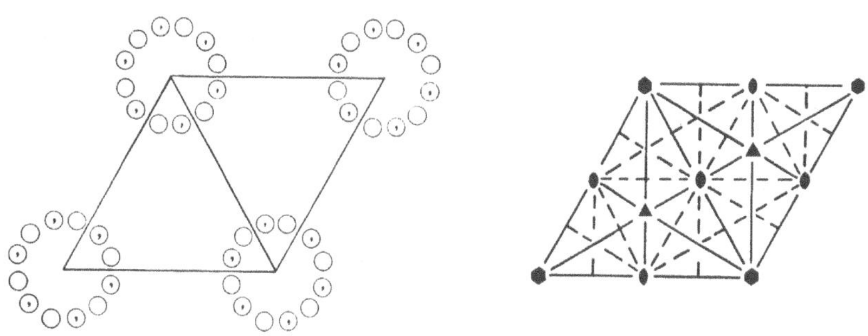

Fig. S5.3
Plane group $6mm$, showing the general equivalent positions and symmetry elements.

The coordinates of the general equivalent positions are:

$$\pm(x,\ y;\ \ \bar{y},\ x-y;\ \ y-x,\ \bar{x};\ \ \bar{y},\ \bar{x};\ \ y-x,\ y;\ \ x,\ x-y)$$

[There should now be no problem with plane groups $p6$, $p3m1$ and $p31m$, as long as the distinction between the last two is recalled.]

5.7 (a)

			Origin on $\bar{1}$	Limiting conditions
4	(e)	1	$x, y, z; x, \frac{1}{2} - y, \frac{1}{2} + z$	hkl : None
			$\bar{x}, \bar{y}, \bar{z}; \bar{x}, \frac{1}{2} + y, \frac{1}{2} - z$	$h0l$: $l = 2n$
				$0k0$: $k = 2n$
2	(d)	$\bar{1}$	$\frac{1}{2}, 0, \frac{1}{2}; \frac{1}{2}, \frac{1}{2}, 0$	As above + hkl : $k + l = 2n$
2	(c)	$\bar{1}$	$0, 0, \frac{1}{2}; 0, \frac{1}{2}, 0$	
2	(b)	$\bar{1}$	$\frac{1}{2}, 0, 0; \frac{1}{2}, \frac{1}{2}, \frac{1}{2}$	
2	(a)	$\bar{1}$	$0, 0, 0; 0, \frac{1}{2}, \frac{1}{2}$	
(100) $p2gg$: $b' = b, c' = c$			(010) $p2$: $a' = a/2, c' = c$	(001) $p2gm$: $a' = a, b' = b$

(b) Space group $P2_1/c$ on the (010) plane is shown in Fig. S5.4.

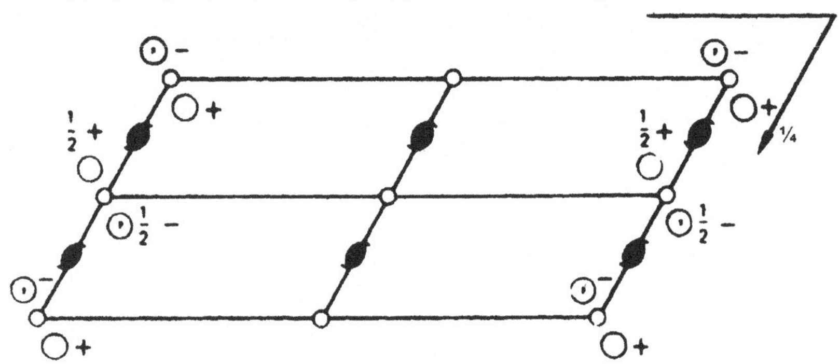

Fig S5.4
Space group $P2_1/c$ in projection on to the (010) plane.

(c) The two molecules of biphenyl in the unit cell lie on one set of the Wyckoff (a)–(d) special positions, with the centre of the C–C′ bond on $\bar{1}$. Hence, the molecule is centrosymmetric and planar, and this planarity imposes a conjugation on the molecule, including the central C–C′ bond. This result is supported by the bond lengths C–C′ ≈ 1.49 Å and C_{arom}–C_{arom} ≈ 1.40 Å. [*In the free molecule state, the rings are rotated about the C–C′ bond to the energetically favourable conformation with their planes at approximately 45° to each other. This result highlights the fact that the conformation of a species obtained by X-ray diffraction may not equate to that in its free state.*]

5.8 Each pair of sites forms two vectors, between the origin and the points: $\pm\{(x_2 - x_1), (y_2 - y_1), (z_2 - z_1)\}$.
Thus, there is a single vector at each of the positions

$$2x, 2y, 2z;\quad 2\bar{x}, 2\bar{y}, 2\bar{z};\quad 2x, 2\bar{y}, 2z;\quad 2\bar{x}, 2y, 2\bar{z}$$

and two superimposed vectors at each of the positions:

$$2x,\ 1/2,\ 1/2 + 2z;\quad 0,\ 1/2 + 2y,\ 1/2;\quad 2\bar{x},\ 1/2,\ 1/2 - 2z;\quad 0,\ 1/2 - 2y,\ 1/2$$

Note that the term $-(2x, 1/2, 1/2 + 2z)$ is crystallographically equivalent to $2\bar{x}, 1/2, 1/2 - 2z$.

5.9 Refer to the following scheme to elucidate the orientations of the symmetry elements:

$$\begin{array}{ccc}
x, y, z & \xrightarrow{\quad b \quad} & 2p - x, \frac{1}{2} + y, z \\
\downarrow \bar{1} & & \downarrow a \\
\bar{x}, \bar{y}, \bar{z} & & \\
2p - x, 2q - y, 2r - z & \xleftarrow{\quad n \quad} & \frac{1}{2} + 2p - x, 2q - \frac{1}{2} - y, z
\end{array}$$

Since $\bar{x}, \bar{y}, \bar{z}$ and $2p - x, 2q - y, 2r - z$ are one and the same point, $p = q = r = 0$, so that the three symmetry planes intersect in a centre of symmetry at the origin. Otherwise, by applying the half-translation, $T = a/2 + b/2 + a/2 + b/2 \equiv 0$. Hence, the centre of symmetry lies at the intersection of the three symmetry planes.

5.10 Fig. S5.5 shows space group *Pbam*.

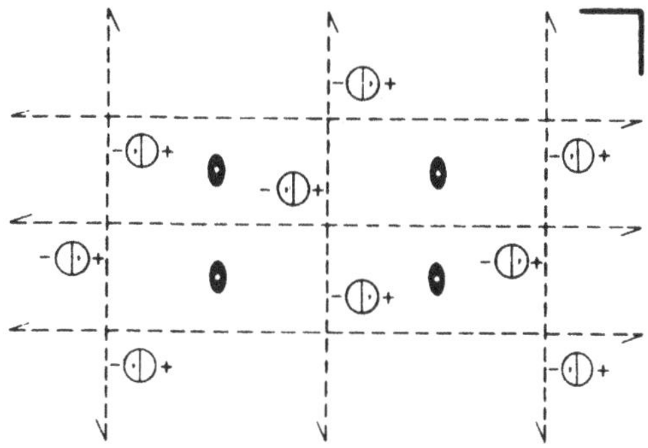

Fig. S5.5
Space group *Pbam* in projection on (001).

Coordinates of general equivalent positions:

$$x, y, z;\quad \tfrac{1}{2} - x, \tfrac{1}{2} - y, z;\quad \tfrac{1}{2} + x, \bar{y}, z;\quad \bar{x}, \tfrac{1}{2} + y, z;$$
$$x, y, \bar{z};\quad \tfrac{1}{2} - x, \tfrac{1}{2} - y, \bar{z};\quad \tfrac{1}{2} + x, \bar{y}, \bar{z};\quad \bar{x}, \tfrac{1}{2} + y, \bar{z}$$

Coordinates of centres of symmetry:

¼, ¼, 0; ¼, ¾, 0; ¾, ¼, 0; ¾, ¾, 0; ¼, ¼, 0; ¼, ¾, 0; ¾, ¼, 0; ¾, ¾, 0;
¼, ¼, ½; ¼, ¾, ½; ¾, ¼, ½; ¾, ¾, ½ ¼, ¼, ½; ¼, ¾, ½; ¾, ¼, ½; ¾, ¾, ½

Change of origin to $\frac{1}{4}, \frac{1}{4}, 0$:
Subtract $\frac{1}{4}, \frac{1}{4}, 0$ from the above set of coordinates of general equivalent positions. Then, let $x_0 = x - \frac{1}{4}$, $y_0 = y - \frac{1}{4}$ and $z_0 = z$. After making all substitutions, drop the subscript, and rearrange to give:

$$\pm\{x, y, z;\quad \bar{x}, \bar{y}, z;\quad \tfrac{1}{2} + x, \tfrac{1}{2} - y, z;\quad \tfrac{1}{2} - x, \tfrac{1}{2} + y, z\}$$

This result may be confirmed by redrawing the space group with the origin on $\bar{1}$.

5.11 Figure S5.6 shows two adjacent unit cells of space group *Pn* on the (010) plane. In the transformation to *Pc*, only the *c* spacing is changed; the new value *c* for *Pc* is given by:

$$\mathbf{c}' = -\mathbf{a} + \mathbf{c}$$

Hence, $Pn \equiv Pc$. By interchanging the labels of the *x* and *z* axes, which are not constrained by the twofold symmetry, $Pc \equiv Pa$. Note that it is necessary then to negate **b** in order to preserve right-handed axes. The translation $a/2$ in the *C* unit cell in *Cm* means that $Ca \equiv Cm$. Since there is no translation along *c* in *Cm*, *Cm* is not equivalent to *Cc*, although *Cc* is equivalent to *Cn*. If the *x* and *z* axes in *Cc* are interchanged, with due attention to **b**, the symbol *Cc* becomes *Aa*. The standard symbols among these groups are *Pc*, *Cm*, and *Cc*.

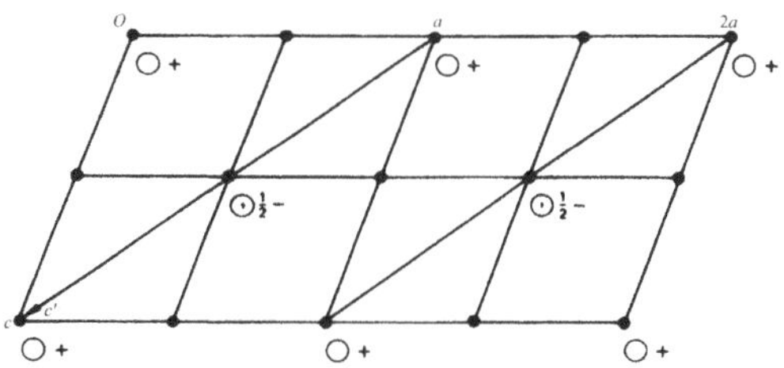

Fig. S5.6
Equivalence of space groups *Pn* and *Pc*.

5.12 $P2/c$: (a) $2/m$, monoclinic. (b) Primitive unit cell, c glide plane $\perp$ b, twofold axis $\parallel$ b. (c) $h0l$: $l = 2n$.
$Pca2_1$: (a) $mm2$, orthorhombic. (b) Primitive unit cell, c glide plane $\perp$ a, a glide plane $\perp$ b, 2_1 axis $\parallel$ c. (c) $0kl$: $l = 2n$; $h0l$: $h = 2n$.
$Cmcm$: (a) mmm, orthorhombic. (b) C face centered unit cell, m plane $\perp$ a, c glide plane $\perp$ b, m plane $\perp$ c. (c) hkl: $h + k = 2n$; $h0l$:$l = 2n$.
$P\bar{4}2_1c$: (a) $\bar{4}2m$, tetragonal. (b) Primitive unit cell, $\bar{4}$ axis $\parallel$ c, 2_1 axes $\parallel$ a and b, c-glide planes $\perp$ [110] and [1$\bar{1}$0]]. (c) hhl: $l = 2n$; $h00$: $h = 2n$.
$P6_122$: (a) 622, hexagonal. (b) Primitive unit cell, 6_1 axis $\parallel$ c, twofold axes $\parallel$ a, b, and u, twofold axes 30° to a, b and u, and in the (0001) plane. (c) $000l$: $l = 6n$. (A similar result obtains for $P6_522$.)
$Pa3$: (a) $m3$, cubic. (b) Primitive unit cell, a glide plane $\perp$ b (Equivalent statements are b glide plane $\perp$ c; c glide plane $\perp$ a), threefold axes $\parallel$ <[111>. (c) $0kl$: $k = 2n$; (Equivalent statements are $h0l$: $l = 2n$, $hk0$: $h = 2n$.)

5.13 (a) The plane group is $p11m$. (b) Plane group $p2$ would result, with the unit cell repeat along b halved, and γ with the special value of 90°.

5.14 (a) Refer to Fig. 4.14a, for a cubic P unit cell, without the A centring points; the unit cell vectors **a**, **b**, and **c** are equal in magnitude. (b) Tetragonal P: $\mathbf{a}_P = \mathbf{b}/2 + \mathbf{c}/2$, $\mathbf{b}_P = -\mathbf{b}/2 + \mathbf{c}/2$, $\mathbf{c}_P = \mathbf{a}$. (c) Monoclinic C: $\mathbf{a}_C = \mathbf{c}$, $\mathbf{b}_C = -\mathbf{b}$, $\mathbf{c}_C = \mathbf{a}$. (d) Triclinic P: $\mathbf{a}_T = \mathbf{a}$, $\mathbf{b}_T = \mathbf{b}/2 + \mathbf{c}/2$, $\mathbf{c}_T = -\mathbf{b}/2 + \mathbf{c}/2$. Other transformations may be acceptable for (d), provided that right-handed axes are maintained.

5.15 The matrices are:

$$\underset{\mathbf{R}_1}{\overset{\bar{4}\text{ along }z}{\begin{pmatrix} 0 & 1 & 0 \\ \bar{1} & 0 & 0 \\ 0 & 0 & \bar{1} \end{pmatrix}}} \quad \underset{\mathbf{R}_2}{\overset{m\perp y}{\begin{pmatrix} 1 & 0 & 0 \\ 0 & \bar{1} & 0 \\ 0 & 0 & 1 \end{pmatrix}}}$$

From Section 4.8.4, $\mathbf{R}_2\mathbf{R}_1\mathbf{h} = \mathbf{h}'$. Form first $\mathbf{R}_3 = \mathbf{R}_2\mathbf{R}_1$, remembering the order of multiplication, and then evaluate

$$\underset{\mathbf{R}_3}{\begin{pmatrix} 0 & \bar{1} & 0 \\ \bar{1} & 0 & 0 \\ 0 & 0 & \bar{1} \end{pmatrix}} \cdot \underset{\mathbf{h}}{\begin{pmatrix} h \\ k \\ l \end{pmatrix}} = \underset{\mathbf{h}'}{\begin{pmatrix} \bar{k} \\ \bar{h} \\ \bar{l} \end{pmatrix}}$$

that is, $\mathbf{R}_3\mathbf{h} = \mathbf{h}'$, so that $\boldsymbol{h}' = \bar{k}\,\bar{h}\,\bar{l}$; $\mathbf{R}_3$ represents a twofold rotation axis about [1$\bar{1}$0]. The point group is $\bar{4}m2$, which becomes the standard symbol $\bar{4}2m$ by a rotation of the symmetry elements, or the reference axes, by 45° in the x, y plane.

5.16 The matrices are multiplied in the usual way, and the components of the translation vectors are added, resulting in

$$\begin{pmatrix} \bar{1} & 0 & 0 \\ 0 & \bar{1} & 0 \\ 0 & 0 & 1 \end{pmatrix} + \begin{pmatrix} 1/2 \\ 1/2 \\ 1/2 \end{pmatrix}$$

which corresponds to a 2_1 axis along $\left[1/4,\ 1/4,\ z\right]$. The space group symbol is $Pna2_1$ (determined uniquely); class, rhombic disphenoid or, colloquially, $mm2$.

5.17 (a) From a hexagonal stereogram (Fig. 3.12), or otherwise, the matrix for a 6_3 rotation about [0, 0, z] is

$$\begin{pmatrix} 1 & \bar{1} & 0 \\ 1 & 0 & 0 \\ 0 & 0 & 1 \end{pmatrix} + \begin{pmatrix} 0 \\ 0 \\ 1/2 \end{pmatrix}$$

and that for the c glide normal to y and passing through the origin is

$$\begin{pmatrix} 1 & 0 & 0 \\ 1 & \bar{1} & 0 \\ 0 & 0 & 1 \end{pmatrix} + \begin{pmatrix} 0 \\ 0 \\ 1/2 \end{pmatrix}$$

(b) The matrix for the combination $\boldsymbol{c6_3}$ is

$$\begin{pmatrix} 1 & \bar{1} & 0 \\ 0 & \bar{1} & 0 \\ 0 & 0 & 1 \end{pmatrix} + \begin{pmatrix} 0 \\ 0 \\ 0 \end{pmatrix}$$

(c) Since the translation vector is zero, this matrix represents an m plane perpendicular to [120}, that is, along x as the plane $(x, 0, z)$. Thus, the space group symbol is $P6_3cm$.

(d) Refer to Fig. S5.7.

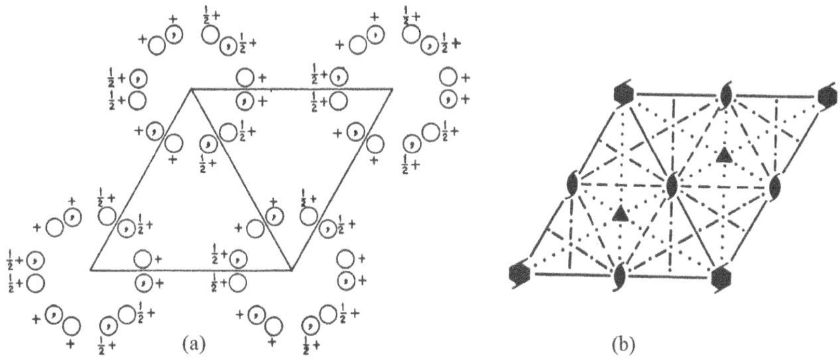

Fig. S5.7
Space group $P6_3cm$: (a) General equivalent positions, and (b) Symmetry elements. [Reproduced by courtesy of IUCr (1).]

(e) The general equivalent positions are:

12 d 1 $x, y, z;\ x-y, x, {}^1\!/_2+z;\ \bar{y}, x-y, z;\ \bar{x}, \bar{y}, {}^1\!/_2+z;\ y-x, \bar{x}, z;\ y, y-x, {}^1\!/_2+z;$
$\bar{y}, \bar{x}, {}^1\!/_2+z;\ \bar{x}, y-x, z;\ y-x, y, {}^1\!/_2+z;\ y, x, z;\ x, x-y, {}^1\!/_2+z;\ x-y, \bar{y}, z.$

(f) There are three sets of special equivalent positions:
6 c m $x, 0, z\ x, x, {}^1\!/_2+z;\ 0, x, z;\ \bar{x}, 0, {}^1\!/_2+z;\ \bar{x}, \bar{x}, z;\ 0, \bar{x}, {}^1\!/_2+z$
4 b 3 ${}^1\!/_3, {}^2\!/_3, z;\ {}^2\!/_3, {}^1\!/_3, z;\ {}^1\!/_3, {}^2\!/_3, {}^1\!/_2+z;\ {}^2\!/_3, {}^1\!/_3, {}^1\!/_2+z$
2 a $3m$ $0, 0, z;\ 0, 0, {}^1\!/_2+z$

(g) Limiting conditions

Wyckoff site	Condition
d	$hkil$ None
	$hh\overline{2h}l$ None
	$h\bar{h}0l$ $l=2n$
c	As above
b	As above, and $hkil$: $l=2n$
a	As for site b

5.18 From Fig. 4.15,

$$\mathbf{a}_R = 2\mathbf{a}_H/3 + \mathbf{b}_H/3 + \mathbf{c}_H/3$$
$$\mathbf{b}_R = -\mathbf{a}_H/3 + \mathbf{b}_H/3 + \mathbf{c}_H/3$$
$$\mathbf{c}_R = -\mathbf{a}_H/3 - 2\mathbf{b}_H/3 + \mathbf{c}_H/3$$

Following Section 2.2.3, $\mathbf{a}_R \cdot \mathbf{a}_R = (2\mathbf{a}_H/3 + \mathbf{b}_H/3 + \mathbf{c}_H/3) \cdot (2\mathbf{a}_H/3 + \mathbf{b}_H/3 + \mathbf{c}_H/3) = 3a^2/9 + c^2/9 = 12$ Å^2 so that $a_R = 1/3\sqrt{3a^2+c^2} = 3.464_1$ Å. Similarly, $\cos\alpha_R = (2\mathbf{a}_H/3 + \mathbf{b}_H/3 + \mathbf{c}_H/3) \cdot (-\mathbf{a}_H/3 + \mathbf{b}_H/3 + \mathbf{c}_H/3)/a_R^2 = 7.5/3.4641^2$ so that $\alpha_R = 51.32°$. (Remember that $a = b = c$ and $\alpha = \beta = \gamma$ in a rhombohedral unit cell.). By an alternative manipulation for the rhombohedral angle, it can be shown that $\sin(\alpha/2) = \frac{3}{2\sqrt{3+(c/a)^2}}$.

5.19 The transformation matrix $\mathbf{S}$ for $R_{hex} \rightarrow R_{obv}$ is given, from the solution to Problem 2.19 by

$$\mathbf{S} = \begin{pmatrix} {}^2\!/_3 & {}^1\!/_3 & {}^1\!/_3 \\ \overline{{}^1\!/_3} & {}^1\!/_3 & {}^1\!/_3 \\ \overline{{}^1\!/_3} & \overline{{}^2\!/_3} & {}^1\!/_3 \end{pmatrix}$$

and its inverse is

$$\mathbf{S}^{-1} = \begin{pmatrix} 1 & \bar{1} & 0 \\ 0 & 1 & \bar{1} \\ 1 & 1 & 1 \end{pmatrix}$$

so that its transpose becomes

$$(\mathbf{S}^{-1})^{\mathrm{T}} = \begin{pmatrix} 1 & 0 & 1 \\ \bar{1} & 1 & 1 \\ 0 & \bar{1} & 1 \end{pmatrix}$$

Hence $(13{*}4)_{\text{hex}}$ is transformed to $(32\bar{1})_{\text{obv}}$, and $[1\bar{2}{*}3]_{\text{hex}}$ to $[405]_{\text{obv}}$.

5.20 Figure S5.8 illustrates the reflection of the point X, Y, Z across the plane $(q\bar{q}z)$, a plane of the type $(h\bar{h}l)$, where the plane intersects the x and y axes at the distances q and $-q$, respectively, from the origin O, making angles of 45° with the reference axes. From the geometry of the figure, it is clear that the coordinates after reflection are $X' = q - (-Y) = q + Y$, and $-Y' = -q - X$, or $Y' = q + X$; Z remains unchanged (see also Section 5.4.12.2). [The change of origin procedure could also be used here.]

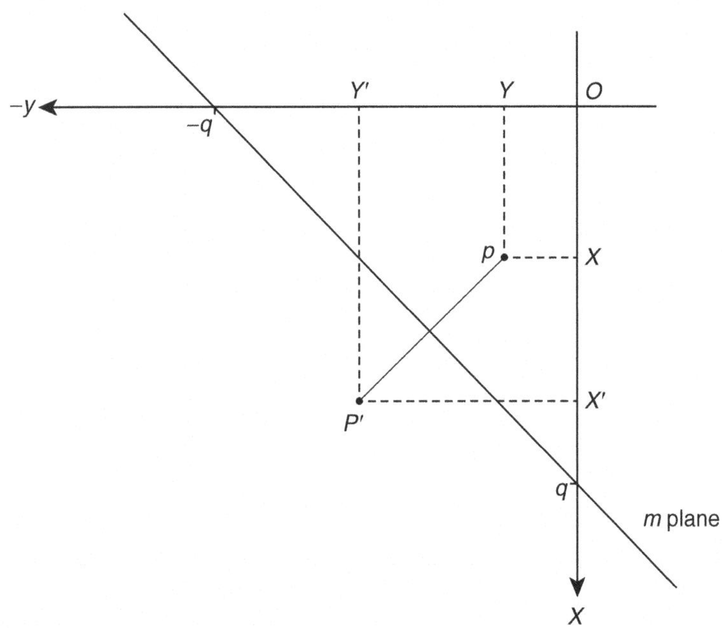

Fig. S5.8
Diagonal plane $(h\bar{h}l)$ in the tetragonal system intersecting the x and y axes at distances q and $-q$ respectively from the origin.

5.21 (a) From a drawing similar to Fig. 4.17, the matrix $\begin{pmatrix} \tfrac{1}{2} & 0 & 0 \\ 0 & 1 & 0 \\ 0 & 0 & 1 \end{pmatrix}$ can be developed. Then, (210) becomes (110) and may be confirmed by drawing to a scale. (b) From a new drawing, the matrix $\begin{pmatrix} 1 & 0 & 0 \\ 0 & \tfrac{1}{2} & 0 \\ 0 & 0 & 1 \end{pmatrix}$ can be developed, whence (210) becomes (410) after clearing the fraction. A scale drawing shows that the original (210) is now the second plane from the origin in the (410) family of planes; $d(410)_{\text{new}} = d(210)_{\text{old}}/2$ under the given transformation. In each case, the Miller index corresponding to the unit cell halving is also halved.

5.22 (a) In *Cmm*2, the polar, twofold axis is normal to the centred plane, but parallel to it in *Amm*2 (≡ *Bmm*2). *Cmmm* and *Ammm* are equivalent by interchange of axes, so that they are not two distinct arrangements of points. (b) (i) Applying the standard setting transforms *Abma* to *Cmca*. (ii) The glide vectors in *Cmca* related by *C*, lie normal to *z*, so the symbol can be written as *Cmce* (Fig. S5.9). The *e* glide generates the points $\frac{1}{2}+x,\ y,\ \frac{1}{2}-z$ and $x,\ \frac{1}{2}+y,\ \frac{1}{2}-z$ from $x, y,\ \frac{1}{2}-z$ (see also Section 5.4.5.) [It may be noted that the current International Tables, Volume A, does not list space groups with an *e* notation.]

5.23 The program LEPAGE gives the conventional unit cell as orthorhombic *F*, with $a' = 8.854$, $b' = 10.597$, $c' = 21.933$ Å, $\alpha' = 90.00°$, $\beta' = 90.00°$, $\gamma' = 90.00°$. The value for β' is 90° within 0.01°. The transformation matrix is

$$\begin{pmatrix} \bar{1} & 0 & 0 \\ 0 & \bar{1} & 0 \\ 1 & 0 & 2 \end{pmatrix}.$$

5.24 If **a** and **b** are fixed initially and a new value **c**′ defined as –**a** + **c** (to ensure an obtuse β angle), an *A* centred cell is obtained; then, **a** and **c**′ must be interchanged; it is necessary also to reverse **b** so as to obtain right-handed axes. Hence, $\mathbf{a}' = -\mathbf{a} + \mathbf{c}$; $\mathbf{b}' = -\mathbf{b}$; $\mathbf{c}' = \mathbf{a}$. Hence, $a' = 1.275$, $b' = 0.509$, $c' = 0.620$ nm; $\beta' = \cos^{-1}(-a/a') = 119.1°$; space group $C\frac{2}{c}$.

5.25

P$3_1$12

$$\begin{pmatrix} 0 & \bar{1} & 0 \\ 1 & \bar{1} & 0 \\ 0 & 0 & 1 \end{pmatrix}\begin{pmatrix} 0 \\ 0 \\ 1/3 \end{pmatrix} \qquad \begin{pmatrix} 1 & 0 & 0 \\ 1 & \bar{1} & 0 \\ 0 & 0 & \bar{1} \end{pmatrix}\begin{pmatrix} 0 \\ 0 \\ 0 \end{pmatrix}$$

3_1 along $[0, 0, z]$ $\qquad 2 \perp x, y, u$

$x, y, z;\ \bar{y}, x-y, 1/3+z;\ y-x, \bar{x}, 2/3+z;$

$x, x-y, \bar{z};\ y-x, y, 1/3-z;\ \bar{y}, \bar{x}, 2/3-z.$

P$3_1$21

$$\begin{pmatrix} 0 & \bar{1} & 0 \\ 1 & \bar{1} & 0 \\ 0 & 0 & 1 \end{pmatrix}\begin{pmatrix} 0 \\ 0 \\ 1/3 \end{pmatrix} \qquad \begin{pmatrix} \bar{1} & 1 & 0 \\ \bar{1} & 0 & 0 \\ 0 & 0 & \bar{1} \end{pmatrix}\begin{pmatrix} 0 \\ 0 \\ 0 \end{pmatrix}$$

3_1 along $[0, 0, z]$ $\qquad$ 2 along x, y, u

$x, y, z;\ \bar{y}, x-y, 1/3+z;\ y-x, \bar{x}, 2/3+z;$

$y, x, \bar{z};\ \bar{x}, y, -x, 1/3-z;\ x-y, \bar{y}, 2/3-z.$

It is clear that the two arrangements of sites differ, because of the different position of the twofold rotation axis with respect to the references axes.

5.26 *P*$4_2$32: Origin at 23. Coordinates of general equivalent positions:

x, y, z	$\bar{x}, \bar{y}, z$	$\bar{x}, y, \bar{z}$	$x, \bar{y}, \bar{z}$
z, x, y	$z, \bar{x}, \bar{y}$	$\bar{z}, \bar{x}, y$	$\bar{z}, x, \bar{y}$
y, z, x	$\bar{y}, z, \bar{x}$	$y, \bar{z}, \bar{x}$	$\bar{y}, \bar{z}, x$
$y+\frac{1}{2}, x+\frac{1}{2}, \bar{z}+\frac{1}{2}$	$\bar{y}+\frac{1}{2}, \bar{x}+\frac{1}{2}, \bar{z}+\frac{1}{2}$	$y+\frac{1}{2}, \bar{x}+\frac{1}{2}, z+\frac{1}{2}$	$\bar{y}+\frac{1}{2}, x+\frac{1}{2}, z+\frac{1}{2}$
$x+\frac{1}{2}, z+\frac{1}{2}, \bar{y}+\frac{1}{2}$	$\bar{x}+\frac{1}{2}, z+\frac{1}{2}, y+\frac{1}{2}$	$\bar{x}+\frac{1}{2}, \bar{z}+\frac{1}{2}, \bar{y}+\frac{1}{2}$	$x+\frac{1}{2}, \bar{z}+\frac{1}{2}, y+\frac{1}{2}$
$z+\frac{1}{2}, y+\frac{1}{2}, \bar{x}+\frac{1}{2}$	$z+\frac{1}{2}, \bar{y}+\frac{1}{2}, x+\frac{1}{2}$	$\bar{z}+\frac{1}{2}, y+\frac{1}{2}, x+\frac{1}{2}$	$\bar{z}+\frac{1}{2}, \bar{y}+\frac{1}{2}, \bar{x}+\frac{1}{2}$

General limiting conditions: $h00$: $h = 2n$, and cyclic permutations.

5.27 Following the procedure discussed in Chapter 5 leads to the following results.

(a)	(i)	(ii)	(iii)
*P*4*mm*	*p*4*mm*	*p*1*m*1	*p*1*m*1
(b)			
*P*4*bm*	*p*4*g*1	*p*1*m*1	*p*1*m*1
(c)			
$I\frac{4_1}{a}md$	*p*4*mm*	*c*2*mm*	*c*2*mm*
(d)	(iv)	(v)	(vi)
P$2_1$3	*p*2*gg*	*p*3	*p*1*g*1

(The plane groups of the projections of the space groups may be found in reference [5].)

5.28 Since 432 is a point group, the origin in $P432$ is at 432. Hence, the coordinates of the general positions are as listed below, together with the numbers appropriate to Fig. P5.3:

(1) x, y, z	(2) $\bar{x}, \bar{y}, z$	(3) $\bar{x}, y, \bar{z}$	(4) $x, \bar{y}, \bar{z}$
(5) z, x, y	(6) $z, \bar{x}, \bar{y}$	(7) $\bar{z}, \bar{x}, y$	(8) $\bar{z}, x, \bar{y}$
(9) y, z, x	(10) $\bar{y}, z, \bar{x}$	(11) $y, \bar{z}, \bar{x}$	(12) $\bar{y}, \bar{z}, x$
(13) $y, x, \bar{z}$	(14) $\bar{y}, \bar{x}, \bar{z}$	(15) $y, \bar{x}, z$	(16) $\bar{y}, x, z$
(17) $x, z, \bar{y}$	(18) $\bar{x}, z, y$	(19) $\bar{x}, \bar{z}, \bar{y}$	(20) $x, \bar{z}, y$
(21) $z, y, \bar{x}$	(22) $z, \bar{y}, x$	(23) $\bar{z}, y, x$	(24) $\bar{z}, \bar{y}, \bar{x}$

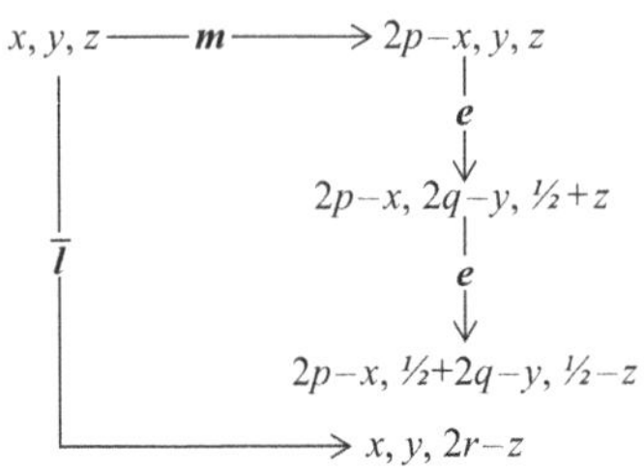

Fig. S5.9
Scheme for forming space group *Cmce*, using *these* symmetry elements.

5.29 Let the *e* glide plane be chosen as (x, y, r) with ½ translation along y; m is (p, y, z), c is (x, q, z). and $\bar{1}$ is set at 0, 0, 0. Now, the scheme in Fig. S5.9 can be devised. It is necessary to use only *one component* of the *e* glide in the scheme: the use of ½ translation along both x and y would be tantamount to applying the C centring condition; the ½ translation could equally well have been chosen along x instead of along y. Thus, the unknowns in the setting become $p = 0$, $q = r = \frac{1}{4}$. On drawing the space group and comparing with *Cmca* in ITA, No. 64, or elsewhere, the symmetry elements are seen to be identical. The symbol *Cmce* is just a way of emphasizing the translational components along $x(a)$ and $y(b)$ arising from the C centring.

Solutions 6

6.1 Inserting values of the fundamental constants into (6.1) gives $\lambda = 1.24/V$, where the units of V are kV. Hence, $\lambda_{\min} = 0.0413$ nm.

6.2 Since $I/I_0 = \exp(-\mu t)$, the fraction transmitted is 0.415

6.3 Using the reciprocal lattice, $\mathbf{d}^*_{123} \cdot \mathbf{d}^*_{123} = (d^*_{123})^2 = (\mathbf{a}^* + 2\mathbf{b}^* + 3\mathbf{c}^*) \cdot (\mathbf{a}^* + 2\mathbf{b}^* + 3\mathbf{c}^*) = 14(a^*)^2 = 14/a^2$, since $a = b = c$. Hence, $d_{123} = a/\sqrt{14} = 0.1678$ nm. From the Bragg equation, $\theta = \sin^{-1}[\lambda/(2d)]$; hence, $\theta_{123} = 27.35°$.

6.4 (a) From Eq. (6.3), $(\sin\theta/\lambda)^2 = 1/4d^2$, and in the cubic system $1/d^2 = (h^2 + k^2 + l^2)/a^2$, so that $(\sin\theta/\lambda)^2 = N/4a^2$, where $N = h^2 + k^2 + l^2$. The calculation is easily programmed for convenience, with the following results for $I(hkl)$.

hkl	111	200	311	400	331	420	422	511
NaCl	317	6668	87	2383	72	1867	1496	71
KCl	31	11547	31	3994	13	3114	2496	5

(b) The weak intensities for KCl where the indices are all odd integers, compared to NaCl under the same condition, is striking, and led to the original assumption that KCl was P cubic. (c) The sum $h^2 + k^2 + l^2$ is the same for both (511) and (333).

6.5 (a) The symmetry shown by the Laue photograph is $4mm$. (b) From Table 6.1, the beam direction is along one of the directions <100>. (c) The possible point groups are 432, $\bar{4}3m$ or $m\bar{3}m$.

6.6 From Eqs. (6.8)–(6.9),

$$A(hkl) = \sum_{j=1}^{N/2}\left[\cos 2\pi(hx_j + ky_j + lz_j) + \cos 2\pi\left(hx_j + ky_j + lz_j + \frac{k+l}{2}\right)\right]$$
$$= 2\sum_{j=1}^{N/2}\cos 2\pi\left(hx_j + ky_j + lz_j + \frac{k+l}{4}\right)\cos 2\pi\left(\frac{k+l}{4}\right).$$

Similarly,

$$B(hkl) = \sum_{j=1}^{N/2}\left[\sin 2\pi(hx_j + ky_j + lz_j) + \sin 2\pi\left(hx_j + ky_j + lz_j + \frac{k+l}{2}\right)\right]$$
$$= 2\sum_{j=1}^{N/2}\sin 2\pi\left(hx_j + ky_j + lz_j + \frac{k+l}{4}\right)\cos 2\pi\left(\frac{k+l}{4}\right).$$

It follows that $A = B = 0$ if $k + l = 2n + 1$. Thus, the limiting condition for A centring is hkl: $k + l = 2n$.

6.7 Using Eq. (6.5) here:

$$\mathrm{F}(hkl) = \sum_{j=1}^{N/2} f_{j,\theta}\left\{\exp[\mathrm{i}2\pi(hx_j + ky_j + lz_j)] + \exp\left[\mathrm{i}2\pi\left(-hx_j - ky_j + lz_j + \frac{h+l}{2}\right)\right]\right\}.$$

In order to proceed further, set $h = k = 0$; then

$$\mathrm{F}(hkl) = \sum_{j=1}^{N/2} f_{j,\theta}\exp(\mathrm{i}2\pi lz_j)(1 + \exp(\mathrm{i}\pi l).$$

As $\exp(\mathrm{i}n\pi) = 1$ for n even and 0 for n odd, the systematic absences for a 2_1 axis parallel to c are $00l$: $l = 2n + 1$.

6.8 Here, the atoms occur in pairs x, y, z and $\frac{1}{2} + x$, $\frac{1}{2} + \frac{1}{2} + y, -z$. Applying Eq. (6.5):

$$\mathrm{F}(hkl) = \sum_{j=1}^{N/2} f_{j,\theta}\left\{\exp[\mathrm{i}2\pi(hx_j + ky_j + lz_j)] + \exp\left[\mathrm{i}2\pi\left(hx_j + ky_j - lz_j + \frac{h+k}{2}\right)\right]\right\}.$$

Set $l = 0$; then, this equation can be simplified:

$$\mathrm{F}(hkl) = \sum_{j=1}^{N/2} f_{j,\theta}\exp[\mathrm{i}2\pi(hx_j + ky_j)]\{1 + \exp[\mathrm{i}\pi(h+k)]\}.$$

Hence, the limiting condition for the n glide plane normal to the z axis is: $hk0$: $h + k = 2n$.

6.9 (a) $P2_1$, $P2_1/m$. (b) $C2$, Cm, $C2/m$. (c) $Pbm2$ ($\mathbf{b\,a\,\bar{c}}$ setting of $Pma2$), $Pb2_1m$ ($\mathbf{b\,c\,a}$ setting of $Pmc2_1$), $Pbmm$ ($\mathbf{c\,a\,b}$ setting of $Pmma$). (d) $Ibca$. (e) $I\frac{4_1}{a}$. (f) $P4cc$, $P\frac{4}{m}cc$. (g) The two molecules could be placed either in general positions of space group $P2_1$ or in special positions, symmetry m or $\bar{1}$, of space group $P2_1/m$. Since a steroid molecule has no symmetry compatible with m or $\bar{1}$, the space group is $P2_1$. (h) hkl: $h + k + l = 2n$; $0kl$: $k = 2n$. ($l = 2n$); $h0l$: $l = 2n$, ($h = 2n$), $hk0$: $h = 2n$, ($k = 2n$); $h00$: ($h = 2n$); $0k0$: ($k = 2n$); $00l$: ($l = 2n$).

6.10 The coordinates of the general equivalent positions are (0, 0, 0; ½, ½, 0) + $x, y, z;\ x, \bar{y}, \tfrac{1}{2}+z$. Considering first the C centring: it is straightforward to show that the multiplicative term $[1+\exp(in\pi)]$, where $n=(h+k)$ exists under the summation in Eq. (6.5). This term is 2 or 0, according as the sum $(h+k)$ is even or odd. Thus, proceeding with Eqs. (6.9)–(6.10):

$$\begin{aligned} A(hkl) &= \sum[\cos 2\pi(hx+ky+lz)+\cos 2\pi(hx-ky+lz+l/2)] \\ &= 2\cos 2\pi(hx+lz+l/4)\cos 2\pi(ky-l/4) \end{aligned}$$

Expanding (Appendix A7):

$$\begin{aligned} A(hkl) &= 2\{[\cos 2\pi(hx+lz)\cos 2\pi l/4-\sin 2\pi(hx+lz)\sin 2\pi l/4] \\ &\quad [\cos 2\pi ky\cos 2\pi l/4+\sin 2\pi ky\sin 2\pi l/4]\} \\ &= 2\{[\cos 2\pi(hx+lz)\cos 2\pi ky\cos^2 2\pi l/4 \\ &\quad -\sin 2\pi(hx+lz)\sin 2\pi ky\sin^2 2\pi l/4]\} \end{aligned}$$

since the cross terms involve $\cos 2\pi l/4\ \sin 2\pi l/4 = \tfrac{1}{2}\sin 4\pi l/4 = 0$ for all k, since k is integral. In a similar manner:

$$\begin{aligned} B(hkl) &= 2\{[\sin 2\pi(hx+lz)\cos 2\pi ky\cos^2 2\pi l/4 \\ &\quad -\cos 2\pi(hx+lz)\sin 2\pi ky\sin^2 2\pi l/4]\} \end{aligned}$$

Separating $A(hkl)$ and $B(hkl)$ for even and odd values of l and remembering the factor of 2 from the C centring:

$l=2n$: $A(hkl) = 4\cos 2\pi(hx+lz)\cos 2\pi ky$

$B(hkl) = 4\sin 2\pi(hx+lz)\sin 2\pi ky$

$l=2n+1$: $A(hkl) = -4\sin 2\pi(hx+lz)\sin 2\pi ky$

$B(hkl) = 4\cos 2\pi(hx+lz)\sin 2\pi ky$

Thus the limiting conditions are: $\begin{cases} hkl: h+k+2n \\ h0l: l=2n \end{cases}$

6.11 The change will be reflected in the geometrical structure factor component $A(hkl)$; $B(hkl)$ remains zero. By change of origin

$$\begin{aligned} A(hkl)_{\frac{1}{2},0,\frac{1}{2}} &= \sum_{j=1}^{N}\cos 2\pi[hx+ky+lz-(h+l)/2] \\ &= \sum_{j=1}^{N}\{\cos 2\pi[hx+ky+lz]\cos 2\pi(h+l)/2 \\ &\quad +\sin 2\pi[hx+ky+lz]\sin 2\pi(h+l)/2\} \end{aligned}$$

Since $\sin 2\pi(h+l)/2$ is always zero,

$$\begin{aligned} A(hkl)_{\frac{1}{2},0,\frac{1}{2}} &= \sum_{j=1}^{N}\cos 2\pi[hx+ky+lz-(h+l)/2] \\ &= \sum_{j=1}^{N}\cos 2\pi[hx+ky+lz]\cos 2\pi(h+l)/2] \\ &= \sum_{j=1}^{N}\cos 2\pi[hx+ky+lz](-1)^{h+l} \end{aligned}$$

Thus, $A(hkl)_{1/2,0,1/2} = (-1)^{h+l}A(hkl)_{0,0,0}$, and the phase, which is here the sign of $A(hkl)$, will be modified by the term $(-1)^{h+l}$ from what it would be with $\overline{1}$ at 0, 0, 0, although it remains either 0 or π.

Solutions 7

7.1 $\boldsymbol{C}_2$ commutes with $\boldsymbol{\sigma}$, $\boldsymbol{E}$ commutes with all symmetry operators.

7.2 (a)

C_{2h}	E	C_2	i	σ_h
E	E	C_2	i	σ_h
C_2	C_2	E	σ_h	i
i	i	σ_h	E	C_2
σ_h	σ_h	i	C_2	E

(b)

D_{2d}	E	S_4	S_4^3	C_2	C_2'	C_2''	σ_d	σ_d'
E	E	S_4	S_4^3	C_2	C_2'	C_2''	σ_d	σ_d'
S_4	S_4	C_2	E	S_4^3	σ_d	σ_d'	C_2''	C_2'
S_4^3	S_4^3	E	C_2	S_4	σ_d'	σ_d	C_2'	C_2''
C_2	C_2	S_4^3	S_4	E	C_2''	C_2'	σ_d'	σ_d
C_2'	C_2'	σ_d'	σ_d	C_2''	E	C_2	S_4^3	S_4
C_2''	C_2''	σ_d	σ_d'	C_2'	C_2	E	S_4	S_4^3
σ_d	σ_d	C_2'	C_2''	σ_d'	S_4	S_4^3	E	C_2
σ_d'	σ_d'	C_2''	C_2'	σ_d	S_4^3	S_4	C_2	E

C_{2h} is Abelian.

7.3

G_r	E	r	r^2	r^3
E	E	r	r^2	r^3
r	r	r^2	r^3	E
r^2	r^2	r^3	E	r
r^3	r^3	E	r	r^2

$\boldsymbol{r}^2$ is its own inverse as $\boldsymbol{r}^2\boldsymbol{r}^2 = \boldsymbol{E}$.

7.4 (a) D_3

(b)

D_3	E	C_3	C_3^2	C_2	C_2'	C_2''
E	E	C_3	C_3^2	C_2	C_2'	C_2''
C_3	C_3	C_3^2	E	C_2'	C_2''	C_2
C_3^2	C_3^2	E	C_3	C_2''	C_2	C_2'
C_2	C_2	C_2''	C_2'	E	C_3^2	C_3
C_2'	C_2'	C_2	C_2''	C_3	E	C_3^2
C_2''	C_2''	C_2'	C_2	C_3^2	C_3	E

(c) C_3, C_2 and C_1.

(d) Deleting $\boldsymbol{C}_2$ from the table requires the deletion also of $\boldsymbol{C}_2'$ and $\boldsymbol{C}_2''$, leaving C_3 ; C_3 is invariant with respect to D_3. The deletion of $\boldsymbol{C}_3$ requires the deletion of $\boldsymbol{C}_3^2$ and two $\boldsymbol{C}_2$ operators, and there is more than one way in which this can be done: C_2 is not invariant with respect to D_3. C_l is also invariant with respect to D_3.

7.5

S_4	E	S_4	C_2	S_4^3
E	E	S_4	C_2	S_4^3
S_4	S_4	C_2	S_4^3	E
C_2	C_2	S_4^3	E	S_4
S_4^3	S_4^3	E	S_4	C_2

In S_4, deletion of any one of the operators $\boldsymbol{S}_4$, $\boldsymbol{C}_2$ or $\boldsymbol{S}_4^3$ requires the simultaneous deletion of the other two, leaving the identity group C_1. Thus, point group S_4 cannot be produced by any combination of point groups.

S_6	E	C_3	C_3^2	i	S_6^5	S_6
E	E	C_3	C_3^2	i	S_6^5	S_6
C_3	C_3	C_3^2	E	S_6^5	S_6	i
C_3^2	C_3^2	E	C_3	S_6	i	S_6^5
i	i	S_6^5	S_6	E	C_3	C_3^2
S_6^5	S_6^5	S_6	i	C_3	C_3^2	E
S_6	S_6	i	S_6^5	C_3^2	E	C_3

Deletion of $\boldsymbol{C}_3$ requires the deletion of $\boldsymbol{C}_3^2$, $\boldsymbol{S}_6^5$ and S_6. Deletion of i requires the deletion of $\boldsymbol{S}_6^5$ and S_6; $i\boldsymbol{C}_3 = \boldsymbol{C}_3 i = \boldsymbol{S}_6^5$. The subgroups of S_4 are C_2 and C_1, and both are invariant with respect to S_4. The subgroups of S_6 are C_3, C_i and C_1, and are all invariant with respect to S_6.

7.6 $[PtCl_4]^{2-}$ ion, point group D_{4h}.

$$E,\ 2C_4,\ C_2,\ 2C_2',\ 2C_2'',\ i,\ 2S_4,\ \sigma_h,\ 2\sigma_v,\ 2\sigma_d;\ \text{order} = 16.$$

Z-1,2-dichloroethene, point group C_{2v}

$$E,\ C_2,\ \sigma_v,\ \sigma_v';\ \text{order} = 4.$$

7.7 By definition of conjugation $B = X^{-1}AX$ and $C = Y^{-1}BY$, where X and Y are any members of the same group. By substitution, $C = Y^{-1}X^{-1}AXY = (XY)^{-1}AXY$. Hence, C is conjugate with A, since $Y^{-1}X^{-1} = (XY)^{-1}$, and XY and $(XY)^{-1}$ are members of the group.

7.8 (a) The right and left cosets are formed first with $H\{E, C_3, C_3^2\}$ and σ_v. Left coset: $\sigma_v E$, $\sigma_v C_3$, $\sigma_v C_3^2$ which, from the C_{3v} group multiplication table, is equivalent to σ_v, σ_v'', σ_v'. By a similar analysis, the right coset is σ_v, σ_v', σ_v''. These two cosets are identical: thus, the subgroup H is invariant with respect to C_{3v}, and the point group may be generated by the combination of H with either the left or right coset. With H' and C_3, the right coset with C_3 is EC_3, $\sigma_v C_3 \equiv C_3$, σ_v', and the left coset is EC_3, $\sigma_v C_3 \equiv C_3$, σ_v''. The two cosets are not identical; hence, H' is not invariant with respect to C_{3v}, and this point groups is not generated by the combination of either coset with H'. There are two other subgroups like H', involving σ_v' and σ_v''.

7.9 (a) C_i at the intersection of C_2 and σ_h; C_{2h} (2/*m*). (b) A second σ_v plane, perpendicular to the first, with two C_2 axes normal to S_4 and midway between the σ_v planes; D_{2d} ($\bar{4}2m$). (c) The S_6 axis is coincident with the C_3 axis; S_6 ($\bar{3}$). (d) Two more C_2 axes exist, perpendicular to the S_6 axis and at 120° from the first, and three σ_d planes midway between the C_2 axes; D_{3d} ($\bar{3}m$).

7.10 (a) $\mathbf{B} = \mathbf{A}^{\mathrm{T}}$ (b) $\mathbf{B} = \mathbf{A}^*$ (c) $\mathbf{B} = \text{Cofactor}\,\mathbf{A}$ (d) $\mathbf{B} = \mathbf{A}^\dagger$. There is no inverse of $\mathbf{A}$ in (c) because the determinant is negative.

7.11 Follow the argument of Section 7.5.2.

$$\mathbf{D}(C_6) = \begin{pmatrix} 1/2 & \overline{\sqrt{3}/2} & 0 \\ \sqrt{3}/2 & 1/2 & 1 \\ 0 & 0 & 1 \end{pmatrix} \qquad \mathbf{D}(C_2') = \begin{pmatrix} \bar{1} & 0 & 0 \\ 0 & 1 & 0 \\ 0 & 0 & \bar{1} \end{pmatrix}$$

$$\mathbf{D}(C_6)\mathbf{D}(C_2') = \begin{pmatrix} \overline{1/2} & \overline{\sqrt{3}/2} & 0 \\ \overline{\sqrt{3}/2} & 1/2 & 1 \\ 0 & 0 & \bar{1} \end{pmatrix} \qquad \mathbf{D}(C_2')\mathbf{D}(C_6) = \begin{pmatrix} \overline{1/2} & \sqrt{3}/2 & 0 \\ \sqrt{3}/2 & 1/2 & 1 \\ 0 & 0 & \bar{1} \end{pmatrix}$$

$\mathbf{D}(C_6)\mathbf{D}(C_2')$ represents a C_2' operator in the *x*, *y* plane 30° anticlockwise from +*y*; $\mathbf{D}(C_2')\mathbf{D}(C_6)$ represents a C_2' operator in the *x*, *y* plane 30° clockwise from +*y*. The three C_2' operators are related by C_6.

7.12 (i) Use the matrix $\begin{pmatrix} \cos\theta & \overline{\sin\theta} & 0 \\ \sin\theta & \cos\theta & 0 \\ 0 & 0 & 1 \end{pmatrix}$ with $\theta = 240°$ (or twice with $\theta = 120°$).

(ii). Obtain $\mathbf{D}(C_3)$ and invert the matrix.

7.13

C_{2v}	E	C_2	σ_v	σ_v'
Γ_{red}	9	−1	1	3

Applying the reduction formula, together with the character table for C_{2h}:

$$a_{A_1} = \frac{1}{4}\{(9 \times 1 \times 1) + [1 \times (-1) \times 1] + (1 \times 1 \times 1) + (3 \times 1 \times 1)\} = 3\}.$$

Hence, $3A_1$ is one element of the reducible representation. Continuing in this way: $\Gamma_{\text{irred}} = 3A_1 + A_2 + 2B_1 + 3B_2$.

7.14 From the C_{2h} character table, the following movements suggest themselves: (i) Translation: $A_1 + B_1 + B_2$. (ii) Rotation: $A_2 + B_1 + B_2$. Hence, vibration corresponds to $2A_1 + B_2$.

7.15 Use the procedure discussed for reducing representations, as in the previous problem. (a) $A_2 + 2B_2$. (b) $A_1 + 2A_2 + E$. (c) $A_2'' + E''$. (d) $2A_1 + B_1 + E$. (e) $B_{1g} + E_{1g} + A_{2u} + E_{2u}$. (f) $A_{1g} + E_g + T_{2g} + T_{2u}$.

7.16 The irreducible representations in C_3 are shown in its character table:

C_3	E	C_3	C_3^2		$\varepsilon = \exp(i2\pi/3)$
A	1	1	1	$z,\ R_z$	$x^2 + y^2,\ z^2$
E	$\left\{\begin{matrix}1\\1\end{matrix}\right.$	$\begin{matrix}\varepsilon\\\varepsilon^*\end{matrix}$	$\left.\begin{matrix}\varepsilon^*\\\varepsilon\end{matrix}\right\}$	$(x,\ y),\ (R_x,\ R_y)$	$(x^2 - y^2,\ xy),\ (yz,\ xz)$

For A with itself: $\frac{1}{3}[(1 \times 1) + (1 \times 1) + (1 \times 1)] = 1$

For A with itself E: $\frac{1}{3}[(1 \times 1) + (1 \times \varepsilon) + (1 \times \varepsilon^*)] = 0$

For E with itself: $\frac{1}{3}[(1 \times 1) + (\varepsilon \times \varepsilon^*) + (\varepsilon^* \times \varepsilon)] = 1$

[The product of a character of one irreducible representation with itself is the character multiplied by its conjugate: the conjugate of 1 is 1, but the conjugate of ε, or $\exp(i2\pi/3)$, is ε^*, or $\exp(-i2\pi/3)$, so that $\varepsilon\varepsilon^* = \varepsilon^*\varepsilon = 1$, and for C_3 $(\varepsilon + \varepsilon^*) = -1$.]

7.17 Following Section 7.10, $\Gamma_{\text{red}} = 18\ 0\ 2\ 0\ 0$. Applying the standard reduction procedure:

$$\Gamma_{\text{irred}} = A_1 + A_2 + 2E_1 + 2T_1 + 2T_2.$$

7.18 Review Section 7.10, as necessary. For D_{3d}, $h = 12$. The rules allow the table to be written in the form:

D_{3d}	E	$2C_3$	$3C_2'$	i	$2S_6$	$3\sigma_d$
A_{1g}	1	1	1	1	1	1
A_{2g}	1	1	–1	1	1	–1
E_g	2	a	b	2	a	b
A_{1u}	1	1	1	–1	1	–1
A_{2u}	1	1	–1	–1	1	1
E_u	2	a	b	–2	$-a$	$-b$

From A_{1g} and E_g, $(1/12)\{(1 \times 1 \times 2) + (2 \times 1 \times a) + 3b + (1 \times 1 \times 2) + (2 \times 1 \times a) + 3b\} = 0$; hence, $2a + 3b = -2$, to which the only sensible solution is $a = -1$ and $b = 0$. Hence, the character table for D_{3d}:

D_{3d}	E	$2C_3$	$3C_2'$	i	$2S_6$	$3\sigma_d$
A_{1g}	1	1	1	1	1	1
A_{2g}	1	1	–1	1	1	–1
E_g	2	–1	0	2	–1	0
A_{1u}	1	1	1	–1	1	–1
A_{2u}	1	1	–1	–1	1	1
E_u	2	–1	0	–2	1	0

7.19 $C_3 C_i = S_3$ [$\varepsilon = \exp(i2\pi/3)$]. For $C_i \otimes C_3$, the symmetry operations are $\{E, i\}$ $\{E, C_3, C_3^2\} = \{E, C_3, C_3^2, i, S_6^5, S_6\}$, which corresponds to point group S_6. *In extenso*:

C_i	E	i
E	1	1
i	1	-1

$\otimes$

C_3	E	C_3	C_3^2
A	1	1	1
E	$\{1$	ε	$\varepsilon^*\}$
	$\{1$	ε^*	$\varepsilon\}$

$=$

S_6	E	C_3	C_3^2	i	S_6^5	S_6
Ag	1	1	1	1	1	1
E_g	$\{1$	ε	ε^*	1	ε	$\varepsilon^*\}$
	$\{1$	ε^*	ε	1	ε^*	$\varepsilon\}$
A_u	1	1	1	-1	-1	-1
E_u	$\{1$	ε	ε^*	-1	$-\varepsilon$	$-\varepsilon^*\}$
	$\{1$	ε^*	ε	-1	$-\varepsilon^*$	$-\varepsilon\}$

7.20 Equation (7.42) for $\alpha = \beta$ is:

$$\sum_{R} \mathbf{D}_\alpha^*(\boldsymbol{R})_{i,i}\, \mathbf{D}_\alpha(R)_{p,p} = \frac{h}{n}\, \delta_{i,p} \quad (\delta_{i,p} = \delta_{j,q}).$$

The sums over all i and p may be written as:

$$\sum_{i=1}^{n_\alpha} \sum_{p=1}^{n_\beta} \mathbf{D}_\alpha^*(\boldsymbol{R})_{i,i}\, \mathbf{D}_\beta(\boldsymbol{R})_{p,p} = \frac{h}{\sqrt{n_\alpha n_\beta}} \delta_{\alpha,\beta} \sum_{i=1}^{n_\alpha} \sum_{p=1}^{n_\beta} \delta_{i,p}$$

The left hand side sums over i and p lead to:

$$\sum_{i=1}^{n_\alpha} \sum_{p=1}^{n_\beta} \mathbf{D}_\alpha^*(\boldsymbol{R})_{i,i}\, \mathbf{D}_\beta(\boldsymbol{R})_{p,p} = \chi_\alpha^*\, \chi_\beta$$

because χ is the trace of a **D**-matrix. Hence, summing now over $\boldsymbol{R}$, and since $\alpha = \beta$,

$$\sum_{R} \chi_\alpha^*(\boldsymbol{R})\, \chi_\beta(\boldsymbol{R}) = \frac{h}{n} \delta_{i,p}$$

which is non-zero only for $i = j$. The left hand term is equal to $\sum_R |\chi(\boldsymbol{R})|^2$; hence,

(a) $1(1)^2 + 2(-1)^2 + 2(1)^2 + 1(-1)^2 + 3(-1)^2 + 3(1)^2 = 12$. ($h = 12$, $n = 1$, $h/n = 12$)

(b) $1(1)^2 + 1(-i \times i) + 1(-1)^2 + 1(-i \times i)^2 + 1(1)^2 + 1(-i \times i) + 1(-1)^2 + 1(-i \times i)^2 = 8$. $h = 8$, $n = 1$, $h/n = 8$)

[E_g^* evaluates as E_g. In this particular example, if the two parts of E_g are added first to give 2 0 –2 0 –2 0 2 0, and h/n =4; but then the equation would have summed to 16, which is erroneous.]
(c) For $i \neq j$, the sum is zero.

7.21 (a) Follow Section 7.7. From the character table for C_{4v}, the direct product of B_1 with itself is formed and summed over all ***R***; hence, $(1\times1\times1)+[(-1)\times(-1)\times2]+(1\times1\times1)+[(-1)\times(-1)\times2]+(1\times1\times2)=8$. In a similar manner, (b) 0, (c) 12, (d) 24.

7.22 *Translational displacements*. About z: ***E*** and $\boldsymbol{C_2}$ leave a vector along z unchanged, whereas ***i*** and $\boldsymbol{\sigma_h}$ reverse it. Hence, the characters are, in order, 1 1 –1 –1. About x: ***E*** and $\boldsymbol{\sigma_h}$ leave x unchanged, whereas $\boldsymbol{C_2}$ and ***i*** reverse it. Hence, the characters are, in order, 1 –1 –1 1. About y: y behaves as does x. *Rotational displacements*: About z: None of the symmetry operations changes the sign of R_z, as can be judged from the figure. Hence, the characters are 1 1 1 1. About x: ***E*** and ***i*** leave x unchanged, whereas $\boldsymbol{C_2}$ and $\boldsymbol{\sigma_h}$ reverse it. Hence, the characters are 1 –1 1 –1. About y: y behaves as does x. *Product terms*: The character for the product terms may be obtained by multiplication. Thus, for x^2, it is (1×1), $(-1)^2$, $(-1)^2$ (1×1), or 1 1 1 1. For yz, it is 1^2, $(-1)\times1$, $(-1)^2$, $1\times(-1)$, or 1 –1, 1 –1. Thus, the assignments in character table C_{2v} are confirmed.

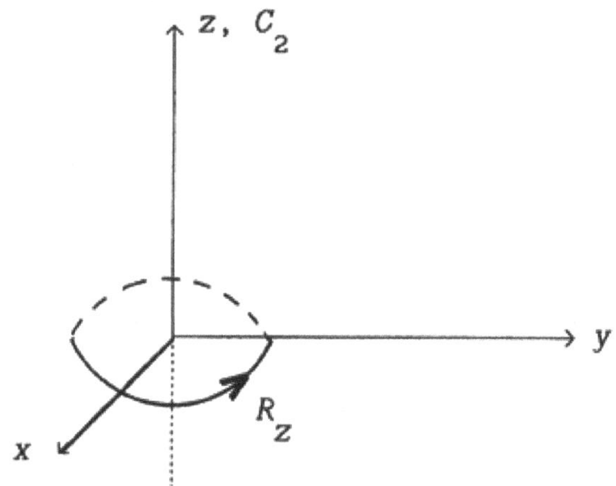

Solutions 8

8.1 (a) From function 1, $3\cos^2\theta-1=3\cos^2\theta-(\cos^2\theta+\sin^2\theta)=2\cos^2\theta-\sin^2\theta$. Using Appendix A8, $\sin^2\theta\cos^2\phi=(x/r)^2$ and $\sin^2\theta\sin^2\phi=(y/r)^2$. Thus, $(x^2+y^2)/r^2=\sin^2\theta(\sin^2\phi+\cos^2\phi)=\sin^2\theta$. Hence, the original function may be written

$$3\cos^2\theta-1=(2/r^2)z^2-(1/r^2)(x^2+y^2)=c(2z^2-x^2-y^2)$$

where c is a constant. Thus, the d orbital is named $d_{2z^2-(x^2+y^2)}$. In most point groups, z^2 and x^2+y^2 transform in the same way, so that $2z^2-x^2-y^2$ will transform like z^2 and the shorter notation d_{z^2} is used for the name of this function. The d_{z^2} angular function is normalized according the the equation

$$N^2\int_0^{2\pi}\mathrm{d}\phi\int_0^{\lambda}(9\cos^4\theta-6\cos^2\theta+1)\sin\theta\,\mathrm{d}\theta=1$$

where the integral over ϕ is 2π. The integrand is now $9\cos^4\theta\sin\theta - 6\cos^2\theta\sin\theta + \sin\theta$. From standard integral tables, $\int\cos^m\theta\sin\theta\,d\theta = -\frac{\cos^{m+1}\theta}{m+1}$, and the integral in θ becomes

$$\left(-\frac{9}{5}\cos^5\theta\,|_0^\pi\right) - \left(-\frac{6}{3}\cos^3\theta\,|_0^\pi\right) + \left(-\cos\theta\,|_0^\pi\right) = \frac{18}{5} - 4 + 2 = \frac{8}{5}.$$

Hence,

$$1 = 2\pi N^2(8/5), \text{ and } N = {}^1\!/\!_4\sqrt{5/\pi}.$$

(b) In a like manner, the function $\sin^2\theta\cos 2\phi$ can be shown to equal $\sin^2\theta(\cos 2\phi - \sin^2\phi)$, which by manipulation similar to that above becomes $c'(x^2 - y^2)$, where c' is another constant. Thus, this d orbital function is named $d_{x^2-y^2}$. The normalization constant evolves from the equation $1 = N^2\sin\theta\int_0^{2\pi}\cos^2 2\phi\,d\phi\int_0^\pi(\sin^5\theta)\,d\theta$; where the integral in ϕ is now π. Then, from integral tables,

$$\int\sin^5 x\,dx = -\frac{1}{5}\sin^4 x\cos x + \frac{4}{5}\int\sin^3 x\,dx$$

$$= -\frac{1}{5}\sin^4 x\cos x + \frac{4}{5}\left(-\frac{1}{3}\cos x(\sin^2 x + 2)\right).$$

Thus the integral in θ becomes $-\frac{1}{5}\sin^4\theta\cos\theta|_0^\pi + \frac{4}{5}(-\frac{1}{3}\cos\theta(\sin^2\theta + 2))|_0^\pi = \frac{8}{15}\cos\theta\,|_0^\pi = 16/15$. Thus, the normalization constant for the $d_{x^2-y^2}$ function is ${}^1\!/\!_4\sqrt{15/\pi}$.

8.2 Working along the line used in the text for boron trifluoride, the representation

C_{3v}	E	$2C_3$	$3\sigma_v$
Γ	3	0	1

is obtained; Γ reduces to $A_1 + E$.

8.3 Under C_{3v}, the atomic orbitals for A_1 are s, p_z and d_{z^2}, for E, p_x, p_y, d_{xy} and d_{yz}. Possible combination leading to three N–H bonds are:

Trigonal planar:	sp^2, p^2d, sd^2, d^3
Trigonal pyramidal:	p^3, pd^2
Unsymmetrical planar:	spd

For an element in the first row of the Periodic Table, contributions from relatively high energy d orbitals can be excluded. Neither p^3 nor sp^2 alone would give the correct bond angles, so a combination of the two is most likely. The linear combination could be written as $\Psi = c_1\psi(sp^2) + c_2\psi(p^3)$, where c_1 and c_2 are adjusted for minimum electronic energy.

8.4 Using the Ni–F bond vectors as a basis, the following representation can be generated:

D_{4h}	E	$2C_4$	C_2	$2C_2'$	$2C_2''$	C_i	$2S_4$	σ_h	$2\sigma_v$	$2\sigma_d$
Γ	4	0	0	2	0	0	0	4	2	0

It reduces to $A_{1g} + B_{1g} + E_u$. The available atomic orbitals are s or d_{z^2}, $d_{x^2-y^2}$, p_x and p_y, so that possible combination is sp^2d or d^2p^2. Further calculations would be needed in order to decide which of these alternatives is correct. [It is known that sp^2d is preferred.]

8.5 The z axis lies through the nitrogen atom and normal to the plane of the three hydrogen atoms; x and y are in the plane of the hydrogen atoms, with x passing through one of them. The matrices for the symmetry operations of C_{3v} are shown in Eq. (7.15). The traces of these matrices are, in order, 3, 0, 0, 1, 1 and 1; hence the representation

C_{3v}	E	$2C_3$	$3\sigma_v$
Γ	3	0	1

which reduces to $A_1 + E$. Thus, p_z spans A_1, and p_x and p_y span E. The three p orbitals that are triply degenerate under the spherical symmetry of the nitrogen atom alone, become split into p_z (non-degenerate) and p_x, p_y (doubly degenerate) under the C_{3v} symmetry of the molecule—an example of ligand-field theory with p orbitals.

8.6 Use the standard reduction procedure. (a) C_{4v}: $A_2 + B_1 + E$. (b) D_{4d}: $E_1 + E_2 + E_3$. (c) D_{6h}: $2A_{1g} + A_{2g} + 2B_{2g} + E_{1g} + 4E_{2g} + A_{2u} + 2B_{1u} + 2B_{2u} + 3E_{1u} + 2E_{2u}$.

8.7 The original symmetry O_h distorts the ion to D_{4h}. The axial bonds no longer couple with the equatorial bonds. Hence, $\Gamma_\sigma = \Gamma_{ax} + \Gamma_{eq}$. Using the two sets of bond vectors the following representation is obtained:

D_{4h}	E	$2C_4$	C_2	$2C_2'$	$2C_2''$	C_i	$2S_4$	σ_h	$2\sigma_v$	$2\sigma_d$
Γ_{ax}	2	2	2	−2	−2	−2	−2	−2	2	2
Γ_{eq}	4	0	0	2	0	0	0	4	2	0

From the standard reduction procedure, $\Gamma_{ax} = 2A_{2u}$; $\Gamma_{eq} = A_{1g} + B_{1g} + E_u$.

8.8 The non-linear N_2O molecule has point group C_{2v} (a) From Section 7.5.3, $\Gamma_{3n} = 9, -1, 1, 3$, which reduces to $3A_1 + A_2 + 2B_1 + 3B_2$. (b) From the character table for C_{2v}, the translational movements span $A_1 + B_1 + B_2$, and rotational movements $A_2 + B_1 + B_2$. Thus, the vibrational degrees are $3A_1 + A_2 + 2B_1 + 3B_2 - (A_1 + B_1 + B_2) - (A_2 + B_1 + B_2)$, or $2A_1 + B_2$, so that there are three degrees of vibrational freedom (normal vibrational modes), in agreement with the $3N - 6$ rule. Since A_1 is totally symmetric, $2A_1$ correspond with the symmetric stretch (a) and symmetric bend (c) in Fig. S8.1; B_2 is the asymmetric stretch (Fig. S8.1) (b). The three vibrations perturb both the dipole moment and polarizability of the molecule, so that all three vibrations are both infrared active and Raman active.

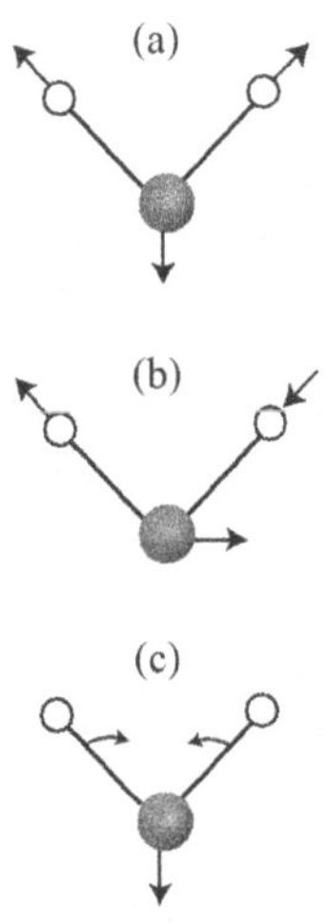

Fig. S8.1
Normal modes for the molecule of N_2O.

8.9 Apply the similarity transformation $\boldsymbol{X}^{-1}\boldsymbol{A}\,\boldsymbol{X}$, where $\boldsymbol{X}$ is any other member of the group other than $\boldsymbol{A}$.
(a) Point group C_4: for $\boldsymbol{A} = \boldsymbol{C}_4$, the four possible products are only $\boldsymbol{C}_4$; for $\boldsymbol{A} = \boldsymbol{C}_4^3$, the products are only $\boldsymbol{C}_4^3$; thus, $\boldsymbol{C}_4$ and $\boldsymbol{C}_4^3$ are in different classes in point group C_4. (b) Point group C_{4v}: for $\boldsymbol{A} = \boldsymbol{C}_4$, the eight possible products, four are $\boldsymbol{C}_4$ and four are $\boldsymbol{C}_4^3$; for $\boldsymbol{A} = \boldsymbol{C}_4^3$, the same products arise; thus, $\boldsymbol{C}_4$ and $\boldsymbol{C}_4^3$ are in one and the same class in point group C_{4v}. [Note also that, if $2\boldsymbol{C}_4$ were split into $\boldsymbol{C}_4$ and $\boldsymbol{C}_4^3$ in C_{4v}, the orthonormal relationship would not hold for these columns (Section 7.6.2).]

8.10 The appropriate direct products are formed, and group multiplication tables (or stereograms) applied as necessary.
(a) $C_i \otimes S_4$: $\{\boldsymbol{E}, \boldsymbol{i}\}\{\boldsymbol{E}, \boldsymbol{S}_4, \boldsymbol{C}_2, \boldsymbol{S}_4^3\} = \{\boldsymbol{E}, \boldsymbol{S}_4, \boldsymbol{C}_2, \boldsymbol{S}_4^3 \mid \boldsymbol{i}, \boldsymbol{C}_4^3, \boldsymbol{\sigma}_h, \boldsymbol{C}_4\}$. Rearranging to standard form: $\{\boldsymbol{E}, \boldsymbol{C}_4, \boldsymbol{C}_2, \boldsymbol{C}_4^3, \boldsymbol{i}, \boldsymbol{S}_4^3, \boldsymbol{\sigma}_h, \boldsymbol{S}_4\}$, which corresponds to point group C_{4h}.

(b) $\sigma_v \otimes S_4$: $\{\boldsymbol{E},\ \boldsymbol{\sigma}_v\}\{\boldsymbol{E},\ \boldsymbol{S}_4,\ \boldsymbol{C}_2,\ \boldsymbol{S}_4^3\} = \{\boldsymbol{E},\ \boldsymbol{S}_4,\ \boldsymbol{C}_2,\ \boldsymbol{S}_4^3,\ \mid \boldsymbol{\sigma}_v,\ \boldsymbol{C}'_{2'},\ \boldsymbol{\sigma}'_v,\ \boldsymbol{C}'_2\}$. Rearranging to standard form: $\{\boldsymbol{E},\ 2\boldsymbol{S}_4,\ \boldsymbol{C}_2,\ \boldsymbol{S}_4^3,\ 2\boldsymbol{C}'_{2'},\ 2\boldsymbol{\sigma}_d\}$, which corresponds to point group D_{2d}; both of these groups emerged from the analysis in Section 8.3.4.

8.11 Refer to Eqs. (8.4)–(8.5); the program MATOPS in the Web Program Suite may be helpful.

$$\left\{\mathbf{4}_{1[100]}\middle|\mathbf{t}_{1/4,0,0}\right\}\begin{pmatrix}x\\y\\z\end{pmatrix} = \begin{pmatrix}x+1/4\\\bar{z}\\y\end{pmatrix}$$

$$\left\{\mathbf{4}_{1[100]}\middle|\mathbf{t}_{1/4,0,0}\right\}^2\begin{pmatrix}x\\y\\z\end{pmatrix} = \left\{\mathbf{4}_{1[100]}\mathbf{4}_{1[100]}\middle|(\mathbf{4}_{1[100]}\mathbf{t}_{1/4,0,0}+\mathbf{t}_{1/4,0,0}\right\}\begin{pmatrix}x\\y\\z\end{pmatrix}$$

$$= \left\{2_{[100]}\middle|\mathbf{t}_{1/2,0,0}\right\}\begin{pmatrix}x\\y\\z\end{pmatrix} = \begin{pmatrix}1/2+x\\\bar{y}\\\bar{z}\end{pmatrix}$$

$$\left\{\mathbf{4}_{1[100]}\middle|\mathbf{t}_{1/4,0,0}\right\}^{-1}\begin{pmatrix}x\\y\\z\end{pmatrix} = \left\{4^{-1}_{1[100]}\middle|-4^{-1}_{1[100]}\mathbf{t}_{1/4,0,0}\right\}\begin{pmatrix}x\\y\\z\end{pmatrix}$$

$$= \begin{pmatrix}1&0&0\\0&0&1\\0&\bar{1}&0\end{pmatrix}\begin{pmatrix}x\\y\\z\end{pmatrix}+\begin{pmatrix}-1/4\\0\\0\end{pmatrix} = \begin{pmatrix}3/4+x\\z\\\bar{y}\end{pmatrix}$$

Thus, the coordinates are $x,\ y,\ z;\ \ 1/4+x,\ \bar{z},\ y;\ \ 1/2+x,\ \bar{y},\ \bar{z};\ \ 3/4+x,\ z,\ \bar{y}$.

8.12

$$\left\{\bar{\mathbf{4}}_{[0,0,z]}\middle|\mathbf{t}_{0,0,0}\right\}\left\{\boldsymbol{c}_{(0,y,0)}\middle|\mathbf{t}_{0,0,1/2}\right\}\mathbf{r} = \left\{\bar{\mathbf{4}}_{[0,0,z]}\,\boldsymbol{c}_{(0,y,0)}\middle|(\bar{\mathbf{4}}_{[0,0,z]}\mathbf{t}_{0,0,1/2}+\mathbf{t}_{0,0,0}\right\}\mathbf{r}$$

$$= \begin{pmatrix}0&1&0\\\bar{1}&0&0\\0&0&\bar{1}\end{pmatrix}\begin{pmatrix}1&0&0\\0&\bar{1}&0\\0&0&1\end{pmatrix}\mathbf{r}+\begin{pmatrix}0\\0\\\overline{1/2}\end{pmatrix}$$

$$= \bar{y},\ \bar{x},\ 1/2-z$$

Thus, $x,\ y,\ z \rightarrow \bar{y},\ \bar{x},\ 1/2-z$ corresponds to a twofold axis along $[1\ \bar{1}\ 1/4]$; the space group is $P\bar{4}\,c2$.

8.13

$$\{\boldsymbol{R}|\mathbf{t}\}^n\mathbf{r} = [\{\boldsymbol{R}|\mathbf{t}\}\{\boldsymbol{R}|\mathbf{t}\}.....\{\boldsymbol{R}|\mathbf{t}\}\{\boldsymbol{R}|\mathbf{t}\}]\{\boldsymbol{R}|\mathbf{t}\}\{\boldsymbol{R}|\mathbf{t}\}\mathbf{r}$$
$$= [\{\boldsymbol{R}|\mathbf{t}\}\{\boldsymbol{R}|\mathbf{t}\}.....\{\boldsymbol{R}|\mathbf{t}\}\{\boldsymbol{R}|\mathbf{t}\}](\boldsymbol{R}\boldsymbol{R}\mathbf{r}+\boldsymbol{R}\mathbf{t}+\mathbf{t})$$
$$= [\{\boldsymbol{R}|\mathbf{t}\}\{\boldsymbol{R}|\mathbf{t}\}.....\{\boldsymbol{R}|\mathbf{t}\}\{\boldsymbol{R}|\mathbf{t}\}](\boldsymbol{R}^2|2\mathbf{t})$$

because $\boldsymbol{R}\mathbf{t} = \mathbf{t}$. Continuing the multiplication process:

$$\{\boldsymbol{R}|\mathbf{t}\}^n\mathbf{r} = [\{\boldsymbol{R}|\mathbf{t}\}\{\boldsymbol{R}|\mathbf{t}\}.....\{\boldsymbol{R}|\mathbf{t}\}\{\boldsymbol{R}|\mathbf{t}\}]\{\boldsymbol{R}^2|2\mathbf{t}\}\mathbf{r}$$
$$= [\{\boldsymbol{R}|\mathbf{t}\}\{\boldsymbol{R}|\mathbf{t}\}.....\{\boldsymbol{R}|\mathbf{t}\}\{\boldsymbol{R}|\mathbf{t}\}](\boldsymbol{R}^3\mathbf{r}+2\boldsymbol{R}\mathbf{t}+\mathbf{t})$$
$$= [\{\boldsymbol{R}|\mathbf{t}\}\{\boldsymbol{R}|\mathbf{t}\}.....\{\boldsymbol{R}|\mathbf{t}\}\{\boldsymbol{R}|\mathbf{t}\}]\{\boldsymbol{R}^3|3\mathbf{t}\}\mathbf{r}$$

Hence, for all n repetitions, $\{\boldsymbol{R}|\mathbf{t}\}^n\mathbf{r} = \{\boldsymbol{R}^n|n\mathbf{t}\}\mathbf{r}$. Thus, n operations of a screw axis n_p, $(p < n)$ are equivalent to the rotation n plus a translational repeat along the direction of the axis. Thus, the possible screw axes are given by the sum of $(n-1)$, for $n = 1, 2, 3, 4$ and 6, that is eleven screw axes in total.

8.14 (a)

$$\{\mathbf{4}_{[001]}|\mathbf{t}_{0,0,0}\}\left\{\boldsymbol{b}_{(1/4,y,z)}|\mathbf{t}_{1/2,1/2,0}\right\} = \left\{\mathbf{4}\boldsymbol{b}|(\mathbf{4}\mathbf{t}_{0,1/2,1/2} + \mathbf{t}_{0,0,0})\right\}$$
$$= \left\{\mathbf{4}\boldsymbol{b}|(\mathbf{t}_{\overline{1/2},1/2,0}\right\}$$
$$= \begin{pmatrix} 0 & \bar{1} & 0 \\ \bar{1} & 0 & 0 \\ 0 & 0 & 1 \end{pmatrix} + \begin{pmatrix} 1/2 \\ 1/2 \\ 0 \end{pmatrix}$$

(b)

$$\left\{\boldsymbol{b}_{(1/4,y,z)}|\mathbf{t}_{1/2,1/2,0}\right\}\{\mathbf{4}_{[001]}|\mathbf{t}_{0,0,0}\} = \left\{\boldsymbol{b}\mathbf{4}|(\boldsymbol{b}\mathbf{t}_{0,0,0} + \mathbf{t}_{1/2,1/2,0})\right\}$$
$$= \left\{\boldsymbol{b}\mathbf{4}|(\mathbf{t}_{1/2,1/2,0}\right\}$$
$$= \begin{pmatrix} 0 & 1 & 0 \\ 1 & 0 & 0 \\ 0 & 0 & 1 \end{pmatrix} + \begin{pmatrix} 1/2 \\ 1/2 \\ 0 \end{pmatrix}$$

This operation transforms x, y, z to $1/2 + y$, $1/2 + x$, z, which is another site in the space group, as may be conformed from the International Tables for space group $P4bm$.

(c) From (a), a point x, y, z is transformed to $1/2 - x$, $1/2 - y$, z, which indicates a diagonal m plane at $1/2$, $1/2$, z, and justifies the scheme used in Section 5.4.12.2. The apparent anomaly of the m plane at $x = 1/2$, $y = 1/2$ instead of the usual half distances of $1/4$, $1/4$, is explained by noting that the repeat distance for the diagonal planes is $d_{[110]}$, so that the intersections of the m plane are at one half of this distance.

8.15 (a) Two molecules per unit cell lie on either $\bar{1}$ or m. Since the molecule cannot be centrosymmetric, both its symmetry and its site symmetry are m. (b) The factor group is C_{2h} ($2/m$), (c) The symmetry of the $Cr(CO)_3$ moiety is C_{3v}. (d) The representation is

C_{2h}	E	C_2	i	σ_h
Γ	6	0	0	2

and Γ reduces to $2A_g + B_g + A_u + 2B_u$. From an inspection of the character table for C_{2h}, infrared activity arises from A_u and $2B_u$, and Raman from $2A_g$ and B_g. (e) They are non-coincident because the crystal is centrosymmetric.

8.16 (a) The Ba and Ti species occupy Wyckoff (a) and (b) sites. Let (a), 0, 0, 0, be chosen for Ba; then Ti occupies (b), $1/2$, $1/2$, $1/2$. The three O species could occupy Wyckoff (c) or (d) sites, but (d) is rejected because it would require a Ba–O–Ba distance of at least *ca.* 4.23 Å, which is longer than the unit cell edge. Therefore, the three O species occupy sites (c). [If Ti had been place at 0, 0, 0 with Ba at $1/2$, $1/2$, $1/2$, an equivalent arrangement, then the three O would occupy sites (d).]

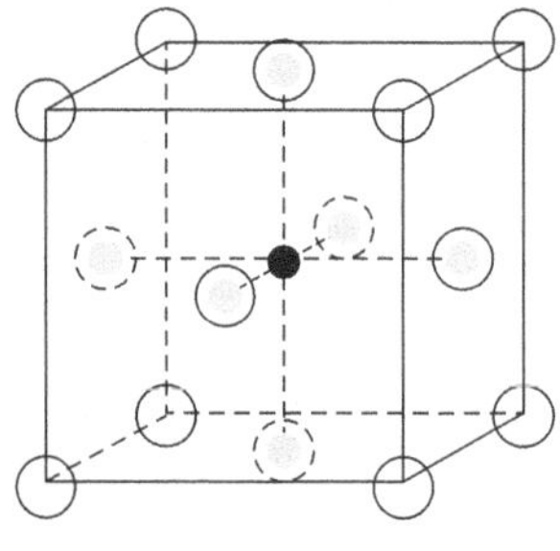

White = Ba, Black = Ti, Shaded = O

(b) The threefold rotation axes and two fourfold rotation axes are degraded, as well as some diagonal m planes, and the resulting structure is tetragonal, space group $P\frac{4}{m}mm$.

(c) Further degradation of symmetry leads to the tetragonal space group $P4mm$.

(d) Of the resulting structures in (b) and (c), that in (c) can be pyroelectric because the z axis is now a polar direction. It could not arise for (b) because that structure is centrosymmetric.

8.17 An F cubic unit cell of side a has twelve lattice points distant $a\sqrt{2}$ from any given lattice point. Bisecting these lengths by plane so as to make a closed polyhedron leads to the required Wigner–Seitz cell:

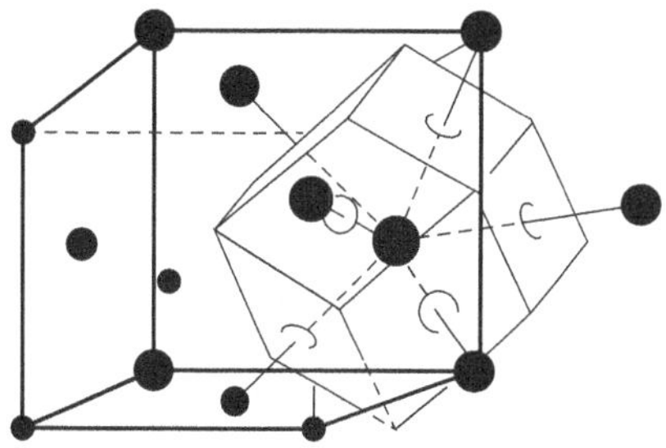

It is a rhombic dodecahedron of symmetry also $m\bar{3}m$.

8.18 Point group D_{6h} is centrosymmetric, so no one mode can be active in both infrared and Raman. From the character table, the relevant integrals may be written:

$$\int \Psi_{A_{1g}} \mu_i \Psi_{A_{1g}} \mathrm{d}\tau = 0 \ (i = x,\ y,\ z)$$

$$\int \Psi_{A_{1g}} \alpha_{ij} \Psi_{A_{1g}} \mathrm{d}\tau \neq 0 \ (ij = x^2,\ y^2,\ z^2)$$

where μ_i and α_{ij} are components of the dipole moment operator and polarizability operator respectively. Hence, A_{1g} is infrared forbidden but Raman active for $i = j = z\ (x, y)$.

8.19 From the analysis in Section 8.3.2, $\Gamma_{\text{vib}} = A_1' + A_2'' + 2E'$. From direct products, A_2'' and E' are infrared active, and A_1' and E' are Raman active, polarized and depolarized respectively. The stretches normally have the higher frequencies, with the symmetric lower than the asymmetric. Out-of-plane bends are generally of higher frequency than in-plane bends. Thus, the probable assignment is:

888 cm^{-1}	A_1'	$\bar{\nu}_1$	Symmetric B–F stretch
1454 cm^{-1}	E'	$\bar{\nu}_3$	Asymmetric B–F stretches
692 cm^{-1}	A_2''	$\bar{\nu}_2$	Out-of-plane bend
480 cm^{-1}	E'	$\bar{\nu}_4$	In-plane bends

8.20 Chlorobenzene belongs to point group C_{2v}; from a cyclic interchange of axes, the molecular plane becomes σ_v. From unshifted atom contributions, the following representation obtains:

C_{2v}	E	C_2	σ_v	σ_v'
Γ_{3n}	36	–4	12	4

By reduction, $\Gamma_{3n} = 12A_1 + 4A_2 + 12B_1 + 8B_2$, so that $\Gamma_{vib} = 11A_1 + 3A_2 + 10B_1 + 6B_2$. From direct products, the infrared active modes are A_1, B_1 and B_2, and the Raman active A_1, A_2, B_1 and B_2. Summarizing:

	C_6H_6	C_6H_5Cl
Infrared	A_{2u}, $3E_{1u}$	$11A_1$, $10B_1$, B_2
Raman	$2A_{1g}$, E_{1g}, $4E_{2g}$	$11A_1$, $10B_1$, B_2 all coincident with the infrared; $3A_2$

The mutual exclusion rule applies to benzene because it is centrosymmetric.

8.21 (a) 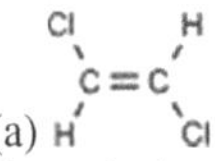

Point group $2/m$. (b) The $+z$ axis is normal to the plane and upwards through the centre of the double bond, the $+x$ axis is along the double bond, left to right, and the y axis is perpendicular to both x and z, such that x, y and z form a right handed axial set. (c) Considering how many orbitals remain unchanged for each operation, taking –1 for each change of sign, leads to the following table:

$2/m$	1	2	$\bar{1}$	m
Γ_{1s}	2	0	0	2
Γ_{2p}	6	0	0	2

(d) Reduction gives $A_g + B_g$ for Γ_{1s}, and $2A_g + A_u + B_g + 2B_u$ for Γ_{2p}.

8.22 Refer to the character table for point group C_{2v}. taking $B_g \otimes B_u$ as an example:

C_{2h}	E	C_2	i	σ_h
B_g	1	$\bar{1}$	1	$\bar{1}$
B_u	1	$\bar{1}$	$\bar{1}$	1
$B_g \otimes B_u = A_u$	1	1	$\bar{1}$	$\bar{1}$

Evaluating all possible products:

$A_g \otimes A_g = A_g$ $\quad A_u \otimes A_u = A_g$ $\quad B_g \otimes B_u = A_u$

$A_g \otimes A_u = A_u$ $\quad A_u \otimes B_g = B_u$ $\quad B_u \otimes B_u = A_g$

$A_g \otimes B_g = B_g$ $\quad A_u \otimes B_u = B_g$

$A_g \otimes B_u = B_u$ $\quad B_g \otimes B_g = A_g$

8.23 The point group for the water molecule is C_{2v}, and a representation Γ is generated on the O–H bonds. If a bond does not move under a symmetry operation of the group, Γ take the value 1 for that operation. The reference axes are shown in Fig. 7.5. Hence,

C_{2v}	E	C_2	σ_v	σ'_v
Γ	2	0	0	2

which reduces to $A_1 + B_2$. There will be two SALCs ψ_1 and ψ_2, so the projection operators $\boldsymbol{P}_{A_1}$ and $\boldsymbol{P}_{B_2}$ are applied to these functions:

C_{2v}	E	C_2	σ_v	σ'_v
$\boldsymbol{P}_{A_1}$	ψ_1	ψ_2	ψ_1	ψ_2
$\boldsymbol{P}_{B_2}$	ψ_1	$-\psi_2$	ψ_1	$-\psi_2$

which produces the SALCs $2\psi_1 + 2\psi_2$ and $2\psi_1 - 2\psi_2$, or simply, $\psi_1 + \psi_2$ and $\psi_1 - \psi_2$. These results must be normalized: as usual, the normalizing factor is $1/\sqrt{\sum_1 c_1^2}$, which is $1/\sqrt{2}$ because the coefficients c_i are unity. Hence, the required SALCs are

$$\Psi_1 = \left(1/\sqrt{2}\right)(\psi_1 + \psi_2)$$

$$\Psi_2 = \left(1/\sqrt{2}\right)(\psi_1 - \psi_2)$$

General bibliography

Barlow, W. (1894) *Zeit. für Kristallogr.* **23**, 1.
Bradley, C. J. and Cracknell, A. P. (1972) *The Mathematical Theory of Symmetry in Solids*, Clarendon.
Burns, G. and Glazer, A. M. (2013) *Space Groups for Solid State Scientists*, Third edition, Elsevier.
Carter, Nathan C. (2009) *Visual Group Theory*, M. A. A. Publishers.
Cotton, F. A. (1990) *Chemical Applications of Group Theory*, 3rd Edition, John Wiley and Sons, Inc.
Dresselhaus, M. S. *et al.* (2008) *Group Theory: Application to the Physics of Condensed Matter*, Springer.
Fyodorov, Y. S. (1891) *Zap. Mineral. Obch.*, **28**, 1; *also* English translation: Harker, D. and Harker, K. (1971) *Amer. Cryst. Assoc. Monograph*, No. 7.
Goodman-Strauss, C., Conway, J. and Burgiel, H. (2008) *The Symmetries of Things*, A K Peters/CKC Press (2008).
Hargittai, I. (1994) *Symmetry: a unifying concept*, Shelter Publications.
Hargittai, M and Hargittai, I. (2009) *Symmetry through the Eyes of a Chemist*, Springer.
Hall, L. H. (1969) *Group Theory in Chemistry*, McGraw-Hill.
Henry, N. F. M. and Lonsdale, K. [Editors] (1965) *International Tables for X-ray Crystallography*, Vol I, Kynoch Press.
Kettle, S. F. A. (1995) *Symmetry and Structure*, John Wiley & Sons Ltd.
McWeeny, R. (1964) *Symmetry*, Pergamon Press.
Nussbaum, A. (1971) *Applied Group Theory for Chemists, Physicists and Engineers*, Prentice-Hall.
Schönflies, A. M. (1891) *Kristallsysteme und Kristallstruktur*, Leipzig.
Streitwolf, H.-W. (1967) *Gruppentheorie in der Festkörperphysik*, Akad. Ver. Geest and Portig K.-G; *also* English translation: Sykes, J. B. (1971) as *Group Theory in Solid-State Physics*, Macdonald.
Weyl, H. (1962) *Symmetry*, Princetown University Press; *also* on-line at: http://ia600809.us.archive.org/11/items/Symmetry_482/Weyl-Symmetry.pdf

Index

The manufacturer's authorised representative in the EU for product safety is Oxford University Press España S.A. of El Parque Empresarial San Fernando de Henares, Avenida de Castilla, 2 - 28830 Madrid (www.oup.es/en or product.safety@oup.com). OUP España S.A. also acts as importer into Spain of products made by the manufacturer.
Printed and bound by CPI Group (UK) Ltd, Croydon, CR0 4YY
14/09/2026
02216285-0001